HANDBOOK OF PLANT VIRUS DISEASES

Handbook of Plant Virus Diseases

Editor:

Aleksei Anatoliyovych Stepanenko, Ph.D.

Delve Publishing

www.delvepublishing.com

Handbook of Plant Virus Diseases

Editor: **Aleksei Anatoliyovych Stepanenko, Ph.D.**

708 3rd Avenue, 6th Floor
New York, NY 10017
United States of America
www.delvepublishing.com

ISBN: 978-1-68095-755-6

Library of Congress Control Number: 2016951511

About the Editor

Aleksei Anatoliyovych Stepanenko, Ph.D.

Aleksei A. Stepanenko received his Masters's degree in Biochemistry from V. N. Karazin Kharkiv National University and PhD in Molecular biology from Institute of Molecular Biology and Genetics, where he is currently working as a Research Fellow. His scope of interests is genomic and phenotypic evolution of tumor cells under chemotherapeutic stresses and viro-immunological approaches as an alternative perspective in cancer therapy.

Contents

List of Contributors

Robert R. Martin
USDA-ARS Horticultural Crops Research Laboratory, Corvallis, OR 97330, USA

James J. Polashock
USDA-ARS, GIFVL, 125A Lake Oswego Rd. Chatsworth, NJ 08019, USA

Ioannis E. Tzanetakis
Department of Plant Pathology, Division of Agriculture, University of Arkansas, Fayetteville, AR 72701, USA

Jiri Sochor
Department of Chemistry and Biochemistry, Faculty of Agronomy, Mendel University in Brno, Zemedelska 1, CZ-613 00 Brno, Czech Republic
Department of Natural Drugs, Faculty of Pharmacy, University of Veterinary and Pharmaceutical Sciences Brno, Palackeho 1-3, CZ-612 42, Czech Republic
Central European Institute of Technology, Brno University of Technology, Technicka 3058/10, CZ- 616 00 Brno, Czech Republic

Petr Babula
Department of Chemistry and Biochemistry, Faculty of Agronomy, Mendel University in Brno, Zemedelska 1, CZ-613 00 Brno, Czech Republic
Department of Natural Drugs, Faculty of Pharmacy, University of Veterinary and Pharmaceutical Sciences Brno, Palackeho 1-3, CZ-612 42, Czech Republic
Central European Institute of Technology, Brno University of Technology, Technicka 3058/10, CZ- 616 00 Brno, Czech Republic

Vojtech Adam
Department of Chemistry and Biochemistry, Faculty of Agronomy, Mendel University in Brno, Zemedelska 1, CZ-613 00 Brno, Czech Republic
Central European Institute of Technology, Brno University of Technology, Technicka 3058/10, CZ- 616 00 Brno, Czech Republic

Boris Krska
Department of Fruit Growing, Faculty of Horticulture, Mendel University in Brno, Valticka 337, CZ-691 44 Lednice, Czech Republic
Rene Kizek
Department of Chemistry and Biochemistry, Faculty of Agronomy, Mendel University in Brno, Zemedelska 1, CZ-613 00 Brno, Czech Republic
Central European Institute of Technology, Brno University of Technology, Technicka 3058/10, CZ- 616 00 Brno, Czech Republic

Paolla M. V. Abreu
Núcleo de Biotecnologia, Universidade Federal do Espírito Santo, Vitória 29040090, Espírito Santo, Brazil

Tathiana F. S. Antunes
Núcleo de Biotecnologia, Universidade Federal do Espírito Santo, Vitória 29040090, Espírito Santo, Brazil

Anuar Magaña-Álvarez
Núcleo de Biotecnologia, Universidade Federal do Espírito Santo, Vitória 29040090, Espírito Santo, Brazil
Laboratorio GeMBio, Centro de Investigación Científica de Yucatán A.C., Mérida 97200, Yucatán, Mexico

Daisy Pérez-Brito
Laboratorio GeMBio, Centro de Investigación Científica de Yucatán A.C., Mérida 97200, Yucatán, Mexico

Raúl Tapia-Tussell
Laboratorio GeMBio, Centro de Investigación Científica de Yucatán A.C., Mérida 97200, Yucatán, Mexico

José A. Ventura
Núcleo de Biotecnologia, Universidade Federal do Espírito Santo, Vitória 29040090, Espírito Santo, Brazil
Instituto Capixaba de Pesquisa, Assistência Técnica e Extensão Rural, Vitória 29050790, Espírito Santo, Brazil

Antonio A. R. Fernandes
Núcleo de Biotecnologia, Universidade Federal do Espírito Santo, Vitória 29040090, Espírito Santo, Brazil

Patricia M. B. Fernandes
Núcleo de Biotecnologia, Universidade Federal do Espírito Santo, Vitória 29040090, Espírito Santo, Brazil

Won Kyong Cho
Department of Agricultural Biotechnology, Plant Genomics and Breeding Institute, Institute for Agriculture and Life Sciences, College of Agriculture and Life Sciences, Seoul National University, Seoul 151-921, Korea

Yeonhwa Jo
Department of Agricultural Biotechnology, Plant Genomics and Breeding Institute, Institute for Agriculture and Life Sciences, College of Agriculture and Life Sciences, Seoul National University, Seoul 151-921, Korea

Kyoung-Min Jo
Department of Agricultural Biotechnology, Plant Genomics and Breeding Institute, Institute for Agriculture and Life Sciences, College of Agriculture and Life Sciences, Seoul National University, Seoul 151-921, Korea

Kook-Hyung Kim
Department of Agricultural Biotechnology, Plant Genomics and Breeding Institute, Institute for Agriculture and Life Sciences, College of Agriculture and Life Sciences, Seoul National University, Seoul 151-921, Korea

Muhammad Tahir
Plant Biotechnology Department, Atta-ur-Rahman School of Applied Biosciences, National University of Sciences and Technology, Sector H-12, Islamabad 44000, Pakistan

Imran Amin
Agricultural Biotechnology Division, National Institute for Biotechnology and Genetic Engineering, Jhang Road, Faisalabad 38000, Pakistan

Muhammad Saleem Haider
School of Biological Sciences, University of the Punjab, New Campus, Lahore 54590, Pakistan

Shahid Mansoor
Agricultural Biotechnology Division, National Institute for Biotechnology and Genetic Engineering, Jhang Road, Faisalabad 38000, Pakistan

Rob W. Briddon
Agricultural Biotechnology Division, National Institute for Biotechnology and Genetic Engineering, Jhang Road, Faisalabad 38000, Pakistan

Frédéric Péréfarres
CIRAD, UMR PVBMT, Pôle de Protection des Plantes, 97410 Saint-Pierre, Ile de La Réunion, France
Université de La Réunion, UMR PVBMT, Pôle de Protection des Plantes, 97410 Saint-Pierre, Ile de La Réunion, France

Magali Thierry
CIRAD, UMR PVBMT, Pôle de Protection des Plantes, 97410 Saint-Pierre, Ile de La Réunion, France
Université de La Réunion, UMR PVBMT, Pôle de Protection des Plantes, 97410 Saint-Pierre, Ile de La Réunion, France

Nathalie Becker
Muséum National d'Histoire Naturelle, Département Systématique et Evolution, USM 601, CNRS UMR 5202 Origine, Structure et Evolution de la Biodiversité, 57 rue Cuvier, CP 50, 75005 Paris, France

Pierre Lefeuvre
CIRAD, UMR PVBMT, Pôle de Protection des Plantes, 97410 Saint-Pierre, Ile de La Réunion, France

Bernard Reynaud
CIRAD, UMR PVBMT, Pôle de Protection des Plantes, 97410 Saint-Pierre, Ile de La Réunion, France

Hélène Delatte
CIRAD, UMR PVBMT, Pôle de Protection des Plantes, 97410 Saint-Pierre, Ile de La Réunion, France

Jean-Michel Lett
CIRAD, UMR PVBMT, Pôle de Protection des Plantes, 97410 Saint-Pierre, Ile de La Réunion, France

José Luis Cruz-Jaramillo
Departamento de Biotecnología y Bioingeniería, Centro de Investigación y de Estudios Avanzados del Instituto Politécnico Nacional Av. IPN 2508, Zacatenco 07360, México D.F., Mexico

Roberto Ruiz-Medrano
Departamento de Biotecnología y Bioingeniería, Centro de Investigación y de Estudios Avanzados del Instituto Politécnico Nacional Av. IPN 2508, Zacatenco 07360, México D.F., Mexico

Lourdes Rojas-Morales
LaNSE, Centro de Investigación y de Estudios Avanzados del IPN Av. IPN 2508, Zacatenco 07360, México D.F., Mexico

José Abel López-Buenfil
Departamento de Biotecnología y Bioingeniería, Centro de Investigación y de Estudios Avanzados del Instituto Politécnico Nacional Av. IPN 2508, Zacatenco 07360, México D.F., Mexico
Servicio Nacional de Sanidad Inocuidad y Calidad Agroalimentaria, Guillermo Pérez Valenzuela 127, Coyoacán 04100, México D.F., Mexico

Oscar Morales-Galván
Servicio Nacional de Sanidad Inocuidad y Calidad Agroalimentaria, Guillermo Pérez Valenzuela 127, Coyoacán 04100, México D.F., Mexico

Claudio Chavarín-Palacio
Servicio Nacional de Sanidad Inocuidad y Calidad Agroalimentaria, Guillermo Pérez Valenzuela 127, Coyoacán 04100, México D.F., Mexico

José Abrahán Ramírez-Pool
Departamento de Biotecnología y Bioingeniería, Centro de Investigación y de Estudios Avanzados del Instituto Politécnico Nacional Av. IPN 2508, Zacatenco 07360, México D.F., Mexico

Beatriz Xoconostle-Cázares
Departamento de Biotecnología y Bioingeniería, Centro de Investigación y de Estudios Avanzados del Instituto Politécnico Nacional Av. IPN 2508, Zacatenco 07360, México D.F., Mexico

Nicole Mielke-Ehret
Biocentre Klein Flottbek, University of Hamburg, Ohnhorststrasse 18, Hamburg 22609, Germany

Hans-Peter Mühlbach
Biocentre Klein Flottbek, University of Hamburg, Ohnhorststrasse 18, Hamburg 22609, Germany

Marcos G. Celli
Instituto de Patología Vegetal (IPAVE – CIAP – INTA) Camino. 60 cuadras Km 5,5, Córdoba, X5020ICA, Argentina

Maria C. Perotto
Instituto de Patología Vegetal (IPAVE – CIAP – INTA) Camino. 60 cuadras Km 5,5, Córdoba, X5020ICA, Argentina
Consejo Nacional de Investigaciones Científicas y Técnicas (CONICET), Camino. 60 cuadras Km 5,5, Córdoba, X5020ICA, Argentina

Julia A. Martino
Consejo Nacional de Investigaciones Científicas y Técnicas (CONICET), Camino. 60 cuadras Km 5,5, Córdoba, X5020ICA, Argentina

Ceferino R. Flores
Estación Experimental Agropecuaria Yuto (INTA), Ruta Nacional 34 Km 1286, Yuto, 4518, Jujuy, Argentina

Vilma C. Conci
Instituto de Patología Vegetal (IPAVE – CIAP – INTA) Camino. 60 cuadras Km 5,5, Córdoba, X5020ICA, Argentina
Consejo Nacional de Investigaciones Científicas y Técnicas (CONICET), Camino. 60 cuadras Km 5,5, Córdoba, X5020ICA, Argentina

Patricia Rodriguez Pardina
Instituto de Patología Vegetal (IPAVE – CIAP – INTA) Camino. 60 cuadras Km 5,5, Córdoba, X5020ICA, Argentina

Chengke Liu
Plant Biology Division, The Samuel Roberts Noble Foundation, Inc., Ardmore, OK, USA

Richard S. Nelson
Plant Biology Division, The Samuel Roberts Noble Foundation, Inc., Ardmore, OK, USA

Grace Armijo
Laboratorio de Biología Molecular y Biotecnología Vegetal, Departamento de Genética Molecular y Microbiología, Facultad de Ciencias Biológicas, Pontificia Universidad Católica de Chile, Santiago, Chile

Rudolf Schlechter
Laboratorio de Biología Molecular y Biotecnología Vegetal, Departamento de Genética Molecular y Microbiología, Facultad de Ciencias Biológicas, Pontificia Universidad Católica de Chile, Santiago, Chile

Mario Agurto
Laboratorio de Biología Molecular y Biotecnología Vegetal, Departamento de Genética Molecular y Microbiología, Facultad de Ciencias Biológicas, Pontificia Universidad Católica de Chile, Santiago, Chile

Daniela Muñoz
Laboratorio de Biología Molecular y Biotecnología Vegetal, Departamento de Genética Molecular y Microbiología, Facultad de Ciencias Biológicas, Pontificia Universidad Católica de Chile, Santiago, Chile

Constanza Nuñez
Laboratorio de Biología Molecular y Biotecnología Vegetal, Departamento de Genética Molecular y Microbiología, Facultad de Ciencias Biológicas, Pontificia Universidad Católica de Chile, Santiago, Chile

Patricio Arce-Johnson
Laboratorio de Biología Molecular y Biotecnología Vegetal, Departamento de Genética Molecular y Microbiología, Facultad de Ciencias Biológicas, Pontificia Universidad Católica de Chile, Santiago, Chile

I. U. Mohammed
Natural Resources Institute, University of Greenwich, Chatham Maritime, Kent ME4 4TB, UK

M. M. Abarshi
Natural Resources Institute, University of Greenwich, Chatham Maritime, Kent ME4 4TB, UK

B. Muli
Food Crops Programme, Kenya Agricultural Research Institute, Mtwapa, Kenya

R. J. Hillocks
Natural Resources Institute, University of Greenwich, Chatham Maritime, Kent ME4 4TB, UK

M. N. Maruthi
Natural Resources Institute, University of Greenwich, Chatham Maritime, Kent ME4 4TB, UK

Chiara Lico
Laboratory of Biotechnology, ENEA, Rome, Italy

Eugenio Benvenuto
Laboratory of Biotechnology, ENEA, Rome, Italy

Selene Baschieri
Laboratory of Biotechnology, ENEA, Rome, Italy

Valérie Nicaise
Fruit Biology and Pathology, Virology Laboratory, Institut National de la Recherche Agronomique, University of Bordeaux, UMR 1332, Villenave d'Ornon, France

Preface

The aim of this book is to highlight the importance of plant viruses that are causing devastating plant diseases with a significant impact on agronomy and economics, focusing not only on well-known viruses but also on those that are emerging as serious threats for the production of socially important crops around the world. The book includes studies covering plant virus structures and molecular biology of virus infection/replication mechanisms, disease symptoms, host range, phylogenetic relationships, virus detection methods, available viral control strategies, plant immunity against viruses and the development of virus resistant plants, among other studies. Specifically, the book is focused on Tobacco mosaic virus, Cassava Brown Streak Viruses, Potato Virus X, blueberry and cranberry viruses, grapevine viruses, Papaya meleira virus, the causal agent of papaya sticky disease, Plum pox virus, the causative agent of sharka disease, Ageratum enation virus, a begomovirus of weeds with the potential to infect crops, begomoviruses infecting chia, Dichorhavirus associated with the citrus leprosis disease, Tomato yellow leaf curl begomovirus, Chrysanthemum stunt viroid and Chrysanthemum chlorotic mottle viroid, which infect chrysanthemums, one of the most popular flowers in the world, Emaraviruses, causing ringspot disease of European Mountain Ash, rose rosette disease, High Plains disease, raspberry leaf blotch disorder, fig mosaic disease and others.

Editor
Aleksei Anatoliyovych Stepanenko, Ph.D.

CHAPTER 1

NEW AND EMERGING VIRUSES OF BLUEBERRY AND CRANBERRY

Robert R. Martin[1], James J. Polashock[2], and Ioannis E. Tzanetakis[3]

[1]USDA-ARS Horticultural Crops Research Laboratory, Corvallis, OR 97330, USA

[2]USDA-ARS, GIFVL, 125A Lake Oswego Rd. Chatsworth, NJ 08019, USA

[3]Department of Plant Pathology, Division of Agriculture, University of Arkansas, Fayetteville, AR 72701, USA

ABSTRACT

Blueberry and cranberry are fruit crops native to North America and they are well known for containing bioactive compounds that can benefit human health. Cultivation is expanding within North America and other parts of the world raising concern regarding distribution of existing viruses as well as the appearance of new viruses. Many of the known viruses of these crops are latent or asymptomatic in at least some cultivars. Diagnosis and detection procedures are often non-existent or unreliable. Whereas new viruses can move into cultivated fields from the wild, there is also the threat that devastating viruses can move into native stands of*Vaccinium* spp. or other native plants from cultivated fields. The aim of this paper is to highlight the importance of blueberry and cranberry viruses, focusing not only on those that are new but also those that are emerging as serious threats for production in North America and around the world.

INTRODUCTION

The genus *Vaccinium* belongs to the *Ericaceae* (heath family). The family has more than 3,500 species growing in all latitudes, from the tropics to the polar regions. A common feature among members of the family is that they grow in acidic, many times, nutrient-deprived soils [1]. Of the more than 400 species in the genus *Vaccinium*, highbush, lowbush and rabbiteye blueberries (*V. corymposum* L.; *V. augustifolium* Ait. and*V. ashei* Reade, respectively) and cranberry (*V. macrocarpon* Ait.) are of high economic importance. Largely due to recent press regarding the health benefits of fresh fruit consumption [2,3,4], including blueberries and cranberries, the industry has been rapidly expanding and blueberries in particular are now grown across the globe [5]. This change has come with the cost of introducing and spreading virus diseases and driving disease epidemics. The cost of virus diseases is massive in perennial fruit crops such as blueberry and cranberry that require a large initial investment to prepare, plant, and establish fields to the point of maturity and full productivity. Even when mature, intensive maintenance is needed to sustain productivity, but once established, fields can remain productive for many years. Some cultivated cranberry beds in New Jersey and Massachusetts are reported to be over 100 years old [6]. Blueberry fields are generally not as long-lived, but can easily be maintained for 30 years or more.

Citation: Robert R. Martin, James J. Polashock, and Ioannis E. Tzanetakis, New and Emerging Viruses of Blueberry and Cranberry, doi: 10.3390/v4112831.

Like most plants, *Vaccinium* spp. are known to harbor viruses belonging to several different families. Virus diseases are accompanied by a variety of symptoms ranging from being visually asymptomatic to plant death. In addition, symptoms can vary between cultivars, regions, production systems and years. Viruses can also be latent for years or not express at all in a nursery setting, further allowing widespread distribution. Virus diseases, particularly of blueberry, have become a major problem in recent years not only because of the expansion of the industry and the change in climate patterns, but also because of lack of grower awareness of the problems that can emerge when propagating non-certified material. Because of the rapid increase in blueberry plantings in recent years, in many cases growers choose to propagate from field plants that could be harboring asymptomatic infections at the time cuttings were taken. This has led to the emergence of several diseases in new areas or to major increases in virus incidence where there was minimal virus presence [7].

This review not only aims to provide up-to-date knowledge of the viruses that infect blueberry and cranberry but also give insight into the measures that need to be taken to control the spread of virus diseases and avoid virus epidemics, so as to be able to grow these crops in a sustainable and profitable environment. As is the case with virus diseases of any vegetatively propagated crop, the most critical control measure is the production and planting of stocks free of targeted pathogens. In the U.S., the programs for producing certified planting stocks of vegetative fruit crops has received a tremendous boost with the development of the National Clean Plant Network (NCPN). NCPN has as its' mission the production and maintenance of the initial starting materials for certification programs that are free of targeted pathogens. Thus, the program aims to introduce cultivars of plants that are needed or desired for fruit production, including new materials developed outside of the U.S. to protect current production and the environment from the introduction of potentially damaging pathogens. The structure of this review is primarily based on the mode of transmission of the viruses as this is the most important factor for the development of an efficient and effective management strategy for the viruses and the diseases they cause.

NEPOVIRUSES—NEMATODE AND/OR POLLEN-BORNE

Blueberry Latent Spherical Virus

Blueberry latent spherical virus (BLSV) was isolated recently from highbush blueberry in northern Japan [8]. Although the virus is readily transmissible to herbaceous hosts, and causes symptoms on indicator species that include *Chenopodium quinoa* and *Nicotiana benthamiana* it appears to be asymptomatic in the nine highbush cultivars evaluated [8]. BLSV is most closely related to *Peach rosette mosaic virus*, a subgroup C*Nepovirus* that also infects blueberry. The genome organization of BLSV is typical of members of subgroup C, with the replication-associated polyprotein in RNA 1, and the movement and coat proteins of the virus encoded in the single polyprotein of RNA2. The only difference in BLSV, as compared to other subgroup C viruses is that it has a serine instead of a cysteine protease encoded in RNA 1 [8]. Given the recent discovery of the virus, its geographic distribution and modes of transmission are still unknown, whereas detection protocols are under development.

Blueberry Leaf Mottle Virus

Blueberry leaf mottle disease was first described in 1977 in Michigan and its presence is limited to this state and parts of eastern Canada [9] although extended surveys have been conducted in Arkansas, Oregon, Washington and British Columbia, Canada [10]. Affected leaves are smaller, distorted and show mottling symptoms (Figure 1). Affected bushes develop stem dieback, are stunted, and yield only a fraction of their healthy counterparts [11]. Symptom intensity is cultivar-dependent with northern highbush blueberry developing most severe symptoms [12]. Ramsdell and Stace-Smith [13] purified a virus from infected material. The virus proved to be the causal agent of the disease and was given the name*Blueberry leaf mottle virus* (BLMoV). The physicochemical properties of the virus and the weak cross-reactivity of *Grapevine Bulgarian latent virus* (GBLV) to BLMoV antisera pointed to the relationship of the virus to members of the genus *Nepovirus* [13]. Partial sequence of the two RNAs of the virus placed BLMoV in subgroup C of the genus [14,15]. Although the virus belongs to the genus *Nepovirus*, nematode or aphid transmissions studies have proven unsuccessful. The virus is readily transmissible by pollen, primarily via honeybee movement and seed [16] similar to another virus in the C subgroup, *Cherry leaf roll virus* [15]. For many years there has been confusion on the identity of BLMoV and GBLV. This has been cleared up after the molecular characterization of GBLV [17]. The two viruses are closely related but are clearly different species. BLMoV can be detected serologically and by RT-PCR; however, there is no data on its population structure, a concern for reliable testing. These data indicate that BLMoV is thus far only found in North America, knowledge that has obvious implications in quarantine and certification schemes.

Figure 1: Symptoms of *Blueberry leaf mottle virus*: (A) small, pale green, rosetted and narrow leaves; (B) pale green, stunted Jersey bush (on left) (Photos courtesy Annemiek Schilder, Michigan State University).

Peach Rosette Mosaic Virus

Rosette mosaic disease of peach was first described in the early 1900s [18]. It took several decades before a virus was associated with the disease [19]. The virus, *Peach rosette mosaic virus* (PRMV), early on was only found in stone fruit trees, grapevine and a few weed species such as dandelion. It was first reported in highbush blueberry in 1981 when northern highbush blueberry was planted into fields previously occupied by PRMV-diseased grapevines [20]. Symptoms include uneven leaf distribution, leaves that are smaller than usual with obvious deformations. No data are available on the effect of the virus on other types of blueberries or yield. Similar to BLSV and BLMoV, the virus belongs to subgroup C of the genus *Nepovirus* [15,21]. PRMV is unique among all nematode transmitted viruses as it can be transmitted by different nematode genera, namely *Xiphinema* and *Longidorus* [22,23]. The distribution of the virus is limited to the area around the Great Lakes of North America (Michigan, Ontario and New York) where the virus has only been found in a few fields. As in the case of BLMoV, PRMV can be detected serologically and by RT-PCR; however there is only information available for a single virus isolate. Given the unknown of virus diversity, extra precautions need to be taken when interpreting detection results. Reports of the virus in Europe and the Middle East need to be further evaluated as the virus was only detected using a single technique and the readings appeared very weak [24,25]. For certification and quarantine purposes, the virus should only be considered present in the areas aforementioned.

Necrotic Ringspot Disease [Tobacco Ringspot Virus (TRSV) and Tomato Ringspot Virus (ToRSV)]

Necrotic ringspot disease was first reported in blueberry in 1960 in New Jersey [26] and was associated with TRSV in 1963 [27] and later reported in Arkansas, Connecticut, Illinois, Michigan, New York, Oregon, New Brunswick and Chile [28,29,30]. ToRSV was first reported in blueberry in Washington in 1972 [28] and subsequently identified in blueberries, in Indiana, Michigan, New York, Oregon, Pennsylvania, Canada, and Chile [9,28,29,30]. Necrotic ringspot disease (Figure 2) has been used synonymously for TRSV infection in blueberry, as this was the virus first associated with the disease. However, symptoms of ToRSV (Figure 2) are very similar to those caused by TRSV and the only way to differentiate the viruses in blueberry is to carry out diagnostic assays. TRSV and ToRSV are vectored by the same nematode, *Xiphinema americanum*, and disease symptoms in the field often appear as oval foci typical for nematode transmitted diseases. Both viruses cause a slow, steady decline in bush productivity in susceptible cultivars, including 'Collins', 'Elliott', 'Jersey', 'Pemberton', 'Rubel', and in several halfhigh blueberry clones in New Brunswick [9]. In some cultivars, infection with TRSV or ToRSV can lead to plant death. 'Bluecrop' appears to become infected very slowly with ToRSV, but it is highly susceptible to TRSV [28]. Necrotic ringspot symptoms of TRSV or ToRSV have not been reported in lowbush or rabbiteye blueberry.

Figure 2: Symptoms of necrotic ringspot disease in blueberry: (A) *Tomato ringspot virus* in blueberry showing necrotic spots and leaf distortion; (B) Chlorotic mosaic symptoms on plant infected with ToRSV; and (C) Mosaic on young leaves of plant infected with *Tobacco ringspot virus*.

TRSV and ToRSV are typical of members in subgroups A and C, respectively, in the genus *Nepovirus* [14]. They contain two positive-sense genomic RNA molecules that are encapsidated in spherical virions of ~28 nm. Both have broad host ranges, are pollen- and seed-transmitted in some hosts with varying degrees of efficiency up to 100%, features that allow them to persist in the environment. TRSV and ToRSV can be detected serologically with ELISA and by RT-PCR; however, there are significant strain differences in both viruses, and one must take care to ensure that an appropriate test is used. TRSV is very common in *Rubus*spp. [31] in the southeastern U.S. and one would expect it to become an emerging problem in blueberry in that region as production expands there.

TRSV and ToRSV are often difficult to detect in blueberry due to uneven distribution in the plants [29], seasonal differences in virus titer in leaf tissues and varying titers of virus in different parts of the plant. These viruses spread very slowly from an infection source and can be controlled by removing affected bushes, plus an additional 3–4 bushes beyond the symptomatic plants, taking care to remove as much of the root system as possible. The area should then be treated with a nematicide prior to replanting. If planting into a site that has widespread distribution of TRSV or ToRSV and the vector nematode present, then control will require removal of the entire field, followed by fumigation and replanting, fallowing the field and keeping it weed free, or planting a crop that is a non-host for these viruses [32]. This latter situation can happen if a site is not tested for the presence of *X. americanum* and the viruses prior to planting, since the viruses and nematode have broad host ranges and are symptomless in many hosts.

ILARVIRUSES—POLLEN-BORNE

Blueberry Shock Virus (BlShV)

Blueberry shock disease was first observed in Washington in 1987 and initially confused with blueberry scorch caused by *Blueberry scorch virus* [33]. However, plants affected with shock produced a second flush of leaves after flowering and the plants appeared normal by late summer except for the lack of fruit. Also, after 1–3 years the affected plants flowered and fruited normally and did not exhibit any additional symptoms [34]. However, during the long cool spring conditions of 2010, 2011 and 2012 plants of several cultivars that had recovered from the shock symptoms exhibited a leaf reddening in the Pacific Northwest (PNW; Oregon, Washington and British Columbia) (Figure 3C). The impact on yield, if any, with the leaf reddening symptoms is still unknown. The virus is pollen-borne and infection only occurs during the bloom period, as determined with trap plants placed bi-weekly in diseased fields throughout two growing seasons [35]. The year after infection, bushes exhibit a "shock reaction" where the flowers and foliage blight in the early spring just as the plant is in full bloom (Figure 3A,B). The blighted tissues fall from the bush and a new flush of leaves develops during the summer. By harvest time, infected bushes look nearly normal except for the absence of fruit. Fruit loss is correlated with the extent of blighting in the spring. In some bushes, only one or a few branches will show symptoms whereas in others the entire bush blights. Bushes, where only partial blighting occurs, will usually show symptoms the following year on previously symptomless wood. BlShV has been identified throughout the PNW, and more recently in California, Nova Scotia, Canada, Pennsylvania, New York [36] and Michigan [37].

BlShV is a member of subgroup 3 of the genus *Ilarvirus*. It is detected readily by ELISA or RT-PCR from flower buds early in the season and leaf tissue through August in the PNW. Control should be focused on not introducing the virus to new production areas on nursery stock. Once the virus is present in a field, removal of infected plants based on symptoms or diagnostics will slow the spread of the virus but not completely prevent further spread. There are several reasons for this: (1) The virus is unevenly distributed in blueberries the first year or two after infection and field testing can easily result in false negatives; and (2) The virus is pollen-borne and the flowers open enough to start releasing pollen before symptoms appear, thus transmission occurs before it is obvious a plant is infected. The recommendation in the PNW is to let the virus run its course through a field realizing that there will be a 1–2 year crop loss in the process. This is less of a loss than removal of a field and replanting, where it may be 4–6 years before a planting is back in full production. It is unknown whether plants will recover in a similar manner in other production areas where the growing conditions are less favorable. Thus, an approach similar to that used in Michigan, where the entire field was removed as soon as the virus was detected, is recommended if the virus is detected in a new production area.

Figure 3: *Blueberry shock virus* symptoms: (A) Leaf and flower necrosis in cv. 'Liberty' showing many symptomatic and asymptomatic bushes in a production field; (B) Close up of leaf necrosis in cv. 'Bluegold'; and (C) Leaf reddening in recovered plants that has been observed in several cultivars in years with prolonged cool conditions in Oregon.

Tobacco Streak Virus

Tobacco streak virus (TSV) was first reported in cranberry in 2001 [38]. The virus was discovered during a routine screen of plant material imported into Scotland. The material originally tested was shipped from New Jersey and the source plots in New Jersey remain infected. It was later noted that vine material in question was brought to New Jersey from Wisconsin. Subsequent limited testing of vines in Wisconsin indicated varying degrees of infection [39] and the prevalence of virus in Wisconsin is still unknown. The infected material showed no symptoms and it is unclear whether the virus has any deleterious effects under certain conditions or in certain cultivars. TSV belongs to subgroup 1 and is the type member of the *Ilarvirus* genus in the family *Bromoviridae*. Although there are both immunological and molecular tests available for the detection of the virus, the extreme diversity observed among isolates needs to be taken into consideration when choosing a detection protocol. TSV has been reported to be thrips transmitted through incidental mechanical inoculation as the insects feed on the plants and infected pollen is introduced into the wounds [40]. This mode of transmission has not been demonstrated in cranberry. Seed transmission may occur at very low incidence, but thus far, seedlings from crosses using infected pollen have tested negative [39].

APHID-BORNE

Blueberry Scorch Virus

A previously undescribed blight of highbush blueberry was observed since the early 1970s in the Sheep Pen Hill area of Burlington County, New Jersey [41]. The symptoms were described as a blighting of both flowers and new vegetative growth just prior to full bloom. The cause at the time was unknown, but it was noted to be actively spreading to all cultivars grown in the area. The disease, known locally as 'Sheep Pen Hill Disease', was noted to be of economic importance as it could lead to total fruit loss and it was cautioned that the disease could potentially spread through plants propagated for sale. A disease with similar symptoms was described in 1980 as affecting the cultivar Berkeley near Puyallup, Washington [42]. Both diseases were shown to be graft transmissible and virus particles were found in the leaves of affected plants [42,43]. The disease etiology was characterized in both states and was shown to be caused by distinct strains of new carlavirus [44,45,46]. The disease became known as Blueberry scorch and the causal agent was thus named*Blueberry scorch virus* (BlScV). The virus particles are flexuous rods approximately 14 × 650 nm encapsidating the monopartite positive-sense single-stranded RNA genome of about 8.5 Kb [47,48].

Symptoms vary depending on virus strain and cultivar from asymptomatic to severe blighting of flowers, young leaves and twig dieback. In addition to these symptoms, leaves of infected bushes sometimes show marginal chlorosis or a red line pattern in late summer and fall (Figure 4). The blighted flowers often fall off soon after blighting, but can remain on the bush throughout the season and through the next dormant season (Figure 4B). Plants with severe blighting bear little or no fruit and take on a scorched appearance (Figure 4A). Virus infections can be latent for many years, depending on cultivar, and severity of expression can vary from year to year. Affected bushes of some cultivars can be productive for many years, whereas others decline quickly over a few years and eventually die. This variation in symptom expression has made field diagnosis difficult and growers are reluctant to remove infected bushes that still bear fruit. The disease has been reported in New Jersey, Massachusetts, Connecticut, Michigan, Oregon, Washington, and British Columbia, Canada [49]. The disease has also been reported in Germany [50], Italy and found in a single plant in the Netherlands [51]. The virus is spread primarily by aphids in a non-persistent manner [52]. BlScV has also been reported from American cranberry (*V. macrocarpon*) and black huckleberry (*V. membranaceum*), but those infections were asymptomatic [53,54]. An aphid control program can help limit field spread. Aphids can also transmit the virus to *Chenopodium quinoa* and *C. amaranticolor* [48] as well as *N. occidentalis*[55]. Mechanical transmission has been successful using infectious transcripts [56]. Blueberries are typically asexually propagated by rooted cuttings. The method of pruning plants to the ground and allowing regrowth for cutting wood production does not allow symptom expression in infected plants. Thus, distribution through infected nursery stock is also an important mode of dissemination [57]. It is imperative that all plants be purchased from certified nurseries that test material with state-of-the-art protocol, able to detect all known virus strains so as to minimize the danger of infected material and the movement of virus and disease into areas where it is absent.

Blueberry Shoestring Virus

The disease was first described in 1950 [58]. Expanding twigs develop red streaking in the spring, discoloration that normally disappears as the season progresses. Symptoms on leaves vary significantly from narrow, pointy leaves to distorted, fully developed blades that may develop an oakleaf pattern (Figure 5). There is normally red discoloration among the veins, symptoms that may affect one or more areas of the leaf, to the point that it covers the majority of the blade. Symptoms vary in intensity depending on the environment and there have been reports were bushes can be asymptomatic for several seasons [59]. Symptoms often are only observed on parts of a plant rather than on all leaves. In severe cases, the disease leads to extensive losses, because of yield reduction and production of unmarketable fruit [60].

Figure 4: *Blueberry scorch virus* symptoms: (A) Blueberry cv. 'Elliott', necrotic flowers and leaves in foreground, healthy bushes in background; (B) Blueberry cv. 'Berkeley' showing flower necrosis and retention, silvery flowers are from previous year, brown flowers from current year; (C) Line and oakleaf patterns; (D) Chlorotic leaf margins in cv. 'Stanley'; E. Leaf reddening in cv. 'Concord'.

Figure 5: *Blueberry shoestring virus* symptoms: (A) Narrow strapped leaves that may be reddish in color, note not all leaves exhibit symptoms; and (B) Oakleaf pattern of reddening on normal shaped leaves (Photos courtesy Mark Longstroth Michigan State University).

Varney [59] provided the first evidence that shoestring disease is caused by a graft-transmissible agent. A virus, *Blueberry shoestring virus* (BSSV), was proven to be the causal agent of the disease after Lesney *et al.*[61] purified the virus and inoculated healthy plants that developed typical diseased symptoms a few months post inoculation. Although BSSV was confirmed as the causal agent was described 35 years ago, it is still understudied. Ramsdell [58] studied the physicochemical properties of the virus, including size, composition and molecular mass of the coat protein. Those properties indicate that BSSV is member of the genus*Sobemovirus* and it is recognized as a member of that genus by the International Committee on Taxonomy of Viruses [62]. BSSV is the only approved member of the *Sobemovirus* genus that is transmitted by an aphid in an efficient manner that can reach 28% in inoculation studies with 30 aphids (*Illinoia pepperi*) that had an inoculation access period of one hour [63]. All types of blueberry are susceptible to the virus but infection rates appear to be higher in certain northern highbush cultivars [64]. As in several of the viruses described here, detection tests available today are based on a single virus isolate and additional research is needed to improve the detection spectrum to as many isolates as possible.

VECTOR UNKNOWN

Blueberry Latent Virus

A new disorder was observed in 2002 in Oregon and Washington in the U.S. and British Columbia, Canada. Highbush blueberry set but dropped virtually all fruit when about 5 mm in diameter. The possibility that a virus was associated with the disorder was examined and dsRNA was isolated from more than 20 affected and several asymptomatic plants. A single molecule of about 3.5 kb was present in the majority of affected plants and controls [65]. The molecule was not associated with the disorder but its presence in the vast majority of the plants tested led to its further characterization. The dsRNA belongs to a virus, provisionally named Blueberry latent virus (BBLV), and will probably be the type member of a new taxon of dsRNA viruses [66,67]. The genome organization of the virus, lacking a movement protein, suggested that it can only move by cell division. Another interesting feature of BBLV is its presence in high percentages in all blueberry germplasm tested from both the east and west coast of the United States as well as Japan [66,68]. Almost 50 isolates from the U.S. and Japan have been sequenced, partially or completely, and the population structure appears extremely stable as diversity does not exceed 0.5% among all isolates. Aerial and soil transmission studies have been performed without success. These data, in combination with the genome structure and extreme genome stability of BBLV, advances the idea that its movement is limited to seed and/or pollen and experiments with three northern highbush and one rabbiteye blueberry cultivar showed 100% seed transmission. It is expected, and given the high incidence of BBLV in all tested germplasm, that the virus is present wherever North American germplasm is grown. The prevalence of the virus should not be of major concern though as no symptoms have been observed on several highbush cultivars that have been monitored for almost 10 years.

Blueberry Mosaic Virus

Blueberry mosaic disease was first reported in the 1950s when experiments proved that it is not a genetic disorder but is rather caused by a graft-transmissible agent [69]. Mosaic is the most noticeable of all blueberry diseases. Symptoms include mottling and mosaic patterns of bright yellow to pink to red (Figure 6). Usually symptoms are apparent on only a few leaves on a bush, although, and depending on the season, they may be absent or cover the majority of the bush. The effect of the disease has not been studied in detail but some reports indicate that infected material has noticeable reduction in yield, and in addition the fruit is of poor quality and may ripen late [69]. Many northern highbush (*V. corymbosum*) cultivars develop symptoms [70]. No symptoms have been observed in other *Vaccinium* species other than *V. vacillans* (a lowbush dryland blueberry; [69]. It is important to note that there was no detection test available until recently [71] that would help determine whether the agent does not infect other blueberry species or causes asymptomatic infections.

There have been several attempts to identify the causal agent including reports of a viroid-like agent found in diseased plants [72]. The viroid was never proven to be the causal agent of the disease and only recently was the putative causal agent fully characterized [73]. All symptomatic plants collected from Arkansas, Michigan, New Jersey, Kentucky and Oregon were infected with a new member of the negative-sense, single-stranded RNA genus *Ophiovirus*.

The disease seems to be static in most areas but in recent years there have been observations of movement to adjacent bushes, suggesting that the vector is soil-borne as proven in the case of other ophioviruses [74]. The disease has been observed in several areas in North America whereas there are reports of limited distribution in New Zealand, Europe, South Africa, Argentina and Chile [75]. With the development of detection methods, it will be feasible to access the presence of the virus in blueberry production areas around the world and determine its effect on production.

Figure 6: Blueberry mosaic symptoms in cultivars chlorosis, reddening and mosaic symptoms observed in many cultivars: (A) 'Bluecrop'; (B) 'Blueray'; and (C) 'Legacy'.

Blueberry Necrotic Ring Blotch Virus

A new disease named blueberry necrotic ring blotch, was first observed on southern highbush blueberry in 2006 in Georgia in the U.S., and has since been observed in the southeastern quadrant of the U.S. including the states of Florida, Mississippi, North Carolina and South Carolina. The disease has not been observed in the northern highbush blueberries (*V. corymbosum*) or native rabbiteye blueberries (*V. virgatum*) in the region. The symptoms initially appear as distinct necrotic rings with green centers (Figure 7A), but as the rings coalesce they can be confused with symptoms caused by several blueberry fungal pathogens. In severe cases the disease progresses to premature defoliation of bushes and initially was thought to be caused by septoria leaf spot, which can also cause premature defoliation. The necrotic rings are observed on the upper and lower surface of the leaves, in contrast to BBRV, which causes rings that usually are only observed on upper leaf surfaces and occasionally on young green shoots.

Figure 7: Blueberry necrotic ring blotch: (A) Virus symptoms in blueberry cultivar 'Star', upper row, symptomatic leaves showing range of necrotic rings and spots, lower row, healthy leaves; and (B) dsRNA extracted from symptomatic plants, the two bands between 1.7 kb and 2.5 kb, represent Totiviruses.

Investigations into the possibility of a viral etiology began with extraction of dsRNA from symptomatic and healthy leaf tissues. Six dsRNA bands were observed consistently from more than 10 sources of diseased tissue (Figure 7B), four of these were absent in unaffected tissue. The four dsRNAs were sequenced and found to have several unique features [76]. The most surprising was the presence of two helicases, which is unique among all known viruses. Additionally, these two helicases represent very different lineages suggesting that Blueberry necrotic ring blotch virus (BNRBV) is the result of a recombination event during a mixed infection in some host, not necessarily blueberry [76]. Additional support for creation of this virus by a recombination event is unique genome organization compared to its most closely related relative, *Citrus leprosis virus* (CiLV). CiLV has two RNAs compared with the four present in BNRBV. It appears that RNA 1 of CiLV is divided into RNA 1 and RNA 2 of BNRBV whereas RNA 2 of CiLV is divided into RNA 3 and RNA 4 of BNRBV. Thus, BNRBV likely represents a new virus genus [76]. *Hibiscus green spot virus*also shows similarities to BNRBV but has three genomic segments and triple gene block type of movement proteins in contrast to the 30K-superfamily type of CiLV and BNRBV.

The virus is detected readily by RT-PCR as there was a perfect correlation between symptomatic and asymptomatic tissue and RT-PCR positive and negative results, respectively, in over 60 isolates tested from multiple states. Future plans include testing that will address symptomless infections in rabbiteye, northern highbush, and additional cultivars of southern highbush blueberries.

Based on aa homologies with CiLV it is likely that BNRBV is transmitted by an eriophyid mite, and experiments to test this hypothesis are underway in Florida and Georgia [77,78]. In greenhouse trials, plants that tested-free of BNRBV developed symptoms in less than 3 weeks when placed adjacent to infected plants, whereas plants of the same cohort maintained in a greenhouse without any infected plants did not develop symptoms [78].

There is strong evidence for host resistance or tolerance, as some southern highbush blueberry cultivars are often found to exhibit extensive symptoms (*i.e.*, 'Star', 'O'Neal', and 'FL 86-19'), whereas others on the same farm or field can be symptomless. Although spread through vegetative propagation appears to be limited, suspect plants should not be utilized for propagation through hard- or softwood cuttings. Plants produced in tissue culture from source plants that have tested free of BNRBV may help to reduce the initial introduction of this disease since they are protected from transmission by an aerial vector during much of the plant increase propagation cycles.

Blueberry Red Ringspot Virus

Red ringspot was first described as a disease of highbush blueberry in New Jersey in 1950 [79]. The disease was shown to be graft transmissible and presumed to be caused by a virus. The symptoms appear as red rings on green stems (Figure 8B) and as red rings with pale green centers 2–3 mm in diameter or round red spots that can coalesce into blotches on older leaves in late summer (Figure 8C). The rings on leaves were traditionally thought to be visible only on the upper surface of the leaves and this was used as a diagnostic character, but some cultivars exhibit the rings on both sides of the leaves. Occasionally, reddish rings appear on the developing green fruit, but are usually not apparent when fruit is fully ripe. Infected bushes of the cultivar 'Ozarkblue' exhibit deformed fruit that are not marketable [80]. Many infected cultivars appear to bear a full crop. However, a limited study in Michigan reported a

25% crop loss in infected plants of the cultivar 'Blueray', but impact on yield in other cultivars is to be documented [81]. A similar disease was reported to occur in American cranberry [82], but in cranberry the leaf symptoms are limited to small red splotches and the rings on the fruit are light colored on the reddening berries late in the season (Figure 8D). The causal agent in blueberry was confirmed to be a virus in the family*Caulimoviridae* in the *Soymovirus* genus and designated *Blueberry red ringspot virus* (BRRV). Virus particles are 42–46 nm in diameter and the circular double-stranded DNA genome is 8.3 Kb [83]. The virus is easily detected by PCR from symptomatic tissue. Detection in asymptomatic leaves is unreliable, but scrapings of green bark from current-year stems can provide reliable detection. Antibodies for this virus are not yet available. Cranberry is infected by distinct strains of the virus [84]. Vectors for the BRRV, as with all members of the genus *Soymovirus*, are yet to be identified, although the virus is actively moving in infected fields in the eastern U.S. The disease has been reported in blueberry in New Jersey, Michigan, North Carolina, Georgia, New York, and Connecticut, [28]. The disease has also been reported in Japan [85], Korea [86], Poland [87], Slovenia [88], and Czech Republic [89]. Since symptoms on soft wood cuttings are variable among cultivars and symptoms on hardwood cuttings may not be visible, infected plants may inadvertently be used as source material for propagation. Thus, spread through propagation of infected plants is also a source of distribution. Significant work on the virus population structure in the U.S. and areas around the world has led to the development of sensitive and reliable detection protocols that can be used for virus screening and minimize the mayhem that can be caused by the propagation and distribution of symptomless material [84].

Figure 8: Symptoms of *Blueberry red ringspot virus*: (A) Early season leaf spots; (B) Red ringspots on young twigs; (C) Late season ring spots on leaves; and (D) Ringspots on cranberry fruit.

Blueberry Virus A

A new disorder named blueberry bronze leaf curl was observed in 2009 in Michigan in the U.S. Symptoms were observed in cultivars 'Jersey', 'Bluecrop', 'Elliott', 'Duke' and 'Pemberton' [90]. Plants exhibiting the disorder developed a bronze to red leaf color in mid-summer with some leaves curling upward (Figure 9). DsRNA extracted from symptomatic leaves was cloned and partially sequenced suggesting a novel virus in the genus *Closterovirus* was present in affected plants. Detection primers were developed, yielding amplicons of the new virus only from symptomatic plants [91]. A closterovirus, designated Blueberry virus A, from highbush blueberries in Japan recently has been sequenced and the entire genomic sequence is available in GenBank (accession # NC_018519.1, Isogai and Yoshikawa). Sequence comparisons between the U.S. and Japanese isolates showed 99% identity at the amino acid level, suggesting the same virus is present in Michigan and Japan. In Japan, the virus has been detected in the cultivars 'Spartan', 'Sierra', 'Bluecrop' and 'Coville' and there were no symptoms associated with the virus infection [92]. This suggests the possibility that the symptoms observed in Michigan may be due to mixed infections.

Figure 9: Symptoms of blueberry bronze leaf curl disease: A. Leaf exhibiting mid season reddening, healthy leaf on left, symptomatic leaf on the right; B. General bronzing with some upward curling of leaves (Photos Courtesy Annemiek Schilder, Michigan State University).\

CONCLUSIONS

It has been more than 15 years since the last review of blueberry viruses [27]. In this period there have been new viruses and diseases described whereas additional information on the epidemiology of known viruses has accumulated, knowledge that we aimed to capture in this review. As with the case of other berry crops, blueberry production has expanded to new areas, not only in North America, the cradle of vaccinium production but also around the world. Now, blueberry is grown on all continents except Antartica, from Europe, Asia, Australasia, Africa and South America. The major shift and expansion of blueberry production has led to new disease challenges. Blueberry is a prime candidate for problems caused by new and reemerging viruses for several reasons: (1) Many of the viruses that infect blueberry have a limited distribution compared to the areas where blueberry production occurs; (2) Production of this crop is expanding rapidly worldwide; (3) Plants are in high demand and in many cases cuttings for propagation are taken from fruiting fields without knowledge of their virus status; (4) As plantings are established in new areas the crop is being exposed to new viruses and vectors; (5) The reaction of new cultivars, which are deployed rapidly and widely, lack information on their susceptibility and reaction to viruses and other pathogens; and (6) Many countries do not have certification programs in place for this crop. All of the above reasons suggest that efforts to minimize the movement of viruses in planting stocks through effective certification programs will go a long way to safeguarding the blueberry industry. Plants are often sold across production regions, countries and continents, providing many opportunities for virus spread over long distances in planting stock. Development of certification programs to enhance the quality of the planting stock in terms of virus infection should be a high priority and where possible harmonization of the certification programs will facilitate safe and efficient movement of blueberry plants. There are a limited number of viruses that impact blueberries, but as the crop is grown in more regions it is expected that additional viruses will be detected in blueberries. A recent example of this is the occurrence of BNRBV in the southeastern U.S., which occurred only a few years after the industry there began to expand rapidly. Also, once introduced into the southeastern U.S., *Blueberry red ringspot virus* appears to spread much more rapidly there than in other regions where it has been reported, suggesting the presence of a more efficient vector.

Changes in cultural practices, changes in climate and reduced or altered pesticide use may all lead to increased virus incidence through indirect impacts on vector populations. As new cultivars are released, they may be more sensitive to virus infections and exhibit more severe symptoms and reduced yields. Alternatively, they may be tolerant of virus infection and facilitate movement of viruses in plants simply because they are symptomless carriers of a virus. Changes in climate likely will result in altered ranges of many virus vectors, such as, whiteflies, aphids, thrips, mealybugs, eriophyid mites *etc.* leading to changes in virus distribution. With the large scale of plant (ornamental and crop plants) movement internationally, introduction of new vectors into a production region can dramatically change the epidemiology of a virus disease (*i.e.*, vine mealybug introduction to California greatly increased the impact of Grapevine leafroll viruses; expansion of the glassy winged sharpshooter into California increased the impact of Xylella on grape production). Once a virus is introduced into a new region eradication efforts can be very expensive (*i.e.*,*Plum pox virus* eradication efforts after its introduction in Pennsylvania has cost the state more than 53 million dollars [93].

Blueberries are an expensive crop to establish, requiring extensive site preparation, irrigation, pruning and 3–5 years to bring into production. The cost of certified plants is relatively small compared to all of the other costs associated with getting a planting into production, yet using virus infected plants could lead to a completely unproductive planting, or at best a planting that takes longer to get into production and will have reduce yields over the life of the planting. Given the importance of the quality of mother plants and certification schemes that will minimize the movement of infected material, our group is working to provide material of the highest quality possible to the end user. Through the National Clean Plant Network we are working to eliminate all known viruses from propagation material, while developing sensitive and reliable tests for all blueberry viruses based on several isolates collected from multiple states in the U.S. and countries around the world. In addition, we study emerging diseases attributed to graft transmissible agents. This work aims to identify the agents and develop detection protocols that would allow for early detection and minimize movement within the indigenous area but also reduce the possibility of movement into areas where diseases are absent. As a systems-based approach to blueberry virus diseases we are also developing harmonized certification schemes among all blueberry production states in the U.S. that will allow for the movement of the highest quality propagation material to all areas where blueberry is grown.

ACKNOWLEDGMENTS

The authors acknowledge the importance of the USDA-APHIS National Clean Plant Network for funding research focusing on graft-transmissible agents of blueberry (Award Numbers 10-8100-1572, 11-8100-1572, 10-8100-1567-IA). We also acknowledge funding from the Southern Region Small Fruit Consortium, Oregon Blueberry Commission, Washington Blueberry Commission, British Columbia Blueberry Industry Development Council)

REFERENCES AND NOTES

1. Walters D.R., Keil D.J. Vascular Plant Taxonomy. 4th. Kendal/Hunt Publishing Company; Dubuque, IA, USA: 1996. Ericaceae; pp. 340–343.
2. Cho E., Seddon J.M., Rosner B., Willett W.C., Hankinson S.C. Prospective study of intake of fruits, vegetables, vitamins, and carotenoids and risk of age-related maculopathy. Arch. Ophthalmol.2004;122:883–892. doi: 10.1001/archopht.122.6.883.
3. Kalt W., Howell A.B., MacKinnon S.L., Goldman I.L. Selected bioactivities of *Vaccinium* berries and other fruit crops in relation to their phenolic contents. J. Sci. Food Agric. 2007;87:2279–2285. doi: 10.1002/jsfa.2985. [Cross Ref]
4. Rimando A.M., Nagmani R., Feller D.R., Yokoyama W. Pterostilbene, a newagonist for the peroxisome proliferator-activated receptor a-isoform, lowers plasma lipoproteins and cholesterol in hypercholesterolemic hamsters. J. Agric. Food Chem. 2005;53:3403–3407. doi: 10.1021/jf0580364.
5. Villata M. Trends in world blueberry production. [(accessed on 31 August 2012)]. Available online:http://www.growingproduce.com/article/26272/trends-in-world-blueberry-production/
6. Oudemans P.V., Caruso F., Stretch A. Cranberry fruit rot in the Northeast: A complex disease. Plant Dis.1998;82:1176–1184. doi: 10.1094/PDIS.1998.82.11.1176.
7. Martin R.R., Tzanetakis I.E., Caruso F.L., Polashock J.J. Emerging and reemerging virus diseases of blueberry and cranberry. Acta Hortic. 2009;810:299–304.
8. Isogai M., Tatuto N., Ujiie C., Watanabe M., Yoshikawa N. Identification and characterization of blueberry latent spherical virus, a new member of subgroup C in the genus Nepovirus. Arch. Virol.2012;157:297–303. doi: 10.1007/s00705-011-1177-7.
9. Jaswal A.S. Occurrence of blueberry leaf mottle, blueberry shoestring-, tomato ringspot and tobacco ringspot viruses in eleven halfhigh blueberry clones grown in New Brunswick, Canada. Can. Plant Dis. Surv.

1990;70:113–117.

10. Martin R.R., Tzanetakis I.E. USDA-ARS, Corvallis, OR, USA and University of Arkansas, Fayetteville, AR, USA. Unpublished data, 2011.
11. Ramsdell D.C., Stace-Smith R. Blueberry leaf mottle, a new disease of highbush blueberry. Acta Hortic.1979;95:37–45.
12. Sandoval C.R., Ramsdell D.C., Hancock J.F. Infection of wild and cultivated *Vaccinium* spp. with blueberry leaf mottle nepovirus. Ann. Appl. Biol. 1995;126:457–464. doi: 10.1111/j.1744-7348.1995.tb05380.x.
13. Ramsdell D.C., Stace-Smith R. Physical and chemical properties of the particles and ribonucleic acid of blueberry leaf mottle virus. Phytopathology. 1981;71:468–472.
14. Bacher J.W., Warkentin D., Ramsdell D., Hancock J.F. Sequence analysis of the 3' termini of RNA1 and RNA2 of blueberry leaf mottle virus. Virus Res. 1994;33:145–156. doi: 10.1016/0168-1702(94)90051-5.
15. Sanfaçon H., Iwanami T., Karasev A.V., van der Vlugt R., Wellink J., Wetzel T., Yoshikawa N. Secoviridae. In: King A.M.Q., Adams M.J., Carstens E.B., Lefkowitz E.J., editors. Virus Taxonomy, Ninth Report of the International Committee on Taxonomy of Viruses. Academic Press; San Diego, CA, USA: 2011. pp. 881–899.
16. Childress A.M., Ramsdell D.C. Bee-mediated transmission of blueberry leaf mottle virus via infected pollen in highbush blueberry. Phytopathology. 1987;77:167–172. doi: 10.1094/Phyto-77-167.
17. Elbeaino T., Digiaro M., Fallanaj F., Kuzmanovic S., Martelli G.P. Complete nucleotide sequence and genome organisation of grapevine Bulgarian latent virus. Arch. Virol. 2011;156:875–879. doi: 10.1007/s00705-010-0908-5.
18. Hildebrand E.M. A new case of rosette mosaic of peach. Phytopathology. 1941;31:353–355.
19. Dias H.F., Cation D. The characterization of a virus responsible for peach rosette mosaic and grape decline in Michigan. Can. J. Bot. 1976;54:1228–1239.
20. Ramsdell D.C., Gillet J.M. Reach rosette mosaic virus in highbush blueberry. Plant Dis. 1981;65:755–758.
21. Lammers A.H., Allison R.F., Ramsdell D.C. Cloning and sequencing of peach rosette mosaic virus RNA 1. Virus Res. 1999;65:57–73. doi: 10.1016/S0168-1702(99)00104-5.
22. Eveleigh E.S., Allen W.R. Description of *Longidorus diadecturus* n.sp. (Nematoda: Longidoridae), a vector of the peach rosette mosaic virus in peach orchards in southwestern Ontario, Canada. Can. J. Zool.1982;60:112–115. doi: 10.1139/z82-014.
23. Klos E.J., Fronec F., Knierim J.M., Cation D. Peach rosette mosaic transmission and control studies.Mich. Agric. Exp. Station Q. Bull. 1967;49:287–293.
24. Awad M.A.E., Ibrahem L.M., Aboul-Ata A.E., Ziedan M., Mazyad H.M., Abdel-Aziz E., Mansour N. Virus-free plum and peach mother plant production in Egypt. Acta Hortic. 1998;472:531–536.
25. Azerý T., Çýçek Y. Detection of virus diseases affecting almond nursery trees in Western Anatolia (Turkey) EPPO Bull. 1997;27:547–550. doi: 10.1111/j.1365-2338.1997.tb00682.x.
26. Varney E.H., Raniere L.C. Necrotic ringspot, a new virus disease of cultivated blueberry.Phytopathology. 1960;50:241.
27. Lister R.M., Raniere L.C., Varney E.H. Relationships of viruses associated with ringspot diseases of blueberry. Phytopathology. 1963;53:1031–1035.
28. Caruso F.L., Ramsdell D.C. Compendium of Blueberry and Cranberry Diseases. APS Press; St. Paul, MN, USA: 1995. p. 87.
29. Fuchs M., Abawi G.S., Marsella-Herrick P., Cox R., Cox K.D., Carroll J.E., Martin R.R. Occurrence of*Tomato ringspot virus* and *Tobacco ringspot virus* in highbush blueberry in New York State. J. Plant Pathol.2010;92:451–459.
30. Medina C., Matus J.T., Zúñiga M., San-Martin C., Arce-Johnson P. Occurrence and distribution of viruses in commercial plantings of *Rubus*, *Ribes* and *Vaccinium* species in Chile. Cienc. Investig. Agrar.2006;33:23–28.
31. Tzanetakis I.E. University of Arkansas, Fayetteville, AR, USA. Unpublished data, 2010.
32. Pinkerton J.N., Martin R.R. Management of Tomato ringspot virus in red raspberry with crop rotation.Int. J. Fruit Sci. 2005;5:55–67. doi: 10.1300/J492v05n03_06.
33. MacDonald S.G., Martin R.R., Bristow P.R. Characterization of an ilarvirus associated with a necrotic shock reaction in blueberry. Phytopathology. 1991;81:210–214. doi: 10.1094/Phyto-81-210.
34. Bristow P.R., Windom G.E., Martin R.R. Recovery of plants infected with Blueberry shock virus (BlShV) Acta Hortic. 2002;574:85–89.
35. Bristow P.R., Martin R.R. Transmission and the role of honeybees in field spread of Blueberry shock ilarvirus, a pollen borne virus ofhighbush blueberry. Phytopathology. 1999;89:124–130. doi: 10.1094/PHYTO.1999.89.2.124.
36. Martin R.R. USDA-ARS, Corvallis, OR, USA. Unpublished data, 2011.
37. Schilder A. Blueberry shock virus, Pest Alert Factsheet. Michigan State University Extension, East Lansing, MI, USA, 2009.
38. Jones A.T., McGavin W.J., Dolan A. Detection of Tobacco streak virus isolates in North American cranberry (*Vaccinium macrocarpon* Ait.) Plant Health Progr. 2011
39. Vorsa N. Rutgers University, Chatsworth, NJ, USA. Personal communication, 2012.
40. Sdoodee R., Teakle D.S. Transmission of tobacco streak virus by *Thrips tabaci* a new method of plant virus transmission. Plant Pathol. 1987;36:377–380. doi: 10.1111/j.1365-3059.1987.tb02247.x.
41. Stretch A.W. A previously undescribed blight disease of highbush blueberry in New Jersey.Phytopathology. 1983;73:375.

42. Bristow P.R., Martin R.R. A virus associated with a new blight disease of highbush blueberry.Phytopathology. 1987;77:1721–1722.

43. Podleckis E.V., Davia R.F., Stretch A.W., Schulz C.P. Flexuous rod particles associated with Sheep Pen Hill Disease of highbush blueberries. Phytopathology. 1986;76:1065.

44. Martin R.R., Bristow P.R. A Carlavirus associated with blueberry scorch disease. Phytopathology.1988;78:1636–1640.

45. Martin R.R., MacDonald S.G., Podleckis E.V. Relationships between blueberry scorch and Sheep Pen Hill viruses of highbush blueberry. Acta Hortic. 1992;308:131–139.

46. Podleckis E.V., Davis R.F. Infection of highbush blueberries with a putative carlarvirus. Acta Hortic.1989;241:338–343.

47. Bernardy M.G., Dubeau C.R., Braun A., Harlton C.E., Bunckle A., Lowery D.T., French C.J., Wegener L. Molecular characterization and phylogenetic analysis of two distinct strains of Blueberry scorch virus from western Canada. Can. J. Plant Pathol. 2005;27:581–591. doi: 10.1080/07060660509507259.

48. Cavileer T.D., Halpern B.T., Lawrence D.M., Podleckis E.V., Martin R.R., Hillman B.I. Nucleotide sequence of the carlavirus associated with blueberry scorch and similar diseases. J. Gen. Virol. 1994;75:711–720. [PubMed]

49. Martin R.R., Bristow P.R., Wegener L.A. Scorch and Shock: Emerging virus diseases of highbush blueberry and other *Vaccinium* spp. Acta Hortic. 2006;715:463–467.

50. Richert-Pöggeler K.R., Turhal A.K., Schuhmann S., Maaß C., Blockus S., Zimmermann E., Eastwell K.C., Martin R.R., Lockhart B. Carlavirus biodiversity in horticultural host plants: Efficient virus detection and identification combining electron microscopy and molecular biology tools. Acta Hortic. 2013 in press.

51. Ciuffo M., Pettiti D., Gallo S., Masenga V., Turina M. First report of *Blueberry scorch virus* in Europe.Plant Pathol. 2005;54:565. doi: 10.1111/j.1365-3059.2005.01198.x.

52. Bristow P.R., Martin R.R., Windom G.E. Transmission, field spread, cultivar response, and impact on yield in highbush blueberry infected with blueberry scorch virus. Phytopathology. 2000;90:474–479. doi: 10.1094/PHYTO.2000.90.5.474.

53. Wegener L.A., Punja Z.K., Martin R.R. First Report of *Blueberry scorch virus* in cranberry in Canada and the United States. Plant Dis. 2004;88:427.

54. Wegener L.A., Punja Z.K., Martin R.R. First Report of *Blueberry scorch virus* in black huckleberry in British Columbia. Plant Dis. 2007;91:328.

55. Lowery D.T., French C.J., Bernardy M. *Nicotiana occidentalis*: A new herbaceous host for Blueberry scorch virus. Plant Dis. 2005;89:205.

56. Lawrence D.M., Hillman B.I. Synthesis of infectious transcripts of blueberry scorch carlavirus *in vitro*. J. Gen. Virol. 1994;75:2509–2512. doi: 10.1099/0022-1317-75-9-2509.

57. Oudemans P.V., Hillman B.I., Linder-Basso D., Polashock J.J. Visual inspections of nursery stock fail to protect new plantings from *Blueberry scorch virus* infection. Crop Prot. 2011;30:871–875. doi: 10.1016/j.cropro.2011.02.013.

58. Ramsdell D.C. Physical and chemical properties of Blueberry shoestring virus. Phytopathology.1979;69:1087–1091.

59. Varney E.H. Mosaic and shoestring virus diseases of cultivated blueberry in New Jersey. Phytopathology.1957;47:307–309.

60. Ramsdell D.C. Blueberry Shoestring. In: Converse R.H., editor. Virus Diseases of Small Fruits. US Department of Agriculture, Agriculture Handbook No. 631, US Government Printing Office; Washington, D.C., USA: 1987. pp. 103–105.

61. Lesney M.S., Ramsdell D.C., Sun M. Etiology of blueberry shoestring disease and some properties of the causal virus. Phytopathology. 1978;68:295–300.

62. Truve E., Fargette D. Sobemovirus. In: King A.M.Q., Adams M.J., Carstens E.B., Lefkowitz E.J., editors. Virus Taxonomy, Ninth Report of the International Committee on Taxonomy of Viruses. Academic Press; San Diego, CA, USA: 2011. pp. 1185–1189.

63. Morimoto K.M., Ramsdell D.C., Gillett J.M., Chaney W.G. Acquisition and transmission of Blueberry shoestring virus by its aphid vector *Illinoia pepperi*. Phytopathology. 1985;75:709–712. doi: 10.1094/Phyto-75-709.

64. Acquaah T., Ramsdell D.C., Hancock J.F. Resistance to blueberry shoestring virus in southern highbush and rabbiteye cultivars. HortScience. 1995;30:1459–1460.

65. Martin R.R., Tzanetakis I.E., Sweeney M., Wegener L. A virus associated with blueberry fruit drop disease. Acta Hortic. 2006;715:497–501.

66. Martin R.R., Zhou J., Tzanetakis I.E. Blueberry latent virus: An amalgam of the *Partitiviridae* and the*Totiviridae*. Virus Res. 2011;155:175–180. doi: 10.1016/j.virusres.2010.09.020.

67. Sabanadzovic S., Valverde R.A., Brown J.K., Martin R.R., Tzanetakis I.E. Southern tomato virus: The link between the families *Totiviridae* and Partitiviridae. Virus Res. 2009;140:130–137. doi: 10.1016/j.virusres.2008.11.018.

68. Isogai M., Nakamura T., Ishii K., Watanabe M., Yamagishi N., Yoshikawa N. Histochemical detection of Blueberry latent virus in highbush blueberry plant. J. Gen. Plant Pathol. 2012;77:304–306.

69. Ramsdell D.C., Stretch A.W. Blueberry Mosaic. In: Converse R.H., editor. Virus Diseases of Small Fruits. US Department of Agriculture, Agriculture Handbook No. 631; Washington, D.C., USA: 1987. pp. 119–120.

70. Martin R.R. USDA-ARS, Corvallis, OR, USA. Unpublished data, 2012.

71. Tzanetakis I.E., Martin R.R. University of Arkansas, Fayetteville, AR, USA and USDA-ARS, Corvallis, OR, USA. Unpublished data, 2012.

72. Zhu S.F., Gillet J.M., Ramsdell D.C., Hadidi A. Isolation

and characterization of viroid-like RNAs from mosaic diseased blueberry tissue. Acta Hortic. 1995;385:132–140.

73. Thekke-Veetil T., Martin R.R., Tzanetakis I.E. University of Arkansas, Fayetteville, AR, USA (T.-V.T. and T.I.E.) and USDA-ARS, Corvallis, OR, USA (M.R.R.). Unpublished data, 2012.

74. Vaira A.M., Gago-Zachert S., Garcia M.L., Guerri J., Hammond J., Milne R.G., Moreno P., Morikawa T., Natsuaki T., Navarro J.A., et al. Ophiovirus. In: King A.M.Q., Adams M.J., Carstens E.B., Lefkowitz E.J., editors. Virus Taxonomy, Ninth Report of the International Committee on Taxonomy of Viruses. Academic Press; San Diego, CA, USA: 2011. pp. 743–748.

75. Martin R.R. USDA-ARS, Corvallis, OR, USA. Personal observations, 2006.

76. Quito-Avila D.F., Martin R.R. Blueberry necrotic ring blotch virus represents a unique genus of plant RNA viruses. Phytopathology. 2012;102:S4.96.

77. Burkle C., Olmstead J.W., Harmon P.F. A potential vector of Blueberry necrotic ring blotch virus and symptoms on various host genotypes. Phytopathology. 2012;102:S4.17.

78. Robinson T.S., Brannen P.M., Deom C.M. Blueberry necrotic ring blotch: A new disorder of southern highbush blueberries. Phytopathology. 2012;102:S4.101.

79. Hutchinson M.T., Varney E.H. Ringspot—A virus disease of cultivated blueberry. Plant Dis. Rep.1954;38:260–262.

80. Cline W.O., Ballington J.R., Polashock J.J. Blueberry red ringspot observations and findings in North Carolina. Acta Hortic. 2009;810:305–312.

81. Gillett J.M. Physical and chemical properties of blueberry red ringspot virus. M.S. Thesis, Michigan State University, East Lansing, MI, USA, 1988.

82. Boone D.M. Ringspot disease of cranberry. Plant Dis. Rep. 1966;50:543–545.

83. Glasheen B.M., Polashock J.J., Lawrence D.M., Gillett J.M., Ramsdell D.C., Vorsa N., Hillman B.I. Cloning, sequencing, and promoter identification of Blueberry red ringspot virus, a member of the family Caulimoviridae with similarities to the "Soybean chlorotic mottle-like" genus. Arch. Virol. 2001;147:2169–2186. [PubMed]

84. Polashock J.J., Ehlenfeldt M.K., Crouch J.A. Molecular detection and discrimination of blueberry red ringspot virus strains causing disease in cultivated blueberry and cranberry. Plant Dis. 2009;93:727–733.

85. Isogai K., Ishii M., Umemoto S., Watanabe M., Yoshikawa N. First report of blueberry red ringspot disease caused by *Blueberry red ringspot virus* in Japan. J. Gen. Plant Pathol. 2009;75:140–143. doi: 10.1007/s10327-009-0145-5.

86. Cho I.S., Chung B.N., Cho J.D., Choi G.S. First report of *Blueberry red ringspot virus* infecting highbush blueberry in Korea. Plant Dis. 2012;96:1074.

87. Kalinowska E., Paduch-Cichal E., Chodorska M., Sala-Rejczak K. First report of *Blueberry red ringspot virus* in highbush blueberry in Poland. J. Plant Pathol. 2011;93:S4.73.

88. Pleško I.M., Marn M.V., Koron D. Detection of *Blueberry red ringspot virus* in highbush blueberry cv. 'Coville' in Slovenia. Julius-Kuhn-Archives. 2010;427:204–205.

89. Přibylová J., Špak J., Kubelková D., Petrzik K. First Report of *Blueberry red ringspot virus* in Highbush Blueberry in the Czech Republic. Plant Dis. 2010;94:1071.

90. Schilder A.C. Michigan State University, East Lansing, MI, USA. Personal communication, 2009.

91. Martin R.R., Schilder A.C. USDA-ARS, Corvallis, OR, USA and Michigan State University, East Lansing, MI, USA. Unpublished data, 2011.

92. Isogai M. Iwate University, Morioka, Iwate, Japan. Personal communication, 2012.

93. Welliver R.A. Plum pox virus case study: The eradication road is paved in gold. Phytopathology2012, 102, S4.154.

CHAPTER 2

SHARKA: THE PAST, THE PRESENT AND THE FUTURE

Jiri Sochor[1, 2, 3], Petr Babula[1, 2, 3], Vojtech Adam[1, 3], Boris Krska[4], and Rene Kizek[1, 3]

[1]Department of Chemistry and Biochemistry, Faculty of Agronomy, Mendel University in Brno, Zemedelska 1, CZ-613 00 Brno, Czech Republic

[2]Department of Natural Drugs, Faculty of Pharmacy, University of Veterinary and Pharmaceutical Sciences Brno, Palackeho 1-3, CZ-612 42, Czech Republic

[3]Central European Institute of Technology, Brno University of Technology, Technicka 3058/10, CZ-616 00 Brno, Czech Republic

[4]Department of Fruit Growing, Faculty of Horticulture, Mendel University in Brno, Valticka 337, CZ-691 44 Lednice, Czech Republic

ABSTRACT

Members the *Potyviridae* family belong to a group of plant viruses that are causing devastating plant diseases with a significant impact on agronomy and economics. *Plum pox virus* (PPV), as a causative agent of sharka disease, is widely discussed. The understanding of the molecular biology of potyviruses including PPV and the function of individual proteins as products of genome expression are quite necessary for the proposal the new antiviral strategies. This review brings to view the members of *Potyviridae* family with respect to *plum pox virus*. The genome of potyviruses is discussed with respect to protein products of its expression and their function. *Plum pox virus* distribution, genome organization, transmission and biochemical changes in infected plants are introduced. In addition, techniques used in PPV detection are accentuated and discussed, especially with respect to new modern techniques of nucleic acids isolation, based on the nanotechnological approach. Finally, perspectives on the future of possibilities for nanotechnology application in PPV determination/identification are outlined.

INTRODUCTION

Sharka or pox or plum pox disease is one of the most devastating viral diseases of stone fruits, with a significant impact on agronomy and economics. The disease is caused by *plum pox virus* (PPV), a member of *Potyvirus* genus in *Potyviridae* family. It is able to damage not only plums (especially *Prunus domestica*L.) [1,2], but also apricots (*Prunus armeniaca* L.), peaches (*Prunus persica* (L.) Batsch.) [3], nectarines (*Prunus persica* var. *nucipersica* (Borkh.) C.K. Schneid., hybrids) [4], almonds (*Prunus dulcis* (Mill.) D. A. Webb, syn. *P. amygdalus* (L.) Batsch., *Amygdalus communis* L., *Amygdalus dulcis* Mill.) [5,6,7], sweet cherries (*Prunus avium* (L.)) [8,9] and tart cherries (*Prunus cerasus* L.) [10,11,12]. PPV is able to infect also important ornamental and wild *Prunus* L species, including those used in traditional medicine: myrobalan (*P. cerasifera* Ehrh.) [13,14,15], American plum (*P. americana* Marsh.) [16,17], dwarf flowering almond (*P. glandulosa* Thunb.) [18,19], and blackthorn (*P. spinosa* L.) [17,20,21,22]. In addition, almost 60 species in eight families have been recognized as experimental host plants [23,24,25,26,27]. Nevertheless, *plum pox virus* is able to infect naturally occurring plants, which are not members of genus *Prunus*. Infected

Citation: Jiri Sochor, Petr Babula, Vojtech Adam, Boris Krska, and Rene Kizek, Sharka: The Past, the Present and the Future, doi: 10.3390/v4112853.

walnut trees (*Juglans regia* L.) were reported in Slovakia in 1996 by Baumgartnerova [28]. This information was based on the presence of diffuse spots on the leaves of infected plants. Despite the fact that PPV infection was detected by ELISA and the virus transmitted to the host plants, the presence of PPV-infected *Juglans regia* trees was not later confirmed, see Polak [29]. On the other hand, ELISA, ISEM and biological tests confirmed presence of PPV in *Euonymus europae* L. (*Celastraceae* R. Br.) and *Ligustrum vulgare* L. (*Oleaceae* Hffg. et Link) [29]. PPVis one of the most important members of family *Potyviridae* and the results of the infection are shown in Figure 1A-D. Family *Potyviridae* includes flexuous filamentous rod-shaped viruses of lengths of 650-950 nm and diameter of 12-15 nm, more exactly ssRNA non-encapsulated viruses with almost one-third plant viruses. The family includes seven genera as *Brambyvirus*, *Bymovirus*,*Ipomovirus*, *Macluravirus*, *Potyvirus*, *Rymovirus* and *Tritimovirus*. The most important members, their characteristics including vector(s), host plant(s) and distribution have been summarized in a plenty of extensive reviews, such as Gibbs and Ohshima [30], Gibbs *et al.* [31], Revers *et al.* [32], Riechmann *et al.*[33], and Ward and Shukla [34].

Potyviruses are transmitted predominantly by insects. Genera *Potyvirus* and *Macluravirus* are transmitted by aphids (*Aphis* Linnaeus, *Brachycaudus* van der Goot, *Brevicoryne* van der Goot, *Macrosiphum* Passerini, *Myzus* Passerini, *Phorodon* Passerini, and *Rhopalosiphum* Koch) [35,36,37,38]. In addition, eriophyid mites spread members of *Rymovirus* and *Tritimovirus* (*Abacarus* Keifer, *Aceria* Keifer) [39,40,41] and whiteflies (*Bemicia* Quaintance et Baker) spread members of *Ipomovirus* [42,43,44]. Only members of *Bymovirus* are transmitted by plasmodiophoroid fungus (*Polymyxa graminis* Ledingham, *Plasmodiophoridae*) [45,46,47,48]. It has been established that potyviruses are generally transmitted by non-circulative manner, which means that virus articles do not cross the vector cell membranes. However, not all potyviruses require specific associations with aphids. This exception is represented by *Bymovirus*, where species are restricted to*Poaceae* (*Graminae*) and are transmitted in soil by zoospores of *Polymyxa graminis* Ledingham. In this case, it is very difficult to study the epidemiology of diseases caused by *Bymovirus*, because they are not transmitted mechanically, purified preparations demonstrate only low infectivity, and the vector is an obligate parasite of roots that inhabits the individual cells of rhizodermis and cortex [49]. In addition, they frequently occur in a mixture with furoviruses. A new soilborne virus - *soybean leaf rugose mosaic virus* (SLRMV) - has been isolated and characterized by Kuroda *et al.* [50]. This virus is closely related to bymoviruses, however, further characterization is necessary. However, *Polymyxa graminis* is obligate parasite of monocot plants (*Glycine max* (L.) Merr. is dicot) and soybeans became diseased when grown in virus-infested soil [50]. With exception of *Bymovirus* with bipartite genome, where open reading frame is divided between two genomic RNAs, all potyviruses have monopartite genome with a genome-linked protein attached to the 5′ end (VPg) and a polyadenosine tail at the 3′ end of the genome. Their genome was characterised also as monocistronic for many years because of the presence of only one functional ORF. However, bioinformatic evidence and experimental verification of the evidence for overlapping coding sequences within the P3 cistron of potyviruses have been presented in the work of Chung *et al.* in 2008 [51].

Figure 1: (A) Peach blossoms showing typical speckling of Plum Pox infected trees. Note the darker pink stripes on petals. (B) Chlorotic rings and blotches develop in peach leaves. (C) Yellow rings on a yellow-fleshed peach cultivar. (D) Ring patterns visible on the pit of an apricot.

In a Turnip mosaic virus, *pipo* (Pretty Interesting *Potyviridae* ORF) consists of 60 codons in the +2 reading frame relative to polyprotein, which places it within the P3 cistron. Disruption of *pipo* leads to the no changes in the polyprotein amino acid sequence; however, these mutants are not able to establish systemic infection. The ability of *pipo* mutants to establish systemic infection is still being discussed. Whereas Wen and Hajimorad in their mutational analysis of the putative pipo of *soybean mosaic virus* suggested that the disruption of PIPO protein impedes movement [52], the same authors together with Maroof later established that not PIPO but P3 is responsible for the changes in the virulence of the *soybean mosaic virus* [53]. A cap-independent translation is mediated by the 5′UTR [54]. The genome is expressed as a single polypeptide (polyprotein) precursor, which is proteolytically processed to mature virus proteins. However, due to fact that*pipo* that is predicted to encode a 7kDa protein was discovered to overlap with the P3 coding regions in all potyviruses. Immunodetection of the *pipo*-encoded protein in Turnip mosaic virus infected cells revealed a 25 kDa polypeptide, consistent with expression of *pipo* as a translational fusion with the N-terminus of P3 [51]. This fact must be carefully considered in the characterization of potyviruses genome organisation. The potyvirus genome consists of a positive-sense single messenger-polarity RNA molecule of the size of about 10 kb (generally 8.5 – 12 kb). The genome organization of potyviruses indicates nine cleavage sites and 11 mature proteins such as P1, HCpro, P3, P3N-PIPO, 6K1, CI, 6K2, NIa (respectively VPg and NIa-Pro), NIb and CP. In the light of above-mentioned fact, PIPO protein is expressed as a trans-frame protein consisting of the amino-terminal half of P3 fused to PIPO and a distinct C-terminus as a result of translational frameshift (P3N-PIPO). There are some exceptions in the potyviruses genome organisation [55]. *Ipomovirus* represents exception in the absence of HCpro and presence of two copies of P1 protein (P1a and P1b), which is shown in Figure 2. The nascent polyprotein precursor is cleaved by three viral proteinases (they are endopeptidases) as P1 (P1/HCpro cleavage site, serine proteinase with *cis*-cleavage activity), HCpro (HCpro/P3 cleavage site, HCpro cysteine proteinase with *cis*-cleavage activity), and NIa, respectively C-terminal proteinase domain NIa-Pro (NIa-Pro cysteine proteinase, all other cleavage sites) [56]. Potyviruses encode specific nonstructural protein called helper component (HC) that is obligatory for virus transmission by aphids. Its role probably consists in binding both to virions and to the cuticular lining of aphid mouthparts with retaining the virions within the maxillary food canal of the aphid stylets. This short-term reversible binding is most likely necessary and sufficient for successful virus transmission from one plant to other plants. However, above-mentioned facts are hypothetical and they have been not fully demonstrated. Functional proteins are generated after action of virus-encoded proteases. P1 proteinase is located at the very beginning of the viral genome of potyviruses. Due to the fact that the serine-like peptidase domain in the P1 coding region was identified in the C-terminal, highly conserved portion for all serine peptidases, associated with proteolytic activity and containing the Gly-Xaa-Ser-Gly motif, P1 is serine peptidase [57]. It varies in the length as well as in an amino acid sequence [58]. P1 poses RNA-binding activity and the possible influence in cell-to-cell viral spreading. Its role in defining the host plant is studied too [59]. Work of Shi *et al.* that showed interactions between P1 proteinase of *soybean mosaic virus* and Rieske Fe/S soybean protein are probably responsible for development of the symptoms of viral disease [60]. This N-terminal "hypervariable" protein is the most divergent protein among all potyviruses [61]. The *Tritimovirus* P1 functions as a suppressor of RNA silencing and an enhancer of disease symptoms were investigated by Young *et al.* [62]. However, P1 function must be further studied [63]. Potyviral helper component-proteinase (HCpro) has been recognized to be involved in the processes necessary for life, respectively infection cycle, especially in interactions with host proteins as well as other viral proteins. Thus, the presence of HC-pro is essential for all potyviruses [64]. These processes include not only interactions with host proteins, but also polyprotein processing and suppression of antiviral RNA silencing [65,66]. In addition, possible interactions between HCpro and host proteins must be carefully considered. This fact has been demonstrated in work of Shen *et al.,* who observed interactions between HCpro and full-length papaya calcireticulin, the multifunctional protein that regulates intracellular calcium(II) ions levels and protein folding in the endoplasmic reticulum [67]. In conclusion, HCpro has multiple functions. Despite the HCpro multiple function, the understanding of its function is still poor. The work of Guo *et al.* revealed the crystal function of cysteine protease domain of HCpro from *turnip mosaic virus* [68]. HCpro cleavages only dipeptide Gly-Gly at its C terminus. A cleaved C domain remains tightly bond at the active site cleft to prevent *trans* activity. Generally, the structure adopts a compact alpha/beta-fold. The catalytic cysteine and histidine residues constitute an active site and HCpro recognizes a consensus sequence YXGVV around the cleavage site between two glycine residues [68]. There are at least six cleavage sites in *cis-/trans*-arrangement in the viral polyprotein recognized by NIa proteinase (NIa-Pro). NIa-Pro represents C-terminal domain of NIa and is shown to be analogous to picornavirus 3C protease [69]. The N-terminal domain of NIa is designated as VPg and will be further discussed. The active – catalytic – sites of NIa-Pro are defined by a consensus heptapeptide sequence surrounding each cleavage site containing triad of His-Asp-Cys, which are conserved among the potyviruses [70]. The mutation of the catalytic residues His46, Asp81, and Cys151 resulted in complete loss of activity [71]. In addition, NIa proteinase possesses structural motifs shared with cellular serine proteases with the substitution of serine by a cysteine as the active site nucleophile. The specificity of enzyme and control its activity is discussed in some recently published papers [72,73,74], which are focused on the investigation of recombinant viruses and their host interactions. Mutation in the Lys(27) of NIa-Pro has established the role of this enzyme in the determination of host specificity. On the other hand, Lys(27) mutation did not affect the protease activity of NIa-Pro [75]. The NIa-Pro is a sequence-specific proteinase required for processing of viral polyprotein in the cytoplasm. This fact has been demonstrated in many potyviruses, such as *pepper vein banding virus* (PVBV), and is generally relevant for all known potyviruses [76]. Its accumulation in nuclei of infected cells manifested as a formation of "inclusion bodies" and has been demonstrated in some papers, for example see work of Anindya *et al.* [71] or Restrepo *et al.* [77]. This fact has been established by Knuhtsen, who observed tobacco etch virus-induced intranuclear inclusions [78]. However, this author is not the first to demonstrate the presence of nuclear inclusions in the nuclei of infected cells. In 1968, Shepard used electron microscopy for the characterization of nuclear (and cytoplasmic as well) inclusions in tobacco etch virus-infected*Nicotiana tabacum* L. cv. Havana 425 [79], and before him in 1941, Sheffied observed formation of nuclear inclusions in infected cells [80]. The NIa-Pro nuclear localization is still investigated. Accumulation of Nia-Pro in nuclei is observable mainly at the end of viral infection cycle. It seems that NIa-Pro executes DNA cleavage activity, respectively nonspecific double-stranded DNA degradation. This fact indicates the NIa-Pro role in viral infection cycle with subsequent degradation of the host DNA [71]. Potential use of NIa-Pro in biotechnologies is indicated in the paper by Fellers *et al.,* where three genes, each consisting of two NIa encoding regions (*tobacco vein mottling virus, tobacco etch virus* and *potato virus Y*) were introduced into*Nicotiana tabacum* L. cv. Burley 21. The authors indicate that "results of above-mentioned paper showed that different potyvirus NIa-Pro genes can be used for a protection against potyvirus infection and may protect plants against more than one potyvirus" [81]. Generally, transformation of plants with translatable sequences corresponding to the structural proteins, such as CP, or non-structural proteins such as the NIb or the P1 of potyviruses can make the plants highly resistant to infection with the corresponding virus [82,83,84]. Pathogen-derived resistance to viruses in transgenic plants is typically based on RNA silencing. This fundamental cytoplasmic antiviral defense system can be efficiently directed against viruses by producing homologous RNA from a transgene in plant cells.

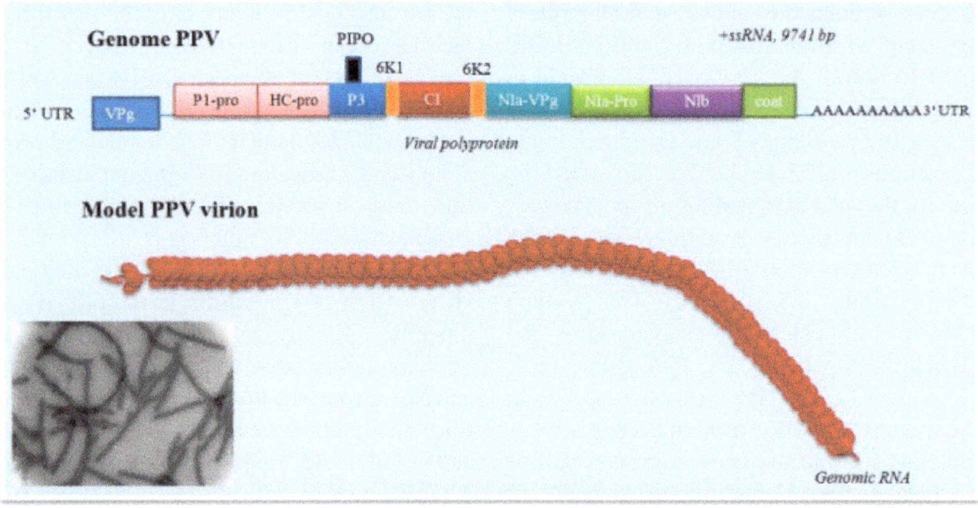

Figure 2: Potyvirus genome. CP: capsid protein; P1: P1 proteinase; HC-pro: helper component proteinase; P3: protein P3; PIPO: protein; CI: cytoplasmic inclusion protein; Nia-VPg: viral genome-linked protein; NIa-Pro: nuclear inclusion protein a; Nib: nuclear inclusion protein N (RNA-directed RNA polymerase).

Plant potyviruses encode two membrane proteins 6K and P3. The role of P3 protein is only poorly understood. The role of P3 in viral infection cycle and as a symptom determinant is shown in the paper by Jenner *et al.* [85]. Development of symptoms of the disease may be closely connected with interactions between the P3 protein and proteins of host plant. The paper by Lin *et al.* demonstrates the possibility of interactions between the P3 protein encoded by *shallot yellow stripe virus* onion isolate and small subunit of RubisCo (ribulose-1,5-bisphosphate carboxylase oxygenase), enzyme, which is essentially involved in carbon fixation [86]. Point mutagenesis in the P3 *soybean mosaic virus* proved its role in virulence in Rsv4 genotype soybean. A single amino acid substitution (Q1033K) in the P3 led to the resistance towards Rsv4-mediated resistance [87]. The P3 protein role in virulence was confirmed by recently published papers [88,89,90,91]. The study of Cui *et al.* was focused on determination of the P3 protein role in *Nicotiana benthamiana* Domin (*Solanaceae*) leaf cells infected by *tobacco etch virus* (TEV) [92]. Authors carried out TEV subcellular localization. The P3 protein was localized in endoplasmic reticulum and formed punctate inclusions in association with Golgi apparatus. These structures originated directly from the endoplasmic reticulum exit side. Localization of P3 (and, in addition P3-6K intermediate) in the endoplasmic reticulum was previously reported by Eiamtanasate *et al.* [93]. Finally, the P3 punctate structure was found to traffic along the actin filaments and colocalize with the W-containing replication vesicles. Deletion analyses demonstrated the irreplaceable role of P3 domains in the retention of the P3 at the Golgi [92]. The 6K1 protein is generally quite insufficiently examined. In addition, data about its function is almost missing. The 6K1 protein is probably responsible for the cell-to-cell movement. This fact may be confirmed by the paper by Hong *et al.,* who observed the 6K1 protein at the cell periphery of infected leaves of *Pinellia ternata*(Thunb.) Ten. ex Breitenb. (*Araceae*) [94] and the paper by Saenz *et al.,* who determined 6K1 protein as the protein inducing viral replication vesicles [89]. However, the lack of a transmembrane domain suggests that it does not conform to recently recognized patterns of vital movement proteins [94]. The potyvirus cylindrical inclusion (CI) protein (or cylindrical inclusion helicase) represents a RNA helicase required for genome replication. The role of CI is discussed; however, this protein is necessary in facilitation cell-to-cell movement and replication. *Tobacco etch virus* (TEV) CI-mutants were replication-defective in protoplast inoculation assay, some mutants possessed cell-to-cell or long distance defects in the paper by Carrington *et al.* [95]. Ultrastuctural analyses revealed interactions between the CI protein and plasmodesmata and capsid protein-containing ribonucleoprotein complexes to make cell-to-cell and long-distance movement possible [96]. The presence of CI is connected with an apparent transient reduction in callose near the plasmodesmata [97]. Structural studies on *Nicotiana benthamiana* plants infected by *plum pox virus* revealed interactions between CI and the photosystem I PSI-K protein as the product of *psaK* gene. Coexpression of CI and PPV was shown to be responsible for a decrease of PSI-K accumulation as a result of *psaK* expression down-regulation. This effect is closely connected with PPV accumulation in *N. benthamiana* plants and confirms the role of CI in PPV infection [98]. In the recent papers, the role of CI in the breaking of resistance of plants against potyviruses is discussed [99]. The role of CI in the resistance against *lettuce mosaic virus* (LMV) is demonstrated in the paper by Tavert-Roudet *et al.* [100]. The C terminus of the CI together with VPg is directly involved in the overcoming *mo1*, the gene responsible for recessive resistance of plants to *lettuce mosaic virus*, resistance. *Mo1* gene encodes the eukaryotic translation initiation factor 4E (eIF 4E). This factor is a component of the eIF 4F complex and binds the cap structure of cellular mRNAs. Interactions between C-terminal part of the LMV CI and eIF 4E was proved [100]. However, further studies are necessary. Above-mentioned facts indicated the role of the viral protein linked to the genome (VPg), which represents the N-terminal domain of NIa. Its role, especially in resistant plants, is discussed in many published papers [101,102,103,104]. The VPg protein covalently links the 5′ end of the viral RNA via tyrosine residue. The *potato virus A* Vpg consists of 189 amino acids with a molecular mass of 21.9 kDa [105]. Its model suggests an elongated structure with a hydrophobic core composed of antiparallel beta-sheets surrounded by helices and a positively charged contact surface where most of the known activities are localized [106]. This protein also acts important roles in viral life cycle and enables systemic invasion. The VPg protein accumulation in companion cells (cells closely associated with sieve tube members, phloem) of infected leaves at the beginning of the systemic infection indicates the VPg as a "phloem" protein and its role in facilitating the virus unloading [107]. However, the questions focused on role of phloem, respectively sieve tube members, in plant viruses long-distance transport, must be further discussed [108]. A single amino acid substitution (His118Tyr) in VPg leads to the overcome of the strain-specific resistance to systemic infection and confirms the role of central part of VPg for systemic invasion of the potyviruses [109]. A Tyr63Ala mutation did not prevent replication completely but blocked spreading of the virus [106]. This protein was found to be phosphorylated as a part of the virus particle by the cellular kinase activity, which means that VPg is accessible to protein-protein interactions [110]. Interactions with HCPro were shown in the papers of Yambalo *et al.* and Roudet-Tavert *et al.* [111,112].

The potyvirus-interacting protein (PVIP), a plant-specific protein with some homologues has been identified in many plants. PVIP function is still discussed, however, as it probably serves as a control in chromatin remodelling and as an ancillary factor to support potyvirus movement in plants [113]. Co-immunoprecipitation and bimolecular fluorescence complementation (BiFC) assays revealed that P3N-PIPO interacts with host protein PCaP1, a cation-binding protein that attaches to the plasma membrane via myristoylation. PIPO domain of P3N-PIPO binds PCaP1; in this process myristoylation of PCaP1 is unnecessary for interaction with P3N-PIPO. It seems that PCaP1 links a complex of viral proteins and genomic RNA to the plasma membrane by binding P3N-PIPO, enabling localization to the plasmodesmata and cell-to-cell movement [55]. Phosphorylation itself probably plays important role in the VPg-mediated functions during the infection cycle of potyviruses [110]. Interactions between viral RNA polymerase and VPg support the VPg role in a viral RNA synthesis [114]. VPg displays still investigated roles in nuclei of host plants [115,116]. However, the nuclear VPg localization is not the only localization in infected cells. Using green fluorescent protein technologies, VPg was localized in endoplasmic reticulum with the probable role in viral RNA translation [117]. Mutants in the VPg domain demonstrate inhibition of nuclear transport debilitated viral genome amplification. Interactions between VPg and eIF 4E that indicate its role in the initiation of translation of the viral RNA has been demonstrated and represent a critical element for virus production [118,119,120,121,122]. VPg is probably an efficient modulator of eIF 4E biochemical function [123,124] and directs eIF 4E to promote viral RNA expression [125]. Interactions of VPg with next molecules, such as the phosphatidylserine of biomembranes [126]. However, the significance of these interactions must be further investigated. The NIb acts as a RNA-dependent RNA polymerase and is generally localized in the nuclei of infected plants [127]. Interactions between NIb and VPg have been demonstrated in some papers [128,129,130,131]. NIb role in uridylylation of VPg protein was investigated in*potato virus A* model and it seems that has important function in the regulation of RNA synthesis [131]. NIb introduction into plants (transgenic plants) may pose one of the directions in progress in the field of potyvirus-resistant plants [132,133].

The coat protein (CP) represents the multifunctional protein [134]. It is important for transmission of*Potyviridae* and, in addition, the degree of the identity between coat protein sequences can be used for the determination of relationship within the *Potyviridae* [135,136,137,138,139,140,141,142,14 3]. The role of CP and its binding with the helper component (HC) in aphid transmission is demonstrated [144,145,146]. CP-HC interactions are highly specific and require a Cys-Cys-Cys motif for specific interaction with CP Asp-Ala-Gly (DAG) motif [147]. Changes in DAG motif are the cause of systemic infection failure, but they do not affect replication and production of virions [148,149]. Further, N terminus of CP demonstrates next factor in the specificity for HC. This fact was confirmed by the work of Dombrovski *et al.*, who compared the differences between two potyviruses - *zucchini yellow mosaic virus* (ZYMV) and *turnip mosaic virus*(TuMV) with the same DAG domain [150]. Amino acid at the position 47 of the CP in the *pea seed-borne mosaic potyvirus* is highly important for ability of potyviruses to induce systemic infection [151]. However, due to the CP N-terminal region hypervariability, individual cases must be considered separately. CP is able to interact with NIb protein. However, the role of this interaction in the functioning of both proteins remains unknown [114]. Different plant viruses have been demonstrated to induce the expression of *HSP70*. This fact has also been established in some potyviruses, where up-regulation of *HSP70* (*heat shock protein* gene)via a cytoplasmic unfolded protein response was established [152,153]. In addition, CP plays important role in these processes – increased amount of aggregated viral CP was observed under *tobacco mosaic virus*infection. Hafren *et al.* observed CP-mediated defect associated with replication under *HSP70* down-regulation. In addition, authors also propose the role of cochaperone CPIP in these processes [154]. The full-length *vanilla necrosis potyvirus* (VNV) coat protein (CP) gene was introduced into *Nicotiana benthamiana*plants via *Agrobacterium tumefaciens*-mediated transformation in the paper by Wang *et al.* [155]. Plants containing a full-length sense CP gene were highly resistant to virus infection. Similar papers demonstrating protective effect of CP gene expression in plants were demonstrated later [156,157]. However, the resistance is RNA-mediated. This circumstance is of high importance in biotechnology and engineering the virus-resistant plants [158,159,160,161,162]. However, virus resistance obtained by expressing the regions of a plant viruses genome in transgenic plants may be suppressed by the infection with another virus, which indicates the specificity in virus resistance [163]. On the other hand, the paper by Shand *et al.* points at the extensive cellular changes including disruption of the normal cell morphology, mitochondrial and chloroplast internal structures reorganization and formation of the atypical lipid deposits [164].

The 6K2 protein is as well as 6K1 protein only poorly understood. This transmembrane protein has no enzymatic function and enables anchoring the viral replication complex to membranes. It induces the formation of vesicles from endoplasmic reticulum membranes of the host leading to the formation of active replication compartments that contain all components required for virus replication. This fact has been confirmed especially by the papers of Restrepo-Hartwig *et al.,* who investigated a role of 6K2 protein in regulation of transport of the NIa via 6K2-NIa interactions to the nucleus [115,165,166]. Deletion or insertion of six histidine residues into various positions of membrane-bound 6K2 protein of *potato virus A* inhibited systemic infection independently in a host-specific manner [103,167]. Characterization and function of above-discussed proteins are summarized in Table 2.

PLUM POX VIRUS

The *plum pox virus* was firstly observed on plums in 1915 in Bulgaria and the disease was described in 1932 by Atanasoff [184]. Therefore, Eastern Europe was the localization of the first PPV epidemic. Later, PPV was reported on apricots (1933, by the same author who described the first PPV observation), on peaches (1961) and on sour cherries in the 1980′s [185]. Between 1932 and 1960, the disease caused by PPV moved from Bulgaria into Yugoslavia, Hungary, Romania, Albania, Czechoslovakia, Germany and Russia. Progression of disease in Western Europe was recorded after Word War II from Germany and Austria to The Netherlands, Switzerland, Greece, England and Turkey (1960s), then France, Italy, and Belgium (early 1970s), Spain and Portugal (early 1980s), Egypt, Syria and Cyprus (late 1980s). After 1990s, the disease was recorded in Chile, USA (Adams County, Pennsylvania, in 1999), Jordan, India and Canada. However, new reports about PPV occurrence (including individual PPV strains) still appear [186,187,188,189,190,191]. In Kazakhstan, PPV was firstly detected on plum and apricot trees in 2004 [192]. In 2005, *plum pox virus* has been found on apricot trees in China (Hunan Province of China, plants with typical yellow rings and diffused chlorotic spots) [193], in 2006, sharka disease in *Prunus* species was demonstrated in Argentina [194] and Pakistan [195]. One of the latest reports indicates detection of PPV in commercial Japanese apricot trees in Tokyo [196]. For the disease distribution and status, see Table 3.

PPV as a member of *Potyviridae,* which is a RNA virus with flexuous filamentous particles approximately 750 × 15 nm. It is widely distributed in Europe, North Africa, Asia and both Americas. As a natural host, PPV is restricted only to members of genus *Prunus* L. with above-mentioned exception. However, many experimental host plants have been identified. PPV is transmitted by several aphid species (see Table 1) [240]; however, it is also graft-transmissible to susceptible *Prunus* species and sap-transmissible to a wide range of herbaceous species. Symptoms of "sharka" highly depend on sensitivity of host plant and environmental conditions, especially on the weather conditions and age of the trees [241].

PPV usually affects both the leaves (leaf symptoms) and the fruit (fruit symptoms) of the plants. The intensity of fruit symptoms is usually significantly increased by the age of infected plants. The symptoms include chlorotic spots or rings, oak-leave patterns and vein clearing on leaves (plum), chlorotic pale-green rings and lines on leaves (apricot), small chlorotic blotches and distortion of the leaves (peach) or pale green patters and rings (cherry) [242,243]. Shallow rings and arabesque depressions, sometimes with brownish or reddish necrotic flesh (plums), light colored depressed rings (apricot), pale rings with diffuse bands on the epidermis (peach) or chlorotic and necrotic rings (cherry) are the main symptoms in the fruits [242]. In addition, fruits may drop prematurely [241]. Infection is usually symptomless in almond. On the other hand, *in vitro* explants lack the presence of symptoms typical for PPV infection; only typical interveinal chlorosis produced by PPV is visible [244]. This fact must be discussed in the light of the composition of cultivation medium, which represents complex matrix composed of macro- and microelements, as well as source of carbon (sugar(s)) and plant growth regulators - phytohormones. Just phytohormones play crucial role in the regulation of physiological processes as well as processes directing to the programmed cell death [245]. Most of the observable symptoms are probably dependent on a complex of virus-host interactions, where several or all viral genes could be involved in some way. On the other hand, the work of Clemente-Moreno *et al.* brings interesting findings and shows the possible protective role of ferrulic acid [244]. The work of Nagyova *et al.*revealed the role the 3′ proximal part of the *plum pox virus P1* gene in the virus-host interactions resulting in various pathotypes and demonstrated a different relative importance of particular PPV genes for symptom manifestation in different herbaceous host plant species [246]. There was an effort of several working groups to classify PPV in accordance with various characteristics, especially on the symptoms caused in experimental conditions of different plant hosts [27,247].

Table 1: Characterization of *Potyviridae* proteins including their function and subcellular localization

Protein (size)	Function	Cellular localization	Reference(s)
P1 (32 – 64 K)	serine proteinase with *cis*-cleavage activity, P1/HCpro cleavage site. Responsible for **symptomatolology**. Usable for investigation of potyviruses phylogeny and relationship	Crystalline inclusions in cytoplasm and nuclei.	[56,61,63,168-170]
HCpro (56 – 58 K)	cysteine proteinase with *cis*-cleavage activity, HCpro/P3 cleavage site. **Aphid transmission, systemic movement, suppression of gene silencing, symptoms development**.	Cytoplasm, aggregates along the endoplasmic reticulum.	[82,171-175]
P3 (37 K)	plant pathogenicity – **virulence, viral infection cycle**.	Crystalline inclusions in cytoplasm and nuclei. Endoplasmic reticulum (punctuate inclusions in association with Golgi)	[86,91,92,169,176,177]
6K1 (6 K?)	cell-to-cell movement?, role in virulence?	The cell periphery.	[94,178,179]
CI (70 K)	RNA helicase required for genome replication. **Cell-to-cell movement. Long-distance movement. Development of symptoms. Breaking the host resistance**.	Cytoplasm, conical structures attached to plasmodesmata.	[95,97-99,180]
6K2	anchoring the viral replication complex to membranes. **Long-distance movement. Development of systemic infection**.	Crystalline inclusions in cytoplasm and nuclei.	[103,167,169]
VPg	covalently links the 5′ end of the viral RNA via tyrosine residue. **Viral cycle. Systemic infection. Overcoming the eIF4E-based recessive resistance.**	Endoplasmic reticulum, nucleus, nucleolus.	[100,102,104,114,117]
NIa-Pro	NIa-pro cysteine proteinase. Host specificity. Host DNA cleavage activity.	Nuclei of infected cells in the form of inclusion bodies.	[71,75]
NIb	RNA-dependent RNA polymerase. **Genome/viral replication**.	Nuclei of infected cells in the form of inclusion bodies. Cytoplasm.	[116,127,131]
CP	**Aphid transmission. Cell-to-cell and long-distance movement. Virion assembly**.	Plasmodesmatal connections between infected leaf mesophyll cells. Cytoplasm of infected cells.	[96,144-146,181-183]

Table 2: Sharka disease status in different countries. * - there are no data since 2003

Disease status	Country	Reference(s)
Widespread (high level)	Croatia, Greece, Bulgaria, Germany, Hungary, Poland, Romania, Russia, Serbia, Slovakia	[197-210]
Restricted distribution (heterogeneous levels)	Albania, Argentina, Austria, Canada, Cyprus, Czech Republic, Egypt, France, Italy, Iran, Kazakhstan, Luxembourg, Moldova, Norway, Pakistan, Portugal, Southern Russia, Slovenia, Spain, Syria, Turkey, Ukraine, United Kingdom, United States	[192,195,197,208,211-229]
Introduced, established	Azores, Bosnia-Herzegovina, Chile, Some states of former USSR including Central Asia, India, Jordan, Lithuania, The Netherlands, Switzerland, Tunisia	[197,208,218,230-235]
Introduced, eradicated	Belgium, Denmark	[197,208]
Not present	Estonia, Finland, Ireland, Israel, Lebanon, Malta, Morocco, Palestine, Sweden	[197,208,236,237]
Unknown status	China*, Libya	[208,238,239]

Based on different experimental plant hosts and symptoms, different strains have been described. Classification based on *Chenopodium foetidum* Lam. as an experimental plant host ranks PPV isolates to yellow, intermediate and necrotic strains [248]. Classification based on molecular-biological data initially brings the recognition of six strains of *plum pox virus*, see Table 4. Nevertheless, new isolates still occur and there is an effort for the establishment new PPV strains (for example "PPV-T, PPV-Penn, *etc.*"; however, these are still only isolates, for example PPV-Penn isolate(s) belongs to the PPV-D strain) [249,250]. On the other hand, PPV-EA isolate demonstrates 79–80 %, 80 %, 77 %, and 77 % homology with isolates of strains D/M, Rec, C, and W, respectively) and is classified as a strain [251,252]. The same situation is in the case of PPV-W that represents a new strain that occurs in Canada and Latvia [253].

Table 3: Classification of PPV strains, their first identification, characterization and distribution. * - isolate, ** - sour cherry isolate with unclear taxonomic relationship, in table as a strain. In addition, isolates are indicated in italics in the table. ...

Strain/Isolate*	First identification	Characterization	Distribution	References
PPV-M (Marcus)	Peach in northern Greece	Common strain in southern, eastern and central Europe. Spread rapidly by aphids, epidemic strain of PPV. Apricot, peach, plum.	Austria, Bosnia and Herzegovina, Bulgaria, Croatia, Czech Republic, France, Germany, Hungary, Iran, Italy, Jordan, Kosovo, Serbia, Slovakia, Slovenia, Spain, Romania, Turkey	[10,14,18,199,250,254-271]
PPV-D (Dideron)	Apricot in Southern France	The most common in western Europe, epidemic strain of PPV. Apricot, peach (only poorly), plum.	Bosnia and Herzegovina, Bulgaria, Canada, Chile, Croatia, Czech Republic, France, Hungary, Japan, Kosovo, Lithuania, Romania, Slovakia, Spain, Turkey, Ukraine	[18,19,196,218,254,258,265,268,271-279]
PPV-rec (recombinant)	Recombination between PPV-M and PPV-D	Widespread in several central and eastern Europe countries. Efficiently transmitted by aphids. Plums.	Bosnia and Herzegovina, Bulgaria, Croatia, Czech republic, Germany, Hungary, Italy, Kosovo, Serbia, Slovakia, Slovenia	[14,265,267,271,277-281]
PPV-EA (El Amar)	Apricot from Egypt	Only Egypt. Limited data about distribution. Apricot, peach, plum trees.	Egypt	[282-286]
PPV-C** (Cherry)	Cherry from Moldova	Eastern and central European countries, Italy (probably eradicated). Transmitted efficiently by aphids. Sweet (PPV-SwC) and sour cherry (PVP-SoC).	Hungary, Italy, Moldova, Romania	[287-290]
PPV-T (Turkey)*	*Plum from Ancara region, Turkey*	*Turkey, data still limited.*	*Turkey*	[250]
PPV-Penn (Pennsylvania)*	*Three different growers. Peach, nectarine, Plum.*	*USA, firstly in Adams County, Pensylvania in 1999.*	*USA*	[291,292]
PPV-W (Winona)	Plum from Canada	Canada, found in Latvia (The LV-141pl and LV-145bt isolates appear to be representatives of the "ancestral" PPV-W strain)	Canada, Latvia	[253,293]

PLUM POX VIRUS GENOME ORGANIZATION

PPV virions are long, flexuous and rod-shaped, approximately 750 nm (660-750 nm) in length and 15 nm (12.5 -20 nm) in width [294]. PPV as a member of *Potyviridae* has the structure typical for all potyviruses with exception of *Bymovirus*. The molecule of ssRNA is of positive polarity. The RNA of PPV has a VPg protein linked to its 5′end and a long poly(A) tail, which is heterogenous in its size at its 3′end [295]. In the case of an aphid non-transmissible PPV, excluding a 3′terminal poly(A) sequence the ssRNA is 9741 nucleotides in length. The 3′noncoding region is 220 nucleotides in length without the poly(A) tail [296]. PPV genome contains a long open reading frame starting at the first AUG codon, nucleotide 36, that is translated upon infection, starting at the second AUG codon as nucleotide 147, into a large polyprotein of 355.5 kDa [70,295,297]. There are still unanswered questions about initiation of potyviruses translation via recognition of the specific viral sequences [125,298]. Because potyviruses including PPV have relatively short 5′non-coding regions with a low content of cytosine and guanine, they avoid a stable secondary structures and lack nonfunctional intraleader opening reading frames [299,300]. Cap-independent internal initiation of translation has been proposed for four members of the genus *Potyvirus* [301,302]. Work of Simón-Buela *et al.* presents different *in vitro* and *in vivo* evidence of cap-independent leaky scanning as the mechanism of translation initiation of

PPV genomic RNA [303]. Originated polyprotein precursor is co- and post- translationally cleaved by three virus-encoded proteinases into 11 mature proteins – P1, HCpro, P3, P3N-PIPO, 6K1, CI, 6K2, NIa (respectively VPg and NIa-Pro), NIb and CP. CP protein of about 36 kDa forms helically arranged "coat". Similarly to other potyviruses, the PPV P1 proteinase represents together with P3/6K1, 6K2, NIa/VPg and N-terminal domain of CP highly variable protein [285,304,305]. This variability enables the determination of PPV isolates and their further characterization [285,304,305,306]. The untranslated region of PPV consists of 147 nucleotides, starting with four adenine residues. It seems that intact 5′end is not essential for PPV replication. PPV 5′ untranslated region that is essential for virus replication is confined to the first 35 residues. Deletion of a long sequence between nucleotide 39 and 145 did not affect the rate of infection and viral accumulation, but affected process of pathogenesis [307,308].

PLUM POX VIRUS TRANSMISSION AND CYTOLOGICAL, HISTOLOGICAL AND BIO-CHEMICAL CHANGES IN INFECTED PLANTS

The PPV transmission is widely discussed in the two directions as aphid and non-aphid transmission. PPV is transmitted over short-distances in a non-persistent manner by *Aphis fabae* Scopoli (*Aphididae*), *Aphis gossypii* Glover(*Aphidae*), *Aphis spiraecola* Patch (*Aphididae*), *Brachycaudus persicae* Passerini (*Aphididae*), *Myzus persicae* Sulzer (*Aphididae*) and next members of *Aphididae* [309,310,311,312,313].Long-distance transmission is based on non-aphid transmission, this means on the spreading of infected plants and infected plant parts. The grafting may represent important risk in PPV spreading [24,314]. Milusheva *et al.* indicates the possibility of PPV transmission by infected seeds in a cultivar-dependent manner [315], however, some previously published papers disproves the possibility of PPV transmission by seeds [316,317]. The next fate of PPV and symptoms expression in infected plants is discussed. In stems of infected plants, PPV is localized in xylem and pith, which may indicate its possible spreading via xylem part of vascular tissue of infected plants [318]. This xylem transport does not suppose the transport for long distances, but for short distances (cell-to-cell) in the horizontal direction, which may be responsible for the localization of PPV in xylem and pith. This type of transport is provided by parenchyma cells. In addition, these cells are interconnected with other parenchyma cells; strict compartmentation is not possible. Later, this localization has been made more accurate. Hoffmann *et al.* proved the PPV presence in ray and axial parenchyma of vascular tissue of infected plants [319]. In petioles, PPV was demonstrated in epidermis and parenchyma cells of ground tissue, not in xylem [318]. These findings enable to express hypotheses about PPV spreading in plants [320,321]. On the other hand, there are significant differences between resistant and susceptible cultivars. Whereas the long-distance PPV spreading is highly limited in resistant cultivars, PPV susceptible cultivars allow long-distance PPV transport via xylem and the development of symptoms of sharka disease [320,322]. The differences between cell-to-cell (short-distance) and long-distance PPV transport between individual plant cultivars have been recorded, however, there are still unanswered questions [323]. *Plum pox virus* infection leads to the changes on different levels, *i.e.* cell and tissue as well as biochemical [1,324]. Whereas the changes in subcellular and cellular levels are connected with subcellular compartmentation of individual PPV proteins [320], changes in biochemistry of infected plants and resistant/susceptible plants may have practical impact in utilization of enzymes/proteins as markers of sharka disease. The paper by Hernandez *et al.* described the response of differently sensitive apricot (*Prunus armeniaca* L.) cultivars to *plum pox virus* infection [325]. This paper is interesting due to choice of cultivars, both resistant (Goldrich) and sensitive (Real Fino), and comparison of responses at antioxidant enzymes levels [325]. The most significant changes were observed in the case of superoxide dismutase(s) (SODs) and ascorbate peroxidase (APX) enzymes. Compared to sensitive cultivar, the significant decrease of APX activity in resistant cultivar was determined. Modifications in peroxidase activity were demonstrated also in*Nicotiana clevelandii* Gray PPV-infected plants, where the role of gaseous phytohormon ethylene in PPV-induced senescence has been partially revealed [326], and in *Chenopodium foetidum* Lam. [327]. On the other hand, levels of SODs, glutathione reductase (GR), dihydroascorbate reductase (DHAR) and monodehydroascorbate reductase (MDAR) of resistant cultivar were significantly increased compared to sensitive cultivar. Levels of catalase (CAT) remained unaltered. These results indicate the role of hydrogen peroxide, which is generated by the SOD activity, in response to PPV on the biochemical level. The experimental work continues in the second article of Hernandez *et al.*, in this case focused on the effect of PPV on photosynthesis and again on antioxidant enzymes [328]. These results were confirmed in pea plants (*Pisum sativum* L.), where PPV infection led to the accumulation of reactive oxygen species in chloroplasts under affecting of photosynthetic processes [329]. This fact is in agreement with papers focused on interactions between potyviral proteins and proteins of photosynthetic apparatus [98].The above-mentioned facts may be a reaction on its damage. In the third study, the same authors investigated response to long-term*plum pox virus* infection in peach plants (*Prunus persica* (L.) Batch cv. GF305 with the great susceptibility to PPV) with the focus on oxidative stress [328]. In this study, an oxidative stress and antioxidant mechanisms imbalance in accordance with the progress of PPV infection was determined. This fact has been confirmed by the following study of the authors [330]. Diaz-Vivancos *et al.* described an oxidative stress as a result of PPV infection in the apoplastic space of only susceptible apricot plants [331,332]. Whereas all above-mentioned papers were focused on oxidative stress caused by PPV infection, Escalettes *et al.* determined differential gene expression ins PPV-infected apricot cultivar (cv. Goldrich) [333]. The CH4 and CH22 fragments coding for a putative myosin and kinesin, were over-expressed in cv. Goldrich. Both myosin and kinesin are closely associated with cytokinesis, where they serve as motor proteins in the organization of phragmoplast microtubules. In the PPV-infected cv. Goldrich, the transketolase analogue CH15 was over-expressed, while it presented a very similarly modified expression in the susceptible genotype. Expression of the ankyrin-like CH27 was obviously enhanced in PPV-inoculated partially resistant apricot and in the susceptible Prunus genotype. The clear repression of CH29 transcript encoding a putative class III chitinase in the partially resistant genotype suggests a virus-mediated repression of this gene in the

Goldrich cultivar [333]. *Arabidopsis thaliana* (L.) Heynh. was used as a model plant for investigation of gene expression alteration due PPV infection in the following study [334]. Genes associated with soluble sugar, starch and amino acid, intracellular membrane/membrane-bound organelles, chloroplast, and protein fate were up-regulated, while genes related to development/storage proteins, protein synthesis and translation, and cell wall-associated components were down-regulated. These gene expression changes were closely associated with PPV infection and symptom development [334]. The paper by Wang *et al.* revealed that genes involved in defense, cellular transport, development, protein synthesis, proteins with binding function in the PPV-infected peach leaf tissue are more active than those in PPV-free leaves [335]. The changes of expression of genes (increase of the following gene transcripts: including beta-1,3-glucanase, cytochrome-P450-like protein, cytochrome P450 monooxygenase, heat-shock protein 70, thioredoxin, alcohol dehydrogenase, catalase, cysteine protease inhibitor, translation factor EF-1 alpha, and pathogenesis-related protein (PRI)) in*Pisum sativum* L. PPV-infected plants support the ability of PPV to induce common stress responses in susceptible plants [335]. The recent studies focused on the changes in growth characteristics and yield of infected plants [336,337]. Similar results were demonstrated by Garcia-Ibarra *et al.*, who investigated changes in protective mechanisms including antioxidant enzymes in peach infected by *apple chlorotic leaf spot virus* (ACLSV, *Flexiviridae*). The results show increases in the APX, dehydroascorbate reductase (DHAR), superoxide dismutase (SOD) and glutathione S-transferase (GST) activities, whereas POX suffered a decrease of about 34 % [338]. Induction of oxidative stress by plant viruses is demonstrated in some works, such as Amari *et al.* [339], Clarke *et al.* [340], Farkas *et al.* [341], Fodor *et al.* [342], and Kiraly *et al.* [343]. In the lights of these findings, changes in markers of oxidative stress may be useful in the detection of PPV infection. However, further characterization of infected plants is necessary, because changes in antioxidant mechanisms under plant virus infection are indistinctive.

DETECTION OF PPV

The detection of PPV undergoes rapid development, from the usage of very simple methods based on host plants to molecular-biological methods and techniques. However, significant limitations for methods used in routine testing of plants for PPV must be carefully considered. These limitations are based especially on irregular distribution and translocation of PPV in plants in accordance with growth characteristics. Biological techniques are based on the infection and especially mechanical inoculation of susceptible plants - hosts, including both herbaceous and woody plants, from more than nine plant families (*Amaranthaceae* incl.*Chenopodiaceae*, *Cannabidaceae*, *Caryophyllaceae*, *Compositae* (*Asteraceae*), *Brassicaceae* (*Cruciferae*),*Fabaceae* (*Leguminosae*), *Ranunculaceae*, *Rosaceae* and *Solanaceae* [27,344,345]. Diagnostically important plant host species and symptoms are summarized in Table 4.

Table 4: Susceptible *Plum pox virus* (PPV) plant host species used in PPV detection and determination

Plant host species	Symptoms	References
Chenopodium foetidum Lam. (*Amaranthaceae*)	yellow spots, some with necrotic centres or necrotic spots; not systemic	[327,346-348]
Nicandra physalodes (L.) P. Gaertn. (*Solanaceae*)	black-brown necrotic local lesions	[344,345]
Prunus domestica L. cv. Italian Prune and cv. Pozegaca (*Rosaceae*)	chlorotic rings, bands and spots	[242,349,350]
Prunus japonica Thunb. (*Rosaceae*)	diffused chlorotic spots	[18,351]
Prunus maritima Marsh. (*Rosaceae*)	chlorotic spots, vein necrosis	[351]
Prunus persica L. (Batsch) cv. GF305	vein necrosis, malformation of leaves	[352,353]
Prunus sibirica L. (*Rosaceae*)	green spots and faint line pattern	[351,354]
Prunus tomentosa Thunb. (*Rosaceae*)	epinasty and malformation of young leaves, chlorotic spots, necrotic spots of older leaves	[23,355,356]
Sorbus domestica L. (*Rosaceae*)	yellow spots, leaf chlorosis	[208,357]

The first, but highly important step in progress of PPV diagnostics was based on the serological assay, respectively on the enzyme-linked immunosorbent assay (ELISA), which takes advantages from antibodies and their subsequent detection. One of the first record about the use of antibodies was published in 1980 [358]. ELISA method underwent progress with some modifications and now, it is one of the most popular and used techniques for PPV detection [359,360,361,362]. Whereas the usage of polyclonal antibodies, which recognize multiple epitopes on any one antigen, is controversial due to problems with specificity and consequently with sensitivity (serum contains a mixture of antibodies of different affinity), monoclonal antibodies, which detect only one epitope on the antigen, helped to overcome above-mentioned problems and are still widely used in PPV diagnostics [363,364,365,366,367]. Monoclonal antibodies are usually obtained after immunization of BALB/c mice with purified PPV isolates [367,368]. Despite the fact that commercial kits had been available and used before the introduction of 5B-IVIA monoclonal antibody, universal and specific detection of any PPV isolate was achieved by the use of 5B-IVIA monoclonal antibody [23,369]. Generally, this antibody enabled the production and use of commercially available kits for PPV detection, which enable specific identification of respective PPV strain. Antibodies are produced especially against non-structural PPV proteins as P3 [169], 6K2 [169], CP and CI [181,370,371], but also HCPro [181,372], NIb [169,181,373], and P1 [169]. However, non-structural proteins antibodies have not found the application in the practice and are used only for scientific purposes. As a secondary antibody, peroxidase-labelled, biotin-streptavidin system or fluorescence probe-labelled secondary antibodies are used [358]. Double Antibody Sandwich Indirect ELISA (DASI-ELISA) and Triple Antibody Sandwich ELISA (TAS-ELISA) have been recently recommended in PPV detection [13,23,266,269,360,374]. They are used especially for the determination of individual PPV strains [23,365,375,376]. Antibodies may be useful for the investigation of subcellular compartmentation of PPV mature proteins, or, in addition, may be used for subcellular localization of some members of *Potyviridae* family [181]. In this case, gold-labelled antibodies are used [169,377]. During the years, improved ELISA-based techniques have been proposed and developed. One of the most recent technique is impedimetric immunosensor, which is based on gold electrodes modified with 1,6-hexanedithiol, gold nanoparticles, anti-PPV IgG polyclonal antibody and bovine serum albumin. The proposed technique displays very good detection limit (10 pg/ml) and is able to differentiate the samples from healthy plants and the samples containing 0.01 % of infected plant extract [378]. The PCR-ELISA, respectively RT-PCR-ELISA, which is based on immunoenzymatic detection of PCR products, represents next possibility of ELISA/PCR modification with significant increase of sensitivity compared to both ELISA and PCR methods [374,379,380].

Hybridization techniques are based on an establishing a non-covalent, sequence-specific interaction between two or more complementary strands of nucleic acids into a single hybrid. The revolutionary were two works of Wetzel *et al.* published in 1991. Authors described molecular cloning and partially described the nucleotide sequence of the genomic RNA of PPV-EA. In addition, they compared this sequence to the corresponding sequence of previously sequenced PPV strains [381]. Finally, a sensitive, polyvalent assay based on the polymerase chain reaction (PCR) was developed for *plum pox potyvirus* (PPV) detection. This technique was adapted for a single tube, the chemical denaturation and reverse transcription of the viral RNA followed by the PCR reaction yielding a 243-base-pair product [382]. These two works started the large-scale application of PCR in the practice. Molecular hybridization techniques as well as different PCR-based assays have been developed to detect the PPV RNA presence in the sample [383,384,385,386]. PCR is widely used to amplify specific nucleic acid regions; thus, there is a necessity to have primers of known nucleotide sequence. In the past, different primers have been proposed and used, mainly on the HCPro [387], C-terminal part of NIb [265,388], C-terminal part of CP [249,265,389], N-terminal part of CP [390] or 3′-noncoding region [272]. The request to known nucleotide sequence is absolutely necessary also in hybridization techniques, which are popular due to relative simplicity. The paper by Pasquini *et al.* describes the strategy, development and use of 70-mer oligonucleotide probes specific for determination and genotyping (identification) the individual PPV isolates [391]. Despite the above-mentioned facts, the RNA isolation is the crucial step in these techniques. There are some difficulties connected with PPV RNA isolation from plant tissues, especially higher rate of polyphenolics and polysaccharides, which affect RNA isolation due to formation of complexes with both RNA and proteins. In addition, these compounds are able affect also activity of enzymes (including reverse transcriptase necessary for RNA transcription into cDNA form) used in PCR. One-Step RT-PCR as well as Two-Step RT-PCR is widely used in PPV RNA detection [265,392,393,394,395,396]. Products of PCR may be visualized by the electrophoretic separation of fragments in agarose gel with subsequent labelling by ethidium bromide or SYBR Green [397,398,399], on the nitrocellulose membrane [400], eventually by immunoenzymatic procedure [379]. There are some techniques, which are based on PCR technique and pose the improvement, especially significant enhancement, of sensitivity. Co-operational PCR (Co-PCR) has been described for sensitive detection of plant viruses and bacteria [401,402,403]. This technique, carried out in a single reaction, minimizes contamination risks and has a level of sensitivity similar to nested PCR and real-time PCR. In addition, it can be coupled with dot blot hybridization, making it possible to characterize the nucleotide sequence [401]. A highly sensitive assay, based on the polymerase chain reaction amplification of cDNA synthesized from the viral RNA of antibody-captured viral particles was developed by Wetzel *et al.* in 1992 [404]. The immunocapture step, by allowing the use of large sample volumes and by the viral particle pre-purification it achieves, dramatically increased the sensitivity of the assay. Recently, modifications of this technique are widely used. An immunocapture reverse transcription-polymerase chain reaction (IC-RT-PCR) based assay for the detection and identification of plant viruses represents technique using clarified plant extracts with degenerate primers, without necessity of isolation of total nucleic acids. This technique was used for detection and determination of *papaya ringspot virus* [405,406], *zantedeschia mild mosaic virus* [407], and*sugarcane streak mosaic virus* [408]. In addition, this technique was used in identification of PPV isolates [14,199,265,266,267,269]. A nucleic acid sequence-based amplification method coupled with flow-through hybridization (NASBA-FH) was developed for *plum pox virus* (PPV) detection with the detection limit 1000 times higher compared to conventional RT-PCR [409,410]. Despite the fact that ELISA followed by PCR techniques is the most frequently used technique in the PPV detection, loop-mediated isothermal amplification (LAMP) will probably become the most frequently applied approach in PPV detection. The great advantage of LAMP is its enormous rate of amplification paired with a very high specificity and low artefact susceptibility, which means great specificity [411]. Spot real-time RT-PCR is a method for detection of *plum pox virus* using conventional ELISA plant crude extracts immobilized on paper. This method has been developed to overcome the need of RNA isolation [412].

The field of nanotechnology is focused on the study and control of phenomena and materials and length scale below 100 nm, it means 1-100 nm [413,414]. This definition is not the only one, which tries to define nanotechnology. One of the most important criterions consists in accentuation the special properties of nanomaterials due to their nanoscaled proportions, which open their unique properties and features [415]. Nanotechnology found its application especially in industry and medicine, especially in electronics, imaging and drug delivery system for targeted therapy [416,417,418,419,420,421,422,423,424,425,426]. On the other hand, their application in the field of plant biology and phytopathology is limited, especially for cell imaging and manipulation [427]. There are works describing nanotechnology application in agriculture in the crop production in improving the yield and product quality [428,429,430,431,432]. On the other hand, plant viruses are considered as used in the nanotechnology, especially due to highly organized protein capsids, which are useful as scaffolds for building nanomaterials [433,434,435]. *Tobacco mosaic virus* has found applications as a building block for nanoelectronics as a template for metal deposition, mineralization and the deposition of the silica, such as nanowires and conductive films, as well as in light-harvesting systems [436,437,438,439,440]. *Cowpea mosaic virus* with its icosahedral protein coat shape can be utilized in non-invasive imaging, biosensors, and in vaccines [434,439,441]. Plant virus nanoparticles may be used in medicine in *in vivo* targeting and tumor targeting [442,443]. Based on the nanotechnology applications in the field of analytical chemistry [444,445,446,447,448], they represent the principal improvement of serological and immunofluorescence techniques in plant virus isolation and determination. In the field of PPV isolation and identification, there is an important requirement for PPV RNA of the highest quality. Therefore, methods of nanotechnology take advantage of nanoparticles on the basis of surface modifications possibility, which can significantly improve the quality of isolated RNA. Magnetic nanoparticles have been used for isolation and purification of nucleic acids; the use of magnetic nanoparticles provides very rapid and simple isolation of nucleic acids [449,450,451,452,453,454,455,456]. In addition, magnetic nanoparticled surface may be variously modified (these modifications include for example the introduction of silanol, epoxide, diol and carboxyl groups respectively). In comparison with traditional methods, the solid-phase process based on the interaction of nucleic acids with chemically-modified surface of magnetic nanoparticles is characterized by convenience, speed, timesaving, and being amenable to automation [457,458]. Using automation together with simple and rapid PPV RNA isolation under high specificity may represent great progress, especially due to the possibility of analyzing many samples in a short time period [459,460,461]. Generally, in the field of PPV detection, there is only limited number of publications focused on the use of nanoparticles in different isolation/detection steps. A paper by Byzova *et al.* introduces the possibility of using monoclonal antibodies as gold nanoparticle conjugates (26 nm in diameter) [462]. Using these conjugates with optimal ratio, an express immunochromatographic assay of PPV with a detection limit of 3 ng/ml and duration of 10 min. was developed [462]. Colloidal gold nanoparticles (5 – 60 nm) as a carrier system conjugated with corresponding antibody are used in the paper by Safenkova *et al.* [463]. This paper demonstrates correlation between gold nanoparticles, respectively conjugate size and affinity. An increase of conjugate size leads to the increase in its affinity [463]. In conclusion, not only improvement of those methods used (ELISA, PCR) and the establishment of new detection techniques, but also the development of new, effective methods usable in PPV RNA isolation with subsequent routine detection, will bring new opportunities to routine PPV detection, and thereby provide the possibility of eradicating plum pox disease as an one of the most devastating viral diseases of stone fruits.

ACKNOWLEDGMENTS

Financial support from the grants NAZV QI 91A032 and CEITEC CZ.1.05/1.1.00/02.0068 is highly acknowledged.

REFERENCES AND NOTES

1. Albrechtova L. Investigations on the distribution of sharka virus (plum pox virus) in tissue of prunus-domestica. J. Plant Dis. Prot. 1986;93:190–201.
2. Neumuller M., Hartmann W. The hypersensitivity of european plum (prunus domestica l.) against the plum pox virus; Proceedings of the twentieth international symposium on virus and virus–like diseases of temperate fruit crops – fruit tree diseases, 2008; pp. 273–279.
3. Badenes M.L., Asins M.J., Carbonell E.A., Glacer G. Genetic diversity in apricot, prunus armeniaca, aimed at improving resistance to plum pox virus. Plant Breed. 1996;115:133–139.
4. Balan V., Ivascu A., Toma S. Xvith international symposium on fruit tree virus diseases. 1995. Susceptibility of apricot, nectarine and peach cultivars and hybrids to plum-pox virus; pp. 299–305.
5. Martinez-Gomez P., Rubio M., Dicenta F., Gradziel T.M. Utilization of almond as source of plum pox virus resistance in peach breeding; Proceedings of the xixth international symposium on virus and virus-like diseases of temperate fruit crops: Fruit tree diseases, 2004; pp. 289–293.
6. Pribek D., Gaborjanyi R., Palkovics L. Molecular characterization of plum pox virus almond isolate; Proceedings of the xixth international symposium on virus and virus–like diseases of temperate fruit crops: Fruit tree diseases, 2004; pp. 289–293.
7. Rubio M., Martinez-Gomez P., Dicenta F. Resistance of almond cultivars to plum pox virus (sharka) Plant Breed. 2003;122:462–464. doi: 10.1046/j.1439-0523.2003.00872.x.
8. Crescenzi A., Nuzzaci M., Levy L., Piazzolla P., Hadidi A. Xvith international symposium on fruit tree virus diseases. 1995. Plum pox virus (ppv) in sweet cherry; pp. 219–225.
9. Fanigliulo A., Comes S., Crescenzi A. Evaluation of cherry cultivars for their response to infection by plum pox virus sweet cherry strain; Proceedings of the xixth international symposium on virus and virus–like diseases of temperate fruit crops: Fruit tree diseases, 2004; pp. 309–316.
10. Boeglin M., Quiot J.B., Labonne G. Risk assessment of contamination of cherry trees by plum pox virus in france; Proceedings of the xixth international symposium on virus and virus–like diseases of temperate fruit crops: Fruit tree diseases, 2004; pp. 221–224.
11. Damsteegt V.D., Scorza R., Stone A.L., Schneider W.L., Webb K., Demuth M., Gildow F.E. Prunus host range of plum pox virus (ppv) in the united states by aphid

and graft inoculation. Plant Dis. 2007;91:18–23. doi: 10.1094/PD-91-0018.

12. Nemchinov L., Hadidi A. Characterization of the sour cherry strain of plum pox virus. Phytopathology.1996;86:575–580. doi: 10.1094/Phyto-86-575.
13. Elibuyuk I.O. Detection of plum pox virus in ornamental prunus cerasifera. Phytoparasitica.2006;34:347–352. doi: 10.1007/BF02981020.
14. Kamenova I. Prunus cerasifera as a host of plum pox virus in bulgaria. J. Plant Pathol. 2008;90:15–18.
15. Rubio M., Garcia-Ibarra A., Dicenta F., Martinez-Gomez P. Plum pox virus (sharka) sensitivity in prunus salicina and prunus cerasifera cultivars against a dideron-type isolate. Plant Breed. 2011;130:283–286.
16. Kalinina A., Brown D.C.W., Ravelonandro M. Susceptibility of ornamental prunus to plum pox potyvirus infection; Proceedings of the international symposium on biotechnology of temperate fruit crops and tropical species, 2007; pp. 601–605.
17. Kegler H., Fuchs E., Gruntzig M., Krczal G., Wegener B. Susceptibility of genotypes of the genus prunus to plum pox potyvirus. J. Plant Dis. Prot. 1996;103:255–261.
18. Sebestyen D., Nemeth M., Hangyal R., Krizbai L., Ember I., Nyerges K., Kolber M., Kiss E., Bese G. Ornamental prunus species as new natural hosts of plum pox virus and their importance in the spread of the virus in hungary. J. Plant Pathol. 2008;90:57–61.
19. Stobbs L.W., Van Driel L., Whybourne K., Carlson C., Tulloch M., Van Lier J. Distribution of plum pox virus in residential sites, commercial nurseries, and native plant species in the niagara region, ontario, canada.Plant Dis. 2005;89:822–827. doi: 10.1094/PD-89-0822.
20. Ilbagi H., Citir A., Bostan H. Prunus spinosa l. - a natural wild host of some important fruit viruses in tekirdag,turkey; Proceedings of the twentieth international symposium on virus and virus–like diseases of temperate fruit crops – fruit tree diseases, 2008; pp. 33–36.
21. Polak J. Diagnosis and identification of plant pathogens. 1997. The role of prunus spinosa l in epidemiology of plum pox virus in the czech republic; pp. 527–530.
22. Polak J., Oukropec I. The determination of sources of resistance to plum pox virus suitable for peach; Proceedings of the twentieth international symposium on virus and virus–like diseases of temperate fruit crops – fruit tree diseases, 2008; pp. 269–272.
23. Cambra M., Capote N., Olmos A., Bertolini E., Gorris M.T., Africander N.L., Levy L., Lenardon S.L., Clover G., Wright D. Proposal for a new international protocol for detection and identification of plum pox virus: Validation of the techniques; Proceedings of the twentieth international symposium on virus and virus–like diseases of temperate fruit crops – fruit tree diseases, 2008; pp. 181–191.
24. Damsteegt V.D., Scorza R., Gildow F.E., Schneider W.L., Stone A.L., Luster D.G. Comparative host range of us isolates of plum pox virus among prunus and other woody plant species following graft inoculation or aphid transmission. Phytopathology. 2004;94:S24–S24.
25. Monsion M., Briard P., Glasa M., Ravelonandro M. Comparative techniques to perform koch's postulates with plum pox virus; Proceedings of the twentieth international symposium on virus and virus–like diseases of temperate fruit crops – fruit tree diseases, 2008; pp. 221–225.
26. Ravelonandro M., Briard P., Monsion M. Strategy to identify and to characterize new isolates of plum pox virus; Proceedings of the xixth international symposium on virus and virus–like diseases of temperate fruit crops: Fruit tree diseases, 2004; pp. 171–175.
27. Salvador B., Delgadillo M.O., Saenz P., Garcia J.A., Simon-Mateo C. Identification of plum pox virus pathogenicity determinants in herbaceous and woody hosts. Mol. Plant-Microbe Interact. 2008;21:20–29. doi: 10.1094/MPMI-21-1-0020.
28. Baumgartnerova H. First findings of plum pox virus in walnut trees (juglans regia l) Acta Virol.1996;40:59–60.
29. Polak J. Hosts and symptoms of plum pox virus: Woody species other than fruit and ornamental species of prunus. EPPO Bull. 2006;36:225–226. doi: 10.1111/j.1365-2338.2006.00977.x.
30. Gibbs A., Ohshima K. Potyviruses and the digital revolution. Annu. Rev. Phytopathol. . 2010;48:205–223.
31. Gibbs A.J., Mackenzie A.M., Wei K.J., Gibbs M.J. The potyvirusesofaustralia.Arch.Virol.2008;153:1411–1420. doi: 10.1007/s00705-008-0134-6.
32. Revers F., Le Gall O., Candresse T., Maule A.J. New advances in understanding the molecular biology of plant/potyvirus interactions. Mol. Plant-Microbe Interact. 12:367–376.
33. Riechmann J.L., Lain S., Garcia J.A. Highlights and prospects of potyvirus molecular-biology. Journal of General Virology. 1992;73:1–16. doi: 10.1099/0022-1317-73-1-1.
34. Ward C.W., Shukla D.D. Taxonomy of potyviruses - current problems and some solutions. Intervirology.1991;32:269–296.
35. Fereres A., Kampmeier G.E., Irwin M.E. Aphid attraction and preference for soybean and pepper plants infected with potyviridae. Ann. Entomol. Soc. Am. 1999;92:542–548.
36. Manachini B., Casati P., Cinanni L., Bianco P. Role of myzus persicae (hemiptera : Aphididae) and its secondary hosts in plum pox virus propagation. J. Econ. Entomol. 2007;100:1047–1052. doi: 10.1603/0022-0493(2007)100[1047:ROMPHA]2.0.CO;2.
37. Srinivasan R., Alvarez J.M. Effect of mixed viral infections (potato virus y-potato leafroll virus) on biology and preference of vectors myzus persicae and macrosiphum euphorbiae (hemiptera : Aphididae) J. Econ. Entomol. 2007;100:646–655. doi: 10.1603/0022-0493(2007)100[646:EOMVIP]2.0.CO;2.
38. Symmes E.J., Perring T.M. Intraspecific variation in zucchini yellow mosaic virus transmission by

myzus persicae and the impact of aphid host plant. J. Econ. Entomol. 2007;100:1764–1772. doi: 10.1603/0022-0493(2007)100[1764:IVIZYM]2.0.CO;2.

39. Choi I.R., French R., Hein G.L., Stenger D.C. Fully biologically active in vitro transcripts of the eriophyid mite-transmitted wheat streak mosaic tritimovirus. Phytopathology. 1999;89:1182–1185. doi: 10.1094/PHYTO.1999.89.12.1182.

40. Stephan D., Moeller I., Skoracka A., Ehrig F., Maiss E. Eriophyid mite transmission and host range of a brome streak mosaic virus isolate derived from a full-length cdna clone. Arch. Virol. 2008;153:181–185. doi: 10.1007/s00705-007-1065-3.

41. Sanchez-Sanchez H., Henry M., Cardenas-Soriano E., Alvizo-Villasana H.F. Identification of wheat streak mosaic virus and its vector aceria tosichella in mexico. Plant Dis. 2001;85:13–17. doi: 10.1094/PDIS.2001.85.1.13.

42. Adkins S., Webb S.E., Achor D., Roberts P.D., Baker C.A. Identification and characterization of a novel whitefly-transmitted member of the family potyviridae isolated from cucurbits in florida. Phytopathology.2007;97:145–154. doi: 10.1094/PHYTO-97-2-0145.

43. Adkins S., Webster C.G., Kousik C.S., Webb S.E., Roberts P.D., Stansly P.A., Turechek W.W. Ecology and management of whitefly-transmitted viruses of vegetable crops in florida. Virus Res. 2011;159:110–114. doi: 10.1016/j.virusres.2011.04.016.

44. Valverde R.A., Sim J., Lotrakul P. Whitefly transmission of sweet potato viruses. Virus Res.2004;100:123–128. doi: 10.1016/j.virusres.2003.12.020.

45. Dessens J.T., Meyer M. Identification of structural similarities between putative transmission proteins of polymyxa and spongospora transmitted bymoviruses and furoviruses. Virus Genes. 1996;12:95–99. doi: 10.1007/BF00370006.

46. Chen J.P. Progress and prospects of studies on polymyxa graminis and its transmitted cereal viruses in china. Progress Nat. Sci. 2005;15:481–490. doi: 10.1080/10020070512331342440.

47. McGrann G.R.D., Adams M.J. Investigating resistance to barley mild mosaic virus. Plant Pathol.2004;53:161–169. doi: 10.1111/j.0032-0862.2004.00998.x.

48. Thompson J.P., Clewett T.G., Jennings R.E., Sheedy J.G., Owen K.J., Persley D.M. Detection of polymyxa graminis in a barley crop in australia. Austral. Plant Pathol. 2011;40:66–75. doi: 10.1007/s13313-010-0015-9.

49. Campbell R.N. Fungal transmission of plant viruses. Annu. Rev. Phytopathol. 1996;34:87–108. doi: 10.1146/annurev.phyto.34.1.87.

50. Kuroda T., Nabata K., Hori T., Ishikawa K., Natsuaki T. Soybean leaf rugose mosaic virus, a new soilborne virus in the family potyviridae, isolated from soybean in japan. Journal of General Plant Pathology.2010;76:382–388. doi: 10.1007/s10327-010-0272-z.

51. Chung B.Y.W., Miller W.A., Atkins J.F., Firth A.E. An overlapping essential gene in the potyviridae.Proc. Natl. Acad. Sci. U. S. A. 2008;105:5897–5902.

52. Wen R.H., Hajimorad M.R. Mutational analysis of the putative pipo of soybean mosaic virus suggests disruption of pipo protein impedes movement. Virology. 2010;400:1–7. doi: 10.1016/j.virol.2010.01.022.

53. Wen R.H., Maroof M.A.S., Hajimorad M.R. Amino acid changes in p3, and not the overlapping pipo-encoded protein, determine virulence of soybean mosaic virus on functionally immune rsv1-genotype soybean. Molecular Plant Pathology. 2011;12:799–807. doi: 10.1111/j.1364-3703.2011.00714.x.

54. Kneller E.L.P., Rakotondrafara A.M., Miller W.A. Cap-independent translation of plant viral rnas. Virus Research. 2006;119:63–75. doi: 10.1016/j.virusres.2005.10.010.

55. Vijayapalani P., Maeshima M., Nagasaki-Takekuchi N., Miller W.A. Interaction of the trans-frame potyvirus protein p3n-pipo with host protein pcap1 facilitates potyvirus movement. PLoS Pathog. 2012;8:1–15.

56. Adams M.J., Antoniw J.F., Beaudoin F. Overview and analysis of the polyprotein cleavage sites in the family potyviridae. Mol. Plant Pathol. 2005;6:471–487. doi: 10.1111/j.1364-3703.2005.00296.x.

57. Verchot J., Koonin E.V., Carrington J.C. The 35-kda protein from the n-terminus of the potyviral polyprotein functions as a 3rd virus-encoded proteinase. Virology. 1991;185:527–535. doi: 10.1016/0042-6822(91)90522-D.

58. Yoshida N., Shimura H., Yamashita K., Suzuki M., Masuta C. Variability in the p1 gene helps to refine phylogenetic relationships among leek yellow stripe virus isolates from garlic. Arch. Virol. 2012;157:147–153. doi: 10.1007/s00705-011-1132-7.

59. Salvador B., Saenz P., Yanguez E., Quiot J.B., Quiot L., Delgadillo M.O., Garcia J.A., Simon-Mateo C. Host-specific effect of p1 exchange between two potyviruses. Mol. Plant Pathol. 2008;9:147–155. doi: 10.1111/j.1364-3703.2007.00450.x.

60. Shi Y., Chen J., Hong X., Adams M.J. A potyvirus p1 protein interacts with the rieske fe/s protein of its host. Mol. Plant Pathol. 2007;8:785–790. doi: 10.1111/j.1364-3703.2007.00426.x.

61. Valli A., Lopez-Moya J.J., Garcia J.A. Recombination and gene duplication in the evolutionary diversification of p1 proteins in the family potyviridae. J. Gen. Virol. 2007;88:1016–1028. doi: 10.1099/vir.0.82402-0.

62. Young B.A., Stenger D.C., Qu F., Morris T.J., Tatineni S., French R. Tritimovirus p1 functions as a suppressor of rna silencing and an enhancer of disease symptoms. Virus Res. 2012;163:672–677. doi: 10.1016/j.virusres.2011.12.019.

63. Rohozkova J., Navratil M. P1 peptidase - a mysterious protein of family potyviridae. J. Biosci.2011;36:189–200. doi: 10.1007/s12038-011-9020-6.

64. Stenger D.C., Hein G.L., Gildow F.E., Horken K.M., French R. Plant virus hc-pro is a determinant of eriophyid mite transmission. J. Virol. 2005;79:9054–9061. doi: 10.1128/JVI.79.14.9054-9061.2005.

65. Endres M., Mlotshwa S., Schauer S., Bowman L., Vance V. Hc-pro suppression of rna silencing: Towards a mechanism. Phytopathology. 2005;95:S123–S123.

66. Torres-Barcelo C., Daros J.A., Elena S.F. Compensatory molecular evolution of hc-pro, an rna-silencing suppressor from a plant rna virus. Mol. Biol. Evol. 2010;27:543–551. doi: 10.1093/molbev/msp272.

67. Shen W.T., Yan P., Gao L., Pan X.Y., Wu J.Y., Zhou P. Helper component-proteinase (hc-pro) protein of papaya ringspot virus interacts with papaya calreticulin. Mol. Plant Pathol. 2010;11:335–346. doi: 10.1111/j.1364-3703.2009.00606.x.

68. Guo B.H., Lin J.Z., Ye K.Q. Structure of the autocatalytic cysteine protease domain of potyvirus helper-component proteinase. J. Biol. Chem. 2011;286:21937–21943. doi: 10.1074/jbc.M111.230706.

69. Matthews D.A., Smith W.W., Ferre R.A., Condon B., Budahazi G., Sisson W., Villafranca J.E., Janson C.A., McElroy H.E., Gribskov C.L., et al. Structure of human rhinovirus 3c protease reveals a trypsin-like polypeptide fold, rna-binding site, and means for cleaving precursor polyprotein. Cell. 1994;77:761–771. doi: 10.1016/0092-8674(94)90059-0.

70. Riechmann J.L., Cervera M.T., Garcia J.A. Processing of the plum pox virus polyprotein at the p3-6k(1) junction is not required for virus viability. J. Gen. Virol. 1995;76:951–956. doi: 10.1099/0022-1317-76-4-951.

71. Anindya R., Savithri H.S. Potyviral nia proteinase, a proteinase with novel deoxyribonuclease activity. J. Biol. Chem. 2004;279:32159–32169. doi: 10.1074/jbc.M404135200.

72. Han H.E., Sellamuthu S., Shin B.H., Lee Y.J., Song S., Seo J.S., Baek I.S., Bae J., Kim H., Yoo Y.J., et al. The nuclear inclusion a (nia) protease of turnip mosaic virus (tumv) cleaves amyloid-beta. Plos One.2010;5

73. Han J.S., Kang H.J., Song B.D., Choi K.Y. Novel mechanism of activity control of potyvirus nia protease. Faseb J. 2002;16:A905–A905.

74. Puhl A.C., Giacomini C., Irazoqui G., Batista-Viera F., Villarino A., Terenzi H. Covalent immobilization of tobacco-etch-virus nia protease: A useful tool for cleavage: Of the histidine tag of recombinant proteins. Biotechnol. Appl. Biochem. 2009;53:165–174. doi: 10.1042/BA20080063.

75. Chen K.C., Chiang C.H., Raja J.A.J., Liu F.L., Tai C.H., Yeh S.D. A single amino acid of niapro of papaya ringspot virus determines host specificity for infection of papaya. Mol. Plant-Microbe Interact.2008;21:1046–1057. doi: 10.1094/MPMI-21-8-1046.

76. Mathur C., Jimsheena V.K., Banerjee S., Makinen K., Gowda L.R., Savithri H.S. Functional regulation of pvbv nuclear inclusion protein-a protease activity upon interaction with viral protein genome-linked and phosphorylation. Virology. 2012;422:254–264. doi: 10.1016/j.virol.2011.10.009.

77. Restrepo M.A., Freed D.D., Carrington J.C. Nuclear transport of plant potyviral proteins. Plant Cell.1990;2:987–998.

78. Knuhtsen H., Hiebert E., Purciful D. Partial-purification and some properties of tobacco etch virus-induced intranuclear inclusions. Virology. 1974;61:200–209. doi: 10.1016/0042-6822(74)90254-2.

79. Shepard J.F. Electron microscopy of subtilisin-treated tobacco etch virus nuclear and cytoplasmic inclusions. Virology. 1968;36:20. doi: 10.1016/0042-6822(68)90112-8.

80. Sheffield F.M.L. The cytoplasmic and nuclear inclusions associated with severe etch virus. J. Royal Microscop. Soc. 1941;61:30–45. doi: 10.1111/j.1365-2818.1941.tb00883.x.

81. Fellers J.P., Collins G.B., Hunt A.G. The nia-proteinase of different plant potyviruses provides specific resistance to viral infection. Crop Sci. 1998;38:1309–1319. doi: 10.2135/cropsci1998.0011183X003800050030x.

82. Fukuzawa N., Itchoda N., Ishihara T., Goto K., Masuta C., Matsumura T. Hc-pro, a potyvirus rna silencing suppressor, cancels cycling of cucumber mosaic virus in nicotiana benthamiana plants. Virus Genes.2010;40:440–446. doi: 10.1007/s11262-010-0460-0.

83. Giner A., Lopez-Moya J.J., Lakatos L. Rna interference and viruses: Current innovations and future trends. CAISTER ACADEMIC PRESS; WYMONDHAM: 2010. Rna silencing in plants and the role of viral suppressors; pp. 25–46.

84. Reyes C.A., De Francesco A., Pena E.J., Costa N., Plata M.I., Sendin L., Castagnaro A.P., Garcia M.L. Resistance to citrus psorosis virus in transgenic sweet orange plants is triggered by coat protein-rna silencing.J. Biotechnol. 2011;151:151–158.

85. Jenner C.E., Wang X.W., Tomimura K., Ohshima K., Ponz F., Walsh J.A. The dual role of the potyvirus p3 protein of turnip mosaic virus as a symptom and avirulence determinant in brassicas. Mol. Plant-Microbe Interact. 2003;16:777–784. doi: 10.1094/MPMI.2003.16.9.777.

86. Lin L., Luo Z.P., Yan F., Lu Y.W., Zheng H.Y., Chen J.P. Interaction between potyvirus p3 and ribulose-1,5-bisphosphate carboxylase/oxygenase (rubisco) of host plants. Virus Genes. 2011;43:90–92. doi: 10.1007/s11262-011-0596-6.

87. Chowda-Reddy R.V., Sun H.Y., Chen H.Y., Poysa V., Ling H., Gijzen M., Wang A.M. Mutations in the p3 protein of soybean mosaic virus g2 isolates determine virulence on rsv4-genotype soybean. Mol. Plant-Microbe Interact. 2011;24:37–43. doi: 10.1094/MPMI-07-10-0158.

88. Ryslava H., Doubnerova V., Janoskova M., Subr Z., Novakova S., Cerovska N. Influence of viral infection on transgenic plants containing genes for nonstructural potyviral proteins hc-pro and p3. Febs J.2005;272:464–464. doi: 10.1111/j.1742-4658.2004.04486.x.

89. Saenz P., Cervera M.T., Dallot S., Quiot L., Quiot J.B., Riechmann J.L., Garcia J.A. Identification of a pathogenicity determinant of plum pox virus in the sequence encoding the c-terminal region of protein p3+6k(1) J. Gen. Virol. 2000;81:557–566.

90. Suehiro N., Natsuaki T., Watanabe T., Okuda S. An important determinant of the ability of turnip mosaic virus to infect brassica spp. And/or raphanus sativus is in its p3 protein. J. Gen. Virol. 2004;85:2087–2098. doi: 10.1099/vir.0.79825-0.

91. Wen R.H., Maroof M.A.S., Hajimorad M.R. Amino acid changes in p3, and not the overlapping pipo-encoded protein, determine virulence of soybean mosaic virus on functionally immune rsv1-genotype soybean. Mol. Plant Pathol. 2011;12:799–807. doi: 10.1111/j.1364-3703.2011.00714.x.

92. Cui X.Y., Wei T.Y., Chowda-Reddy R.V., Sun G.Y., Wang A.M. The tobacco etch virus p3 protein forms mobile inclusions via the early secretory pathway and traffics along actin microfilaments. Virology.2010;397:56–63. doi: 10.1016/j.virol.2009.11.015.

93. Eiamtanasate S., Juricek M., Yap Y.K. C-terminal hydrophobic region leads prsv p3 protein to endoplasmic reticulum. Virus Genes. 2007;35:611–617. doi: 10.1007/s11262-007-0114-z.

94. Hong X.Y., Chen J., Shi Y.H., Chen J.P. The '6k1' protein of a strain of soybean mosaic virus localizes to the cell periphery. Arch. Virol. 2007;152:1547–1551. doi: 10.1007/s00705-007-0972-7.

95. Carrington J.C., Jensen P.E., Schaad M.C. Genetic evidence for an essential role for potyvirus ci protein in cell-to-cell movement. Plant J. 1998;14:393–400. doi: 10.1046/j.1365-313X.1998.00120.x.

96. RodriguezCerezo E., Findlay K., Shaw J.G., Lomonossoff G.P., Qiu S.G., Linstead P., Shanks M., Risco C. The coat and cylindrical inclusion proteins of a potyvirus are associated with connections between plant cells. Virology. 1997;236:296–306. doi: 10.1006/viro.1997.8736.

97. Roberts I.M., Wang D., Findlay K., Maule A.J. Ultrastructural and temporal observations of the potyvirus cylindrical inclusions (cis) show that the ci protein acts transiently in aiding virus movement. Virology.1998;245:173–181. doi: 10.1006/viro.1998.9132.

98. Jimenez I., Lopez L., Alamillo J.M., Valli A., Garcia J.A. Identification of a plum pox virus ci-interacting protein from chloroplast that has a negative effect in virus infection. Mol. Plant-Microbe Interact.2006;19:350–358. doi: 10.1094/MPMI-19-0350.

99. Abdul-Razzak A., Guiraud T., Peypelut M., Walter J., Houvenaghel M.C., Candresse T., Le Gall O., German-Retana S. Involvement of the cylindrical inclusion (ci) protein in the overcoming of an eif4e-mediated resistance against lettuce mosaic potyvirus. Mol. Plant Pathol. 2009;10:109–113. doi: 10.1111/j.1364-3703.2008.00513.x.

100. Tavert-Roudet G., Abdul-Razzak A., Doublet B., Walter J., Delaunay T., German-Retana S., Michon T., Le Gall O., Candresse T. The c terminus of lettuce mosaic potyvirus cylindrical inclusion helicase interacts with the viral vpg and with lettuce translation eukaryotic initiation factor 4e. J. Gen. Virol.2012;93:184–193. doi: 10.1099/vir.0.035881-0.

101. Ayme V., Souche S., Caranta C., Jacquemond M., Chadoeuf J., Palloix A., Moury B. Different mutations in the genome-linked protein vpg of potato virus y confer virulence on the pvr2(3) resistance in pepper. Mol. Plant-Microbe Interact. 2006;19:557–563. doi: 10.1094/MPMI-19-0557.

102. Charron C., Nicolai M., Gallois J.L., Robaglia C., Moury B.T., Palloix A., Caranta C. Natural variation and functional analyses provide evidence for co-evolution between plant eif4e and potyviral vpg. Plant J.2008;54:56–68. doi: 10.1111/j.1365-313X.2008.03407.x.

103. Rajamaki M.L., Valkonen J.P.T. The 6k2 protein and the vpg of potato virus a are determinants of systemic infection in nicandra physaloides. Mol. Plant-Microbe Interact. 1999;12:1074–1081. doi: 10.1094/MPMI.1999.12.12.1074.

104. Valkonen J.P.T., Rajamaki M.L., Hamalainen J., Kekarainen T., Oruetxebarria I. Incompatible potyvirus-host interactions in resistant plants: The viral genome-linked protein (vpg) of potato virus a as an avirulence determinant; Proceedings of the 7th aschersleben symposium new aspects of resistance research on culitivated plants: Virus diseases; QUEDLINBURG: BUNDESANSTALT FUR ZUCHTUNGS FORSCHUNG KULTURPFLANZEN; 2000. pp. 92–95.

105. Puurand U., Makinen K., Paulin L., Saarma M. The nucleotide-sequence of potato-virus-a genomic rna and its sequence similarities with other potyviruses. J. Gen. Virol. 1994;75:457–461. doi: 10.1099/0022-1317-75-2-457.

106. Rantalainen K.I., Eskelin K., Tompa P., Makinen K. Structural flexibility allows the functional diversity of potyvirus genome-linked protein vpg. J. Virol. 2011;85:2449–2457. doi: 10.1128/JVI.02051-10.

107. Rajamaki M.L., Valkonen J.P.T. Localization of a potyvirus and the viral genome-linked protein in wild potato leaves at an early stage of systemic infection. Mol. Plant-Microbe Interact. 2003;16:25–34. doi: 10.1094/MPMI.2003.16.1.25.

108. Vuorinen A.L., Kelloniemi J., Valkonen J.P.T. Why do viruses need phloem for systemic invasion of plants? Plant Sci. 2011;181:355–363. doi: 10.1016/j.plantsci.2011.06.008.

109. Rajamaki M.L., Valkonen J.P.T. Viral genome-linked protein (vpg) controls accumulation and phloem-loading of a potyvirus in inoculated potato leaves. Mol. Plant-Microbe Interact. 2002;15:138–149. doi: 10.1094/MPMI.2002.15.2.138.

110. Puustinen P., Rajamaki M.L., Ivanov K.I., Valkonen J.P.T., Makinen K. Detection of the potyviral genome-linked protein vpg in virions and its phosphorylation by host kinases. J. Virol. 2002;76:12703–12711.

111. Roudet-Tavert G., Michon T., Walter J., Delaunay T., Redondo E., Le Gall O. Central domain of a potyvirus vpg is involved in the interaction with the host translation initiation factor eif4e and the viral protein hcpro. J. Gen. Virol. 2007;88:1029–1033. doi: 10.1099/vir.0.82501-0.

112. Yambao M.L.M., Masuta C., Nakahara K., Uyeda I. The central and c-terminal domains of vpg of clover

yellow vein virus are important for vpg-hcpro and vpg-vpg interactions. J. Gen. Virol. 2003;84:2861–2869. doi: 10.1099/vir.0.19312-0.

113. Dunoyer P., Thomas C., Harrison S., Revers F., Maule A. A cysteine-rich plant protein potentiates potyvirus movement through an interaction with the virus genome-linked protein vpg. J. Virol.2004;78:2301–2309. doi: 10.1128/JVI.78.5.2301-2309.2004.

114. Hong Y.L., Levay K., Murphy J.F., Klein P.G., Shaw J.G., Hunt A.G. A potyvirus polymerase interacts with the viral coat protein and vpg in yeast-cells. Virology. 1995;214:159–166. doi: 10.1006/viro.1995.9944.

115. Carrington J.C., Haldeman R., Dolja V.V., Restrepohartwig M.A. Internal cleavage and trans-proteolytic activities of the vpg-proteinase (nia) of tobacco etch potyvirus in-vivo. J. Virol. 1993;67:6995–7000.

116. Hajimorad M.R., Ding X.S., Flasinski S., Mahajan S., Graff E., HaldemanCahill R., Carrington J.C., Cassidy B.G. Nla and nlb of peanut stripe potyvirus are present in the nucleus of infected cells, but do not form inclusions. Virology. 1996;224:368–379. doi: 10.1006/viro.1996.0544.

117. Beauchemin C., Boutet N., Laliberte J.F. Visualization of the interaction between the precursors of vpg, the viral protein linked to the genome of turnip mosaic virus, and the translation eukaryotic initiation factor iso 4e in planta. J. Virol. 2007;81:775–782. doi: 10.1128/JVI.01277-06.

118. Borgstrom B., Johansen I.E. Mutations in pea seedborne mosaic virus genome-linked protein vpg alter pathotype-specific virulence in pisum sativum. Mol. Plant-Microbe Interact. 2001;14:707–714. doi: 10.1094/MPMI.2001.14.6.707.

119. Daughenbaugh K.F., Fraser C.S., Hershey J.W.B., Hardy M.E. The genome-linked protein vpg of the norwalk virus binds eif3, suggesting its role in translation initiation complex recruitment. Embo J.2003;22:2852–2859. doi: 10.1093/emboj/cdg251.

120. Leonard S., Plante D., Wittmann S., Daigneault N., Fortin M.G., Laliberte J.F. Complex formation between potyvirus vpg and translation eukaryotic initiation factor 4e correlates with virus infectivity. J. Virol.2000;74:7730–7737. doi: 10.1128/JVI.74.17.7730-7737.2000.

121. Leonard S., Viel C., Beauchemin C., Daigneault N., Fortin M.G., Laliberte J.F. Interaction of vpg-pro of turnip mosaic virus with the translation initiation factor 4e and the poly(a)-binding protein in planta. J. Gen. Virol. 2004;85:1055–1063. doi: 10.1099/vir.0.19706-0.

122. Plante D., Viel C., Leonard S., Tampo H., Laliberte J.F., Fortin M.G. Tumip mosaic virus vpg does not disrupt the translation initiation complex but interferes with cap binding. Physiol. Mol. Plant Pathol.2004;64:219–226. doi: 10.1016/j.pmpp.2004.06.003.

123. Grzela R., Strokovska L., Andrieu J.P., Dublet B., Zagorski W., Chroboczek J. Potyvirus terminal protein vpg, effector of host eukaryotic initiation factor eif4e. Biochimie. 2006;88:887–896. doi: 10.1016/j.biochi.2006.02.012.

124. Michon T., Estevez Y., Walter J., German-Retana S., Le Gall O. The potyviral virus genome-linked protein vpg forms a ternary complex with the eukaryotic initiation factors eif4e and eif4g and reduces eif4e affinity for a mrna cap analogue. Febs J. 2006;273:1312–1322. doi: 10.1111/j.1742-4658.2006.05156.x.

125. Eskelin K., Hafren A., Rantalainen K.I., Makinen K. Potyviral vpg enhances viral rna translation and inhibits reporter mrna translation in planta. J. Virol. 2011;85:9210–9221. doi: 10.1128/JVI.00052-11.

126. Rantalainen K.I., Christensen P.A., Hafren A., Otzen D.E., Kalkkinen N., Makinen K. Interaction of a potyviral vpg with anionic phospholipid vesicles. Virology. 2009;395:114–120.

127. Li X.H., Valdez P., Olvera R.E., Carrington J.C. Functions of the tobacco etch virus rna polymerase (nib): Subcellular transport and protein-protein interaction with vpg/proteinase (nia) J. Virol. 1997;71:1598–1607.

128. Daros J.A., Schaad M.C., Carrington J.C. Functional analysis of the interaction between vpg-proteinase (nia) and rna polymerase (nib) of tobacco etch potyvirus, using conditional and suppressor mutants. J. Virol.1999;73:8732–8740.

129. Guo D.Y., Rajamaki M.L., Saarma M., Valkonen J.P.T. Towards a protein interaction map of potyviruses: Protein interaction matrixes of two potyviruses based on the yeast two-hybrid system. J. Gen. Virol. 2001;82:935–939.

130. Lin L., Shi Y.H., Luo Z.P., Lu Y.W., Zheng H.Y., Yan F., Chen J., Chen J.P., Adams M.J., Wu Y.F. Protein-protein interactions in two potyviruses using the yeast two-hybrid system. Virus Res. 2009;142:36–40. doi: 10.1016/j.virusres.2009.01.006.

131. Puustinen P., Makinen K. Uridylylation of the potyvirus vpg by viral replicase nib correlates with the nucleotide binding capacity of vpg. J. Biol. Chem. 2004;279:38103–38110. doi: 10.1074/jbc.M402910200.

132. Jones A.L., Johansen I.E., Bean S.J., Bach I., Maule A.J. Specificity of resistance to pea seed-borne mosaic potyvirus in transgenic peas expressing the viral replicase (nib) gene. J. Gen. Virol. 1998;79:3129–3137.

133. Sivamani E., Brey C.W., Dyer W.E., Talbert L.E., Qu R.D. Resistance to wheat streak mosaic virus in transgenic wheat expressing the viral replicase (nib) gene. Mol. Breed. 2000;6:469–477. doi: 10.1023/A:1026576124482.

134. Cassidy B.G., Nelson R.S. Differences in protection phenotypes in tobacco plants expressing coat protein genes from peanut stripe potyvirus with or without an engineered atg. Mol. Plant-Microbe Interact.1995;8:357–365. doi: 10.1094/MPMI-8-0357.

135. Crescenzi A., d'Aquino L., Nuzzaci M., Ostuni A., Bavoso A., Comes S., De Stradis A., Piazzolla P. Production of strain specific antibodies against a synthetic polypeptide corresponding to the n-terminal region of the plum pox potyvirus coat protein. J. Virol. Methods. 1997;69:181–189. doi: 10.1016/S0166-0934(97)00158-4.

136. Gotz R., Huth W., Maiss E. Molecular analyses of the coat protein region of different viruses on poaceae belonging to the potyviridae. Agronomie. 1995;15:491–494. doi: 10.1051/agro:19950719.

137. Handley J.A., Smith G.R., Dale J.L., Harding R.M. Sequence diversity in the coat protein coding region of twelve sugarcane mosaic potyvirus isolates from australia, USA and south africa. Arch. Virol.1998;143:1145–1153. doi: 10.1007/s007050050362.

138. Kobayashi K., Rabinowicz P., BravoAlmonacid F., Helguera M., Conci V., Lot H., Mentaberry A. Coat protein gene sequences of garlic and onion isolates of the onion yellow dwarf potyvirus (oydv) Arch. Virol. 1996;141:2277–2287. doi: 10.1007/BF01718630.

139. Ryu K.H., Kim S.J., Park W.M. Nucleotide sequence analysis of the coat protein genes of two korean isolates of sweet potato feathery mottle potyvirus. Arch. Virol. 1998;143:557–562. doi: 10.1007/s007050050311.

140. Valkonen J.P.T., Puurand U., Slack S.A., Makinen K., Saarma M. 3 strain groups of potato-a-potyvirus based on hypersensitive responses in potato, serological properties, and coat protein sequences. Plant Dis.1995;79:748–753. doi: 10.1094/PD-79-0748.

141. Massey B., Cui X., Hiebert E., Elliott M.S., Waipara N., Hayes L., Charudattan R. Partial sequencing of the genomic rna of araujia mosaic virus and comparison of the coat protein sequence with those of other potyviruses. Arch. Virol. 2007;152:2125–2129. doi: 10.1007/s00705-007-1044-8.

142. Pfosser M.F., Baumann H. Phylogeny and geographical differentiation of zucchini yellow mosaic virus isolates (potyviridae) based on molecular analysis of the coat protein and part of the cytoplasmic inclusion protein genes. Arch. Virol. 2002;147:1599–1609. doi: 10.1007/s00705-002-0839-x.

143. Sanchez F., Wang X., Jenner C.E., Walsh J.A., Ponz F. Strains of turnip mosaic potyvirus as defined by the molecular analysis of the coat protein gene of the virus. Virus Res. 2003;94:33–43. doi: 10.1016/S0168-1702(03)00122-9.

144. Andrejeva J., Puurand U., Merits A., Rabenstein F., Jarvekulg L., Valkonen J.P.T. Potyvirus helper component-proteinase and coat protein (cp) have coordinated functions in virus-host interactions and the same cp motif affects virus transmission and accumulation. J. Gen. Virol. 1999;80:1133–1139.

145. Atreya P.L., Lopezmoya J.J., Chu M.H., Atreya C.D., Pirone T.P. Mutational analysis of the coat protein n-terminal amino-acids involved in potyvirus transmission by aphids. J. Gen. Virol. 1995;76:265–270.

146. Blanc S., LopezMoya J.J., Wang R.Y., GarciaLampasona S., Thornbury D.W., Pirone T.P. A specific interaction between coat protein and helper component correlates with aphid transmission of a potyvirus.Virology. 1997;231:141–147. doi: 10.1006/viro.1997.8521.

147. Flasinski S., Cassidy B.G. Potyvirus aphid transmission requires helper component and homologous coat protein for maximal efficiency. Arch. Virol. 1998;143:2159–2172. doi: 10.1007/s007050050449.

148. Lopez-Moya J.J., Pirone T.P. Charge changes near the n terminus of the coat protein of two potyviruses affect virus movement. J. Gen. Virol. 1998;79:161–165.

149. Lopez-Moya J.J., Wang R.Y., Pirone T.P. Context of the coat protein dag motif affects potyvirus transmissibility by aphids. J. Gen. Virol. 1999;80:3281–3288.

150. Dombrovsky A., Huet H., Chejanovsky N., Raccah B. Aphid transmission of a potyvirus depends on suitability of the helper component and the n terminus of the coat protein. Arch. Virol. 2005;150:287–298. doi: 10.1007/s00705-004-0407-7.

151. Andersen K., Johansen I.E. A single conserved amino acid in the coat protein gene of pea seed-borne mosaic potyvirus modulates the ability of the virus to move systemically in chenopodium quinoa. Virology.1998;241:304–311. doi: 10.1006/viro.1997.8967.

152. Aparicio F., Thomas C.L., Lederer C., Niu Y., Wang D.W., Maule A.J. Virus induction of heat shock protein 70 reflects a general response to protein accumulation in the plant cytosol. Plant Physiol.2005;138:529–536. doi: 10.1104/pp.104.058958.

153. Sugio A., Dreos R., Aparicio F., Maule A.J. The cytosolic protein response as a subcomponent of the wider heat shock response in arabidopsis. Plant Cell. 2009;21:642–654. doi: 10.1105/tpc.108.062596.

154. Hafren A., Hofius D., Ronnholm G., Sonnewald U., Makinen K. Hsp70 and its cochaperone cpip promote potyvirus infection in nicotiana benthamiana by regulating viral coat protein functions. Plant Cell.2010;22:523–535. doi: 10.1105/tpc.109.072413.

155. Wang Y.Y., Gardner R.C., Pearson M.N. Resistance to vanilla necrosis potyvirus in transgenic nicotiana benthamiana plants containing the virus coat protein gene. J. Phytopathol. 1997;145:7–15. doi: 10.1111/j.1439-0434.1997.tb00335.x.

156. Dinant S., Maisonneuve B., Albouy J., Chupeau Y., Chupeau M.C., Bellec Y., Gaudefroy F., Kusiak C., Souche S., Robaglia C., et al. Coat protein gene-mediated protection in lactuca sativa against lettuce mosaic potyvirus strains. Mol. Breed. 1997;3:75–86. doi: 10.1023/A:1009671925550.

157. Xu D., Liu H., Li F., Howell B., Tian T., Li R. Biological characterization and complete genomic sequence of carrot thin leaf virus. Phytopathology. 2011;101:S195–S195.

158. Fuchs M., Gal-On A., Raccah B., Gonsalves D. Epidemiology of an aphid nontransmissible potyvirus in fields of nontransgenic and coat protein transgenic squash. Transgenic Res. 1999;8:429–439. doi: 10.1023/A:1008935426211.

159. Higgins C.M., Hall R.M., Mitter N., Cruickshank A., Dietzgen R.G. Peanut stripe potyvirus resistance in peanut (arachis hypogaea l.) plants carrying viral coat protein gene sequences. Transgenic Res.2004;13:59–67. doi: 10.1023/B:TRAG.0000017166.29458.74.

160. Ravelonandro M., Dunez J., Scorza R., Labonne G. 17th international symposium on virus and virus-like diseases of temperate fruit crops: Fruit tree diseases, vols 1 and 2. INTERNATIONAL SOCIETY HORTICULTURAL

SCIENCE; LEUVEN: 1998. Challenging transgenic plums expressing potyvirus coat protein genes with viruliferous aphids; pp. 413–420.

161. Scorza R., Callahan A.M., Levy L., Damsteegt V., Ravelonandro M. 17th international symposium on virus and virus-like diseases of temperate fruit crops: Fruit tree diseases, vols 1 and 2. INTERNATIONAL SOCIETY HORTICULTURAL SCIENCE; LEUVEN 1: 1998. Transferring potyvirus coat protein genes through hybridization of transgenic plants to produce plum pox virus resistant plums (prunus domestica l.) pp. 421–427.

162. Yadav N., Khan J.A. Over expression of narcissus potyvirus coat protein in e. Coli. Indian J. Virol.2009;20:44–44.

163. Savenkov E.I., Valkonen J.P.T. Coat protein gene-mediated resistance to potato virus a in transgenic plants is suppressed following infection with another potyvirus. J. Gen. Virol. 2001;82:2275–2278.

164. Shand K., Theodoropoulos C., Stenzel D., Dale J.L., Harrison M.D. Expression of potato virus y cytoplasmic inclusion protein in tobacco results in disorganization of parenchyma cells, distortion of epidermal cells, and induces mitochondrial and chloroplast abnormalities, formation of membrane whorls and atypical lipid accumulation. Micron. 2009;40:730–736. doi: 10.1016/j.micron.2009.04.011.

165. Restrepohartwig M.A., Carrington J.C. Regulation of nuclear transport of a plant potyvirus protein by autoproteolysis. J. Virol. 1992;66:5662–5666.

166. Restrepohartwig M.A., Carrington J.C. The tobacco etch potyvirus 6-kilodalton protein is membrane-associated and involved in viral replication. J. Virol. 1994;68:2388–2397.

167. Spetz C., Valkonen J.P.T. Potyviral 6k2 protein long-distance movement and symptom-induction functions are independent and host-specific. Mol. Plant-Microbe Interact. 2004;17:502–510. doi: 10.1094/MPMI.2004.17.5.502.

168. Yoshida N., Shimura H., Yamashita K., Suzuki M., Masuta C. Variability in the p1 gene helps to refine phylogenetic relationships among leek yellow stripe virus isolates from garlic. Arch. Virol. 2011;157:147–153.

169. Martin M.T., Gelie B. Non-structural plum pox potyvirus proteins detected by immunogold labelling. Eur. J. Plant Pathol. 1997;103:427–431. doi: 10.1023/A:1008653000005.

170. Arbatova J., Lehto K., Pehu E., Pehu T. Localization of the p1 protein of potato y potyvirus in association with cytoplasmic inclusion bodies and in the cytoplasm of infected cells. J. Gen. Virol.1998;79:2319–2323.

171. Mangrauthia S.K., Jain R.K., Praveen S. Sequence motifs comparisons establish a functional portrait of a multifunctional protein hc-pro from papaya ringspot potyvirus. J. Plant Biochem. Biotechnol. 2008;17:201–204.

172. Plisson C., Drucker M., Blanc S., German-Retana S., Le Gall O., Thomas D., Bron P. Structural characterization of hc-pro, a plant virus multifunctional protein. J. Biol. Chem. 2003;278:23753–23761.

173. Soitamo A.J., Jada B., Lehto K. Hc-pro silencing suppressor significantly alters the gene expression profile in tobacco leaves and flowers. BMC Plant Biol. 2011:11.

174. Wu H.W., Lin S.S., Chen K.C., Yeh S.D., Chua N.H. Discriminating mutations of hc-pro of zucchini yellow mosaic virus with differential effects on small rna pathways involved in viral pathogenicity and symptom development. Mol. Plant-Microbe Interact. 2010;23:17–28. doi: 10.1094/MPMI-23-1-0017.

175. Zheng H.Y., Yan F., Lu Y.W., Sun L.Y., Lin L., Cai L., Hou M.S., Chen J.P. Mapping the self-interacting domains of tumv hc-pro and the subcellular localization of the protein. Virus Genes. 2011;42:110–116. doi: 10.1007/s11262-010-0538-8.

176. Hjulsager C.K., Olsen B.S., Jensen D.M.K., Cordea M.I., Krath B.N., Johansen I.E., Lund O.S. Multiple determinants in the coding region of pea seed-borne mosaic virus p3 are involved in virulence against sbm-2 resistance. Virology. 2006;355:52–61. doi: 10.1016/j.virol.2006.07.016.

177. Langenberg W.G., Zhang L.Y. Immunocytology shows the presence of tobacco etch virus p3 protein in nuclear inclusions. J. Struct. Biol. 1997;118:243–247. doi: 10.1006/jsbi.1997.3856.

178. Hjulsager C.K., Lund O.S., Johansen I.E. A new pathotype of pea seedborne mosaic virus explained by properties of the p3-6k1-and viral genome-linked protein (vpg)-coding regions. Mol. Plant-Microbe Interact.2002;15:169–171. doi: 10.1094/MPMI.2002.15.2.169.

179. Waltermann A., Maiss E. Detection of 6k1 as a mature protein of 6 kda in plum pox virus-infected nicotiana benthamiana. J. Gen. Virol. 2006;87:2381–2386. doi: 10.1099/vir.0.81873-0.

180. Wei T.Y., Zhang C.W., Hong J.A., Xiong R.Y., Kasschau K.D., Zhou X.P., Carrington J.C., Wang A.M. Formation of complexes at plasmodesmata for potyvirus intercellular movement is mediated by the viral protein p3n-pipo. PLoS Pathog. 2010:6.

181. Riedel D., Lesemann D.E., Maiss E. Ultrastructural localization of nonstructural and coat proteins of 19 potyviruses using antisera to bacterially expressed proteins of plum pox potyvirus. Arch. Virol.1998;143:2133–2158. doi: 10.1007/s007050050448.

182. Voloudakis A.E., Malpica C.A., Aleman-Verdaguer M.E., Stark D.M., Fauquet C.M., Beachy R.N. Structural characterization of tobacco etch virus coat protein mutants. Arch. Virol. 2004;149:699–712. doi: 10.1007/s00705-003-0247-x.

183. Hema M., Reddy C.V.S., Savithri H.S., Sreenivasulu P. Assembly of recombinant coat protein of sugarcane streak mosaic virus into potyvirus-like particles. Indian J. Exp. Biol. 2008;46:793–796.

184. Atanasoff D. Yearbook university of sofia. Vol. 11. University of Sofia; 1932. Plum pox. A new virus disease; pp. 49–69.

185. Atanasoff D. Mosaic of stone fruits. Phytopathology. 1935:8.

186. Candresse T., Svanella-Dumas L., Gentit P., Caglayan K., Cevik B. First report of the presence of plum pox virus rec strain in turkey. Plant Dis. 2007;91:331–331.

187. Myrta A., Rwahnih M.A., Savino V. Presence of a recombinant isolate of plum pox virus in apulia. J. Plant Pathol. 2005;87:127–130.

188. Snover-Clift K.L., Clement P.A., Jablonski R., Mungari R.J., Mavrodieva V.A., Negi S., Levy L. First report of plum pox virus on plum in new york state. Plant Dis. 2007;91:1512–1512.

189. Zagrai I., Gaboreanu I., Ferencz B., Zagrai L., Pamfil D., Popescu O., Ravelonandro M., Capote N., Kovacs K. Bulletin of the university of agricultural sciences and veterinary medicine, vol 62, 2006: Animal husbandry and biotechnologies. 2006. First detection and molecular caracterization of plum pox virus recombinant strain in romania; pp. 291–298.

190. Thompson D., Varga A., De Costa H., Birch C., Glasa M., James D. First report of plum pox virus recombinant strain on prunus spp. In canada. Plant Dis. 2009;93:674–674.

191. Kamenova I., Dallot S., Bozkova V., Milusheva S. First report of the plum pox virus recombinant strain on peach in bulgaria. Plant Dis. 2011;95:1320–1321.

192. Cambra M., Capote N. Plum pox virus (ppv) in kazakhstan. EPPO Bull. 2006;36:210–211. doi: 10.1111/j.1365-2338.2006.01064.x.

193. Navratil M., Safarova D., Karesova R., Petrzik K. First incidence of plum pox virus on apricot trees in china. Plant Dis. 2005;89:338–338.

194. Dal Zotto A., Ortego J.M., Raigon J.M., Caloggero S., Rossini M., Ducasse D.A. First report in argentina of plum pox virus causing sharka disease in prunus. Plant Dis. 2006;90:523–523.

195. Capote N., Cambra M. Plum pox virus (ppv) in pakistan. EPPO Bull. 2006;36:212–212. doi: 10.1111/j.1365-2338.2006.01065.x.

196. Maejima K., Hoshi H., Hashimoto M., Himeno M., Kawanishi T., Komatsu K., Yamaji Y., Hamamoto H., Namba S. First report of plum pox virus infecting japanese apricot (prunus mume sieb. Et zucc.) in japan.J. Gen. Plant Pathol. 2010;76:229–231. doi: 10.1007/s10327-010-0233-6.

197. Kolber M., Nemeth M., Chernets A., Kalashian Y., Dulic-Markovic I., Glasa M., Isac M., Kriska B., Malinowski T., Zawadzka B., et al. Current situation of plum pox disease on stone fruit species in middle and eastern europe; Proceedings of the 18th international symposium on virus & virus–like diseases of temperate fruit crops – top fruit diseases, vols 1 and 2; LEUVEN 1: INTERNATIONAL SOCIETY HORTICULTURAL SCIENCE; 2001. pp. 73–78.

198. Mikec I., Kajić V., Križanac I. Plum pox virus (ppv) in croatia. EPPO Bull. 2006;36:207–207.

199. Mikec I., Kajic V., Krajacic M., Skoric D. Occurrence and distribution of plum pox virus in croatia; Proceedings of the twentieth international symposium on virus and virus–like diseases of temperate fruit crops – fruit tree diseases, 2008; pp. 193–196.

200. Isac M., Zagrai I. Plum pox virus (ppv) in romania. EPPO Bull. 2006;36:213–213. doi: 10.1111/j.1365-2338.2006.00963.x.

201. Varveri C. Plum pox virus (ppv) in greece. EPPO Bull. 2006;36:209–210. doi: 10.1111/j.1365-2338.2006.00957.x.

202. Navrátil M. Plum pox virus (ppv) in the czech republic. EPPO Bull. 2006;36:208–208. doi: 10.1111/j.1365-2338.2006.00991.x.

203. Jarausch W. Plum pox virus (ppv) in germany. EPPO Bull. 2006;36:209–209. doi: 10.1111/j.1365-2338.2006.00993.x.

204. Milusheva S., Kamenova I., Stoev A. Plum pox virus (ppv) in bulgaria. EPPO Bull. 2006;36:206–206.

205. Kölber M. Plum pox virus (ppv) in hungary. EPPO Bull. 2006;36:210–210. doi: 10.1111/j.1365-2338.2006.00958.x.

206. Prichodko Y. Plum pox virus (ppv) in russia. EPPO Bull. 2006;36:213–213. doi: 10.1111/j.1365-2338.2006.00964.x.

207. Malinowski T. Plum pox virus (ppv) in poland. EPPO Bull. 2006;36:212–213. doi: 10.1111/j.1365-2338.2006.00962.x.

208. EPPO. Plum pox potyvirus. EPPO Bull. 2004;34:247–256. doi: 10.1111/j.1365-2338.2004.00726.x.

209. Dulic-Markovic I., Jevremovic D. Plum pox virus (ppv) in serbia. EPPO Bull. 2006;36:213–214. doi: 10.1111/j.1365-2338.2006.00965.x.

210. Glasa M. Plum pox virus (ppv) in slovakia. EPPO Bull. 2006;36:214–214. doi: 10.1111/j.1365-2338.2006.00966.x.

211. Ortego J., Dal Zotto A., Caloggero S., Raigón J.M., Gasparini M.L., Ojeda M.E., Ducasse D.A. Current status of ppv worldwide - plum pox virus (ppv) in argentina. EPPO Bull. 2006;36:205–205.

212. Stamo B., Myrta A. Current status of ppv worldwide - plum pox virus (ppv) in albania. EPPO Bull.2006;36:205–205. doi: 10.1111/j.1365-2338.2006.00947.x.

213. Akbas B., Degirmenci K., Ciftci O., Kaya A., Yurtmen M., Uzunogullari N., Celik N., Torkolmez S. Update on plum pox virus distribution in turkey. Phytopathologia Mediterranea. 2011;50:75–83.

214. Di Terlizzi B., Boscia D. Plum pox virus (ppv) in italy. EPPO Bull. 2006;36:210–210. doi: 10.1111/j.1365-2338.2006.00959.x.

215. Ortego J., Zotto A.D., Caloggero S., Raigón J.M., Gasparini M.L., Ojeda M.E., Ducasse D.A. Plum pox virus (ppv) in argentina. EPPO Bull. 2006;36:205–205.

216. Kalashian Y., Chernets A. Plum pox virus (ppv) in moldova. EPPO Bull. 2006;36:211–211. doi: 10.1111/j.1365-2338.2006.00972.x.

217. Youssef S.A., Shalaby A. Plum pox virus (ppv) in egypt. EPPO Bull. 2006;36:208–208. doi: 10.1111/j.1365-2338.2006.00955.x.

218. Norkus T., Staniulis J., Zizyte M., Melnyk M., Yusko L., Snihur H., Budzanivska I., Polischuk V. Molecular

identification of plum pox virus isolates from lithuania and ukraine. Zemdirbyste-Agriculture.2008;95:277–285.

219. Wang A., Sanfacon H., Stobbs L.W., James D., Thompson D., Svircev A.M., Brown D.C.W. Plum pox virus in canada: Progress in research and future prospects for disease control. Can. J. Plant Pathol.-Rev. Can. Phytopathol. 2006;28:182–196.
220. Blystad D.R., Munthe T. Plum pox virus (ppv) in norway. EPPO Bull. 2006;36:212–212. doi: 10.1111/j.1365-2338.2006.00961.x.
221. Zamharir M.G., Bashir N.S., Khakvar R. Plum pox virus (ppv) in iran. EPPO Bull. 2006;36:210–210.
222. Speich P. Plum pox virus (ppv) in france. EPPO Bull. 2006;36:208–209. doi: 10.1111/j.1365-2338.2006.00992.x.
223. Marn M.V., Mavric I. Plum pox virus (ppv) in slovenia. EPPO Bull. 2006;36:214–214.
224. Kondratenko P., Udovychenko V. Plum pox virus (ppv) in ukraine. EPPO Bull. 2006;36:217–217. doi: 10.1111/j.1365-2338.2006.00974.x.
225. Mumford R.A. Plum pox virus (ppv) in the united kingdom. EPPO Bull. 2006;36:217–217. doi: 10.1111/j.1365-2338.2006.00969.x.
226. Ismaeil F. Plum pox virus (ppv) in syria. EPPO Bull. 2006;36:216–216. doi: 10.1111/j.1365-2338.2006.00997.x.
227. Levy L. Plum pox virus (ppv) in the united states of america. EPPO Bull. 2006;36:217–218. doi: 10.1111/j.1365-2338.2006.00998.x.
228. Caglayan K. Plum pox virus (ppv) in turkey. EPPO Bull. 2006;36:216–217. doi: 10.1111/j.1365-2338.2006.00971.x.
229. Cambra M.A., Serra J., Cano A., Cambra M. Plum pox virus (ppv) in spain. EPPO Bull. 2006;36:215–215.
230. Muñoz M., Collao M. Plum pox virus (ppv) in chile. EPPO Bull. 2006;36:207–207.
231. Matić S., Durić G., Myrta A. Current status of ppv worldwide - plum pox virus (ppv) in bosnia and herzegovina. EPPO Bull. 2006;36:205–206.
232. Staniulis J. Plum pox virus (ppv) in lithuania. EPPO Bull. 2006;36:211–211. doi: 10.1111/j.1365-2338.2006.00995.x.
233. Verhoeven J.T.J., Roenhorst J.W., Jongedijk G.P. Plum pox virus (ppv) in the netherlands. EPPO Bull.2006;36:211–212.
234. Boulila M., Ravelonandro M. Plum pox virus (ppv) in tunisia. EPPO Bull. 2006;36:216–216.
235. Ramel M.E., Gugerli P., Bünter M. Plum pox virus (ppv) in switzerland. EPPO Bull. 2006;36:215–216.
236. Lemmetty A. Plum pox virus (ppv) in finland. EPPO Bull. 2006;36:208–208. doi: 10.1111/j.1365-2338.2006.00956.x.
237. Choueiri E. Plum pox virus (ppv) in lebanon. EPPO Bull. 2006;36:211–211. doi: 10.1111/j.1365-2338.2006.00960.x.
238. Navratil M., Safarova D., Karesova R., Petrzik K. Plum pox virus (ppv) in china. EPPO Bull.2006;36:207–207.
239. Reyes F., Fiore N., Reyes M.A., Sepulveda P., Paredes V., Prieto H. Biological behavior and partial molecular characterization of six chilean isolates of plum pox virus. Plant Dis. 2003;87:15–20. doi: 10.1094/PDIS.2003.87.1.15.
240. Manachini B., Casati P., Aliverti I., Cinanni L. Transmission of ppv-m to prunus persica by brachycaudus schwartzi and phorodon humuli (hem. aphididae) J. Appl. Entomol. 2004;128:677–680. doi: 10.1111/j.1439-0418.2004.00908.x.
241. Zawadzka B., Rozpara E., Grzyb Z. Vi international symposium on plum and prune genetics, breeding and pomology. INTERNATIONAL SOCIETY HORTICULTURAL SCIENCE; LEUVEN 1: 1998. The response of some new plum cultivars to plum pox virus (ppv) pp. 81–85.
242. Paprstein F., Karesova R., Navratil M. Evaluation of ppv symptoms on plum fruits; Proceedings of the eighth international symposium on plum and prune genetics, breeding and pomology, 2007; pp. 255–257.
243. Varveri C., Zintzaras E., Dimou D., Di Terlizzi B. Monitoring and spatiotemporal analysis of ppv-m spread in two apricot orchards in southern greece; Proceedings of the 18th international symposium on virus & virus–like diseases of temperate fruit crops – top fruit diseases, vols 1 and 2; LEUVEN 1: INTERNATIONAL SOCIETY HORTICULTURAL SCIENCE; 2001. pp. 129–133.
244. Clemente-Moreno M.J., Piqueras A., Hernandez J.A. Implication of peroxidase activity in development of healthy and ppv-infected micropropagated gf305 peach plants. Plant Growth Regul. 2011;65:359–367. doi: 10.1007/s10725-011-9608-8.
245. Pacheco R., Garcia-Marcos A., Manzano A., de Lacoba M.G., Camanes G., Garcia-Agustin P., Diaz-Ruiz J.R., Tenllado F. Comparative analysis of transcriptomic and hormonal responses to compatible and incompatible plant-virus interactions that lead to cell death. Mol. Plant-Microbe Interact. 2012;25:709–723. doi: 10.1094/MPMI-11-11-0305.
246. Nagyova A., Kamencayova M., Glasa M., Subr Z.W. The 3'-proximal part of the plum pox virus p1 gene determinates the symptom expression in two herbaceous host plants. Virus Genes. 2012;44:505–512.
247. Varveri C., Vassilakos N. Preliminary biological and molecular characterization of mild ppv isolates naturally induced in n-benthamiana plants; Proceedings of the twentieth international symposium on virus and virus–like diseases of temperate fruit crops – fruit tree diseases; LEUVEN 1: INTERNATIONAL SOCIETY HORTICULTURAL SCIENCE; 2008. pp. 249–253.
248. Sutic D., Jordovic M., Rankovic M., Festi H. Comparative studies of some sharka virus isolates. Proceedings of the VIII Symposium sur les Maladies a Virus des Arbres Frutiers. Annales de Phytopathologie, 1971; 1971. pp. 185–194.
249. James D., Varga A., Thompson D., Hayes S. Detection of a new and unusual isolate of plum pox virus in plum (prunus domestica) Plant Dis. 2003;87:1119–1124. doi: 10.1094/PDIS.2003.87.9.1119.

250. Serce C.U., Candresse T., Svanella-Dumas L., Krizbai L., Gazel M., Caglayan K. Further characterization of a new recombinant group of plum pox virus isolates, ppv-t, found in orchards in the ankara province of turkey. Virus Res. 2009;142:121–126. doi: 10.1016/j.virusres.2009.01.022.

251. Matic S., Elmaghraby I., Law V., Varga A., Reed C., Myrta A., James D. Serological and molecular characterization of isolates of plum pox virus strain el amar to better understand its diversity, evolution, and unique geographical distribution. Journal of Plant Pathology. 2011;93:303–310.

252. Candresse T., Macquaire G., Lanneau M., Bousalem M., Wetze T., Quiot-Douine L., Quiot J.B., Dunez J. Detection of plum pox potyvirus and analysis of its molecular variability using immunocapture-pcr1.EPPO Bull. 1994;24:585–594. doi: 10.1111/j.1365-2338.1994.tb01072.x.

253. Glasa M., Malinowski T., Predajna L., Pupola N., Dekena D., Michalczuk L., Candresse T. Sequence variability, recombination analysis, and specific detection of the w strain of plum pox virus. Phytopathology.2011;101:980–985. doi: 10.1094/PHYTO-12-10-0334.

254. Glasa M., Matisova J., Kudela O. Characterization of plum pox virus isolates from slovakia. Acta Virol.1998;42:226–229.

255. Navratil M., Simonova V., Fialova R., Valova P. Molecular variability of czech plum pox virus isolates. Acta Virol. 1998;42:254–256.

256. Navratil M., Simonova V., Paprstein F., Karesova R. Detection and serological identification of plum pox virus isolates in the czech republic; 17th international symposium on virus and virus-like diseases of temperate fruit crops: Fruit tree diseases, vols 1 and 2; 1998. pp. 373–378.

257. Al Rwahnih M., Boscia D., Myrta A., Di Terlizzi B. First record of plum pox virus in jordan; Proceedings of the 18th international symposium on virus & virus–like diseases of temperate fruit crops – top fruit diseases, vols 1 and 2, 2001; pp. 141–144.

258. Kamenova I., Milusheva S., Borisova A., Stoev A., Myrta A. Typing of plum pox virus isolates in bulgaria: Preliminary results. Biotechnol. Biotechnol. Equip. 2002;16:10–13.

259. Bianco P.A., Aliverti I., Casati P., Belli G., Comes S., Fanigliulo A., Crescenzi A. Detection and characterization of plum pox virus isolates in lombardia and veneto regions (italy); Proceedings of the xixth international symposium on virus and virus–like diseases of temperate fruit crops: Fruit tree diseases, 2004; pp. 165–169.

260. Cambra M., Gorris M.T., Capote N., Asensio M., Martinez M.C., Bertolini E., Collado C., de Mendoza A.H., Mataix E., Lopez A. Epidemiology of plum pox virus in japanese plums in spain; Proceedings of the xixth international symposium on virus and virus–like diseases of temperate fruit crops: Fruit tree diseases, 2004; pp. 195–200.

261. Jarausch W., Bassler A., Molla N., Krezal G. First detection and molecular characterisation of ppv-m strains in plum orchards in south-western germany; Proceedings of the xixth international symposium on virus and virus–like diseases of temperate fruit crops: Fruit tree diseases, 2004; pp. 159–164.

262. Marn M.V., Mavric I. Current status of sharka disease in slovenia; Proceedings of the xixth international symposium on virus and virus–like diseases of temperate fruit crops: Fruit tree diseases, 2004; pp. 245–249.

263. Dallot S., Boeglin M., Labonne G. Spatial pattern and genetic structure of ppv-m in a delimited area of stone fruit orchards in southern france; Proceedings of the twentieth international symposium on virus and virus–like diseases of temperate fruit crops – fruit tree diseases, 2008; pp. 235–242.

264. Dallot S., Kamenova I., Glasa M., Pittnerova S., Kominek P., Paunovic S., Jevremovic D., Virscek-Marn M., Plesko I.M., Milusheva S. Prevalence and genetic structure of ppv-m in six european countries; Proceedings of the twentieth international symposium on virus and virus–like diseases of temperate fruit crops – fruit tree diseases, 2008; pp. 227–234.

265. Kajic V., Cerni S., Krajacic M., Mikec I., Skoric D. Molecular typing of plum pox virus isolates in croatia. J. Plant Pathol. 2008;90:9–13.

266. Zagrai L., Zagrai I., Ferencz B., Gaboreanu I., Kovacs K., Petricele I., Popescu O., Pamfil D., Capote N. Serological and molecular typing of plum pox virus isolates in the north of romania. J. Plant Pathol.2008;90:41–46.

267. Zagrai I., Zagrai L., Ferencz B., Petricele I., Pamfil D., Popescu O., Briciu A., Festila A. Serological and molecular typing of plum pox virus isolates in the transylvania, romania. Not. Bot. Horti Agrobot. Cluj-Na. 2009;37:265–272.

268. Capote N., Cambra M.A., Botella P., Gorris M.T., Martinez M.C., Lopez-Quilez A., Cambra M. Detection, characterization, epidemiology and eradication of plum pox virus marcus type in spain. J. Plant Pathol. 2010;92:619–628.

269. Zagrai I., Zagrai L., Preda S., Kelemen B., Petricele I., Popescu O., Pamfil D., Isac M. Genetic diversity of plum pox virus isolates in muntenia, romania. Rom. Biotech. Lett. 2010;15:5303–5309.

270. Dallot S., Glasa M., Jevremovic D., Kamenova I., Paunovic S., Labonne G. Mediterranean and central-eastern european countries host viruses of two different clades of plum pox virus strain m. Arch. Virol.2011;156:539–542. doi: 10.1007/s00705-011-0918-y.

271. Myrta A., Palmisano F., Pulaj B., Susuri L.R., Boscia D. Incidence of plum pox virus and its strains in kosovo. J. Plant Pathol. 2011;93:725–728.

272. Rosales M., Hinrichsen P., Herrera G. 17th international symposium on virus and virus-like diseases of temperate fruit crops: Fruit tree diseases, vols 1 and 2. INTERNATIONAL SOCIETY HORTICULTURAL SCIENCE; LEUVEN 1: 1998. Molecular characterization of plum pox virus isolated from apricots, plums and

peaches in chile; pp. 401–405.

273. Rochon D., Theilmann J., James D., Reade R., Yang L., Upton C. Partial molecular characterization of plum pox virus isolates occurring in canada. Can. J. Plant Pathol.-Rev. Can. Phytopathol. 2003;25:198–208.

274. Marie-Jeanne V., Thebaud G., Peyre J., Labonne G. Spatially arranged diversity of plum pox virus (ppv) isolates in a cluster of orchards in southern france; Proceedings of the xixth international symposium on virus and virus-like diseases of temperate fruit crops: Fruit tree diseases,INTERNATIONAL SOCIETY HORTICULTURAL SCIENCE: LEUVEN 1, 2004; LEUVEN 1: INTERNATIONAL SOCIETY HORTICULTURAL SCIENCE; pp. 213–215.

275. Elibuyuk I.O. Current situation of sharka disease in ankara, turkey. Phytoparasitica. 2004;32:417–420. doi: 10.1007/BF02979855.

276. Theilmann J., Yang L., Rochon D. Sequence analysis of isolates of the canadian plum pox virus, and comparisons to isolates from europe and the united states. Can. J. Plant Pathol.-Rev. Can. Phytopathol.2006;28:144–151.

277. Matic S., Al-Rwahnih M., Myrta A. Diversity of plum pox virus isolates in bosnia and herzegovina.Plant Pathol. 2006;55:11–17. doi: 10.1111/j.1365-3059.2005.01309.x.

278. Gadiou S., Safarova D., Navratil M. Genetic variability of plum pox virus isolates in the czech republic.Eur. J. Plant Pathol. 2008;121:513–517. doi: 10.1007/s10658-008-9272-7.

279. Zagrai I., Zagrai L., Preda S., Kelemen B., Petricele I., Popescu O., Pamfil D., Isac M. Genetic diversity of plum pox virus isolates in muntenia, romania. Rom. Biotech. Lett. 2011;15:5303–5309.

280. Glasa M., Paunovic S., Jevremovic D., Myrta A., Pittnerova S., Candresse T. Analysis of recombinant plum pox virus (ppv) isolates from serbia confirms genetic homogeneity and supports a regional origin for the ppv-rec subgroup. Arch. Virol. 2005;150:2051–2060. doi: 10.1007/s00705-005-0548-3.

281. Navratil M., Safarova D., Gadiou S., Franova J., Kucerova J., Talacko L. The partial molecular characterization of plum pox virus infecting sweet cherry trees in the czech republic; Proceedings of the twentieth international symposium on virus and virus–like diseases of temperate fruit crops – fruit tree diseases; LEUVEN 1: INTERNATIONAL SOCIETY HORTICULTURAL SCIENCE; 2008. pp. 203–208.

282. Wetzel T., Candresse T., Ravelonandro M., Delbos R.P., Mazyad H., Aboulata A.E., Dunez J. Nucleotide-sequence of the 3'-terminal region of the rna of the elamar strain of plum pox potyvirus. J. Gen. Virol. 1991;72:1741–1746. doi: 10.1099/0022-1317-72-7-1741.

283. Abdel-Ghaffar M.H., El-Nasr M.A.A., Hari V. Studies on an apricot strain of plum pox potyvirus isolated from el amar, egypt; 17th international symposium on virus and virus-like diseases of temperate fruit crops: Fruit tree diseases, vols 1 and 2; 1998. pp. 385–391.

284. Aboul-Ela A., Aboul-Ata A.E., Mazyad H.M. International symposium on apricot culture, vols 1 and 2.INTERNATIONAL SOCIETY HORTICULTURAL SCIENCE; LEUVEN 1: 1999. Plum pox potyvirus situation in egyptian stone fruit trees; pp. 745–751.

285. Myrta A., Varga A., James D. The complete genome sequence of an el amar isolate of plum pox virus (ppv) and its phylogenetic relationship to other ppv strains. Arch. Virol. 2006;151:1189–1198. doi: 10.1007/s00705-005-0703-x.

286. Matic S., Elmaghraby I., Law V., Varga A., Reed C., Myrta A., James D. Serological and molecular characterization of isolates of plum pox virus strain el amar to better understand its diversity, evolution, and unique geographical distribution. J. Plant Pathol. 2011;93:303–310.

287. Nemchinov L., Hadidi A., Kolber M., Nemeth M. Molecular evidence for the occurrence of plum pox virus - cherry subgroup in hungary; 17th international symposium on virus and virus-like diseases of temperate fruit crops: Fruit tree diseases, vols 1 and 2; LEUVEN 1: INTERNATIONAL SOCIETY HORTICULTURAL SCIENCE; 1998. pp. 503–508.

288. Nemchinov L., Hadidi A., Maiss E., Cambra M., Candresse T., Damsteegt V. Sour cherry strain of plum pox potyvirus (ppv): Molecular and serological evidence for a new subgroup of ppv strains.Phytopathology. 1996;86:1215–1221. doi: 10.1094/Phyto-86-1215.

289. Crescenzi A., dAquino L., Comes S., Nuzzaci M., Piazzolla P., Boscia D., Hadidi A. Characterization of the sweet cherry isolate of plum pox potyvirus. Plant Dis. 1997;81:711–714. doi: 10.1094/PDIS.1997.81.7.711.

290. Isac M., Plopa C., Calinescu M., Myrta A. Detection of the viral diseases presently with the stone fruit species in romania; Proceedings of the twentieth international symposium on virus and virus–like diseases of temperate fruit crops – fruit tree diseases, 2008; pp. 59–63.

291. Schneider W.L., Damsteegt V.D., Gildow F.E., Stone A.L., Sherman D.J., Levy L.E., Mavrodieva V., Richwine N., Welliver R., Luster D.G. Molecular, ultrastructural, and biological characterization of pennsylvania isolates of plum pox virus. Phytopathology. 2011;101:627–636. doi: 10.1094/PHYTO-09-10-0256.

292. Damsteegt V.D., Stone A.L., Schneider W.L., Luster D.G., Gildow F.E. Potential prunus host range of ppv-penn isolates by aphid transmission; Proceedings of the xixth international symposium on virus and virus–like diseases of temperate fruit crops: Fruit tree diseases; LEUVEN 1: INTERNATIONAL SOCIETY HORTICULTURAL SCIENCE; 2004. pp. 201–205.

293. James D., Varga A. Nucleotide sequence analysis of plum pox virus isolate w3174: Evidence of a new strain. Virus Res. 2005;110:143–150. doi: 10.1016/j.virusres.2005.02.004.

294. Riechmann J.L., Lain S., Garcia J.A. Highlights and prospects of potyvirus molecular-biology. J. Gen. Virol. 1992;73:1–16. doi: 10.1099/0022-1317-73-1-1.

295. Salvador B., García J.A., Simón-Mateo C. Causal agent of sharka disease: Plum pox virus genome and function of gene products. EPPO Bull. 2006;36:229–238. doi: 10.1111/j.1365-2338.2006.00979.x.

296. Teycheney P.Y., Tavert G., Delbos R., Ravelonandro M.,

Dunez J. The complete nucleotide-sequence of plum pox virus-rna (strain-d) Nucleic Acids Res. 1989;17:10115–10116. doi: 10.1093/nar/17.23.10115.

297. Riechmann J.L., Lain S., Garcia J.A. Identification of the initiation codon of plum pox potyvirus genomic rna. Virology. 1991;185:544–552. doi: 10.1016/0042-6822(91)90524-F.

298. Khan M.A., Goss D.J. Poly(a)-binding protein increases the binding affinity and kinetic rates of interaction of viral protein linked to genome with translation initiation factors eifiso4f and eifiso4f. 4b complex. Biochemistry. 2012;51:1388–1395.

299. Alvarez V., Ducasse D.A., Biderbost E., Nome S.F. Sequencing and characterization of the coat protein and 3' non-coding region of a new sweet potato potyvirus. Arch. Virol. 1997;142:1635–1644. doi: 10.1007/s007050050185.

300. Colinet D., Kummert J., Lepoivre P. Evidence for the assignment of two strains of splv to the genus potyvirus based on coat protein and 3' non-coding region sequence data. Virus Res. 1997;49:91–100. doi: 10.1016/S0168-1702(97)01456-1.

301. Carrington J.C., Freed D.D. Cap-independent enhancement of translation by a plant potyvirus 5' nontranslated region. J. Virol. 1990;64:1590–1597.

302. Kneller E.L.P., Rakotondrafara A.M., Miller W.A. Cap-independent translation of plant viral rnas. Virus Res. 2006;119:63–75. doi: 10.1016/j.virusres.2005.10.010.

303. SimonBuela L., Guo H.S., Garcia J.A. Cap-independent leaky scanning as the mechanism of translation initiation of a plant viral genomic rna. J. Gen. Virol. 1997;78:2691–2699.

304. Fanigliulo A., Comes S., Maiss E., Piazzolla P., Crescenzi A. The complete nucleotide sequence of plum pox virus isolates from sweet (ppv-swc) and sour (ppv-soc) cherry and their taxonomic relationships within the species. Arch. Virol. 2003;148:2137–2153. doi: 10.1007/s00705-003-0175-9.

305. Glasa M., Marie-Jeanne V., Moury B., Kudela O., Quiot J.B. Molecular variability of the p3-6k(1) genomic region among geographically and biologically distinct isolates of plum pox virus. Arch. Virol.2002;147:563–575. doi: 10.1007/s007050200006.

306. Poncarova Z., Kominek P. Restriction fragment length polymorphism differentiation of plum pox virus isolates. Acta Virol. 1998;42:268–269.

307. SimonBuela L., Guo H.S., Garcia J.A. Long sequences in the 5' noncoding region of plum pox virus are not necessary for viral infectivity but contribute to viral competitiveness and pathogenesis. Virology.1997;233:157–162. doi: 10.1006/viro.1997.8574.

308. Simon-Buela L., Osaba L., Garcia J.A., Lopez-Moya J.J. Preservation of 5 '-end integrity of a potyvirus genomic rna is not dependent on template specificity. Virology. 2000;269:377–382. doi: 10.1006/viro.2000.0229.

309. Fernandez-Calvino L., Lopez-Abella D., Lopez-Moya J.J., Fereres A. Comparison of potato virus y and plum pox virus transmission by two aphid species in relation to their probing behavior. Phytoparasitica.2006;34:315–324. doi: 10.1007/BF02980959.

310. Fos A., Massonie G. Experimental transmission of plum pox virus by brachycaudus-persicae passerini. Agronomie. 1993;13:515–518. doi: 10.1051/agro:19930608.

311. Gildow F.E., Damsteegt V.D., Stone A.L., Schneider W.L., Luster D.G., Levy L. Transmission of three north american isolates of plum pox virus: Identification of aphid vectors and species-specific transmission from infected stone fruits; Proceedings of the xixth international symposium on virus and virus–like diseases of temperate fruit crops: Fruit tree diseases, 2004; pp. 207–211.

312. Glasa M., Boeglin M., Labonne G. Aphid transmission of natural recombinant plum pox virus isolates to different prunus ssp - a contribution for understanding the epidemiology of an atypical ppv; Proceedings of the xixth international symposium on virus and virus–like diseases of temperate fruit crops: Fruit tree diseases; LEUVEN 1: INTERNATIONAL SOCIETY HORTICULTURAL SCIENCE; 2004. pp. 217–220.

313. Lopezmoya J.J., Canto T., Diazruiz J.R., Lopezabella D. Transmission by aphids of a naturally non-transmissible plum pox virus isolate with the aid of potato-virus-y helper component. J. Gen. Virol.1995;76:2293–2297. doi: 10.1099/0022-1317-76-9-2293.

314. Paprstein F., Karesova R., Navratil M. Vi international symposium on plum and prune genetics, breeding and pomology. INTERNATIONAL SOCIETY HORTICULTURAL SCIENCE; LEUVEN 1: 1998. Results of plum pox virus transmission into germplasm of plums; pp. 279–282.

315. Milusheva S., Gercheva P., Bozhkova V., Kamenova I. Experiments on transmission of plum pox virus through prunus seeds. J. Plant Pathol. 2008;90:23–26.

316. Glasa M., Hriovsky I., Kudela O. Evidence for non-transmission of plum pox virus by seed in infected plum and myrobalan. Biologia. 1999;54:481–484.

317. Schimanski H.H., Gruntzig M., Fuchs E. Non-transmission of the plum pox virus in plum and apricot seed source clones. Zentralblatt Fur Mikrobiologie. 1988;143:121–123.

318. Dicenta F., Audergon J.M. Localization of plum pox virus in apricot stem and petiole tissues by tissue printing onto nitrocellulose membrane. Ann. Appl. Biol. 1995;126:105–110. doi: 10.1111/j.1744-7348.1995.tb05006.x.

319. Hoffmann A., Stockert T., Hartmann W., Stosser R. Localization of plum pox virus (ppv, sharka) in plum (prunus domestica l.) using tissue print immunoblotting (tpib) J. Plant Dis. Prot. 1997;104:411–414.

320. Rubio M., Gosalvez B., Hernandez J.A., Sanchez-Navarro J.A., Pallas V. Localization of coat protein and nucleic acid of plum pox virus in prunus petiole and stem tissues; Proceedings of the xixth international symposium on virus and virus–like diseases of temperate fruit crops: Fruit tree diseases; LEUVEN 1: INTERNATIONAL SOCIETY HORTICULTURAL SCIENCE; 2004. pp.

295–300.

321. Rubio M., Martinez-Gomez P., Dicenta F. Study of long-distance movement of plum pox virus (sharka) as an alternative resistance-evaluation method in prunus. Sci. Hortic. 2008;118:223–227.

322. Dicenta F., Martinez-Gomez P., Rubio M., Audergon J.M. Localisation and movement of plum pox virus in apricot stem tissues. Ann. Appl. Biol. 2003;142:99–105. doi: 10.1111/j.1744-7348.2003.tb00234.x.

323. Quiot J.B., Tab F., Labonne G., Adamolle C., Boeglin M. Location of plum pox virus in peach and apricot trees; Proceedings of the xiith ishs symposium on apricot culture and decline, vols 1 and 2; LEUVEN 1: INTERNATIONAL SOCIETY HORTICULTURAL SCIENCE; 2006. pp. 489–492.

324. Elouaghlidi F., Sautter C. Immunofluorescence localization of plum pox virus and prunus necrotic ringspot virus in tissue-sections of prunus-cerasus and prunus-juxa. J. Plant Dis. Prot. 1988;95:70–76.

325. Hernandez J.A., Talavera J.M., Martinez-Gomez P., Dicenta F., Sevilla F. Response of antioxidative enzymes to plum pox virus in two apricot cultivars. Physiol. Plant. 2001;111:313–321. doi: 10.1034/j.1399-3054.2001.1110308.x.

326. Visedo G., Fernandezpiqueras J., Garcia J.A. Peroxidase isozyme analysis of factors involved in development of symptoms in nicotiana-clevelandii infected by plum pox virus. Physiol. Plant. 1991;83:165–169. doi: 10.1111/j.1399-3054.1991.tb01297.x.

327. Visedo G., Fernandezpiqueras J., Garcia J.A. Comparison among the isozyme profiles associated with ethrel treatments of leaves, and with senescence and plum pox virus-infection in chenopodium-foetidum.Physiol. Plant. 1991;83:159–164. doi: 10.1111/j.1399-3054.1991.tb01296.x.

328. Hernandez J.A., Olmos J.E., Portillo B., Rubio M., Dicenta F., Martinez-Gomez P. Effect of plum pox virus infection on photosynthesis and antioxidant enzyme activity in peach 'gf305'; Proceedings of the xixth international symposium on virus and virus–like diseases of temperate fruit crops: Fruit tree diseases; LEUVEN 1: INTERNATIONAL SOCIETY HORTICULTURAL SCIENCE; 2004. pp. 349–353.

329. Diaz-Vivancos P., Clemente-Moreno M.J., Rubio M., Olmos E., Garcia J.A., Martinez-Gomez P., Hernandez J.A. Alteration in the chloroplastic metabolism leads to ros accumulation in pea plants in response to plum pox virus. J. Exp. Bot. 2008;59:2147–2160. doi: 10.1093/jxb/ern082.

330. Hernandez J.A., Diaz-Vivancos P., Rubio M., Olmos E., Ros-Barcelo A., Martinez-Gomez P. Long-term plum pox virus infection produces an oxidative stress in a susceptible apricot, prunus armeniaca, cultivar but not in a resistant cultivar. Physiol. Plant. 2006;126:140–152. doi: 10.1111/j.1399-3054.2005.00581.x.

331. Diaz-Vivancos P., Rubio M., Mesonero V., Martinez-Gomez P., Hernandez J.A., Periago P.M. Effect of plum pox virus infection on the antioxidant capacity of leaf apoplast from apricot plants; Proceedings of the xiiith international symposium on apricot breeding and culture; LEUVEN 1: INTERNATIONAL SOCIETY HORTICULTURAL SCIENCE; 2006. pp. 119–122.

332. Diaz-Vivancos P., Rubio M., Mesonero V., Periago P.M., Barcelo A.R., Martinez-Gomez P., Hernandez J.A. The apoplastic antioxidant system in prunus: Response to long-term plum pox virus infection. J. Exp. Bot. 2006;57:3813–3824. doi: 10.1093/jxb/erl138.

333. Escalettes V.S.L., Hullot C., Wawrzy'nczak D., Mathieu E., Eyquard J.P., Le Gall O., Decroocq V. Plum pox virus induces differential gene expression in the partially resistant stone fruit tree prunus armeniaca cv. Goldrich. Gene. 2006;374:96–103. doi: 10.1016/j.gene.2006.01.021.

334. Babu M., Griffiths J.S., Huang T.S., Wang A. Altered gene expression changes in arabidopsis leaf tissues and protoplasts in response to plum pox virus infection. Bmc Genomics. 2008:9.

335. Wang A., Chapman P., Chen L., Stobbs L.W., Brown D.C.W., Brandle J.E. A comparative survey, by expressed sequence tag analysis, of genes expressed in peach leaves infected with plum pox virus (ppv) and free from ppv. Can. J. Plant Pathol.-Rev. Can. Phytopathol. 2005;27:410–419.

336. Milosevic T., Milosevic N. Plum pox virus as a stress factor in the one-year-old shoot and fruit growth and yield of plum cv. Stanley. Cereal Res. Commun. 2009;37:241–244.

337. Milosevic T.M., Glisic I.P., Milosevic N.T., Glisic I.S. Plum pox virus as a stress factor in the vegetative growth, fruit growth and yield of plum (prunus domestica) cv. 'Cacanska rodna'. Eur. J. Plant Pathol.2010;126:73–79. doi: 10.1007/s10658-009-9526-z.

338. Garcia-Ibarra A., Clemente-Moreno M.J., Barba-Espin G., Diaz-Vivancos P., Rubio M., Dicenta F., Martinez-Gomez P., Hernandez J.A. Changes in the antioxidative metabolism induced by apple chlorotic leaf spot virus infection in peach [prunus persica (l.) batsch] Environ. Exp. Bot. 2011;70:277–282. doi: 10.1016/j.envexpbot.2010.10.004.

339. Amari K., Diaz-Vivancos P., Pallas V., Amelia Sanchez-Pina M., Antonio Hernandez J. Oxidative stress induction by prunus necrotic ringspot virus infection in apricot seeds. Physiol. Plant. 2007;131:302–310.

340. Clarke S.F., Guy P.L., Burritt D.J., Jameson P.E. Changes in the activities of antioxidant enzymes in response to virus infection and hormone treatment. Physiol. Plant. 2002;114:157–164. doi: 10.1034/j.1399-3054.2002.1140201.x.

341. Farkas G.L., Kiraly Z., Solymosy F. Role of oxidative metabolism in the localization of plant viruses.Virology. 1960;12:408–421. doi: 10.1016/0042-6822(60)90163-X.

342. Fodor J., Gullner G., Adam A.L., Barna B., Komives T., Kiraly Z. Local and systemic responses of antioxidants to tobacco mosaic virus infection and to salicylic acid in tobacco - role in systemic acquired resistance. Plant Physiol. 1997;114:1443–1451.

343. Kiraly L., Barnaz B., Kiralyz Z. Plant resistance to

pathogen infection: Forms and mechanisms of innate and acquired resistance. J. Phytopathol. 2007;155:385–396. doi: 10.1111/j.1439-0434.2007.01264.x.

344. Kröll J. Experiments to verify the plum pox virus (ppv) by communicating it on chenopodium foetidum schrad. Zentralbl. Bakteriol. Parasitenkd. Infektionskr. Hyg. 1975;130:226–231.

345. Kröll J. Verification of the plum pox virus (ppv) using the tray test. Zentralbl. Bakteriol. Naturwiss. 1978;133:59–64.

346. Glasa M., Matisova J., Hricovsky I., Kudela O. Susceptibility of peach gf 305 seedlings and selected herbaceous plants to plum pox virus isolates from western slovakia. Acta Virol. 1997;41:341–344.

347. Polak J. European spindle tree and common privet a new natural hosts of plum pox virus; Proceedings of the 18th international symposium on virus & virus–like diseases of temperate fruit crops – top fruit diseases, vols 1 and 2, 2001; pp. 125–128.

348. Visedo G., Fernandezpiqueras J., Garcia J.A. Isozyme profiles associated with the hypersensitive response of chenopodium-foetidum to plum pox virus-infection. Physiol. Plant. 1990;78:218–224. doi: 10.1111/j.1399-3054.1990.tb02084.x.

349. Di Terlizzi B., Savino V. Sanitary status and sanitation of stone fruit trees in south east italy; Iobc/wprs and ishs international conference on integrated fruit production, proceedings of the meeting, 1996; pp. 326–327.

350. Paprstein F., Karesova R. Long term evaluation of plum cultivars artificially infected by ppv; Proceedings of the 18th international symposium on virus & virus–like diseases of temperate fruit crops – top fruit diseases, vols 1 and 2, 2001; pp. 121–123.

351. James D., Thompson D. Hosts and symptoms of plum pox virus: Ornamental and wild prunus species. EPPO Bull. 2006;36:222–224. doi: 10.1111/j.1365-2338.2006.00976.x.

352. Rubio M., Martinez-Gomez P., Pinochet J., Dicenta F. Evaluation of resistance to sharka (plum pox virus) of several prunus rootstocks. Plant Breed. 2005;124:67–70.

353. Martinez-Gomez P., Rubio M., Gambin M., Dicenta F. Behaviour of six prunus rootstocks in presence of sharka (plum pox potyvirus); Proceedings of the xiith ishs symposium on apricot culture and decline, vols 1 and 2; LEUVEN 1: INTERNATIONAL SOCIETY HORTICULTURAL SCIENCE; 2006. pp. 227–231.

354. Zhebentyayeva T.N., Reighard G.L., Lalli D., Gorina V.M., Krska B., Abbott A.G. Origin of resistance to plum pox virus in apricot: What new aflp and targeted ssr data analyses tell. Tree Genet. Genomes. 2008;4:403–417. doi: 10.1007/s11295-007-0119-8.

355. Damsteegt V.D. 17th international symposium on virus and virus-like diseases of temperate fruit crops: Fruit tree diseases, vols 1 and 2. 1998. The versatility of prunus tomentosa as a bioindicator of viruses; pp. 143–146.

356. Damsteegt V.D., Waterworth H.E., Mink G.I., Howell W.E., Levy L. Prunus tomentosa as a diagnostic host for detection of plum pox virus and other prunus viruses. Plant Dis. 1997;81:329–332. doi: 10.1094/PDIS.1997.81.4.329.

357. Dosba F. Genetics and breeding of tree fruits and nuts. INTERNATIONAL SOCIETY HORTICULTURAL SCIENCE; LEUVEN 1: 2003. Progress and prospects in stone fruit breeding; pp. 35–43.

358. Barnavetro I., Gyorgy B., Schuster V., Varro R. Use of peroxidase labeled antibodies for detection of plum pox virus. Acta Phyto. Acad. Sci. Hung. 1980;15:333–338.

359. Adams A.N., Guise C.M., Crossley S.J. Plum pox virus detection in dormant plum trees by pcr and elisa. Plant Pathol. 1999;48:240–244. doi: 10.1046/j.1365-3059.1999.00336.x.

360. Caglayan K., Sertkaya G., Ulubas C., Kolber M., Krizbai L. Characterization of plum pox potyvirus (ppv) by das-elisa and rt-pcr/rflp analysis in turkey; Proceedings of the xixth international symposium on virus and virus–like diseases of temperate fruit crops: Fruit tree diseases, 2004; pp. 183–188.

361. Dovas C.I., Mamolos A.P., Katis N.I. Fluctuations in concentration of two potyviruses in garlic during the growing period and sampling conditions for reliable detection by elisa. Ann. Appl. Biol. 2002;140:21–28. doi: 10.1111/j.1744-7348.2002.tb00153.x.

362. Thomidis T., Karajiannis I. Using elisa and pcr to test the potential for spread of plum pox virus by seeds of different stone fruit cultivars. N. Z. J. Crop Hortic. Sci. 2003;31:69–72. doi: 10.1080/01140671.2003.9514237.

363. Asensio M., Gorris M.T., Sanz A., Camarasa E., Perez E., Carbonell E.A., Cambra M. Characterization and detection of plum pox virus using monoclonal antibodies; Xvith international symposium on fruit tree virus diseases, 1995; pp. 354–356.

364. Lopezmoya J.J., Sanz A., Cambra M., Gorris M.T., Anaya C., Miguet J.G., Cortes E., Lopezabella D. Production and characterization of monoclonal-antibodies to plum pox virus and their use in differentiation of mediterranean isolates. Arch. Virol. 1994;135:293–304. doi: 10.1007/BF01310015.

365. Orban S., Theilmann J., Rochon D. Production of polyclonal antibodies for the detection of plum pox virus in leaf tissue by a modified das-elisa assay. Can. J. Plant Pathol.-Rev. Can. Phytopathol. 2007;29:86–86.

366. Richter J., Proll E., Rabenstein F., Stanarius A. Serological detection of members of the potyviridae with polyclonal antisera. J. Phytopathol. 1994;142:11–18.

367. Hilgert I., Cikanek D., Kristofova H., Karesova R., Navratil M. Monoclonal-antibodies suitable for plum pox virus determination. Hybridoma. 1993;12:215–220. doi: 10.1089/hyb.1993.12.215.

368. Myrta A., Potere O., Boscia D., Candresse T., Cambra M., Savino V. Production of a monoclonal antibody specific to the el amar strain of plum pox virus. Acta Virol. 1998;42:248–250.

369. Cambra M., Asensio M., Gorris M.T., Pérez E., Camarasa E., Garcia J.A., Moya J.J., Lopez-Abella D., Vela C., Sanz A. Detection of plum pox potyvirus using monoclonal antibodies to structural and non-structural proteins. EPPO Bull. 1994;24:569–577. doi: 10.1111/

j.1365-2338.1994.tb01070.x.

370. Martin M.T., Garcia J.A., Cervera M.T., Goldbach R.W., Vanlent J.W.M. Intracellular-localization of 3 nonstructural plum pox potyvirus proteins by immunogold labeling. Virus Res. 1992;25:201–211. doi: 10.1016/0168-1702(92)90134-U.

371. Purcifull D.E., Hiebert E. Serological relationships involving potyviral nonstructural proteins. Arch. Virol. 1992:97–122. doi: 10.1007/978-3-7091-6920-9_11.

372. Ravelonandro M., Peyruchaud O., Garrigue L., Demarcillac G., Dunez J. Immunodetection of the plum pox virus helper component in infected plants and expression of its gene in transgenic plants. Arch. Virol.1993;130:251–268. doi: 10.1007/BF01309658.

373. Gil M., Esteban O., Garcia J.A., Pena L., Cambra M. Resistance to plum pox virus in plants expressing cytosolic and nuclear single-chain antibodies against the viral rna nib replicase. Plant Pathol. 2011;60:967–976. doi: 10.1111/j.1365-3059.2011.02448.x.

374. Olmos A., Cambra M., Dasi M.A., Candresse T., Esteban O., Gorris M.T., Asensio M. Simultaneous detection and typing of plum pox potyvirus (ppv) isolates by heminested-pcr and pcr-elisa. J. Virol. Methods.1997;68:127–137. doi: 10.1016/S0166-0934(97)00120-1.

375. Kamenova I., Peters D. The differential reactivity of two polyclonal antisera to plum pox virus isolates collected at different stages after immunization. Biotechnol. Biotechnol. Equip. 1999;13:33–39.

376. Subr Z., Matisova J. Preparation of diagnostic monoclonal antibodies against two potyviruses. Acta Virol. 1999;43:255–257.

377. Himmler G., Brix U., Steinkellner H., Laimer M., Mattanovich D., Katinger H.W.D. Early screening for anti-plum pox virus monoclonal-antibodies with different epitope specificities by means of gold-labeled immunosorbent electron-microscopy. J. Virol. Methods. 1988;22:351–358. doi: 10.1016/0166-0934(88)90118-8.

378. Jarocka U., Wasowicz M., Radecka H., Malinowski T., Michalczuk L., Radecki J. Impedimetric immunosensor for detection of plum pox virus in plant extracts. Electroanalysis. 2011;23:2197–2204. doi: 10.1002/elan.201100152.

379. Pollini C.P., Giunchedi L., Bissani R. Specific detection of d- and m-isolates of plum pox virus by immunoenzymatic determination of pcr products. J. Virol. Methods. 1997;67:127–133. doi: 10.1016/S0166-0934(97)00085-2.

380. Youssef S.A., Shalaby A.A., Mazyad H.M., Hadidi A. Detection and identification of prune dwarf virus and plum pox virus by standard and multiplex rt-pcr probe capture hybridization (rt-pcr-elisa) J. Plant Pathol. 2002;84:113–119.

381. Wetzel T., Candresse T., Ravelonandro M., Delbos R.P., Mazyad H., Aboulata A.E., Dunez J. Nucleotide-sequence of the 3'-terminal region of the rna of the elamar strain of plum pox potyvirus. Journal of General Virology. 1991;72:1741–1746. doi: 10.1099/0022-1317-72-7-1741.

382. Wetzel T., Candresse T., Ravelonandro M., Dunez J. A polymerase chain-reaction assay adapted to plum pox potyvirus detection. J. Virol. Methods. 1991;33:355–365. doi: 10.1016/0166-0934(91)90035-X.

383. Subr Z., Glasa M. Plum pox virus variability detected by the advanced analytical methods. Acta Virol.2008;52:75–89.

384. Wetzel T., Tavert G., Teycheney P.Y., Ravelonandro M., Candresse T., Dunez J. Dot hybridization detection of plum pox virus using p-32 labeled rna probes representing nonstructural viral protein genes. J. Virol. Methods. 1990;30:161–171. doi: 10.1016/0166-0934(90)90017-A.

385. Varveri C., Candresse T., Cugusi M., Ravelonandro M., Dunez J. Use of a p-32-labeled transcribed rna probe for dot hybridization detection of plum pox virus. Phytopathology. 1988;78:1280–1283. doi: 10.1094/Phyto-78-1280.

386. Hammond J., Puhringer H., Machado A.D., Machado M.L.D. A broad-spectrum pcr assay combined with rflp analysis for detection and differentiation of plum pox virus isolates; 17th international symposium on virus and virus-like diseases of temperate fruit crops: Fruit tree diseases, vols 1 and 2; LEUVEN 1: INTERNATIONAL SOCIETY HORTICULTURAL SCIENCE; 1998. pp. 483–490.

387. Gadiou S., Safarova D., Navratil M. Differentiation of plum pox virus isolates by single-strand conformation polymorphism and low-stringency single specific primer pcr analysis of hc-pro genome region.Acta Virol. 2009;53:53–56.

388. Subr Z., Pittnerova S., Glasa M. A simplified rt-pcr-based detection of recombinant plum pox virus isolates. Acta Virol. 2004;48:173–176.

389. Spiegel S., Kovalenko E.M., Varga A., James D. Detection and partial molecular characterization of two plum pox virus isolates from plum and wild apricot in southeast kazakhstan. Plant Dis. 2004;88:973–979. doi: 10.1094/PDIS.2004.88.9.973.

390. Szemes M., Kalman M., Myrta A., Boscia D., Nemeth M., Kolber M., Dorgai L. Integrated rt-pcr/nested pcr diagnosis for differentiating between subgroups of plum pox virus. J. Virol. Methods.2001;92:165–175. doi: 10.1016/S0166-0934(00)00284-6.

391. Pasquini G., Barba M., Hadidi A., Faggioli F., Negri R., Sobol I., Tiberini A., Caglayan K., Mazyad H., Anfoka G., et al. Oligonucleotide microarray-based detection and genotyping of plum pox virus. J. Virol. Methods. 2008;147:118–126. doi: 10.1016/j.jviromet.2007.08.019.

392. Olmos A., Bertolini E., Gil M., Cambra M. Real-time assay for quantitative detection of non-persistently transmitted plum pox virus rna targets in single aphids. J. Virol. Methods. 2005;128:151–155. doi: 10.1016/j.jviromet.2005.05.011.

393. Capote N., Gorris M.T., Martinez M.C., Asensio M., Olmos A., Cambra M. Interference between d and m types of plum pox virus in japanese plum assessed by specific monoclonal antibodies and quantitative real-time reverse transcription-polymerase chain reaction.

Phytopathology. 2006;96:320–325. doi: 10.1094/PHYTO-96-0320.

394. Kim W.S., Stobbs L.W., Lehman S.M., James D., Svircev A.M. Direct real-time pcr detection of plum pox virus in field surveys in ontario. Can. J. Plant Pathol.-Rev. Can. Phytopathol. 2008;30:308–317.

395. Capote N., Bertolini E., Olmos A., Vidal E., Martinez M.C., Cambra M. Direct sample preparation methods for the detection of plum pox virus by real-time rt-pcr. Int. Microbiol. 2009;12:1–6.

396. Levy L., Kolber M., Tokes G., Nemeth M., Hadidi A. 3' non-coding region rt-pcr detection and molecular hybridization of plum pox virus in anthers of infected stone fruit; Xvith international symposium on fruit tree virus diseases; LOUVAIN: INTERNATIONAL SOCIETY HORTICULTURAL SCIENCE; 1995. pp. 331–339.

397. Schneider W.L., Sherman D.J., Stone A.L., Damsteegt V.D., Frederick R.D. Specific detection and quantification of plum pox virus by real-time fluorescent reverse transcription-pcr. J. Virol. Methods.2004;120:97–105. doi: 10.1016/j.jviromet.2004.04.010.

398. Varga A., James D. Real-time rt-pcr and sybr green i melting curve analysis for the identification of plum pox virus strains c, ea, and w: Effect of amplicon size, melt rate, and dye translocation. J. Virol. Methods. 2006;132:146–153. doi: 10.1016/j.jviromet.2005.10.004.

399. Varga A., James D. Detection and differentiation of plum pox virus using real-time multiplex pcr with sybr green and melting curve analysis: A rapid method for strain typing. J. Virol. Methods. 2005;123:213–220. doi: 10.1016/j.jviromet.2004.10.005.

400. Korschineck I., Himmler G., Sagl R., Steinkellner H., Katinger H.W.D. A pcr membrane spot assay for the detection of plum pox virus-rna in bark of infected trees. J. Virol. Methods. 1991;31:139–146. doi: 10.1016/0166-0934(91)90152-P.

401. Olmos A., Bertolini E., Cambra M. Simultaneous and co-operational amplification (co-pcr): A new concept for detection of plant viruses. J. Virol. Methods. 2002;106:51–59. doi: 10.1016/S0166-0934(02)00132-5.

402. Bertolini E., Torres E., Olmos A., Martin M.P., Bertaccini A., Cambra M. Co-operational pcr coupled with dot blot hybridization for detection and 16srx grouping of phytoplasmas. Plant Pathol. 2007;56:677–682. doi: 10.1111/j.1365-3059.2007.01601.x.

403. Martos S., Torres E., El Bakali M.A., Raposo R., Gramaje D., Armengol J., Luque J. Co-operational pcr coupled with dot blot hybridization for the detection of phaeomoniella chlamydospora on infected grapevine wood. J. Phytopathol. 2011;159:247–254. doi: 10.1111/j.1439-0434.2010.01758.x.

404. Wetzel T., Candresse T., Macquaire G., Ravelonandro M., Dunez J. A highly sensitive immunocapture polymerase chain-reaction method for plum pox potyvirus detection. J. Virol. Methods. 1992;39:27–37. doi: 10.1016/0166-0934(92)90122-T.

405. Sreenivasulu M., Gopal D. Development of recombinant coat protein antibody based ic-rt-pcr and comparison of its sensitivity with other immunoassays for the detection of papaya ringspot virus isolates from india, (vol 26, pg 25, 2010) Plant Pathol. J. 2010;26:207. doi: 10.5423/PPJ.2010.26.2.207.

406. Sreenivasulu M., Gopal D. Development of recombinant coat protein antibody based ic-rt-pcr and comparison of its sensitivity with other immunoassays for the detection of papaya ringspot virus isolates from india. Plant Pathol. J. 2010;26:25–31. doi: 10.5423/PPJ.2010.26.1.025.

407. Huang C.H., Hu W.C., Yang T.C., Chang Y.C. Zantedeschia mild mosaic virus, a new widespread virus in calla lily, detected by elisa, dot-blot hybridization and ic-rt-pcr. Plant Pathol. 2007;56:183–189.

408. Hema M., Kirthi N., Sreeivasulu P., Savithri H.S. Development of recombinant coat protein antibody based ic-rt-pcr for detection and discrimination of sugarcane streak mosaic virus isolates from southern india.Arch. Virol. 2003;148:1185–1193. doi: 10.1007/s00705-003-0015-y.

409. Olmos A., Bertolini E., Cambra M. Isothermal amplification coupled with rapid flow-through hybridisation for sensitive diagnosis of plum pox virus. J. Virol. Methods. 2007;139:111–115. doi: 10.1016/j.jviromet.2006.09.012.

410. Olmos A., Bertolini E., Cambra M. Isothermal amplification for detection of plum pox virus; Proceedings of the twentieth international symposium on virus and virus–like diseases of temperate fruit crops – fruit tree diseases; LEUVEN 1: INTERNATIONAL SOCIETY HORTICULTURAL SCIENCE; 2008. pp. 209–213.

411. Hadersdorfer J., Neumuller M., Treutter D., Fischer T.C. Fast and reliable detection of plum pox virus in woody host plants using the blue lamp protocol. Ann. Appl. Biol. 2011;159:456–466. doi: 10.1111/j.1744-7348.2011.00510.x.

412. Capote N., Bertolini E., Martinez M.C., Olmos A., Gorris M.T., Cambra M. Spot real-time rt-pcr: A method for direct detection of plum pox virus avoiding rna extraction; Proceedings of the twentieth international symposium on virus and virus–like diseases of temperate fruit crops – fruit tree diseases; LEUVEN 1: INTERNATIONAL SOCIETY HORTICULTURAL SCIENCE; 2008. pp. 215–219.

413. Sanchez F., Sobolev K. Nanotechnology in concrete - a review. Constr. Build. Mater. 2010;24:2060–2071. doi: 10.1016/j.conbuildmat.2010.03.014.

414. Shea C.M., Grinde R., Elmslie B. Nanotechnology as general-purpose technology: Empirical evidence and implications. Tech. Anal. Strat. Manag. 2011;23:175–192. doi: 10.1080/09537325.2011.543336.

415. Banerjee R., Katsenovich Y., Lagos L., McIintosh M., Zhang X., Li C.Z. Nanomedicine: Magnetic nanoparticles and their biomedical applications. Curr. Med. Chem. 2010;17:3120–3141. doi: 10.2174/092986710791959765.

416. Cai S., Yang Q.H., Bagby T.R., Forrest M.L. Lymphatic drug delivery using engineered liposomes and solid lipid nanoparticles. Adv. Drug Deliv. Rev. 2011;63:901–908.

doi: 10.1016/j.addr.2011.05.017.

417. Feng B., Matsui H., Tomizawa K. Nanoparticle-based drug delivery systems for solid brain tumors.Curr. Nanosci. 2011;7:47–54. doi: 10.2174/157341311794480327.

418. Ghaderi S., Ramesh B., Seifalian A.M. Fluorescence nanoparticles "quantum dots" as drug delivery system and their toxicity: A review. J. Drug Target. 2011;19:475–486. doi: 10.3109/1061186X.2010.526227.

419. Chen G.J., Wang L.F. Design of magnetic nanoparticles-assisted drug delivery system. Curr. Pharm. Design. 2011;17:2331–2351. doi: 10.2174/138161211797052574.

420. Malam Y., Lim E.J., Seifalian A.M. Current trends in the application of nanoparticles in drug delivery. Curr. Med. Chem. 2011;18:1067–1078. doi: 10.2174/092986711794940860.

421. Parveen S., Misra R., Sahoo S.K. Nanoparticles: A boon to drug delivery, therapeutics, diagnostics and imaging. Nanomed.-Nanotechnol. Biol. Med. 2012;8:147–166. doi: 10.1016/j.nano.2011.05.016.

422. Paulo C.S.O., das Neves R.P., Ferreira L.S. Nanoparticles for intracellular-targeted drug delivery.Nanotechnology. 2011:22.

423. Prow T.W., Grice J.E., Lin L.L., Faye R., Butler M., Becker W., Wurm E.M.T., Yoong C., Robertson T.A., Soyer H.P., et al. Nanoparticles and microparticles for skin drug delivery. Adv. Drug Deliv. Rev.2011;63:470–491. doi: 10.1016/j.addr.2011.01.012.

424. Su X., Zhan X., Tang F., Yao J.Y., Wu J. Magnetic nanoparticles in brain disease diagnosis and targeting drug delivery. Curr. Nanosci. 2011;7:37–46. doi: 10.2174/157341311794480363.

425. Talekar M., Kendall J., Denny W., Garg S. Targeting of nanoparticles in cancer: Drug delivery and diagnostics. Anti-Cancer Drugs. 2011;22:949–962. doi: 10.1097/CAD.0b013e32834a4554.

426. Yoo J.W., Doshi N., Mitragotri S. Adaptive micro and nanoparticles: Temporal control over carrier properties to facilitate drug delivery. Adv. Drug Deliv. Rev. 2011;63:1247–1256. doi: 10.1016/j.addr.2011.05.004.

427. Eggenberger K., Frey N., Zienicke B., Siebenbrock J., Schunck T., Fischer R., Brase S., Birtalan E., Nann T., Nick P. Use of nanoparticles to study and manipulate plant cells. Adv. Eng. Mater. 2010;12:B406–B412. doi: 10.1002/adem.201080009.

428. Baruah S., Dutta J. Nanotechnology applications in pollution sensing and degradation in agriculture: A review. Environ. Chem. Lett. 2009;7:191–204. doi: 10.1007/s10311-009-0228-8.

429. KnauerK.,BucheliT.Nano-materials-theneedforresearch in agriculture. Agrarforschung.2009;16:390–395.

430. Knauer K., Bucheli T.D. Nano-materials: Research needs in agriculture. Rev. Suisse Agric.2009;41:341–345.

431. Shrivastava S., Dash D. Agrifood nanotechnology: A tiny revolution in food and agriculture. J. Nano Res. 2009;6:1–14. doi: 10.4028/www.scientific.net/JNanoR.6.1.

432. Lokko Y., Mba C., Spencer M., Till B., Lagoda P. Nanotechnology and synthetic biology - potential in crop improvement. J. Food Agric. Environ. 2011;9:599–604.

433. Young M., Willits D., Uchida M., Douglas T. Plant viruses as biotemplates for materials and their use in nanotechnology. In Annu. Rev. Phytopathol. 2008:361–384.

434. Aniagyei S.E., DuFort C., Kao C.C., Dragnea B. Self-assembly approaches to nanomaterial encapsulation in viral protein cages. J. Mater. Chem. 2008;18:3763–3774. doi: 10.1039/b805874c.

435. Srivastava S.K. Use of plant viruses and their genes in the areas of biotechnology and nanotechnology.Indian J. Virol. 2008;19:101–102.

436. Behrens S.S. Synthesis of inorganic nanomaterials mediated by protein assemblies. J. Mater. Chem.2008;18:3788–3798. doi: 10.1039/b806551a.

437. Lee S.Y., Lim J.S., Harris M.T. Synthesis and application of virus-based hybrid nanomaterials.Biotechnol. Bioeng. 2012;109:16–30. doi: 10.1002/bit.23328.

438. Royston E., Ghosh A., Kofinas P., Harris M.T., Culver J.N. Self-assembly of virus-structured high surface area nanomaterials and their application as battery electrodes. Langmuir. 2008;24:906–912. doi: 10.1021/la7016424.

439. Soto C.M., Ratna B.R. Virus hybrids as nanomaterials for biotechnology. Curr. Opin. Biotechnol.2010;21:426–438. doi: 10.1016/j.copbio.2010.07.004.

440. Zhou Y. Recent progress in biomolecule-templated nanomaterials. Curr. Nanosci. 2006;2:123–134. doi: 10.2174/157341306776875785.

441. Flenniken M.L., Uchida M., Liepold L.O., Kang S., Young M.J., Douglas T. Viruses and nanotechnology. 2009. A library of protein cage architectures as nanomaterials; pp. 71–93.

442. Franzen S., Lockney D.M., Wang R.Q., Lommel S., Hauck M. In vivo imaging and tumor targeting using a plant virus nanoparticle. Abstr. Pap. Am. Chem. Soc. 2011:241.

443. Lockney D.M., Guenther R.N., Loo L., Overton W., Antonelli R., Clark J., Hu M., Luft C., Lommel S.A., Franzen S. The red clover necrotic mosaic virus capsid as a multifunctional cell targeting plant viral nanoparticle. Bioconjugate Chem. 2011;22:67–73. doi: 10.1021/bc100361z.

444. Liang M.M., Guo L.H. Application of nanomaterials in environmental analysis and monitoring. J. Nanosci. Nanotechnol. 2009;9:2283–2289. doi: 10.1166/jnn.2009.SE22.

445. Martin C.R. Nanomaterials: Evolving applications in membranes, energy production, and analytical chemistry. Abstr. Pap. Am. Chem. Soc. 2000;219:U727–U727.

446. Martin C.R., Mitchell D.T. Nanomaterials in analytical chemistry. Anal. Chem. 1998;70:322–327. doi: 10.1021/ac9818430.

447. Scida K., Stege P.W., Haby G., Messina G.A., Garcia C.D. Recent applications of carbon-based nanomaterials in analytical chemistry: Critical review. Anal. Chim. Acta. 2011;691:6–17. doi: 10.1016/j.aca.2011.02.025.

448. Valentini F., Palleschi G. Nanomaterials and analytical

chemistry. Anal. Lett. 2008;41:479–520. doi: 10.1080/00032710801912805.

449. Bandyopadhyay A., Chatterjee S., Sarkar K. Rapid isolation of genomic DNA from e. Coli xl1 blue strain approaching bare magnetic nanoparticles. Curr. Sci. 2011;101:210–214.

450. Chiang C.L., Sung C.S., Chen C.Y. Application of silica-magnetite nanocomposites to the isolation of ultrapure plasmid DNA from bacterial cells. J. Magn. Magn. Mater. 2006;305:483–490. doi: 10.1016/j.jmmm.2006.02.088.

451. Sarkar T.R., Irudayaraj J. Carboxyl-coated magnetic nanoparticles for mrna isolation and extraction of supercoiled plasmid DNA. Anal. Biochem. 2008;379:130–132.

452. Zhang H.P., Bai S., Xu L., Sun Y. Fabrication of mono-sized magnetic anion exchange beads for plasmid DNA purification. J. Chromatogr. B. 2009;877:127–133. doi: 10.1016/j.jchromb.2008.11.026.

453. Masarik M., Gumulec J., Sztalmachova M., Hlavna M., Babula P., Krizkova S., Ryvolova M., Jurajda M., Sochor J., Adam V., et al. Isolation of metallothionein from cells derived from aggressive form of high-grade prostate carcinoma using paramagnetic antibody-modified microbeads off-line coupled with electrochemical and electrophoretic analysis. Electrophoresis. 2011;32:3576–3588.

454. Adam V., Huska D., Hubalek J., Kizek R. Easy to use and rapid isolation and detection of a viral nucleic acid by using paramagnetic microparticles and carbon nanotubes-based screen-printed electrodes.Microfluid. Nanofluid. 2010;8:329–339. doi: 10.1007/s10404-009-0464-z.

455. Huska D., Hubalek J., Adam V., Vajtr D., Horna A., Trnkova L., Havel L., Kizek R. Automated nucleic acids isolation using paramagnetic microparticles coupled with electrochemical detection. Talanta.2009;79:402–411. doi: 10.1016/j.talanta.2009.04.007.

456. Masarik M., Huska D., Adam V., Hubalek J., Provaznik I., Trnkova L., Kizek R. Automated pipetting system coupled with micro- and nanopartieles as a new tool for study of nucleic acids. Int. J. Mol. Med.2010;26:S46–S46.

457. Wang J., Hui F., Fang X.W., Vinciguerra J., Hunicke-Smith S., WalkerPeach C.R. Magnetic bead-based viral nucleic acid isolation incorporating a quantitative internal control. J. Mol. Diagn. 2004;6:427–427.

458. Zhang Z.C., Cui Y., Wan Q.H. Surface modification of magnetic silica microspheres and its application to the isolation of plant genomic nucleic acids. Chin. J. Anal. Chem. 2007;35:31–36. doi: 10.1016/S1872-2040(07)60024-3.

459. Miszczak F., Shuck K.M., Lu Z.C., Go Y.Y., Zhang J.Q., Sells S., Vabret A., Pronost S., Fortier G., Timoney P.J., et al. Evaluation of two magnetic-bead-based viral nucleic acid purification kits and three real-time reverse transcription-pcr reagent systems in two taqman assays for equine arteritis virus detection in semen. J. Clin. Microbiol. 2011;49:3694–3696. doi: 10.1128/JCM.01187-11.

460. Sun H.X., Zeng X., Liu M., Elingarami S., Li G.P., Shen B., He N.Y. Synthesis of size-controlled fe3o4@sio2 magnetic nanoparticles for nucleic acid analysis. J. Nanosci. Nanotechnol. 2012;12:267–273. doi: 10.1166/jnn.2012.5170.

461. Yang S.Y., Chieh J.J., Wang W.C., Yu C.Y., Hing N.S., Horng H.E., Hong C.Y., Yang H.C., Chang C.F., Lin H.Y. Magnetic nanoparticles for high-sensitivity detection on nucleic acids via superconducting-quantum-interference-device-based immunomagnetic reduction assay. J. Magn. Magn. Mater. 2011;323:681–685. doi: 10.1016/j.jmmm.2010.10.011.

462. Byzova N.A., Safenkova I.V., Chirkov S.N., Avdienko V.G., Guseva A.N., Mitrofanova I.V., Zherdev A.V., Dzantiev B.B., Atabekov J.G. Interaction of plum pox virus with specific colloidal gold-labeled antibodies and development of immunochromatographic assay of the virus. Biochem.-Moscow.2010;75:1393–1403. doi: 10.1134/S000629791011012X.

463. Safenkova I.V., Zherdev A.V., Dzantiev B.B. Correlation between the composition of multivalent antibody conjugates with colloidal gold nanoparticles and their affinity. J. Immunol. Methods. 2010;357:17–25. doi: 10.1016/j.jim.2010.03.010.

CHAPTER 3

A CURRENT OVERVIEW OF THE PAPAYA MELEIRA VIRUS, AN UNUSUAL PLANT VIRUS

Paolla M. V. Abreu[1], Tathiana F. S. Antunes[1], Anuar Magaña-Álvarez[1, 2], Daisy Pérez-Brito[2], Raúl Tapia-Tussell[2], José A. Ventura[1, 3], Antonio A. R. Fernandes[1], and Patricia M. B. Fernandes[1]

[1]Núcleo de Biotecnologia, Universidade Federal do Espírito Santo, Vitória 29040090, Espírito Santo, Brazil

[2]Laboratorio GeMBio, Centro de Investigación Científica de Yucatán A.C., Mérida 97200, Yucatán, Mexico

[3]Instituto Capixaba de Pesquisa, Assistência Técnica e Extensão Rural, Vitória 29050790, Espírito Santo, Brazil

ABSTRACT

Papaya meleira virus **(PMeV) is the causal agent of papaya sticky disease, which is characterized by a spontaneous exudation of fluid and aqueous latex from the papaya fruit and leaves. The latex oxidizes after atmospheric exposure, resulting in a sticky feature on the fruit from which the name of the disease originates. PMeV is an isometric virus particle with a double-stranded RNA (dsRNA) genome of approximately 12 Kb. Unusual for a plant virus, PMeV particles are localized on and linked to the polymers present in the latex. The ability of the PMeV to inhabit such a hostile environment demonstrates an intriguing interaction of the virus with the papaya. A hypersensitivity response is triggered against PMeV infection, and there is a reduction in the proteolytic activity of papaya latex during sticky disease. In papaya leaf tissues, stress responsive proteins, mostly calreticulin and proteasome-related proteins, are up regulated and proteins related to metabolism are down-regulated. Additionally, PMeV modifies the transcription of several miRNAs involved in the modulation of genes related to the ubiquitin-proteasome system. Until now, no PMeV resistant papaya genotype has been identified and roguing is the only viral control strategy available. However, a single inoculation of papaya plants with PMeV dsRNA delayed the progress of viral infection.**

INTRODUCTION

Plant viruses cause many diseases of international importance and are responsible for huge losses of crop production and quality in all parts of the world. The production is impaired due to the susceptibility of the plants to the viruses.

Citation: Paolla M. V. Abreu, Tathiana F. S. Antunes, Anuar Magaña-Álvarez, Daisy Pérez-Brito, Raúl Tapia-Tussell, José A. Ventural, Antonio A. R. Fernandes, and Patricia M. B. Fernandes, A Current Overview of the Papaya meleira virus, an Unusual Plant Virus, doi: 10.3390/v7041853.

The main producers of papaya are India, Brazil, Indonesia, Dominican Republic, Nigeria and Mexico, and combined they produce approximately 10 million tons per year [1] (Figure 1).

Brazil is one of the main papaya producers; in 2012, the country produced 1.52 million tons, which was approximately 12.2% of the world supply. The production area was 31.3 thousand hectares with an average yield of 48.5 t/ha, and the production value estimated at $ 786 million [1]. Mexico, on the other hand, is the main exporter to the USA. In 2013, it cultivated approximately 16,368 hectares, with a production of 764,514 tons [2]. The economic importance of this crop for both countries, besides the high levels of fruit exportation, is that this crop is the source of income for hundreds of smallholder farmers.

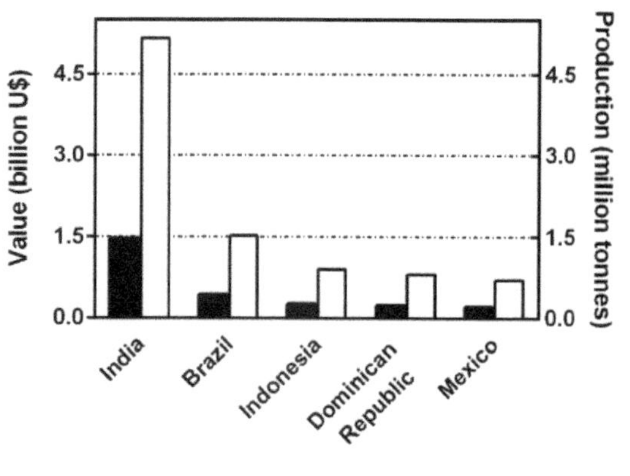

Figure 1: The graph represents the five highest papaya producing countries in 2012. Solid bars correspond to revenue of billions of US dollars and open bars, production in million tonnes.

Papaya viruses cause diseases of international importance with serious reductions in fruit production that may even totally destroy affected orchards. Until now, more than ten different viruses have been reported in papaya worldwide. Nevertheless, the most important viruses that affect papaya are the *Papaya ringspot virus*(PRSV), the *Papaya leaf distortion mosaic virus* (PLDMV), the *Papaya lethal yellowing virus* (PLYV), the *Papaya mosaic virus* (PapMV) and the *Papaya meleira virus* (PMeV), which have been known to cause serious damage to the crop production throughout the world.

PRSV causes the most destructive viral disease of papaya crop, the papaya ringspot, and has been found in many tropical and subtropical areas where papaya is grown, including the USA, South America, Mexico and Japan [3]. PLDMV, was first reported in 1954 on the island of Okinawa, Japan. PLDMV emergence in PRSV-resistant papaya transgenic lines was considered as an emerging threat to papaya culture in China [4]. PYLV is restricted to Brazil and its increasing spread reaching high incidence rates [5]. PapMV was first reported in 1962 in Florida, USA. The disease has spread to other countries, reaching Bolivia, Peru, Venezuela and Mexico. In Mexico, PRSV and PapMV occurred in single or mixed infections and a synergistic interaction between the two independent viruses in the same host can occur and lead to increased symptoms and virus accumulation [6].

In Brazil, the percentage of plants eradicated with papaya ringspot is about 2% per year in well-managed orchards. In traditional orchards, the losses may reach 80% [7]. In Mexico, PRSV causes severe damage in the main papaya production states with crop losses of up to 85% [8]. The PMeV infects at least 20% of the plants during the economic cycle of crop. In some orchards, where rouging of sticky diseased plants was not carried out, an incidence of the disease of up to 100% and cause total yield losses [9].

Little information exists about PYLV and its economic importance for fruit production, but in some orchards the disease may have an incidence of up to 40% of the plants. The disease caused by PLDMV has low economic importance to the papaya production in Brazil [7], and in Mexico this disease has not be found.

Little has been published in the patent databases concerning methods for the detection of the *Papaya meleira virus*, as well as on strategies for eradicating the papaya sticky disease. A search in the Derwent Innovations Index yielded approximately 160 patents related to papaya viruses, but most of these patents involved the*Papaya ringspot virus* or the *Papaya mosaic virus*. In our view, research groups all over the world face a significant opportunity to reduce production losses in papaya cultivation by doing basic research on papaya infestation by PMeV.

PAPAYA MELEIRA VIRUS

PMeV, the causal agent of Papaya sticky disease or "meleira", was reported in the 1980s in Brazil [10], and in 2008, it also appeared in Mexico [11]. Although PMeV is considered one of the major viruses infecting papaya in Brazil and in Mexico, knowledge of the sequence and genomic organization of this virus is poor. PMeV has not been sequenced nor classified by the International Committee on the Taxonomy of Viruses (ICTV). A comparative analysis of a ~560 bp fragment amplified from the PMeV replicase gene from the major Brazilian papaya-producing states isolates indicated that PMeV possesses a similarity with the mycoviruses of the family Totiviridae [12,13]. However, a conclusive taxonomic classification will only be possible when the full genome is sequenced. PMeV is an isometric virus particle that has a double-stranded RNA (dsRNA) genome of 12 Kb [14,15]. Papaya fruits with sticky disease symptoms are commercially unacceptable as the disease compromises their texture and flavor, rendering them unfit for consumption, which prevents their exportation to the international market [9].

ETIOLOGY

Papaya sticky diseased affected plants are characterized by spontaneous exudation of fluid and aqueous latex from the fruit and leaves. The latex oxidizes after atmospheric exposure, resulting in small necrotic lesions on the edges of young leaves (Figure 2A) and a sticky appearance of the fruit from which the name of the disease originates (Figure 2B) [9]. In advanced stages of the disease, irregular light-green areas have been observed on the surface of the infected fruits in Brazil; however, this symptom has not been commonly observed in Mexico. In addition to these symptoms, diseased plants in Mexico show small internal blotches in the pulp of the diseased fruit, necrotic spots in the petiole (Figure 2C) and, in the most severe cases, the presence of latex inside the fruit cavity covering the seeds was also observed (Figure 2D) [11].

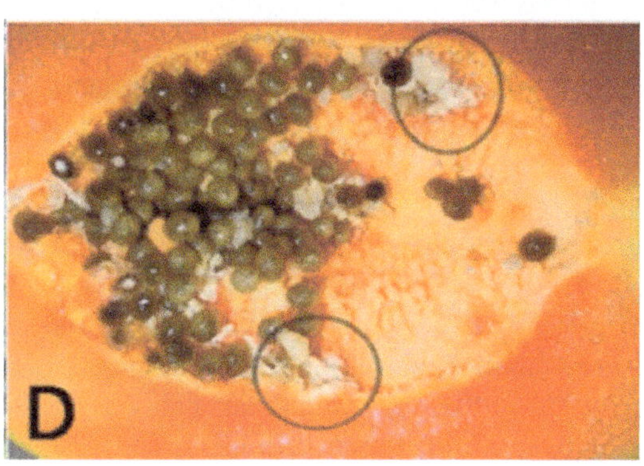

Figure 2: Symptoms of PMeV infection in *Carica papaya*. Papaya plant showing the symptoms of sticky disease (A). Plants spontaneously leak the latex, which oxidizes after atmospheric exposure resulting in small necrotic lesions on the edges of young leaves (B) and ...

Initially, symptoms of the disease were attributed to a disturbance in the calcium and boron absorption that resulted from water stress or an imbalance of these elements in the soil [16]. In addition to the abiotic factors, the involvement of microorganisms was suggested after the isolation of a bacteria of the genus Bartonella from the diseased plants [17].

Monitoring the sticky disease dispersal in commercial orchards indicated a pathogen involvement [10]. The biotic etiology of this disease was verified in preliminary studies where healthy papaya plants that were injected with a latex obtained from the diseased plants developed the anticipated typical symptoms as early as 45 days after inoculation, suggesting that the causative agent was present in the plant latex [10]. Finally, the viral etiology of the disease was confirmed after the purification of viral particles that were present in the latex and that was followed by the inoculation of these particles in the healthy papaya seedlings, which developed typical sticky disease symptoms [15]. Transmission electron microscope studies of the diseased plant leaves and fruit latex samples indicated the presence of a large number of isometric particles of approximately 50 nm in diameter. Ultrafine tissue sections revealed that these particles were restricted to the lactiferous vesicles [14].

The viral nucleic acid with approximately 12,000 base pairs was extracted from purified viral particles and visualized in an agarose gel [14,15]. The genetic material composition was proven to be RNA by its resistance to DNase digestion and susceptibility to RNase A digestion. In order to confirm whether the viral genome was comprised of single-stranded RNA (ssRNA) or double-stranded RNA (dsRNA), it was digested with S1 nuclease that degrades ssRNA. As a result, the viral nucleic acid was resistant to S1 nuclease, and it was concluded that *Papaya meleira virus* is a dsRNA genome virus [15].

Viruses with a dsRNA genome are unusual among plant viruses, which have, in most cases, a genome consisting of single stranded RNA (ssRNA). dsRNA viruses represent a small group among plant viruses, and they are grouped in the Endornaviridae, Partitiviridae and Reoviridae families [18].

Moreover, unusually for a plant virus, PMeV appears to reside primarily in the laticifers, which is an inhospitable environment, where it the modifies potassium levels and the osmotic balance that leads to cellular rupture [19]. The results obtained from the electron microscopy and molecular analyses indicated that the viral particles were localized on [14] and linked to the polymers present in the latex, perhaps acting as a protective mechanism or in assisting the viral transport (Figure 3) [19].

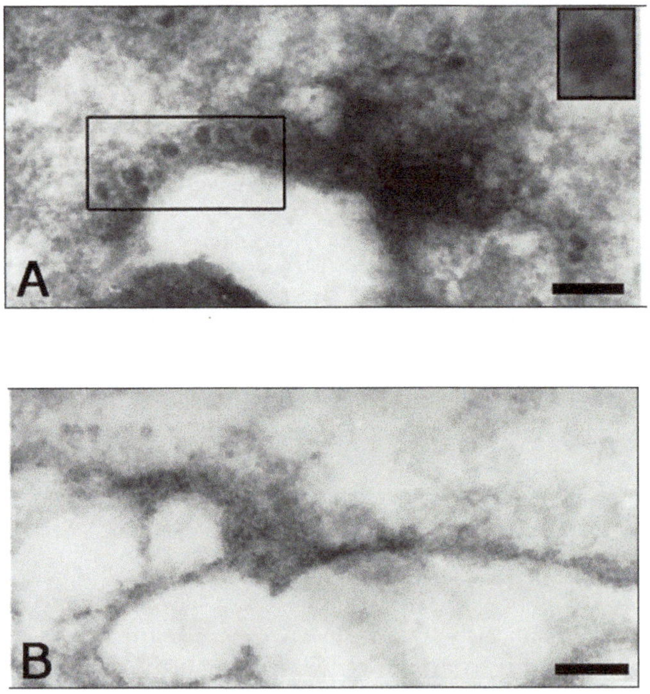

Figure 3: Papaya latex analysis by transmission electron microscopy (TEM). (**A**) Diseased and (**B**) healthy latex samples were studied by TEM, which revealed the PMeV particles (highlight) on the polymers. Inset show the amplified virus particle. The scale bar is 200 ...

EPIDEMIOLOGY

Sticky disease spreads rapidly, and currently it occurs mostly in Northeast Brazil [12]. Outside Brazil, Mexico is the only other country where the disease was reported [11]; at this time, the virus continues its dissemination and it has been detected in all producer states and its incidence in papaya orchards has been increasing.

Agricultural practices are responsible for the spread of the disease within the orchard because the dispersion of the sticky disease often occurs along the array of the crop row [9]. Therefore, laboratory experiments were conducted in order to test five different mechanical inoculation methods that were designed to simulate injuries caused by work tools and the movement of vehicles in the orchard. Among the tested methods, cutting the leaf, cutting the leaf stalk, scraping the leaf surface, scraping the stems and the injection into the stem apex only the injection of infected latex into the stem apex resulted in plant infection [20]. In Mexico, the PMeV spatial distribution was studied in experimental plots over a year, and molecular diagnosis was performed at intervals of 21 days on all of the evaluated plants. The high percentage of infected plants (approximately 78%) per line during and after the harvest pointed to mechanical transmission [21], and this result corroborated the findings of Ventura *et al.* in 2003 [9] (Figure 4). For that reason, we may conclude that among agricultural practices, fruit thinning is responsible for the spreading of sticky disease.

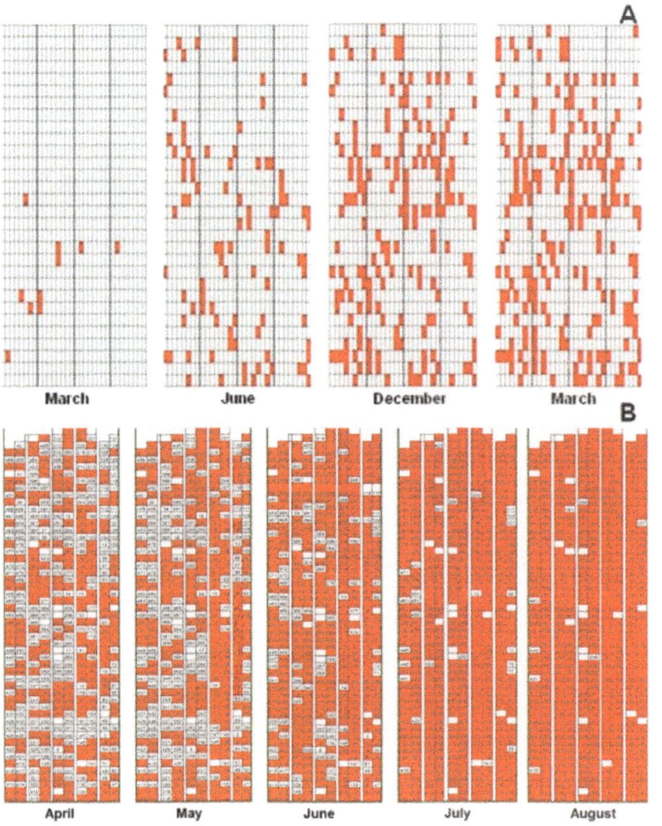

Figure 4: Spatial pattern maps of PMeV on papaya orchards in Brazil (A) and in Mexico (B). Red squares represent diseased plants and white represent healthy ones.

As for the progress of the disease with time, the epidemiological behavior varied according to different factors, such as environmental factors, possible vectors and cultural practices. A positive correlation has been observed between the temperature and precipitation in the course of the disease [21,22]. In the field, the disease occurred mainly in two epidemic patterns according to the Logistic and Gompertz models ($R^2 > 0.98$), which influenced the culture lifespan permanence [22]. In general, the first symptoms of the sticky disease appeared in plants that were 6 to 9 months old, depending upon the papaya variety. The initial onset of the incidence of sticky disease varied according to the source of inoculum, from seeds or alternative hosts and vectors, which played an important role in the quality of the papaya seedlings' health [9,23].

Identifying the mode of transmission of the papaya sticky disease has been an issue. Abreu *et al.* [24] investigated the PMeV transmission by *C. papaya* cultivar Golden seeds. They also developed a method for the diagnosis of the viral infection in seedlings. No PMeV was detected by conventional RT-PCR on the numbers of seedlings that were derived from the seeds of healthy (n = 187) and infected (p = 172) plants. However, another study conducted with seeds from *C. papaya* cultivar Maradol demonstrated that PMeV is seed-borne and that it can be transmitted to the next generation via contaminated seeds. Furthermore, this finding may explain how this virus appeared in Mexico [23]. The frequency of virus transmission by seeds can vary greatly depending on the virus-host interaction, and may be very small in some cases [25,26]. Therefore, it may vary according to the papaya cultivar, being very low for *C. papaya* cultivar Golden that requires a larger number of seedlings in order to be analyzed, and it is very high for *C. papaya* cultivar Maradol accounted for the high incidence of PMeV in Mexico. Moreover, for the seed transmission of plant viruses, the seed might be infected in the generative cells and the virus is maintained in the germ cells or, less often, in the seed coat [27]. Papaya seeds are extremely wrinkled, and each seed is enclosed in a gelatinous membrane, and, with fruits that are infected with the sticky disease, soaked with infected latex. For this reason, it is still necessary to confirm whether PMeV is in the seed embryo or in the outside shell.

It is worth mentioning that it is extremely important to conduct surveys in other papaya producing countries of the region and with the major producers worldwide in order to verify the virus presence in these areas, especially given the increase in the movement of seeds among the producing countries.

The possible involvement of insects as sticky disease vectors has been considered based upon early studies on the field spread pattern of this disease [10,28], especially with evidence of the existence of an aerial vector associated with the disease. It was shown that the presence of dsRNA and sticky disease symptoms occurred six and eight months, respectively, after healthy papaya plants were exposed to PMeV-infected whitefly (*Bemisia tabaci* type B) [29], although the virus was not detected in this possible vector, which is not a papaya pest. On the other hand, adults of another whitefly species, *Trialeurodes variabilis*, a pest to Brazilian papaya and other fruit trees, [30] possess PMeV dsRNA. Nevertheless, despite large populations of the whitefly, no virus infection was observed in those plants, including plants exposed to insects after they fed on PMeV inoculated plants, suggesting that the insect acquires the virus but does not transmit it to papaya plants [31]. A greenhouse study confirmed that *T. variabilis* was unable to transmit PMeV from diseased to healthy papayas, even though the presence of the virus was ascertained in adults and nymphs [20]. Currently, research teams in Brazil and Mexico are undertaking research to identify the virus vector and elucidate the mechanism of transmission.

Sticky disease also interferes with the natural resistance of papaya fruits to fruit flies (Diptera: Tephretidae). Green unripe fruits from PMeV infected plants are susceptible to the Mediterranean fruit fly (Medfly),*Ceratitis capitata* (Wied.), a pest of quarantine importance that usually only infests papaya fruits in the advanced stages of maturity. There is a direct relationship between the time of plant infection with the virus and the infestation of the papaya fruits by the fruit fly *C. capitata* [32].

Another important aspect in plant disease epidemiology is the alternative hosts. In this sense, the presence of dsRNA with a molecular weight similar to that associated with sticky disease was detected in samples of the Gramineae species *Brachiaria decunbens*, which is present in the papaya orchards that are affected by the sticky disease in the south of Bahia [15].

DETECTION METHODS

Aiming for the earlier identification of infected plants, two different molecular approaches have been developed. The first strategy is based upon the extraction of PMeV genomic dsRNA from the plant latex using organic solvents followed by gel electrophoresis to visualization the resultant ~12 kb viral dsRNA band [33]. This method has been used both to analyze symptomatic plants and to confirm the field diagnosis. This has the advantage of relatively low cost and rapid turnover, but requires large quantities of latex and a high titre of the virus, making it unsuitable for the detection at the early stages of infection or when only a little amount of the plant material is available.

The second method applies conventional reverse transcription PCR (RT-PCR) to latex diluted in ammonium citrate. Viral dsRNA was used as a template in RT-RAPD (Reverse Transcription-Random Amplified Polymorphic DNA) to randomly produce DNA sequences from the PMeV genome. RAPD does not require prior DNA sequence information because it uses single, short random primers so it can be applied to generate random DNA sequence data from an uncharacterized virus RNA genome. RT-RAPD fragments were cloned and sequenced. The nucleotide sequences of the viral dsRNA were used to designed specific primers for PMeV. Through RT-PCR, one of these primers (C05-3) amplified a 669-nucleotide fragment that, after sequencing, showed a high similarity to other viral RNA dependent RNA polymerases (RDRP) [13]. This method requires less latex and is more sensitive, but the latex storage and latex proteins in the PCR reaction are problematic.

The efficiency of either molecular diagnosis strategy, the direct PMeV dsRNA extraction or the amplification by RT-PCR, depends upon previous latex processing. Usually papaya producers send fruits, leaves and stems inside plastic bags to diagnostic laboratories. Sometimes the conditions during plant transport makes diagnosis from the latex impracticable as it must be correctly stored and processed to avoid dsRNA destruction and consequent diagnosis failure. Rodrigues and colleagues [33] evaluated different procedures for latex storage prior to PMeV molecular diagnosis. The results showed that PMeV dsRNA was protected for 25 days when the latex was diluted in a citrate buffer pH 5.0 (1:1 v/v) and maintained at −20 °C. At the same temperature, some protection was observed for pure latex or for latex diluted in ultra-pure water. However, the dsRNA was almost completely degraded after 25 days when it was maintained at 25 °C, which indicated the need for freezing [33].

New diagnostic methods have been proposed using conventional and quantitative RT-PCR (qRT-PCR), allowing for the molecular diagnosis in papaya tissues [24]. Two primers pairs, each pair (forward and reverse) was selected specifically for conventional RT-PCR or qRT-PCR reactions, were designed by targeting the PMeV sequence homologous to RDRP [24]. These methods have allowed for the identification of asymptomatic plants infected by PMeV in the field. In Mexico, primers were also developed for PMeV diagnostics using RT-PCR that was capable of detecting both isolates from Mexico and from Brazil and it was sensitive enough to detect the virus in any type of plant tissue and in asymptomatic plants [34].

DISEASE MANAGEMENT

Until now, a papaya genotype that is resistant to PMeV has not been identified despite all of the efforts coming from Brazil and Mexico. Symptoms are triggered only after flowering and, therefore, an infected plant without fruit or symptoms can remain unnoticed for months in the field acting as an inoculum source until it is finally detected and eliminated [7,9]. Thus, roguing of infected plants is the only available strategy to control this viral disease [35,36]. In Brazil, this practice is governed by Normative Instruction number 17, 27 May 2010, of the Brazilian Ministry of Agriculture, Livestock and Supply (MAPA). Weekly inspections are performed in the entire culture crop and plants with symptoms similar to those caused by PMeV are removed. In both countries, with the goals of reducing sticky disease dissemination as well as the prevention or the delaying of its introduction into areas where the disease has not been noted, the following measures have also been recommended: (i) to conduct weekly inspections in orchards and eliminate diseased plants (roguing) as soon as the first symptoms of the sticky disease are detected [7,9]; (ii) to install nurseries and new orchards as far as possible from other orchards, especially those with a history of sticky disease; (iii) not to collect seeds from sticky diseased plants and orchards with PMeV occurrence [23]; (iv) to try to reduce injuries to plants during cultural treatments and disinfect all material used in the process of thinning and harvesting fruits; (v) to manage low-growing vegetation and keep rows clean and the area between rows trimmed to diminish the variety of weed species; (vi) to eliminate all orchards (diseased and healthy) at the end of the economic cycle of production to eliminate sources of inoculum; (vii) to destroy abandoned orchards, mainly those with plants infected by PMeV as one diseased papaya tree can infect the entire orchard and (viii) to implement crop rotation in papaya producing areas.

PAPAYA AND PMEV INTERACTION

Papaya latex is mostly comprised of cysteine proteinases in addition to the carbohydrates, lipids, salts, phenols and glutathione [37]. These proteinases are synthesized as inactive precursors that are converted into mature enzymes after tissue wounding and latex exudation [37]. The PMeV's ability to inhabit such a hostile environment shows an intriguing interaction with papaya.

PMeV particles are found only in papaya laticifers and not in other cell types [14]; therefore, the sticky disease symptoms, spontaneous latex exudation from fruits and leaves, might be a direct effect of the virus [19]. The characterization of the effects of the PMeV on papaya latex demonstrated that the accumulation of calcium oxalate crystals in the infected papaya latex positively correlated with the increased production of hydrogen peroxide (H_2O_2) in the laticifers of the plant, indicating a hypersensitive response (HR) of papaya laticifers against PMeV infection [19].

Latex fluidity and exudation in PMeV infected papaya suggest that the virus uses the host laticifers to move itself through the plant [19]. Electron microscopy and molecular data show the virus in close association with the latex particles. This association may allow the virus to flow more efficiently because the PMeV might use this mechanism to move quickly from the inoculation site to non-infected cells. This study also demonstrated alterations on lipids, phenols, alkaloids and sugar concentration occurring in papaya latex during PMeV infection [19].

Papaya latex is known to be mainly composed of proteins [38,39]; for this reason, the effects of PMeV on the laticifers' regulatory network were addressed through the proteomic analysis of papaya latex. 160 unique papaya latex proteins were identified, where the majority of the proteins were associated with different types of stresses (e.g., cadmium ion, programmed cell death, drought, salinity, and microorganisms) [40]. Thus, papaya laticifers are involved in other plant stress responses in addition to the known herbivorous toxicity [37,41,42] and tissue protection after mechanical wounding [37,43]. Quantitative analysis revealed 10 down regulated proteins in the latex of diseased plants, including 9 cysteine proteases (chymopapain), showing that the PMeV influences the level of the proteases known to be activated during latex exudation [40].

Moreover, a reduction in the proteolytic activity of papaya latex during sticky disease has also been demonstrated, specifically associated with cysteine proteases, showing that these proteins are important in the response of the papaya to stress and that PMeV is able to overcome this process [40]. The reduced proteolytic activity seems to have an inhibitory effect on the latex coagulation by facilitating its flow through the laticifers, and, consequently, the virus spread within the plant [40]. Additionally, the accumulation of H_2O_2 in the laticifers of the infected papaya [19] played a negative regulatory role on the cysteine-protease activity during the sticky disease [37,44], most likely by oxidizing and inactivating the active site of the enzyme. Taken together, these results suggest the negative modulation of cysteine proteases of papaya latex by PMeV infection in an attempt to delay the process of the laticifers' programmed cell death (PCD) [45] and the minimization of the virus particle degradation.

A major defense mechanism in plants is the "hypersensitive response" (HR) whereby cells infected with pathogens, and often surrounding cells as well, are instructed to self-destruct by the host plant [46]. This is thought to deny nutrients to the invading pathogen. It is proposed that in plants, the process is initiated by a reaction between nitric oxide (NO) and hydrogen peroxide (H_2O_2) [47]. The increased H_2O_2 production in the papaya laticifers during PMeV infection has already been demonstrated [19]. To assess the involvement of nitric oxide (NO) in plant defense signaling, papaya seedlings were challenged with a NO donor, sodium nitroprusside (SNP), and PMeV. SNP treatment led to the highest increase in peroxidase and superoxide dismutase activity, also the levels of phenolics and the carbohydrate concentration were also increased [48].

H_2O_2 is a well-known systemic response elicitor, and, as mentioned above, PMeV infected papaya laticifers have an increased production of this molecule. Additionally, necrotic lesions in the plant leaf tip is the first disease symptom [7,9]. Therefore, healthy and PMeV infected papaya leaves were analyzed using global protein expression analytical techniques (2-DE and DIGE) followed by protein identification by MS analysis (MALDI-TOF-MS/MS and LC-IonTrap-MS/MS) [49].

Stress responsive proteins, mostly calreticulin and proteasome-related proteins, are up regulated and proteins related to metabolism are down regulated in sticky diseased papaya leaf tissues [49], showing a major investment in plant defense [37]. In total, 42 up regulated and 33 down regulated protein spots were observed in the infected samples. Of those, 48 spots were identified, 26 being up regulated and 22 being down regulated [49].

The upregulation of proteasome-related proteins supports the assumption that PMeV proteins, or even host proteins, which are essential for the viral infection can be targeted for degradation [49]. Several studies have reported the involvement of the ubiquitin/26S proteasome system (UPS) in the signaling and regulation of the interactions between plants and pathogens [50,51,52], particularly with viral exploitation and interference with the UPS [53,54,55]. Therefore, these proteins are marker candidates useful for an additional analysis of papaya sticky disease to help better understand and control this disease [49].

Noncoding RNAs comprise the majority of transcribed RNA, and they perform a wide range of functions in cellular and developmental processes. As a result, noncoding RNAs are also implicated in the development and pathophysiology of many diseases, and they represent potential targets for therapeutic intervention. MicroRNAs (miRNAs) are one of a number of classes of endogenous, small (21–24 nucleotide), non-coding RNAs found in both animals and plants [56,57].

In the papaya, 24 [58] and 75 [59] conserved miRNAs were identified via the analyses of small RNA deep sequencing data and the genomic sequence. However, the conservation of mature miRNAs between plant species has enabled a computer-based approach to predict the secondary structures of the sequences surrounding miRNA. Actually, 4,251 plant mature miRNA sequences from the Plant miRNAs Database were used to search for hairpin structures in the *C. papaya* genome [60]. 462 known miRNA sequences were detected in 537 hairpins in the *C. papaya* genome; additionally, these miRNAs were classified into 72 miRNAs families [60].

Changes in miRNA expression are associated with viral infection in *Arabidopsis* [61], and viral protein-induced alterations in miRNA expression have been associated with symptom development in *Nicotiana tabacum* [62]. Therefore, in order to understand the unique plant-virus interaction between PMeV and *C. papaya*, and, consequently, the mechanisms involved in the onset of sticky disease symptoms, the miRNA response of the plant to infection was investigated [60]. The MicroRNAs that target the components of the 20S and 26S proteasomal degradation processes and components of other stress response pathways were studied. The expression of each was compared using real-time PCR in healthy and infected papaya leaf tissue [60]. The expression of these miRNAs changed, showing that PMeV can exploit and interfere with the UPS and with other stress response pathways [60]. PMeV modifies the level of several miRNAs involved in the modulation of genes related to the UPS system. The accumulation of these miRNAs in plants with sticky disease symptoms is lower, showing that PMeV takes over the UPS system for its own benefit, increasing its infectivity in papaya [60].

PMeV infection also influences the production of reactive oxygen species (ROS) [19,60]. This was particularly noticeable through the expression of target genes involved in the ROS pathway that played an important signaling role in the plants' controlling processes, such as the response to biotic stimuli, which is lowered during PMeV infection in plants with sticky disease symptoms [60].

Therefore, evidence suggests that the PMeV uses the plant UPS response as a strategy to favor its replication and maintain host viability, and it modifies the accumulation of miRNAs that modulate important defense genes eliciting the onset of sticky disease symptoms [60].

BIOTECHNOLOGICAL STRATEGIES TO PAPAYA MELEIRA *VIRUS* CONTROL

Efforts to control viral papaya diseases are long-standing. The currently used methods are vector control, the use of papaya plants that are resistant to certain viruses, and cross-protection, among others. Chemical or biological vector controls are used as well, and can be very effective considering vectors that need to feed for some time on a crop before the virus is transmitted, but are of much less value when the transmission occurs very rapidly and may already have taken place before the vector succumbs to the pesticide. As the sticky disease vector is still unknown, chemical or biological control cannot be used. The cross-protection strategy involves the use of a mild virus strain to control severe strains of the same virus. However, cross protection depends upon the availability of mild strains that can be used for effective protection against the target virus, and it requires extra agricultural practice and care that limits the benefits of this approach [3]. Until now, PMeV mild strains are unknown and, therefore, cross-protection is not a strategy that can be considered to control this virus.

Considering the lack of effectiveness of the methods mentioned above for PMeV control, the development of resistant papaya plants to viruses is the main goal, and virus-resistant transgenic plants are currently the most effective control of viral diseases.

Pathogen resistant plants can arise naturally or be developed by classic genetics or through the use of transgenic plants. The development of viral resistant varieties through conventional breeding methods has been complicated due to the sexual incompatibility of wild species and cultivated papaya [63].

A promising development in agricultural virus control is the concept that plants can initiate defense responses based upon RNA silencing when challenged by viral RNA. The use of transgenic plants that constituently express viral material has been successfully developed in different crops [64,65]. Genetic modified papaya resistant to certain viruses, for example to PRSV, is marketed in some countries. The concept of pathogen derived resistance (PDR) has been employed for PRSV management through coat protein (CP), RNA silencing, and replicase gene technology [3]. All of these gene technologies depend upon the viral genome sequence knowledge. Therefore, this methodology will only be considered after the complete PMeV genome has been sequenced.

Alternatively, one way of inducing viral resistance in plants is to introduce artificially viral dsRNA molecules that are able to trigger the post-transcriptional gene silencing (PTGS) [66,67]. It has been shown that PMeV infected papaya plants inoculated with the dsRNA viral genome (Figure 5) have a delay in the infection process, suggesting that the plant defense has been elicited that has therefore inhibited viral replication [24]. Therefore, these results are a prospectus for sticky disease control.

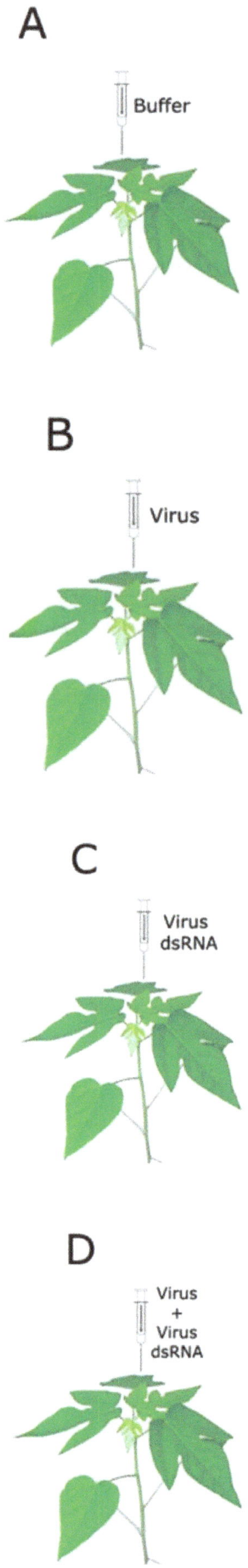

Figure 5: Induced resistance in papaya by viral dsRNA inoculation. Papaya seedlings were inoculated with (A) buffer (control); (B) PMeV; (C) PMeV dsRNA and (D) PMeV and PMeV dsRNA. A delay in the infection process occurred in the plants simultaneously inoculated ...

Plant hormones may play a role in regulating signaling networks involved in plant defense. When a pathogen attack occurs, the plant produces these hormones, and the amounts and compositions vary depending upon the strategy of infection and the type of pathogen. At present, there has been significant progress in identifying the key components and the understanding of the role of phytohormones in plant responses to biotic attacks [68,69]. Studies on the papaya-PMeV interaction, such as those focused on the plant signaling pathways within the defense response, allow for a greater understanding of this mechanism and they open up the possibility for disease control by phytohormone-mediated papaya resistance [70,71].

PERSPECTIVES

Since the first report of the sticky disease in Brazil a few decades ago, many researchers have conducted several studies that have focused on different aspects of the disease; from this research, progress has been made in the molecular diagnosis of PMeV as well as in the knowledge of some of the mechanisms of transmission and spatiotemporal spreading of the virus. More recently, studies have focused on the proteomics of the most cultivated varieties of papaya in Mexico and Brazil, especially with a search for genes that are expressed in the presence of the virus because these genes can be molecular markers useful in searching for resistance to PMeV in papaya plants. Additionally, studies on the possible insect vectors for PMeV transmission have been conducted. More importantly, the virus genome is being sequenced as well.

To date, no variety of papaya has been found to be PMeV resistant. It is very important to focus on obtaining papaya plants that are resistant or tolerant to this virus through breeding programs that are based upon the information provided by the aforementioned studies or by the more effective genetic modification of papaya, which may be the best strategy for the efficient control of PMeV and for long-term protection.

ACKNOWLEDGMENTS

This work was supported by grants from FINEP (Financiadora de Estudos e Projetos), CNPq (Conselho Nacional de Desenvolvimento Científico e Tecnológico), CAPES (Coordenação de Aperfeiçoamento de Pessoal de Nível Superior), FAPES (Fundação de Amparo à Pesquisa e Inovação do Estado do Espírito Santo) and CONACYT (Consejo Nacional de Ciencia e Tecnología).

AUTHOR CONTRIBUTIONS

All authors conceived the review topic, read and approved the final manuscript. P.M.V.A. and T.F.S.A. equally contributed to this work; drafted the manuscript and generated the figures. A.M.A contributed on Biotechnological Strategies and Perspectives. D.P.B., R.T.T. and J.A. V. contributed to writing Epidemiology. A.A.R.F. and P.M.B.F. corrected and edited the manuscript.

REFERENCES

1. FAOSTAT FAOSTAT Production. [(accessed on 19 December 2014)]. Available online:http://faostat.fao.org/site/339/default.aspx.
2. SIAP Sistema de Información Agrícola y Pesquera (SIAP) [(accessed on 19 December 2014)]. Available online: http://www.siap.gob.mx/cierre-de-la-produccion-agricola-por-cultivo/
3. Azad M.A.K., Amin L., Sidik N.M. Gene technology for *Papaya ringspot virus* disease management. Sci. World J. 2014 doi: 10.1155/2014/768038.
4. Bau H.-J., Kung Y.-J., Raja J., Chan S.-J., Chen K.-C., Chen Y.-K., Wu H.-W., Yeh S.-D. Potential threat of a new pathotype of *Papaya leaf distortion mosaic virus* infecting transgenic papaya resistant to *Papaya ringspot virus*. Phytopathology. 2008;98:848–856. doi: 10.1094/PHYTO-98-7-0848.
5. Daltro C.B., PereiraII Á.J., Cascardo R.S., Alfenas-Zerbini P., Bezerra-Junior J.E.A., Lima J.A.A., Zerbini F.M., Andrade E.C. Genetic variability of *Papaya lethal yellowing virus* isolates from Ceará and Rio Grande do Norte States, Brazil. Trop. Plant Pathol. 2012;37 doi: 10.1590/S1982-56762012000100004.
6. Noa-Carrazana J., González-de-León D., Ruiz-Castro B., Piñero D., Silva-Rosales L. Distribution of*Papaya ringspot virus* and *Papaya mosaic virus* in papaya plants (*Carica papaya*) in Mexico. Plant Dis.2006;90:1004–1011. doi: 10.1094/PD-90-1004.
7. Ventura J.A., Costa H., da Silva Tatagiba J. Diseases of Fruits and Vegetables: Volume II. Springer; Berlin/Heidelberg, Germany: 2004. Papaya diseases and integrated control; pp. 201–268.
8. Teliz D., Mora G., Nieto D., Gonsalves D., Garcia E., Matheis L., Avila C. La mancha anular del papayo en México. Rev. Mexicana Fitopatol. 1991;9:64–68.
9. Ventura J.A., Costa H., Tatagiba J.S., Andrade J.S., Martins D.S. Meleira do mamoeiro: Etiologia, sintomas e epidemiologia. In: Martins D.S., editor. Papaya Brasil: Qualidade do Mamão Para o Mercado Interno. Incaper; Vitória, Brazil: 2003. pp. 267–276.
10. Rodrigues C.H., Ventura J.A., Maffia L.A. Distribuição e transmissão da meleira em pomares de mamão no Espírito Santo. Fitopatol. Bras. 1989;14:118.
11. Perez-Brito D., Tapia-Tussell R., Cortes-Velazquez A., Quijano-Ramayo A., Nexticapan-Garcez A., Martín-Mex R. First report of *Papaya meleira virus* (PMeV) in Mexico. Afr. J. Biotechnol. 2012;11:13564–13570. doi: 10.5897/AJB12.1189.
12. Daltro C.B., Abreu E.F.M., Aragão F.J.L., Andrade E.C. Genetic diversity studies of *Papaya meleira virus*. Trop. Plant Pathol. 2014;39:104–108. doi: 10.1590/S1982-56762014000100013.
13. Araújo M.M.M., Tavares E.T., Silva F.R., Marinho V.L.A., Júnior M.T.S. Molecular detection of *Papaya meleira virus* in the latex of *Carica papaya* by RT-PCR. J. Virol. Methods. 2007;146:305–310. doi: 10.1016/j.jviromet.2007.07.022.
14. Kitajima E.W., Rodrigues C.H., Silveira J.S., Alves F., Ventura J.A., Aragao F.J.L., Oliveira L.H.R. Association of isometric virus like particles, restricted to laticifers, with "meleira" ("sticky disease") of papaya (*Carica papaya*) Fitopatol. Bras. 1993;18:118–122.
15. Maciel-Zambolim E., Kunieda-Alonso S., Matsuoka K., de Carvalho M., Zerbini F. Purification and some properties of *Papaya meleira virus*, a novel virus infecting papayas in Brazil. Plant Pathol.2003;52:389–394. doi: 10.1046/j.1365-3059.2003.00855.x.
16. Nakagawa J.T.Y., Suzukama Y. Exudação de Látex Pelo Mamoeiro. Estudo de Ocorrência em Teixeira de Freitas, BA. Congresso Brasileiro de Fruticultura; Brasil: 1987. pp. 555–559.
17. Akiba F. Bactérias pleomórficas, Gram negativas, e de crescimento lento em meio de cultura, isoladas do sistema vascular de diversas espécies de plantas apresentando sintomas de declínio. Fitopatol. Bras.1989;14:110–111.
18. Fauquet C.M., Fargette D. International Committee on Taxonomy of Viruses and the 3142 unassigned species. Virol. J. 2005;2:e64. doi: 10.1186/1743-422X-2-64.
19. Rodrigues S.P., da Cunha M., Ventura J.A., Fernandes P.M.B. Effects of the *Papaya meleira virus* on papaya latex structure and composition. Plant Cell Rep. 2009;28:861–871. doi: 10.1007/s00299-009-0673-7.
20. Rodrigues S.P., Andrade J.S., Ventura J.A., Lindsey G.G., Fernandes P.M.B. *Papaya meleira virus* is neither transmitted by infection at wound sites nor by the whitefly *Trialeurodes variabilis*. J. Plant Pathol.2009;91:87–91.
21. Magaña-Alvarez A. Evaluación de la transmisión de Papaya Meleira Virus (PMeV) por Semillas. Maestría. Centro de Investigación Científica de Yucatán A.C.; Mexico, Mexico: 2013.
22. Cosmi F.C. Epidemiologia da Meleira do Mamoeiro no Norte do estado do Espírito Santo. Mestrado, UFES; Alegre, Espírito Santo, Brazil: 2010.
23. Tapia-Tussell R., Magaña-Alvarez A., Cortes-Velazquez A., Itza-Kuk G., Nexticapan-Garcez A., Quijano-Ramayo A., Martin-Mex R., Perez-Brito D. Seed transmission of *Papaya meleira virus* in papaya (*Carica papaya*) cv. Maradol. Plant Pathol. 2015;64:272–275. doi: 10.1111/ppa.12279.
24. Abreu P., Piccin J.G., Rodrigues S.P., Buss D.S., Ventura J.A., Fernandes P. Molecular diagnosis of*Papaya meleira virus* (PMeV) from leaf samples of *Carica papaya* L. using conventional and real-time RT-PCR. J. Virol. Methods. 2012;180:11–17. doi: 10.1016/j.jviromet.2011.12.003.
25. Forster R.L., Seifers D.L., Strausbaugh C.A., Jensen S.G., Ball E.M., Harvey T.L. Seed transmission of the High Plains virus in sweet corn. Plant Dis. 2001;85:696–699. doi: 10.1094/PDIS.2001.85.7.696.
26. Mikel M.A., Darcy C.J., Ford R.E. Seed transmission of *Maize dwarf mosaic virus* in sweet corn. J. Phytopathol. 1984;110:185–191. doi: 10.1111/j.1439-0434.1984.tb00746.x.

27. Zaitlin M., Palukaitis P. Advances in understanding plant viruses and virus diseases. Annu. Rev. Phytopathol. 2000;38:117–143. doi: 10.1146/annurev.phyto.38.1.117.

28. Vidal C.A., Laranjeira F.F., Nascimento A.S., Habibe T.C. Distribuição espacial da meleira do mamoeiro em zonas de trópico úmido e trópico semi-árido. Fitopatol. Bras. 2004;29:276–281. doi: 10.1590/S0100-41582004000300006.

29. Vidal C.A., Nascimento A.S., Habibe T.C. Transmissão do vírus da meleira do mamoeiro por insetos. In: Martins D.S., editor. Papaya Brasil: Qualidade do Mamão Para o Mercado Interno. Incaper; Vitória, Brazil: 2003. pp. 612–615.

30. Picaço M.C., Moura M.F., Moreira M.D. Biologia, identificação e manejo de moscas-brancas em fruteiras. In: Laercio Z., editor. Manejo Integrado, Produção Integrada, Fruteiras Tropicais, Doenças e Pragas. UVS Pulbisher; Viçosa, Brazil: 2003. pp. 243–284.

31. Andrade J.d.S., Ventura J.A., Rodrigues S.P., Fernandes P.M.B., Tatagiba J.D.S., Costa H. Avaliação de diferentes métodos de inoculação da meleira em plantas jovens de mamão. Fitopatol. Bras. 2003;28:288. doi: 10.1590/S0100-41582003000500001.

32. Martins D.D., Ventura J.A., Lima R.D.A., Culik M.P., Costa H., Ferreira P.S.F. Interaction between*Papaya meleira virus* (PMeV) infection of papaya plants and mediterranean fruit fly infestation of fruits.Crop. Prot. 2012;36:7–10. doi: 10.1016/j.cropro.2012.01.001.

33. Rodrigues S.P.G.O.P., Andrade J.S., Ventura J.A., Fernandes P.M.B. Simplified molecular method for the diagnosis of *Papaya meleira virus* in papaya latex and tissues. Summa Phytopathol. 2005;31:281–283.

34. López-Ochoa L.A., Moreno-Valenzuela O.A., Zamudio-Moreno E. Secuencia parcial y metodo de diagnóstico por RT-PCR del virus de la meleira de la papaya. 2012000081A. Mx Patent. filed 16 Dec 2011, and issue June 2013.

35. Habibe T.C., Dantas J.L.L., do Nascimento A.S., Dantas A.C.V.L. Suscetibilidade de Genótipos de Mamoeiro (Carica Papaya L.) ao Vírus da Meleira, Sob Condições de Trópico Semi-Árido. Embrapa Mandioca e Fruticultura; Cruz Das Almas, Brazil: 2003.

36. Dos S. Martins D. Papaya Brasil: Manejo, Qualidade e Mercado do Mamão. Incaper; Vitória, Brazil: 2007.

37. El Moussaoui A., Nijs M., Paul C., Wintjens R., Vincentelli J., Azarkan M., Looze Y. Revisiting the enzymes stored in the laticifers of *Carica papaya* in the context of their possible participation in the plant defence mechanism. Cell Mol. Life Sci. 2001;58:556–570. doi: 10.1007/PL00000881. [PubMed]

38. Bravo L.M., Hermosilla J., Salas C.E. A biochemical comparison between latex from *Carica candamarcensis* and *C. Papaya*. Braz. J. Med. Biol. Res. 1994;27:2831–2842. [PubMed]

39. Moutim V., Silva L.G., Lopes M.T.P., Fernandes G.W., Salas C.E. Spontaneous processing of peptides during coagulation of latex from *Carica papaya*. Plant Sci. 1999;142:115–121. doi: 10.1016/S0168-9452(98)00226-X.

40. Rodrigues S.P., Ventura J.A., Aguilar C., Nakayasu E.S., Choi H., Sobreira T.J.P., Nohara L.L., Wermelinger L.S., Almeida I.C., Zingali R.B., et al. Label-free quantitative proteomics reveals differentially regulated proteins in the latex of sticky diseased *Carica papaya* L. plants. J. Proteomics. 2012;75:3191–3198. doi: 10.1016/j.jprot.2012.03.021.

41. Konno K. Plant latex and other exudates as plant defense systems: Roles of various defense chemicals and proteins contained therein. Phytochemistry. 2011;72:1510–1530. doi: 10.1016/j.phytochem.2011.02.016.

42. Konno K., Hirayama C., Nakamura M., Tateishi K., Tamura Y., Hattori M., Kohno K. Papain protects papaya trees from herbivorous insects: Role of cysteine proteases in latex. Plant J. 2004;37:370–378. doi: 10.1046/j.1365-313X.2003.01968.x.

43. Azarkan M., Dibiani R., Baulard C., Baeyens-Volant D. Effects of mechanical wounding on *Carica papaya* cysteine endopeptidases accumulation and activity. Int. J. Biol. Macromol. 2006;38:216–224. doi: 10.1016/j.ijbiomac.2006.02.021.

44. Azarkan M., el Moussaoui A., van Wuytswinkel D., Dehon G., Looze Y. Fractionation and purification of the enzymes stored in the latex of *Carica papaya*. J. Chromatogr. B. 2003;790:229–238. doi: 10.1016/S1570-0232(03)00084-9.

45. Solomon M., Belenghi B., Delledonne M., Menachem E., Levine A. The involvement of cysteine proteases and protease inhibitor genes in the regulation of programmed cell death in plants. Plant Cell.1999;11:431–444. doi: 10.1105/tpc.11.3.431.

46. Greenberg J.T., Yao N. The role and regulation of programmed cell death in plant-pathogen interactions. Cell Microbiol. 2004;6:201–211. doi: 10.1111/j.1462-5822.2004.00361.x.

47. Delledonne M., Zeier J., Marocco A., Lamb C. Signal interactions between nitric oxide and reactive oxygen intermediates in the plant hypersensitive disease resistance response. Proc. Natl. Acad. Sci. USA.2001;98:13454–13459. doi: 10.1073/pnas.231178298.

48. Buss D.S., Dias G.B., Santos M.P., Ventura J.A., Fernandes P.M.B. Oxidative stress defence response of*Carica papaya* challenged by nitric oxide, *Papaya meleira virus* and *Saccharomyces cerevisiae*. Open Nitric Oxide J. 2011;3:55–64. doi: 10.2174/1875042701103010055.

49. Rodrigues S.P., Ventura J.A., Aguilar C., Nakayasu E.S., Almeida I.C., Fernandes P., Zingali R.B. Proteomic analysis of papaya (*Carica papaya* L.) displaying typical sticky disease symptoms. Proteomics.2011;11:2592–2602. doi: 10.1002/pmic.201000757.

50. Citovsky V., Zaltsman A., Kozlovsky S.V., Gafni Y., Krichevsky A. Proteasomal degradation in plant-pathogen interactions. Semin. Cell Dev. Biol. 2009;20:1048–1054. doi: 10.1016/j.semcdb.2009.05.012.

51. Dielen A.S., Badaoui S., Candresse T., German-Retana S. The ubiquitin/26S proteasome system in plant-pathogen

interactions: A never-ending hide-and-seek game. Mol. Plant Pathol. 2010;11:293–308. doi: 10.1111/j.1364-3703.2009.00596.x.

52. Marino D., Peeters N., Rivas S. Ubiquitination during plant immune signaling. Plant Physiol.2012;160:15–27. doi: 10.1104/pp.112.199281.

53. Alcaide-Loridan C., Jupin I. Ubiquitin and plant viruses, let's play together! Plant Physiol. 2012;160:72–82. doi: 10.1104/pp.112.201905.

54. Isaacson M.K., Ploegh H.L. Ubiquitination, ubiquitin-like modifiers, and deubiquitination in viral infection. Cell Host Microbe. 2009;5:559–570. doi: 10.1016/j.chom.2009.05.012.

55. Randow F., Lehner P.J. Viral avoidance and exploitation of the ubiquitin system. Nat. Cell Biol.2009;11:U527–U535. doi: 10.1038/ncb0509-527.

56. Park W., Li J.J., Song R.T., Messing J., Chen X.M. CARPEL FACTORY, a Dicer homolog, and HEN1, a novel protein, act in microRNA metabolism in *Arabidopsis thaliana*. Curr. Biol. 2002;12:1484–1495. doi: 10.1016/S0960-9822(02)01017-5.

57. Reinhart B.J., Weinstein E.G., Rhoades M.W., Bartel B., Bartel D.P. MicroRNAs in plants. Gene Dev.2002;16:1616–1626. doi: 10.1101/gad.1004402.

58. Aryal R., Yang X.Z., Yu Q.Y., Sunkar R., Li L., Ming R. Asymmetric purine-pyrimidine distribution in cellular small RNA population of papaya. BMC Genomics. 2012;13:682. doi: 10.1186/1471-2164-13-682.

59. Liang G., Li Y., He H., Wang F., Yu D.Q. Identification of miRNAs and miRNA-mediated regulatory pathways in *Carica papaya*. Planta. 2013;238:739–752. doi: 10.1007/s00425-013-1929-6. [PubMed]

60. Abreu P.M.V., Gaspar C.G., Buss D.S., Ventura J.A., Ferreira P.C.G., Fernandes P.M.B. *Carica papaya*microRNAs are responsive to *Papaya meleira virus* infection. PLoS ONE. 2014;9:e103401. doi: 10.1371/journal.pone.0103401.

61. Tagami Y., Inaba N., Kutsuna N., Kurihara Y., Watanabe Y. Specific enrichment of miRNAs in*Arabidopsis thaliana* infected with *Tobacco mosaic virus*. DNA Res. 2007;14:227–233. doi: 10.1093/dnares/dsm022.

62. Bazzini A.A., Hopp H.E., Beachy R.N., Asurmendi S. Infection and coaccumulation of *Tobacco mosaic virus* proteins alter microRNA levels, correlating with symptom and plant development. Proc. Natl. Acad. Sci. USA. 2007;104:12157–12162. doi: 10.1073/pnas.0705114104. [PMC free article] [PubMed]

63. Gonsalves D., Vegas A., Prasartsee V., Drew R.A., Suzuki J.Y., Tripathi S. Developing papaya to control *Papaya ringspot virus* by transgenic resistance, intergeneric hybridization, and tolerance breeding. In: Janick J., editor. Plant Breeding Reviews. Volume 26. John Wiley & Sons; Hoboken, NJ, USA: 2006. pp. 35–73.

64. Scorza R., Callahan A., Levy L., Damsteegt V., Webb K., Ravelonandro M. Post-transcriptional gene silencing in plum pox virus resistant transgenic European plum containing the plum pox potyvirus coat protein gene. Transgenic Res. 2001;10:201–209. doi: 10.1023/A:1016644823203.

65. Tricoli D.M., Carney K.J., Russell P.F., Mcmaster J.R., Groff D.W., Hadden K.C., Himmel P.T., Hubbard J.P., Boeshore M.L., Quemada H.D. Field evaluation of transgenic squash containing single or multiple virus coat protein gene constructs for resistance to *Cucumber Mosaic Virus*. Biotechnology.1995;13:1458–1465. doi: 10.1038/nbt1295-1458.

66. Tenllado F., Martinez-Garcia B., Vargas M., Diaz-Ruiz J.R. Crude extracts of bacterially expressed dsRNA can be used to protect plants against virus infections. BMC Biotechnol. 2003;3:e3. doi: 10.1186/1472-6750-3-3.

67. Tenllado F.D.-R.J.R. Double-stranded RNA-mediated interference with plant virus infection. J. Virol.2001;75:12288–12297. doi: 10.1128/JVI.75.24.12288-12297.2001. [PMC free article] [PubMed]

68. Nicaise V. Crop immunity against viruses: Outcomes and future challenges. Front. Plant Sci.2014;5:e660. doi: 10.3389/fpls.2014.00660.

69. Robert-Seilaniantz A., Grant M., Jones J.D. Hormone crosstalk in plant disease and defense: More than just jasmonate-salicylate antagonism. Annu. Rev. Phytopathol. 2011;49:317–343. doi: 10.1146/annurev-phyto-073009-114447.

70. Alazem M., Lin K.Y., Lin N.S. The abscisic acid pathway has multifaceted effects on the accumulation of *Bamboo mosaic virus*. Mol. Plant Microbe Interact. 2014;27:177–189. doi: 10.1094/MPMI-08-13-0216-R.

71. Chen L., Zhang L., Li D., Wang F., Yu D. WRKY8 transcription factor functions in the TMV-cg defense response by mediating both abscisic acid and ethylene signaling in *Arabidopsis*. Proc. Natl. Acad. Sci. USA.2013;110:E1963–E1971. doi: 10.1073/pnas.1221347110

CHAPTER 4

A CURRENT OVERVIEW OF TWO VIROIDS THAT INFECT CHRYSANTHEMUMS: CHRYSANTHEMUM STUNT VIROID AND CHRYSANTHEMUM CHLOROTIC MOTTLE VIROID

Won Kyong Cho, Yeonhwa Jo, Kyoung-Min Jo, and Kook-Hyung Kim

Department of Agricultural Biotechnology, Plant Genomics and Breeding Institute, Institute for Agriculture and Life Sciences, College of Agriculture and Life Sciences, Seoul National University, Seoul 151-921, Korea

ABSTRACT

The chrysanthemum (*Dendranthema* X *grandiflorum*) belongs to the family *Asteraceae* and it is one of the most popular flowers in the world. Viroids are the smallest known plant pathogens. They consist of a circular, single-stranded RNA, which does not encode a protein. Chrysanthemums are a common host for two different viroids, the *Chrysanthemum stunt viroid* (CSVd) and the *Chrysanthemum chlorotic mottle viroid*(CChMVd). These viroids are quite different from each other in structure and function. Here, we reviewed research associated with CSVd and CChMVd that covered disease symptoms, identification, host range, nucleotide sequences, phylogenetic relationships, structures, replication mechanisms, symptom determinants, detection methods, viroid elimination, and development of viroid resistant chrysanthemums, among other studies. We propose that the chrysanthemum and these two viroids represent convenient genetic resources for host–viroid interaction studies.

INTRODUCTION

The chrysanthemum (*Dendranthema* X *grandiflorum*) is a member of the family *Asteraceae*, and it is one of the popular flowers in the world. The international market for cut and potted chrysanthemums is increasing, and chrysanthemums in many European and Asian countries are commercially very important for the floral industry [1]. Several pathogens, including viruses, viroids, and phytoplasma, cause serious diseases in chrysanthemums. To date, nine viruses and two viroids are known to infect chrysanthemums [2].

Viroids are the smallest known plant pathogens. They consist of a circular, single-stranded RNA, which does not encode a protein. Viroid RNAs range from 246 to 401 bases [3,4]. Viroids traffic from cell to cell via plasmodesmata [5]. To date, over 30 species of viroids have been reported; these can be divided into two families, the *Pospiviroidae* and the *Avsunviroidae* [6]. The family *Pospiviroidae* includes five genera, such as*Apscaviroid*, *Cocadviroid*, *Coleviroid*, *Hostuviroid*, and *Pospiviroid*. So far, ten species including *Chrysanthemum stunt viroid* (CSVd) and *Potato spindle tuber viroid* (PSTVd) are members of the genus*Pospiviroid* [7]. The genomic RNA of members of the family *Pospiviroidae*, which replicate in the nucleus, assumes rod-like or quasi-

Citation: Won Kyong Cho, Yeonhwa Jo, Kyoung-Min Jo, and Kook-Hyung Kim, A Current Overview of Two Viroids That Infect Chrysanthemums: Chrysanthemum stunt viroid and Chrysanthemum chlorotic mottle viroid, doi: 10.3390/v5041099.

rod-like conformation in which, based on local sequence similarity, five domains have been proposed: the left terminal, pathogenicity, central, variable, and right-terminal domains [8]. Moreover, several conserved regions have been identified in the rod-like conformation, including the central conserved region (CCR), likely involved in replication [9,10,11,12], and the terminal conserved region (TCR) or the terminal conserved hairpin (TCH), which appear mutually exclusive [13,14,15]. The family*Avsunviroidae* is composed of three genera, such as *Avsunviroid*, *Elaviroid*, and *Pelamoviroid*. *Peach latent mosaic viroid* (PLMVd) and *Chrysanthemum chlorotic mottle viroid* (CChMVd) are members of the genus*Pelamoviroid*. They have highly branched structures with self-cleaving ribozymes, which are not present in the family *Pospiviroidae*. These viroids replicate in the host chloroplast [16].

Interestingly, chrysanthemums are a common host for two different viroids, the CSVd and the CChMVd, which are quite different from each other in structure and function (Table 1). Here, we summarize and discuss current research associated with CSVd and CChMVd, and we suggest that chrysanthemums and viroids would serve as convenient genetic resources for host-viroid interaction studies.

Table 1: Characteristics of CSVd and CChMVd viroids

Characteristics	CSVd	CChMVd
Disease	Chrysanthemum stunt	Chrysanthemum chlorotic mottle
Symptoms	Light green young leaves, chlorotic spots, stunting, small leaves and flowers, and decreased rooting ability	Yellow-green mottling, chlorosis, and dwarfed size
Family and genus	*Pospiviroidae*, *Pospiviroid*	*Avsunviroidae*, *Pelamoviroid*
Genome size	354–356 nt	398–401 nt
Replication method	Asymmetric rolling circle mechanism	Symmetric rolling circle mechanism with the hammerhead ribozymes
Replication localization	Nucleus	Chloroplast
Structure	Rod-like structure including central conserved region (CCR)	Branched conformation including hammerhead ribozymes
Transmission	Sap, grafting, and seed	Sap, grafting
Host	Chrysanthemums, *Petunia hybrida*, tomato, *Gynura aurantiaca*, *Ageratum*, dahlia, *Senecio*, *Vinca major*, *Argyranthemum frutescens* and many plants belonging to the families *Solanaceae* and *Asteraceae*	Restricted to chrysanthemums

SYMPTOMS, ISOLATION, NUCLEOTIDE SEQUENCES, AND STRUCTURES

Chrysanthemum stunt disease was reported in the early 1950s [17]. It causes light green young leaves, stunting, small leaves and flowers, and reduced rooting ability. Later, infectious material was first isolated from leaves of stunted chrysanthemum, after nucleic acid extraction, centrifugation in sucrose density gradient and electrophoresis in polyacrylamide gels demonstrated the presence of a low molecular weight RNA, which was different from the known PSTVd [17]. This novel viroid, associated with stunt disease in chrysanthemums, was distinct from other known viruses, and it was named CSVd [18]. The complete nucleotide sequence and secondary structure of CSVd was first determined in Australia in 1981 [19]. The identified CSVd comprised 356 bases and displayed about 69% sequence identity to PSTVd. Subsequently, another CSVd with 354 bases was sequenced [20]. The structures and structural transitions of various viroids, including CSVd, have been determined with thermodynamic, kinetic, and hydrodynamic methods (Figure 1a) [21]. Recently, the sequences of three different CSVd isolates were determined from the US, China, and Australia. Both the US and Australian isolates were composed by several sequence variants, confirming the quasi-species nature of the viroid, but the Chinese isolate consisted of a single variant indicating a low molecular variability for CSVd [22]. Comparative analyses of the nucleotide sequences of 117 CSVd sequence variants revealed nucleotide variations at 103 sites scattered throughout the CSVd genome. However, it is not known whether these nucleotide changes are related to the species specificity of CSVd infections [23]; in addition, possible relationships between nucleotide changes and variant-specific pathogenicity have been recently questioned [23].

A

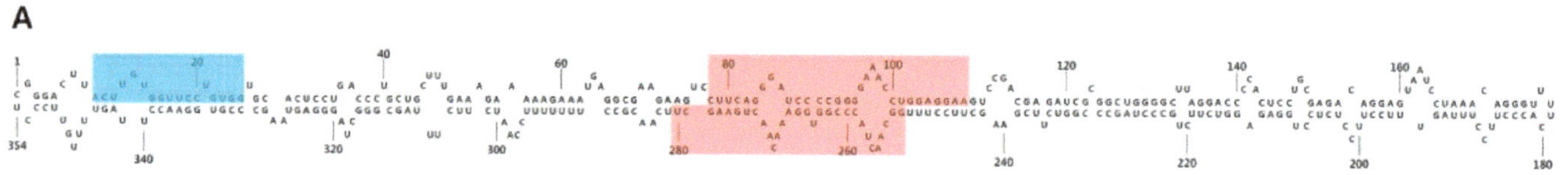

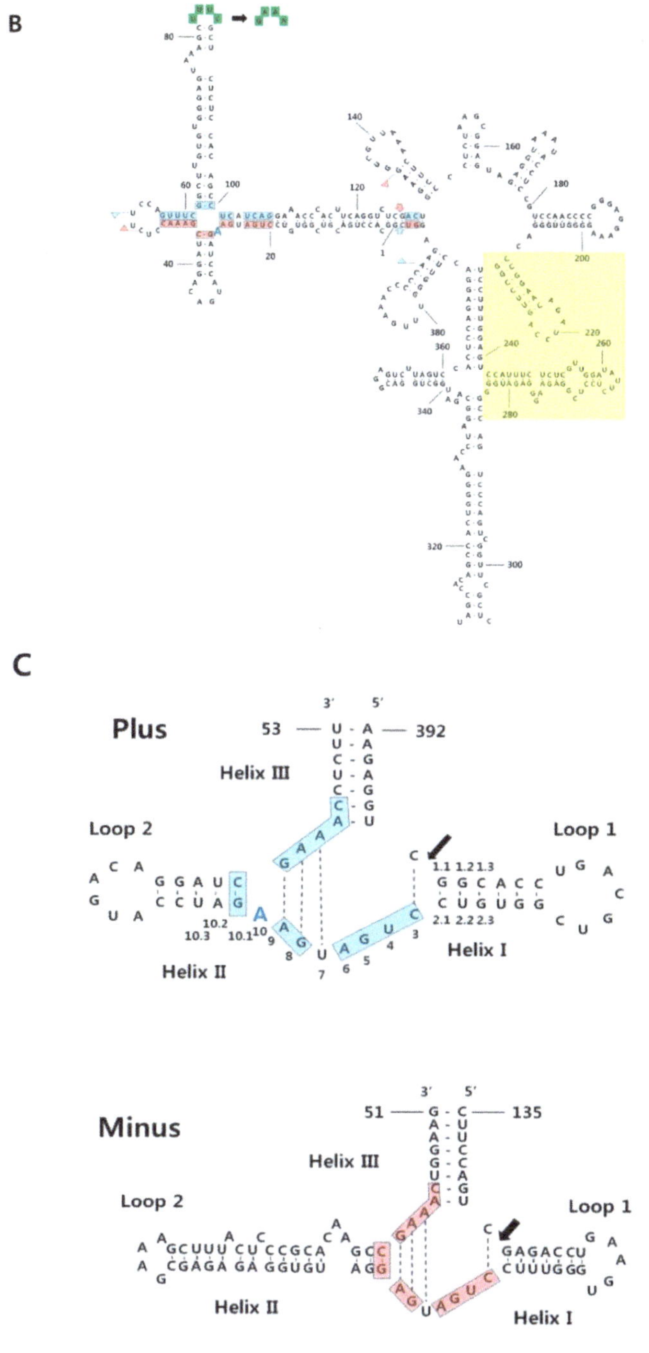

Figure 1: Predicted secondary structures of CSVd and CChMVd. (A) Predicted secondary structure of CSVd was adapted with permission from [22]. The central conserved region (CCR) is indicated by light-red shading and the terminal conserved region is indicated by ...

Chrysanthemum chlorotic mottle disease was first reported in 1967 in the cultivar "Yellow Delaware" in New York State [24]. This disease could be transmitted by grafting one chrysanthemum to another chrysanthemum cultivar. It causes yellow-green mottling, chlorosis, and dwarf symptoms, but some infected cultivars were asymptomatic [25]. Due to the low abundance of CChMVd in infected plants, compared to the abundance of CSVd, the complete, 398–401 nt sequence of CChMVd was only identified in 1997 (Figure 1b) [26]. Transcripts of CChMVd *in vitro* were also infectious and caused chlorotic mottle disease [26]. CChMVd can form hammerhead structures in both plus and minus strands, which self-cleave during *in vitro*and *in vivo* transcription (Figure 1c) [26]. The hammerhead structures of CChMVd display unique features, including an unpaired A residue after the conserved A9 residue in the plus self-cleaving domain, and an unusually long helix II in the minus one (Figure 1c). The predicted secondary structure of CChMVd is a very stable, branched conformation, which is similar to that of PLMVd. In contrast to viroids with a quasi-rod-like conformation, both CChMVd and PLMVd were insoluble in 2 M LiCl [26]. In one recent study, an *in vitro*transcript of CChMVd was inoculated to measure the mutation rate of CChMVd, which resulted in being the highest one ever reported in any biological entity [27].

HOST RANGE

Due to the limited number of studies on viroids, it was thought that CSVd only infected chrysanthemums [31]. Initial reports showed that chrysanthemum plants could serve as host for both CSVd and CChMVd. Moreover, circular and linear forms of CSVd, purified from infected chrysanthemum plants, were infectious when inoculated into *Gynura aurantiaca* plants [32]. Also, a linear CSVd RNA, synthesized by *in vitro*transcription, could infect chrysanthemum plants and other plants that belong to the families *Solanaceae* and*Asteraceae*, without causing any disease symptoms [33]. Under natural conditions, infections of CSVd have been reported in *Petunia hybrida* [34], *Ageratum*, *Dahlia* [35], and *Senecio* (cineraria) [36]. CSVd was also identified in *Vinca major* with reverse transcriptase (RT)-PCR [37]. An infection of CSVd, identified with RT-PCR, was also reported in asymptomatic *Argyranthemum frutescens* (marguerite daisy) [38,39]. In contrast to the quite wide range identified for CSVd, the known host range of CChMVd is very restricted. For example, of 51 species and cultivars tested, CChMVd was infectious in only two chrysanthemum species [40].

In general, CSVd and CChMVd are mechanically transmitted through tools, like knives and scissors, used on infected chrysanthemums and then on healthy material for grafting and flower cutting. A previous study also demonstrated seed-borne transmission of CSVd in chrysanthemums; in that case, the ambient temperature during a cross could influence the rate of CSVd transmission to progeny [41]. However, CSVd could not be transmitted from infested soil [42], and no insect vector has been reported to transmit it.

IDENTIFICATION OF THE TWO VIROIDS

Many large-scale screening studies have been performed to detect CSVd and CChMVd. For instance, a total of 2,480 chrysanthemum samples collected from six Australian states were tested for viroid and virus infections with enzyme-linked, immunosorbent assays (ELISAs) and cDNA probes [43]. As a result, CSVd has been found in mixed infection with other viruses [43].

CSVd is one of several quarantined pathogens in European countries, under order of the European Union's Plant Health Directive (2000/29/EC) [44]. The European Food Safety Authority (EFSA) has released a Scientific Opinion on the assessment of the risk of solanaceous pospiviroids for the EU territory and the identification and evaluation of risk management options [45]. In Italy, to produce healthy chrysanthemum plants, phytosanitary measures, like RT-PCR and hybridization assays, were carried out to test for CSVd infections in 39 seasonal and 93 year-round chrysanthemums

[1]. In Slovenia, 12 out of 200 chrysanthemum plants were found to be infected with CSVd [46]. Recently, a study reported the first identification of CSVd in Turkey; where two infected plants out of 154 chrysanthemums tested positive in an RT-PCR assay [47]. Moreover, CSVd was identified by RT-PCR in infected chrysanthemum plants in Egypt [48].

In Asian countries, CSVd and CChMVd have been reported in Japan, Korea, China, and India. Cultivated and wild chrysanthemums were tested for CSVd infections in Japan with RT-PCR [49]. Out of 89 samples, 80 samples were infected with CSVd. In addition, eight wild chrysanthemum species, which did not exhibit any CSVd symptoms, carried CSVd. Among 21 CSVd isolates, a total of five variants were revealed; these variants showed frequent mutations.

In Akita Prefecture, Japan, CChMVd infections in chrysanthemums caused distinct yellow leaf mottling and necrosis. These symptoms were caused by CChMVd variants containing a UUUC tetraloop (Figure 1b) [50], a structural domain already identified as a specific CChMVd pathogenic determinant [26]. A previous study had detected CChMVd infections in various chrysanthemums grown for three years in Akita Prefecture, revealing that 20% of chrysanthemums were infected with CChMVd [51]. In natural conditions, chrysanthemums that were only infected with CChMVd variants with the UUUC tetraloop did not show any noticeable disease symptoms [51]. In this case, light and temperature may interfere with symptoms induced by variants containing the symptomatic domain. Furthermore, in Kyoto, CChMVd was also detected in cut chrysanthemums from Japan and the Netherlands; this indicated that CChMVd infections have spread over a wide range in Japan [52].

In Korea, CSVd was first identified in chrysanthemum cv. Chunkwang in 2001. The complete sequences for two isolates (K1 and K2) were determined [53]. Later, a total of 64 commercial chrysanthemum cultivars in Korea were tested for CSVd infections, and the infection rate of each cultivar ranged from 9.7% to 66.8% [54]. Furthermore, the complete sequence was reported for the CChMVd-SSHA6 viroid from Korea; this variant caused yellow spots and growth reduction in some infected cultivars [54].

In China, CChMVd was first reported in 2008. In 2010, 13 samples with mild chlorotic spots were collected, and all were infected with CChMVd [55]. In India, chrysanthemum cultivars were screened for CSVd infections with RT-PCR and DNA-RNA hybridization: 70% of cultivars were infected with CSVd [36].

REPLICATION MECHANISMS

Viroids replicate *in vivo* via a rolling circle mechanism, which requires a specific cleavage procedure that converts multimeric viroid copies into monomeric forms [56,57]. Viroids, including CSVd, in the family*Pospiviroidae* replicate in the nucleus via an asymmetric rolling circle mechanism [58]. In contrast, CChMVd and members of the family *Avsunviroidae* replicate via a symmetric rolling circle mechanism in the chloroplast, mediated by the hammerhead ribozyme. The hammerhead ribozyme in the family *Avsunviroidae*determines self-cleavage of oligomeric RNAs during viroid replication; replication is further catalyzed by RNA polymerase and RNA ligase [59].

The hammerhead ribozyme of CChMVd has been intensively studied. The plus hammerhead ribozyme of CChMVd contains an additional A (A10) residue between the conserved A9 and the quasi-conserved G10.1 residues (Figure 1c) [28]. The additional A10 residue causes a moderate decrease in the trans-cleaving rate, while substitution of A10 to C and A10 to G causes major detrimental effects (Figure 1c). In contrast, the A10 to U substitution increased the trans-cleaving rate by 3–4-fold. Some hammerheads display deviations from the consensus sequence because certain residues might be involved in unknown important functions, other than self-cleavage [28]. In addition, high-pressure experiments identified two different types of CChMVd hammerhead conformation that exhibit fast-cleaving and slow cleaving activities [60].

Interactions between loop 1 and loop 2 of CChMVd hammerheads in both polarity strands (and in most natural hammerheads) have been shown to play major roles in self-clearing activity of these ribozymes [30]. Moreover, site-directed mutagenesis, bioassays, and progeny analysis studies showed that a kissing-loop interaction is important for *in vitro* folding and viroid infectivity (Figure 1b) [61]. For instance, the introduction of a single mutation in the kissing loops led to low or no infectivity, but an introduction of second mutation restoring the kissing-loop interaction resulted in an infectious agent [61]. These results indicated that the loop-loop tertiary interactions are important for viroid infectivity and for the folding and catalytic activity of most natural hammerheads [30].

SYMPTOM DETERMINANTS

Of the two viroids discussed here, CSVd is the more serious pathogen. It leads to reductions in the quality and quantity of chrysanthemum production. However, the sequence regions that confer the different symptoms are currently unknown. In contrast, the functional determinants for pathogenicity have been mapped in CChMVd. A previous study identified a nonpathogenic CChMVd strain that could protect against challenge inoculations with a pathogenic strain. Sequence analysis and site-directed mutagenesis have revealed that a substitution from UUUC to GAAA in the pathogenic CChMVd strain led to a nonpathogenic viroid with normal replication ability (Figure 1b). The identified pathogenicity determinant was located in a tetraloop of the predicted, branched conformation of CChMVd [62]. Later, the tetraloop of CChMVd, was further investigated with site-directed mutagenesis, bioassay, and analyses of the progenies [29]. The authors found that substitution of the tetraloop with a triloop or a pentaloop did not affect the infectivity of CChMVd. However, the thermodynamically stable GAAA tetraloop of the nonpathogenic CChMVd strain could not be replaced with other stable tetraloops of the UNCG family without causing functional changes. These data indicate that the sequence was the major factor that preserved the functional tetraloop motif, rather than the structure. After the introduction of a site-directed modified CChMVd variant into chrysanthemum, this nonpathogenic variant eventually evolved into a pathogenic CChMVd one, which acquired the UUUC tetraloop characteristics. This study provided evidence of the stronger biological fitness of the pathogenic CChMVd with the UUUC tetraloop, compared to that of the nonpathogenic CChMVd with the GAAA tetraloop.

DETECTION METHODS

Due to the fact that viroids have very low molecular weights, polyacrylamide gel electrophoresis (PAGE) was first used to detect viroid RNAs [63,64]. Molecular method to detect and quantify CSVd RNA in infected plants based on hybridization assays, where the viroid RNA hybridized to a ^{32}P-labeled cDNA of CSVd, were also developed [65,66]. The ^{32}P-labelled cDNA probe that targeted PSTVd was able to detect both PSTVd and CSVd [67]. In addition, several nonradioactive probes were developed, like photobiotin-labeled DNA probes, [68,69]. Both ^{32}P-labeled and biotin-labeled cDNA probes were used and compared for the detection of CSVd and PSTVd [69]. The most sensitive method for detecting viroids is probably the RT-PCR-based approach. A previous study developed a simple and rapid method for nucleic acid extraction, without tissue homogenization, for detecting CSVd [70]. Typically, two different methods have been applied for detecting CSVd, either a combination of RT-PCR and hybridization, or a combination of RT-PCR and ELISA [39,70].

To identify several viroids or a novel viroid in a single host, several RT-PCR primer sets or digoxigenin-labeled polyprobes have been developed based on consensus sequences identified in all pospiviroids [71,72]. Also, a one-tube/one-step RT-PCR method was developed to detect seven viroids, including CSVd, belonging to four genera [73]. In addition, a multiplex, direct RT-PCR method was developed to detect CSVd and CChMVd in a small amount of plant tissues [74,75]. Furthermore, a previous study developed real-time PCR (TaqMan®) assays for detecting six viroids, including CSVd, infecting solanaceous hosts [44]. Recently, a multiplex RT-PCR primer set was developed to detect viruses and viroids, including*Chrysanthemum virus B*, *Tomato aspermy virus*, CSVd, and CChMVd in chrysanthemums [76]. These assays might be useful for detecting different viroids infecting the same host simultaneously. These methods could be applied to quarantine, certification, and screening purposes.

EFFORTS TO ELIMINATE VIROIDS FROM INFECTED PLANTS

Several attempts have been made to generate viroid-free plants. First, a long-term heat treatment at 35 °C for 14–37 weeks was proposed for treating CSVd-infected plants [77]. In contrast, a cold treatment was shown to be effective for eliminating CSVd in infected chrysanthemum plants [78]. Those authors grew CSVd-infected chrysanthemum plants at 5 °C for six months; subsequently, the meristem tips of cold-treated chrysanthemums were used to generate non-infected chrysanthemum plants [78]. Another study found that promoting vegetative propagation during a low-temperature period could eliminate CSVd from the infected chrysanthemums [42]. However, it was also shown that, after reducing the level of CSVd RNA in the plant during a low-temperature period, CSVd replication could again increase when the infected plants were grown at normal or high temperatures [79].

Recently, a Japanese research group established a new approach for obtaining CSVd-free chrysanthemums with leaf primordium-free, shoot apical meristems (LP-free SAMs). The attachment of LP-free SAMs to the root tips of CSVd-free chrysanthemums or cabbage resulted in the production of 14% and 3% CSVd-free plants, respectively [80]. A previous study generated CSVd-free and CSVd-infected plants by culturing different-sized SAMs dissected from the "Piato" plant; the authors reported that CSVd-infected plants flowered, even in the long-day condition [81]. These data suggested that CSVd might induce autonomous flowering in chrysanthemum, which is known to be a qualitative, short-day flowering plant [81]. In the same way, this research group has regenerated CChMVd-free chrysanthemums [52]. To date, the elimination of viroids from infected plants has remained a challenge.

IDENTIFICATION AND DEVELOPMENT OF VIROID-RESISTANT CHRYSANTHEMUM CULTIVARS

Most cultivated chrysanthemums are known to be infected with CSVd, which even if asymptomatic, could transfer CSVd to other susceptible cultivars in fields. Controlling CSVd with chemical treatments and culturing approaches is currently a challenging endeavor. Therefore, the ideal would be to find chrysanthemum cultivars resistant to CSVd, for developing new CSVd-resistant chrysanthemum lines. In a recent study, 35 chrysanthemum lines, including commercial cultivars, wild species, and interspecific hybrids, were screened to identify resistant plants by exposing uninfected plants to CSVd-infected plants with a grafting system [82]. RT-PCR was routinely performed to check for CSVd infections in the upper leaves of the scion every month after grafting. Of the 35 lines, the "Okayamaheiwa" cultivar showed strong resistance against CSVd. After crossing the "Okayamaheiwa" cultivar with two other susceptible cultivars, "Sei-elza" and "Anri," 76 and 8 F1 progeny individuals were produced. Among the F1 progeny, 14 were not infected with CSVd, thus indicating that CSVd resistance was achieved in the first hybrid generation [82]. CSVd was not present in the SAMs or LPs of resistant chrysanthemum cultivars after grafting. Taking the absence of CSVd in the SAMs and LPs as a marker, 20 cultivars were identified as CSVd resistant out of 85 commercial cultivars tested; however, later grafting experiments showed that CSVd was detected in two of these candidate resistant cultivars [83].

Another previous study examined six cultivars, and identified a cultivar named "Utage," which exhibited a reduced level of CSVd replication. A total of 67 cultivars, obtained from the self-pollination of "Utage," were tested for CSVd resistance by RT-PCR. Of those, three plants (C7, A30, and A27) were found to have strong CSVd resistance [84].

Transgenic chrysanthemum lines were generated where a specific double-stranded RNA ribonuclease gene (*pac1*) of *Schizosaccharomyces pombe* was introduced with Agrobacterium-mediated transformation [85,86]. The generated transgenic plant, which expressed pac1, exhibited reduced levels of viroid infection and accumulation [85]. In addition, these plants were resistant to *Tomato spotted wilt virus* (TSWV). When hybrid lines were crossed with the transgenic line or with a wild chrysanthemum species, the progeny displayed either resistance or high susceptibility to TSWV infections. This study demonstrated the utility of*pac1* for achieving both viroid and virus resistance.

OTHER STUDIES ASSOCIATED WITH CSVD AND CCHMVD

Cross-protection can be defined as the interference in symptom expression generated by a previously inoculated viroid against subsequent infection by other strains of the same or a closely related viroid [87]. Five different viroids have been used for cross-protection studies in tomato and chrysanthemum plants [87]. Each viroid displayed different symptoms in chrysanthemum plants and, interestingly, three of them, CSVd, a mild PSTVd strain, and a severe PSTVd strain, could protect against *Citrus exocortis viroid* (CEVd); in contrast, CChMVd could not protect against the severe PSTV strain, CEVd, or CSVd, even though the challenge-inoculated viroid had replicated. Thus, viroid-infected plants showing attenuated symptoms might induce cross-protection against other viroids. Moreover it has been shown that non-symptomatic strain of CChMVd induces cross-protection against the symptomatic strain of the same viroid species [62].

Virus and viroid infections generate small interfering RNAs that interact with the RNA-induced silencing complex (RISC) of the host. A previous study demonstrated the presence of small RNAs (22 nt) derived from CSVd by northern blot analysis accumulating in similar amounts in different plants, regardless of differences in symptoms [88]. In addition, small interfering RNAs may be involved in the inhibition of enzyme activity; for example, the stunted phenotype of CSVd-infected plants might be associated with decreased levels of the plant growth hormone gibberelic acid (GA), due to disturbed GA 20-oxidase activity [89].

FUTURE DIRECTIONS IN HOST-VIROID INTERACTION STUDIES

Compared to studies on plant viruses, studies on viroids are very limited and mostly restricted to their detection and identification. In particular, host factors involved in viroid replication, pathogenicity, and movement have not been well characterized. For studying host–viroid interactions, chrysanthemum interactions with CSVd and CChMVd have several advantages. First, the two viroids are quite different from each other in replication mechanisms and structure. They belong to two different viroid families, but they do have a common host: chrysanthemums. Second, several protein-RNA interaction techniques will lead to the identification of chrysanthemum proteins that might function for both viroids, while others might be specific for either CSVd or CChMVd. In addition, next-generation sequencing-based approaches might provide information on the chrysanthemum transcriptome and how it is regulated by the two different viroids. Chrysanthemum plants are easily transformed, and creating transgenic plants is not difficult. However, there are several obstacles to using the chrysanthemum as a host, including the finding of a viroid-free, diploid chrysanthemum, because most chrysanthemums are infected with viroids and many are polyploid.

Currently, our research group is working to establish a model system for host-viroid interactions with transcriptomic approaches. We anticipate that host transcript RNA profiling studies will reveal several host candidate genes involved in chrysanthemum-viroid interactions. Characterization of host gene functions will provide important information in understanding viroid disease processes, which may facilitate the development of new methods for their control.

ACKNOWLEDGMENTS

This research was supported, in part, by grants from the National Research Foundation of Korea (NRF), a grant funded by the Korean government (MEST) (No. 2011-0010027); the Next-Generation BioGreen 21 Program (No. PJ00798402), through the Rural Development Administration; and the Vegetable Breeding Research Center (No. 710001-03), through the Agriculture Research Center program from the Ministry for Food, Agriculture, Forestry, and Fisheries. The authors would like to thank the two anonymous reviewers for their helpful comments.

REFERENCES AND NOTES

1. Tomassoli L., Faggioli F., Zaccaria A., Caccia R., Albani M., Barba M. Molecular diagnosis of*Chrysanthemum stunt viroid* for routine indexing. Phytopathol. Mediterr. 2004;43:285–288.
2. Verma N., Sharma A., Ram R., Hallan V., Zaidi A.A., Garg I.D. Detection, identification and incidence of*Chrysanthemum B carlavirus* in chrysanthemum in India. Crop Prot. 2003;22:425–429. doi: 10.1016/S0261-2194(02)00184-9.
3. Ding B. Viroids: Self-replicating, mobile, and fast-evolving noncoding regulatory RNAs. Wiley Interdiscip. Rev. RNA. 2010;1:362–375. doi: 10.1002/wrna.22.
4. Navarro B., Gisel A., Rodio M.E., Delgado S., Flores R., di Serio F. Viroids: How to infect a host and cause disease without encoding proteins. Biochimie. 2012;94:1474–1480. doi: 10.1016/j.biochi.2012.02.020.
5. Ding B., Wang Y. Viroids: Uniquely simple and tractable models to elucidate regulation of cell-to-cell trafficking of RNA. DNA Cell. Biol. 2009;28:51–56. doi: 10.1089/dna.2008.0811.
6. Góra-Sochacka A. Viroids: Unusual small pathogenic RNAs. Acta Biochim. Pol. 2004;51:587–608.
7. Owens R.A., Flores R., Di Serio F., Li S.F., Pallás V., Randles J.W., Sano T., Vidalakis G. Viroids. In: Andrew M.Q.K., Michael J.A., Eric B.C., Elliot J.L., editors. Virus Taxonomy: Ninth Report of the International Committee on Taxonomy of Viruses. Elsevier Academic Press; London, UK: 2012. pp. 1221–1234.
8. Keese P., Symons R.H. Domains in viroids: Evidence of intermolecular RNA rearrangements and their contribution to viroid evolution. Proc. Natl. Acad. Sci. USA. 1985;82:4582–4586. doi: 10.1073/pnas.82.14.4582.
9. McInnes J., Symons R.H. Comparative Structure of Viroids and Their Rapid Detection using Radioactive and Nonradioactive Nucleic Acid Probes. In: Maramorosch K., editor. Viroids and Satellites: Molecular Parasites at the Frontier of Life. CRC Press; Boca Raton, FL, USA: 1991. pp. 21–58.
10. Baumstark T., Schröder A.R., Riesner D. Viroid processing: Switch from cleavage to ligation is driven by a change from a tetraloop to a loop E conformation. EMBO J. 1997;16:599–610. doi: 10.1093/emboj/16.3.599.

11. Gas M.-E., Hernández C., Flores R., Daròs J.-A. Processing of nuclear viroids *in vivo*: An interplay between RNA conformations. PLoS Pathog. 2007;3:e182. doi: 10.1371/journal.ppat.0030182.
12. Gas M.-E., Molina-Serrano D., Hernández C., Flores R., Daròs J.-A. Monomeric linear RNA of citrus exocortis viroid resulting from processing *in vivo* has 5'-phosphomonoester and 3'-hydroxyl termini: Implications for the RNase and RNA ligase involved in replication. J. Virol. 2008;82:10321–10325. doi: 10.1128/JVI.01229-08.
13. Koltunow A.M., Rezaian M.A. Grapevine yellow speckle viroid: Structural features of a new viroid group. Nucleic Acids Res. 1988;16:849–864. doi: 10.1093/nar/16.3.849.
14. Puchta H., Ramm K., Sänger H.L. The molecular structure of hop latent viroid (HLV), a new viroid occurring worldwide in hops. Nucleic Acids Res. 1988;16:4197–4216.
15. Flores R., Di Serio F., Hernández C. Viroids: The Noncoding Genomes, Seminars in Virology, 1997. Elsevier; Waltham, MA, USA: 1997. pp. 65–73.
16. Ding B. The biology of viroid-host interactions. Annu. Rev. Phytopathol. 2009;47:105–131. doi: 10.1146/annurev-phyto-080508-081927.
17. Diener T., Lawson R. Chrysanthemum stunt: A viroid disease. Virology. 1973;51:94–101. doi: 10.1016/0042-6822(73)90369-3.
18. Diener T.O. Viroids: The smallest known agents of infectious disease. Annu. Rev. Microbiol. 1974;28:23–40.
19. Haseloff J., Symons R.H. *Chrysanthemum stunt viroid*: Primary sequence and secondary structure.Nucleic Acids Res. 1981;9:2741–2752.
20. Gross H.J., Krupp G., Domdey H., Raba M., Jank P., Lossow C., Alberty H., Ramm K., Sanger H.L. Nucleotide sequence and secondary structure of citrus exocortis and chrysanthemum stunt viroid. Eur. J. Biochem. 1982;121:249–257. doi: 10.1111/j.1432-1033.1982.tb05779.x.
21. Sieger G., Hofmann H., Förtsch J., Gross H.J., Randies J.W., Sänger H.L., Riesner D. Conformational transitions in viroids and virusoids: Comparison of results from energy minimization algorithm and from experimental data. J. Biomol. Struct. Dyn. 1984;2:543–571. doi: 10.1080/07391102.1984.10507591.
22. Yoon J.Y., Palukaitis P. Sequence comparisons of global *Chrysanthemum stunt viroid* variants: Multiple polymorphic positions scattered through the viroid genome. Virus Genes. 2012;46:97–104.
23. Yoon J., Baek E., Palukaitis P. Are there strains of *Chrysanthemum stunt viroid*? J. Plant Pathol.2012;94:697–701.
24. Dimock A.W., Geissinger C.M. A newly recognized disease of chrysanthemum caused by a graft-transmissible agent. Phytopathology. 1969;59:1024 (abstr.).
25. Dimock A.W., Geissinger C.M., Horst R.K. Chlorotic mottle: A newly recognized disease of chrysanthemum. Phytopathology. 1971;61:415–419. doi: 10.1094/Phyto-61-415.
26. Navarro B., Flores R. *Chrysanthemum chlorotic mottle viroid*: Unusual structural properties of a subgroup of self-cleaving viroids with hammerhead ribozymes. Proc. Natl. Acad. Sci. USA. 1997;94:11262–11267. doi: 10.1073/pnas.94.21.11262.
27. Gago S., Elena S.F., Flores R., Sanjuán R. Extremely high mutation rate of a hammerhead viroid.Science. 2009;323:1308–1308.
28. De la Peña M., Flores R. An extra nucleotide in the consensus catalytic core of a viroid hammerhead ribozyme: Implication for the design of more efficient ribozymes. J. Biol. Chem. 2001;276:34586–34593.
29. De la Pena M., Flores R. *Chrysanthemum chlorotic mottle viroid* RNA: Dissection of the pathogenicity determinant and comparative fitness of symptomatic and non-symptomatic variants. J. Mol. Biol.2002;321:411–421. doi: 10.1016/S0022-2836(02)00629-0.
30. Dufour D., De La Peña M., Gago S., Flores R., Gallego J. Structure-function analysis of the ribozymes of *Chrysanthemum chlorotic mottle viroid*: A loop-loop interaction motif conserved in most natural hammerheads. Nucleic Acids Res. 2009;37:368–381.
31. Runia W.T., Peters D. The response of plant species used in agriculture and horticulture to viroid infections. Eur. J. Plant. Pathol. 1980;86:135–146.
32. Palukaitis P., Symons R.H. Purification and characterization of the circular and linear forms of chrysanthemum stunt viroid. J. Gen. Virol. 1980;46:477–489.
33. Matsushita Y., Penmetcha K.K.R. *In vitro*-transcribed *Chrysanthemum stunt viroid* RNA is infectious to chrysanthemum and other plants. Phytopathology. 2009;99:58–66. doi: 10.1094/PHYTO-99-1-0058.
34. Verhoeven J.T.J., Arts M.S.J., Owens R.A., Roenhorst J.W. Natural infection of petunia by chrysanthemum stunt viroid. Eur. J. Plant. Pathol. 1998;104:383–386.
35. Nakashima A., Hosokawa M., Maeda S., Yazawa S. Natural infection of *Chrysanthemum stunt viroid* in dahlia plants. J. Gen. Plant. Pathol. 2007;73:225–227.
36. Singh D., Pathania M., Ram R., Zaidi A.A., Verma N. Screening of chrysanthemum cultivars for*Chrysanthemum stunt viroid* in an Indian scenario. Arch. Phytopathol. Plant. Protect. 2010;43:1517–1523. doi: 10.1080/03235400802583628.
37. Nie X., Singh R.P., Bostan H. Molecular cloning, secondary structure, and phylogeny of three pospiviroids from ornamental plants. Can. J. Plant Pathol. 2005;27:592–602. doi: 10.1080/07060660509507260.
38. Torchetti E.M., Navarro B., Trisciuzzi V.N., Nuccitelli L., Silletti M.R., Di Serio F. First report of*Chrysanthemum stunt viroid* in *Argyranthemum frutescens* in Italy. J. Plant. Pathol. 2012;94:451–454.
39. Menzel W., Maiss E. Detection of *Chrysanthemum stunt viroid* (CSVd) in cultivars of *Argyranthemum frutescens* by RT-PCR-ELISA. Z. Pflanzenkr. Pflanzenschutz. 2000;107:548–552.

40. Horst R.K. The Viroids, The Viruses series. Volume 11. Plenum Press; New York, NY, USA: 1987. Chrysanthemum chlorotic mottle; pp. 291–295.

41. Chung B.N., Pak H.S. Seed transmission of *Chrysanthemum stunt viroid* in chrysanthemum. Plant Pathol. J. 2008;24:31–35.

42. Sugiura H., Hanada K. *Chrysanthemum stunt viroid*, a disease of large-flowered chrysanthemum [Dendranthema] in Niigata prefecture [Japan] J. Japan. Soc. Hort. Sci. 1998;67:432–438. doi: 10.2503/jjshs.67.432.

43. Hill M.F., Giles R.J., Moran J.R., Hepworth G. The incidence of chrysanthemum stunt viroid, chrysanthemum B carlavirus, tomato aspermy cucumovirus and tomato spotted wilt tospovirus in Australian chrysanthemum crops. Australas. Plant Pathol. 1996;25:174–178.

44. Monger W., Tomlinson J., Booonham N., Marn M.V., Plesko I.M., Molinero-Demilly V., Tassus X., Meekes E., Toonen M., Papayiannis L. Development and inter-laboratory evaluation of real-time PCR assays for the detection of pospiviroids. J. Virol. Methods. 2010;169:207–210.

45. Baker R., Candresse T., Dormannsné Simon E., Gilioli G., Grégoire J.-C., Jeger M.J., Karajova O.E., Lövei G., Makowski D., Manceau C., et al. Scientific opinion on the assessment of the risk of solanaceous pospiviroids for the EU territory and the identification and evaluation of risk management options: Scientific Opinion; EFSA Panel on Plant Health (PLH) EFSA J. 2011;9:2330.

46. Mehle N., Seljak G., Verhoeven J.T.J., Jansen C.C.C., Prezelj N., Ravnikar M. *Chrysanthemum stunt viroid* newly reported in Slovenia. Plant Pathol. 2010;59:1159–1159.

47. Bostan H., Gazel M., Elibuyuk I.O. Occurrence of *Pospiviroid* in potato, tomato and some ornamental plants in Turkey. Afr. J. Biotechnol. 2012;9:2613–2617.

48. El-Dougdoug K.A., Rezk A.A., Rehab D.A., Sofy A.R. Partially nucleotide sequence and secondary structure of *Chrysanthemum stunt viroid* Egyptian isolate from infected-chrysanthemum plants. Int. J. Virol.2012;8:133–150. doi: 10.3923/ijv.2012.133.150.

49. Matsushitali Y. Nucleotide sequences and distribution of *Chrysanthemum stunt viroid* in Japan. J. Japan. Soc. Hort. Sci. 2007;76:333–337.

50. Yamamoto H., Sano T. Occurrence of *Chrysanthemum chlorotic mottle viroid* in Japan. J. Gen. Plant. Pathol. 2005;71:156–157. doi: 10.1007/s10327-004-0179-7.

51. Yamamoto H., Sano T. An epidemiological survey of *Chrysanthemum chlorotic mottle viroid* in Akita Prefecture as a model region in Japan. J. Gen. Plant. Pathol. 2006;72:387–390. doi: 10.1007/s10327-006-0309-5.

52. Hosokawa M., Matsushita Y., Ohishi K., Yazawa S. Elimination of *Chrysanthemum chlorotic mottle viroid* (CChMVd) recently detected in Japan by leaf-primordia free shoot apical meristem culture from infected cultivars. J. Japan. Soc. Hort. Sci. 2005;74:386–391. doi: 10.2503/jjshs.74.386.

53. Chung B.N., Choi G.S., Kim H.R., Kim J.S. *Chrysanthemum stunt viroid* in *Dendranthema grandiflorum*. Plant Pathol. J. 2001;17:194–200.

54. Chung B.N., Lim J.H., Choi S.Y., Kim J.S., Lee E.J. Occurrence of *Chrysanthemum stunt viroid* in chrysanthemum in Korea. Plant Pathol. J. 2005;21:377–382. doi: 10.5423/PPJ.2005.21.4.377.

55. Zhang Z.X., Pan S., Li S.F. First Report of *Chrysanthemum chlorotic mottle viroid* in Chrysanthemum in China. Plant. Dis. 2011;95:1320.

56. Flores R., Daròs J.-A., Hernández C. *Avsunviroidae* family: Viroids containing hammerhead ribozymes. Adv. Virus Res. 2000;55:271–323. doi: 10.1016/S0065-3527(00)55006-4.

57. Flores R., Hernández C., Alba A.E.M.d., Daròs J.-A., Serio F.D. Viroids and viroid-host interactions.Annu. Rev. Phytopathol. 2005;43:117–139. doi: 10.1146/annurev.phyto.43.040204.140243.

58. Branch A.D., Robertson H.D. A replication cycle for viroids and other small infectious RNA's. Science.1984;223:450–455.

59. Flores R., Navarro J.A., Pena M., Navarro B., Ambrós S., Vera A. Viroids with hammerhead ribozymes: Some unique structural and functional aspects with respect to other members of the group. Biol. Chem.1999;380:849–854.

60. Kaddour H., Vergne J., Hervé G., Maurel M.C. High-pressure analysis of a hammerhead ribozyme from*Chrysanthemum chlorotic mottle viroid* reveals two different populations of self-cleaving molecule. FEBS J.2011;278:3739–3747. doi: 10.1111/j.1742-4658.2011.08291.x.

61. Gago S., De la Peña M., Flores R. A kissing-loop interaction in a hammerhead viroid RNA critical for its*in vitro* folding and *in vivo* viability. RNA. 2005;11:1073–1083. doi: 10.1261/rna.2230605.

62. De la Peña M., Navarro B., Flores R. Mapping the molecular determinant of pathogenicity in a hammerhead viroid: A tetraloop within the *in vivo* branched RNA conformation. Proc. Natl. Acad. Sci. USA. 1999;96:9960–9965. doi: 10.1073/pnas.96.17.9960.

63. Mosch W.H.M., Huttinga H., Hakkaart F.A., De Bokx J.A. Detection of chrysanthemum stunt and potato spindle tuber viroids by polyacrylamide gelelectrophoresis. Eur. J. Plant Pathol. 1978;84:85–93.

64. Morris T., Smith E. Potato spindle tuber disease: Procedures for the detection of viroid RNA and certification of disease-free potato tubers. Phytopathology. 1977;67:145–150. doi: 10.1094/Phyto-67-145.

65. Palukaitis P., Symons R.H. Hybridization analysis of *Chrysanthemum stunt viroid* with complementary DNA and the quantitation of viroid RNA sequences in extracts of infected plants. Virology. 1979;98:238–245. doi: 10.1016/0042-6822(79)90541-5.

66. Candresse T., Macquaire G., Monsion M., Dunez J. Detection of *Chrysanthemum stunt viroid* (CSVd) using nick translated probes in a dot-blot hybridization assay.

J. Virol. Methods. 1988;20:185–193. doi: 10.1016/0166-0934(88)90122-X.

67. Macquaire G., Monsion M., Mouches C., Candresse T., Dunez J. Spot hybridization: Application to viroid identification. Ann. Inst. Pasteur. Virol. 1984;135:219–230.

68. McInnes J.L., Habili N., Symons R.H. Nonradioactive, photobiotin-labelled DNA probes for routine diagnosis of viroids in plant extracts. J. Virol. Methods. 1989;23:299–312. doi: 10.1016/0166-0934(89)90162-6.

69. Candresse T., Macquaire G., Brault V., Monsion M., Dunez J. ^{32}P-and biotin-labelled *in vitro* transcribed cRNA probes for the detection of *Potato spindle tuber viroid* and *Chrysanthemum stunt viroid*. Res. Virol.1990;141:97–107. doi: 10.1016/0923-2516(90)90059-R.

70. Nakahara K., Hataya T., Uyeda I. A simple, rapid method of nucleic acid extraction without tissue homogenization for detecting viroids by hybridization and RT-PCR. J. Virol. Methods. 1999;77:47–58. doi: 10.1016/S0166-0934(98)00135-9.

71. Bostan H., Nie X., Singh R.P. An RT-PCR primer pair for the detection of *Pospiviroid* and its application in surveying ornamental plants for viroids. J. Virol. Methods. 2004;116:189–193. doi: 10.1016/j.jviromet.2003.11.014.

72. Torchetti E.M., Navarro B., Di Serio F. A single polyprobe for detecting simultaneously eight pospiviroids infecting ornamentals and vegetables. J. Virol. Methods. 2012;186:141–146. doi: 10.1016/j.jviromet.2012.08.005.

73. Ragozzino E., Faggioli F., Barba M. Development of a one tube-one step RT-PCR protocol for the detection of seven viroids in four genera: *Apscaviroid*, *Hostuviroid*, *Pelamoviroid* and *Pospiviroi*. J. Virol. Methods. 2004;121:25–29. doi: 10.1016/j.jviromet.2004.05.012.

74. Hosokawa M., Shiba H., Kawabe T., Nakashima A., Yazawa S. A simple and simultaneous detection method for two different viroids infecting chrysanthemum by multiplex direct RT-PCR. J. Japan. Soc. Hort. Sci. 2007;76:60–65. doi: 10.2503/jjshs.76.60.

75. Hosokawa M., Matsushita Y., Uchida H., Yazawa S. Direct RT-PCR method for detecting two chrysanthemum viroids using minimal amounts of plant tissue. J. Virol. Methods. 2006;131:28–33. doi: 10.1016/j.jviromet.2005.07.001.

76. Song A., You Y., Chen F., Li P., Jiang J., Chen S. A multiplex RT-PCR for rapid and simultaneous detection of viruses and viroids in chrysanthemum. Lett. Appl. Microbiol. 2013;56:8–13. doi: 10.1111/lam.12007.

77. Hollings M., Stone O.M. Attempts to eliminate chrysanthemum stunt from chrysanthemum by meristem-tip culture after heat-treatment. Ann. Appl. Biol. 1970;65:311–315. doi: 10.1111/j.1744-7348.1970.tb04592.x.

78. Paduch-Cichal E., Kryczyński S. A low temperature therapy and meristem-tip culture for eliminating four viroids from infected plants. J. Phytopathol. 1987;118:341–346. doi: 10.1111/j.1439-0434.1987.tb00465.x.

79. Chung B.N., Huh E.J., Kim J.S. Effect of temperature on the concentration of *Chrysanthemum stunt viroid* in CSVd-infected chrysanthemum. Plant Pathol. J. 2006;22:152–154. doi: 10.5423/PPJ.2006.22.2.152.

80. Hosokawa M., Otake A., Ohishi K., Ueda E., Hayashi T., Yazawa S. Elimination of *Chrysanthemum stunt viroid* from an infected chrysanthemum cultivar by shoot regeneration from a leaf primordium-free shoot apical meristem dome attached to a root tip. Plant Cell Rep. 2004;22:859–863. doi: 10.1007/s00299-004-0770-6.

81. Hosokawa M., Ueda E., Ohishi K., Otake A., Yazawa S. *Chrysanthemum stunt viroid* disturbs the photoperiodic response for flowering of chrysanthemum plants. Planta. 2004;220:64–70. doi: 10.1007/s00425-004-1318-2.

82. Matsushita Y., Aoki K., Sumitomo K. Selection and inheritance of resistance to *Chrysanthemum stunt viroid*. Crop Prot. 2012;35:1–4. doi: 10.1016/j.cropro.2011.12.009.

83. Nabeshima T., Hosokawa M., Yano S., Ohishi K., Doi M. Screening of chrysanthemum cultivars with resistance to *Chrysanthemum stunt viroid*. J. Japan. Soc. Hort. Sci. 2012;81:285–294. doi: 10.2503/jjshs1.81.285.

84. Omori H., Hosokawa M., Shiba H., Shitsukawa N., Murai K., Yazawa S. Screening of chrysanthemum plants with strong resistance to chrysanthemum stunt viroid. J. Japan. Soc. Hort. Sci. 2009;78:350–355. doi: 10.2503/jjshs1.78.350.

85. Ogawa T., Toguri T., Kudoh H., Okamura M., Momma T., Yoshioka M., Kato K., Hagiwara Y., Sano T. Double-stranded RNA-specific ribonuclease confers tolerance against *Chrysanthemum stunt viroid* and*Tomato spotted wilt virus* in transgenic chrysanthemum plants. Breeding Sci. 2005;55:49–55. doi: 10.1270/jsbbs.55.49.

86. Toguri T., Ogawa T., Kakitani M., Tukahara M., Yoshioka M. *Agrobacterium*-mediated transformation of chrysanthemum (*Dendranthema grandiflora*) plants with a disease resistance gene (*pac1*) Plant Biotech.2003;20:121–127. doi: 10.5511/plantbiotechnology.20.121.

87. Niblett C.L., Dickson E., Fernow K.H., Horst R.K., Zaitlin M. Cross protection among four viroids.Virology. 1978;91:198–203. doi: 10.1016/0042-6822(78)90368-9.

88. Chung B.N., Choi G.S., Kim K.T. *Chrysanthemum stunt viroid* induces the accumulation of small RNAs associated with RNA silencing in infected chrysanthemum. Plant Pathol. J. 2006;22:235–238.

89. Huh E.J., Lee S.K., Chung B.N., Lee I.J., Choi S.Y. Changes of growth and gibberellin contents in chrysanthemum by infection of *Chrysanthemum stunt viroid*. Hort. Environ. Biotech. 2006;47:366–370.

CHAPTER 5

AGERATUM ENATION VIRUS—A BEGOMOVIRUS OF WEEDS WITH THE POTENTIAL TO INFECT CROPS

Muhammad Tahir[1], Imran Amin[2], Muhammad Saleem Haider[3], Shahid Mansoor[2], and Rob W. Briddon[2]

[1]Plant Biotechnology Department, Atta-ur-Rahman School of Applied Biosciences, National University of Sciences and Technology, Sector H-12, Islamabad 44000, Pakistan

[2]Agricultural Biotechnology Division, National Institute for Biotechnology and Genetic Engineering, Jhang Road, Faisalabad 38000, Pakistan

[3]School of Biological Sciences, University of the Punjab, New Campus, Lahore 54590, Pakistan

ABSTRACT

Samples of two *Ageratum conyzoides*, one *Sonchus oleraceus* and one turnip (*Brassica rapa* var. rapa) exhibiting virus-like symptoms were collected from Pakistan and Nepal. Full-length begomovirus clones were obtained from the four plant samples and betasatellite clones from three of these. The begomovirus sequences were shown to be isolates of *Ageratum enation virus* (AEV) with greater than 89.1% nucleotide sequence identity to the 26 AEV sequences available in the databases. The three betasatellite sequences were shown to be isolates of Ageratum yellow leaf curl betasatellite (AYLCB) with greater than 90% identity to the 18 AYLCB sequences available in the databases. The AEV sequences were shown to fall into two distinct strains, for which the names Nepal (consisting of isolates from Nepal, India, and Pakistan—including the isolates identified here) and India (isolates occurring only in India) strains are proposed. For the clones obtained from two AEV isolates, with their AYLCB, infectivity was shown by *Agrobacterium*-mediated inoculation to *Nicotiana benthamiana*, *N. tabacum*, *Solanum lycopersicon* and *A. conyzoides*. *N. benthamiana* plants infected with AEV alone or betasatellite alone showed no symptoms. *N. benthamiana*plants infected with AEV with its associated betasatellite showed leaf curl symptoms. The findings show that AEV is predominantly a virus of weeds that has the capacity to infect crops. AYLCB appears to be the common partner betasatellite of AEV and is associated with diseases with a range of very different symptoms in the same plant species. The inability to satisfy Koch's postulates with the cloned components of isolate SOL in *A. conyzoides* suggests that the etiology may be more complex than a single virus with a single betasatellite.

INTRODUCTION

Viruses of the family *Geminiviridae* are important pathogens of cultivated plants throughout the warmer parts of the World and are also increasingly spreading into more temperate areas. Mainly a problem in developing nations of the tropics, some of the viruses also cause significant problems in more developed countries. A good example is *Tomato yellow leaf curl virus* (TYLCV). TYLCV has its origins in the eastern Mediterranean/Middle East but has in the last 20 years spread across southern Europe, North Africa, the Caribbean, parts of the Americas, China, Japan and Australia [1]. Much of our knowledge of the diversity of geminiviruses comes from the study of viruses

Citation: Muhammad Tahir, Imran Amin Muhammad Saleem Haider Shahid Mansoor and Rob W. Briddon, Ageratum enation virus—A Begomovirus of Weeds with the Potential to Infect Crops, doi: 10.3390/v7020647.

found in cultivated plants [2]. Only fairly recently has the focus of investigation also included weed species. These efforts have shown that the diversity of geminiviruses is extremely wide and that weeds may harbour known, agriculturally important virus species, as well as previously unidentified virus species, which could potentially affect cultivated plants in the future [3,4].

Geminiviruses have genomes consisting of circular, single-stranded (ss) DNA that are encapsidated in characteristic twinned, quasi-icosahedral particles. Viruses in the family are assigned to one of seven genera based on their host range, genome arrangement and insect vector [5,6]. The majority of geminiviruses, as well as economically most important, are in the genus *Begomovirus*. Begomoviruses are transmitted exclusively by the whitefly *Bemisia tabaci* to dicotyledonous hosts. The majority of begomoviruses native to the New World have genomes consisting of two ssDNA components of approximately equal size (~2800 nt) that are designated DNA A and DNA B, both of which are required for systemic infection of plants. The DNA A component encodes all viral functions required for viral DNA replication, control of gene expression and transmission between plants whereas the DNA B component encodes two proteins required for inter- and intracellular movement in host plants [7].

Although some bipartite begomoviruses occur in the Old World, most are monopartite, their genomes being a homolog of the DNA A component of bipartite viruses. A small number of these viruses are truly monopartite, but the majority instead associate with a group of ssDNA satellites known collectively as betasatellites [8]. Betasatellites are typically half the size of their helper begomoviruses (~1350 nt) and are required by their helper viruses to infect the plants species from which they were isolated but require the helper virus for replication and spread in plants as well as transmission between plants [9]. All betasatellites identified thus far have a highly conserved structure consisting of a single open reading frame (known as βC1), a region of sequence rich in adenine (A-rich) and a sequence highly conserved between all isolates known as the satellite conserved region (SCR). The SCR contains a predicted hairpin structure with the loop containing the nonanucleotide sequence TAATATTAC with similarity to the origin of virion-strand DNA replication of geminiviruses [10].

The genomes of monopartite begomoviruses encode six genes. The two in the virion-sense encode the coat protein (CP), involved in virus movement within and between plants, and the V2 protein, which is involved in virus movement in plants as well as, for some species, in overcoming host defenses triggered by double-stranded RNA (known as RNA interference [RNAi]) [7,11]. The complementary-sense genes encode the transcriptional activator protein (TrAP; a transcription factor that, for some begomoviruses, up-regulates expression of the virion-sense genes, modulates host gene expression and may be involved in overcoming RNAi), the replication enhancer protein (REn; that is involved in providing a cellular environment suitable for virus replication [12]), the C4 protein (that may be involved in overcoming RNAi and may be a pathogenicity determinant [13,14]) and the replication associated protein (Rep).

Rep is a rolling circle replication (RCR) initiator protein that interferes with host cell-cycle and is the only virus encoded protein required for virus replication [15]). To initiate RCR Rep recognizes and binds short repeated sequences (known as "iterons"), adjacent to the TATA box of the complementary-sense promoter within the non-coding intergenic region (IR) of the virus genome [16,17]. The amino acid sequences of Rep predicted to interact with iterons are known as the "iteron related domain" (IRD) [18].

The analysis presented here concerns a virus, *Ageratum enation virus* (AEV), first identified in Nepal in the late 1990s. We show that AEV has two distinct strains, and to occur across a large geographic area. Although initially isolated from weeds, we show it is also now crossing into cultivated species and could become a more significant problem for agriculture in the future.

MATERIALS AND METHODS

DNA Extraction, PCR Amplification and Cloning

Total nucleic acids were extracted from leaf samples as previously described [19]. For isolates SOL and ACL primer pair WTGF/WTGR [20] were used to PCR amplify an approximately 1500 bp fragment spanning all of the intergenic region and most of the Rep gene (data not shown). A specific pair of abutting primers BGAF/BGAR [21] containing a unique restriction enzyme (*Apa*I) site, to PCR-amplify the complete begomovirus genome. The full-length begomovirus from isolate ABF was PCR-amplified with primer pair BF/BR [22]. The full-length genome of the begomovirus from isolate ACN was PCR-amplified using abutting primers AAGCTTTGATGAGTTCCGCTG/AAGCTTCTCAAGCAGAGAATGGCG containing a unique *Hind*III restriction site designed to a partial sequence obtained with the universal begomovirus primers described previously [23]. Betasatellites were amplified with universal primers beta01/beta02 [24]. Potentially full-length amplification products for were cloned using either the InsT/A clone PCR product cloning kit (Fermentas) or the pGEM-T Easy kit (Promega), as recommended by the manufacturers.

Sequencing and Sequence Analysis

The complete nucleotide sequences of potentially full-length clones were determined by dideoxynucleotide chain-termination sequencing using Genome Lab Dye Terminator Cycle Sequencing kits (Beckman Coulter, Nyon, Switzerland) with reaction products run on a Beckman Coulter automated sequencer (CEQ 8000) or commercially (Macrogen [Seoul, Korea] or Lark Technologies [Takeley, UK]). Sequence information was stored, assembled and analyzed using the Lasergene sequence analysis package (DNAStar Inc., Madison, WI, USA).

Phylogenetic analyses were conducted on matrices of aligned sequences using the neighbor-joining and bootstrap options of Phylip (v. 3.5c) running on an IBM compatible personal computer. Sequence alignments were produced using CLUSTAL X [25]. Phylogenetic dendrograms were viewed, manipulated and printed using Treeview [26]. All sequence alignments to assess the taxonomic status of sequences (viruses/betasatellites) used the Clustal V algorithm (as implemented in MegAlign, Lasergene package), as recommended by the *Geminiviridae* Study Group of the International Committee on Taxonomy of Viruses [27].

Constructs for Infectivity

Partial direct repeat constructs of the begomoviruses and betasatellites cloned from isolates SOL and ACL were produce for the analysis of infectivity. For the begomovirus clones an *Apa*I-*Xba*I fragment of ~1.6 kb, containing the entire intergenic region, was released from the full-length clone and ligated into the binary vector pGreen0029 [28]. The full-length begomovirus insert, released with *Apa*I, was then ligated into the*Apa*I site of the pGreen vector containing the 1.6 kb fragment to yield a 1.6-mer partial tandem repeat construct. A similar strategy, but using restriction endonucleases *Kpn*I and *Cla*I, was followed to produce partial repeat constructs of the betasatellites.

Agrobacterium-Mediated Inoculation

Plants for inoculation were grown in a glasshouse with a 16 h light/8 h dark cycle at 28 °C. The binary vector constructs were electroporated into *Agrobacterium tumefaciens* strain GV3101. *Agrobacterium*cultures for inoculation were grown at 28 °C for 48 h to an optical density at 600 nm of 0.6 with antibiotic selection. The bacterial cells were pelleted (5000 g for 15 min at 4 °C) and resuspended in 10 mM $MgCl_2$and 150 µg/mL of acetosyringone. Cells were incubated on ice for 3 h and then inoculated into plants as previously described [29].

RESULTS

Cloning, Sequencing and Sequence Analysis of Begomoviruses

Leaf samples of field infected weeds *Ageratum conyzoides* (isolate ACL) and *Sonchus oleraceous* (SOL), showing yellow vein symptoms (Figure 1, panels A and B), were collected from Lahore, Pakistan, in 2006 and 2005, respectively. A leaf sample of turnip (*Brassica rapa* var. rapa), exhibiting foliar vein yellowing, was collected from Faisalabad, Pakistan during 2006 (isolate ABF). Additionally a leaf sample originating from an *A. conyzoides* plant, collected in Nepal during the 1990s (isolate ACN), showing a novel symptom was included. This plant appeared superficially healthy (Figure 1, panel C). However, closer inspection of leaves under transmitted light showed vein darkening (Figure 1, panel D) and, on the lower leaf surface, showed raised veins and structures on veins resembling enations (Figure 1, panel E) which on the upper leaf surface were depressions (Figure 1, panel F)—we shall refer to this novel symptom as "dimples".

Figure 1: Symptoms exhibited by the plants from which AEV clones were obtained. *Sonchus oleraceous* with yellow veins (A) *Ageratum conyzoides* with yellow veins (B), *A. conyzoides*

from Nepal (C). A leaf from the *A. conyzoides* plant from Nepal is photographed under

The complete nucleotide sequences of single begomovirus clones from each of the four isolates were determined. The sequences were determined to be 2750 bp, 2749 bp, 2747 bp and 2746 bp, respectively, in length and are available in the databases under accession numbers given in Table 1. Analyses of the sequences shows them, in all respects, to be typical of the genomes of monopartite and DNA A components of bipartite begomoviruses encoding two genes (for the CP and V2 protein) in the virion-sense and four genes (for the Rep, TrAP, REn and C4 protein) in the complementary-sense. The positions and coding capacity of the predicted genes are given in Table 1. Between the virion- and complementary-sense genes lies a non-coding intergenic region which contains a predicted hairpin structure with, within the loop, the conserved (between all geminiviruses) nonanucleotide sequence (TAATATTAC) which forms part of the origin or virion-strand DNA replication.

Table 1: Origins of the virus isolates and features of the begomovirus and betasatellite clones obtained

Isolate	Origin (year)	Plant Species (Symptoms *)	Component	Accession No.	Size (bp)	ORF	Start/Stop Codon (Nucleotide Coordinates)	Predicted Size (no. of Amino Acids)	Predicted Molecular Weight (kDa)
SOL	Lahore, Pakistan (2005)	*Sonchus oleraceus* (YV)	begomovirus	AM261836	2750	V2	135/482	115	13.3
						CP	295/1065	256	29.7
						Rep	2599/1514	361	40.7
						TrAP	1611/1180	143	16.2
						REn	1466/1062	134	15.9
						C4	2442/2185	85	9.3
			betasatellite	AM412239	1368	βC1	605/189	138	16.1
ACL	Lahore, Pakistan (2006)	*Ageratum conyzoides* (YV)	begomovirus	AM698011	2749	V2	134/481	115	13.3
						CP	294/1064	256	29.6
						Rep	2598/1513	361	40.8
						TrAP	1610/1206	134	15.0
						REn	1465/1061	134	15.9
						C4	2441/2184	85	9.4
			betasatellite	AM698010	1355	βC1	596/180	138	16.1
ACN	Nepal (2001)	*Ageratum conyzoides* (LC, EN)	begomovirus	AJ437618	2746	V2	134/481	115	13.3
						CP	294/1064	256	29.5
						Rep	2597/1512	361	40.6
						TrAP	1610/1206	134	15.1
						REn	1465/1061	134	15.9
						C4	2441/2184	85	9.2
			betasatellite	-	-	-	-	-	-
ABF	Faisalabad, Pakistan (2006)	*Brassica rapa* var. rapa (YV)	begomovirus	AM701770	2747	V2	134/481	115	13.3
						CP	294/1064	256	29.6
						Rep	2598/1513	361	40.8
						TrAP	1610/1206	134	15.2
						REn	1465/1061	134	15.8
						C4	2441/2184	85	9.4
			betasatellite	AM701771	1359	βC1	596/180	138	16.1

Despite being isolated from widely different regions (Nepal and Pakistan) and with a 10 year gap, the four sequences show high levels of sequence conservation (between 95.2%–98.7% nucleotide sequence identity). Based on the 89% nucleotide sequence identity demarcation threshold for identification of begomovirus species, this indicates that all four clones are isolates of a single species [27,30].

Comparison of the Sequences Obtained Here with Sequences Available in the Databases

The AEV clone obtained from Nepal was obtained some time ago and was at the time recognized as a distinct species within the genus *Begomovirus* and given the name *Ageratum enation virus* [30]. The sequence of AEV-Nepal thus represents the type isolate of this begomovirus species. The other three sequences, with greater than 89% nucleotide sequence identity to AEV-Nepal, thus are variants of AEV. A Blast search of the GenBank nucleotide sequence database with the four AEV sequences presented here identified 26 additional sequences originating from India with high sequence identity (>89%): GQ268327isolated from *Trichosanthes dioica* (unpublished), EU867513 isolated from *Amaranthus cruentus*(unpublished), KC795968 and KC818421 isolated from tomato (unpublished), JX436472 isolated from tomato (unpublished), HE861940 isolated from soybean (unpublished), JQ911765 and HM149260 isolated from *Papaver somniferum* (unpublished), JX436473 isolated from Fenugreek (unpublished), JQ911767,KJ488990 and KJ488991 isolated from *Ageratum* sp. (unpublished), JF728865 and JF728867 (isolated from carrot [31]), JF728860-JF728864 and JF728866 isolated from *A. conyzoides* [31], JF682242 isolated from*Amaranthus* (unpublished), FN794201 isolated from *Crassocephalum crepidioides* [32], FN543099 isolated from *Zinnia elegans* [33], FJ177031 isolated from *Cleome gynandra* [34] and FN794198 isolated from *A. conyzoides* [32]. Some of these sequences have previously been reported as isolates of AEV. Overall these 30 sequences (including the 4 identified here) show between 88.9% and 100% nucleotide sequence identity (Table S1), with the highest levels of identity to isolates of *Tobacco curly shoot virus* (TbCSV). This confirms the 30 sequences as isolates of a distinct begomovirus species, which has previously been named AEV [30].

A closer analysis of the AEV sequences shows them to fall into three groups. The first group consists of all the AEV sequences characterized here, as well as GQ268327, EU867513, KC818421, KC795968, andFJ177031. These nine sequences show between 93.9% and 100% nucleotide sequence identity (Table S1). The sequences in the second group, consisting of FN543099, FN794198, FN794201, JF728860-64,JF728866, JX436472, JX436473, JQ911765, JF682242, JQ911767, HM149260, KJ488991, KJ488990 andHE861940 show between 94.6% and 100% identity. Between the two groups the identity levels vary between 90.6%. and 92.9%. Based on the presently applicable demarcation threshold for strains within a species (93%; [27]), this would indicate that the two groups represent strains of AEV, for which we propose the names "Nepal" and "India", respectively. The third group of sequences (JF728865 and JF728867) show relatively high sequence identities to both the Nepal (group 1—91.8% to 98.1% identity) and the India (group 2—94.3% to 99.2% identity) strain AEV sequences, making it difficult to assign them to a strain.

Interestingly the four AEV sequences have Rep IRD sequences predicted to be either FKIN (AM261836,AM698011, AM701770) or LKIN (AJ437618). The other three Nepal strain isolates also have LKIN. In contrast, all India strain isolates have an IRD with the predicted sequence FQIY. The corresponding iteron sequences are GGT/AGT for all Nepal strain isolates and either GGTG/AC/A, or possibly GTACT, for all India strain isolates and both problematic isolates. The alignment of all AEV sequences suggested that the differences between the Nepal and India strain isolates may be due to recombination across the origin of replication, possibly with *Papaya leaf crumple virus* (PaLCrV) as the donor of the ori for the India strain (results not shown). PaLCrV occurs in India and has IRD and iteron sequences identical to those of India strain AEV sequences.

A phylogenetic dendrogram based upon an alignment of the four sequences obtained here, the 26 other AEV sequences available in the databases and selected other begomovirus sequences available in the databases is shown in Figure 2A. This shows the 30 AEV sequences to form a group distinct from all other sequences in the tree but to be most closely related to TbCSV. The division of AEV sequences into two distinct groups, for which we have proposed the names Nepal and India strains, are well supported by bootstrapping. The two problematic AEV isolates (JF728865 and JF728867) group with, but are distinct from and basal to the Nepal strain isolates, despite having overall more sequence identity to isolates of the India strain. JF728867has a Nepal strain IRD (FKIN), whereas JF728865 has an Indian strain IRD (FQIY). It is for this reason that we proposed, at least provisionally, to include these two isolates under the Nepal strain, awaiting the availability of further sequences and analyses.

A

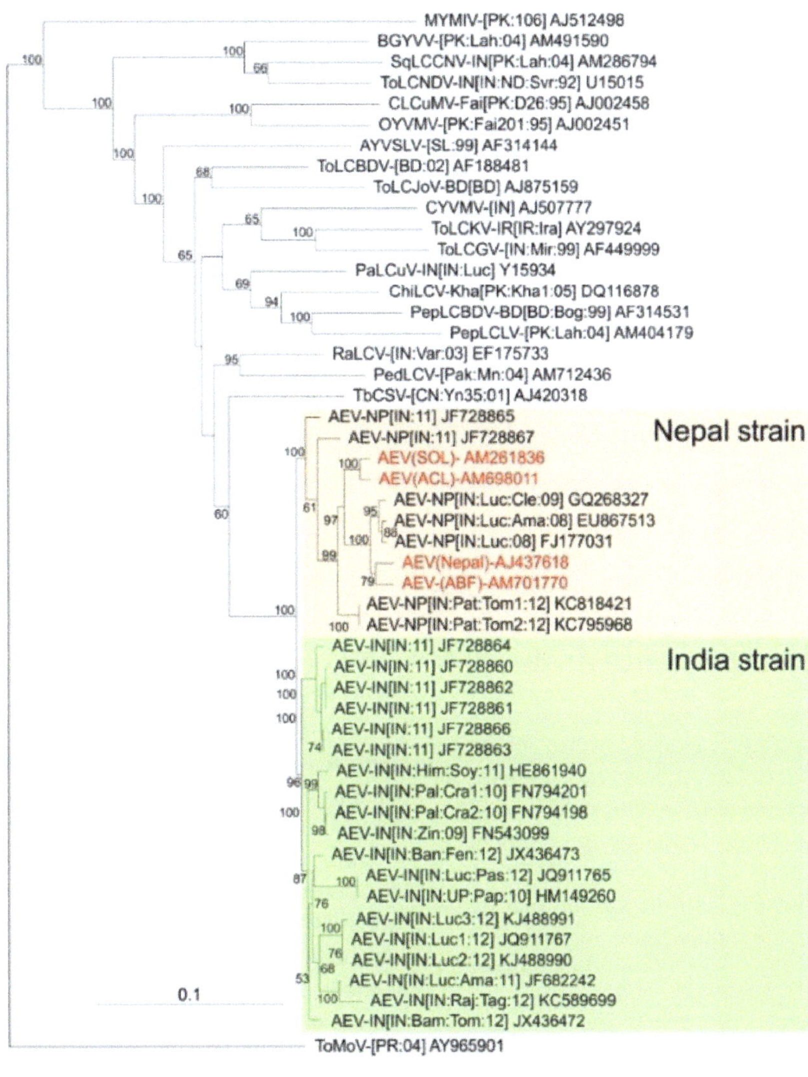

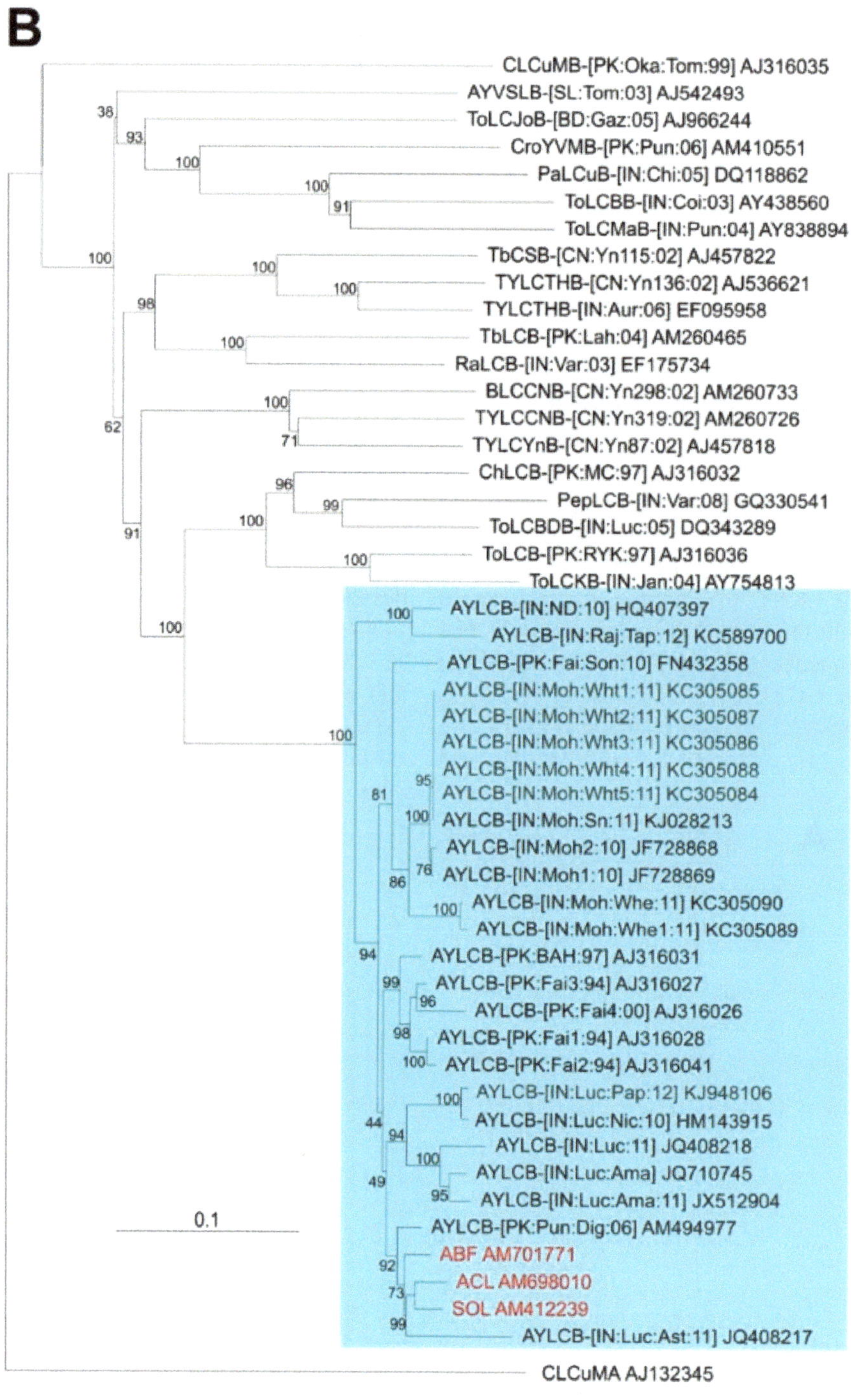

Figure 2: Phylogenetic analyses of begomovirus and betasatellite sequences. Phylogenetic dendrograms based upon alignments of the complete nucleotide sequences of the genomes (or DNA A genomic components of) begomoviruses (A) and betasatellites (B) identified here

Cloning, Sequencing and Sequence Analysis of Betasatellites

Three full-length betasatellite clones were obtained. Unfortunately the satellite for isolate ACN was not obtained for comparison, although the presence of a betasatellite with this isolate has been shown previously [35]. The three sequences have a length typical of betasatellites (between 1355 to 1368 bp, Table 1), being approximately half the size of their helper begomoviruses. Analysis of the sequences show them to have an arrangement typical of this class of satellites consisting of a single open reading frame in the complementary-sense encoding the βC1 protein (Table 1), a region of sequence rich in adenine and a sequence highly conserved between all betasatellites—the satellite conserved region [35].

Comparison of the three betasatellite sequences obtained here to sequences available in the databases shows them to be most similar to Ageratum yellow leaf curl betasatellite (AYLCB; with between 90.6% and 99.8% nucleotide sequence identity to the 18 other sequences available in the databases). Six defective AYLCB are available in the databases (JQ408217, KC305086-90) ranging in size from 891 bp to 1270 bp, which are included in the analysis but not included in the calculated identity values. To all other betasatellite sequences available in the databases the percentage nucleotide sequence identity was less than 62% (not shown) with the highest sequence identity levels (57.5%–61.3%) to an isolate of Tomato leaf curl Bangladesh betasatellite (AJ542489).

A phylogenetic dendrogram, based upon an alignment of the full-length sequences of the 3 betasatellites obtained here, all available AYLCB sequences in the databases and selected other betasatellite sequences from the databases is shown in Figure 2B. This shows AYLCB to form a clade with Chili leaf curl betasatellite, Pepper leaf curl betasatellite, Tomato leaf curl Bangladesh betasatellite, Tomato leaf curl betasatellite and Tomato leaf curl Karnataka betasatellite—a grouping well supported by bootstrapping. It is interesting to note that all AYLCB isolates from Pakistan, with the exception of FN432358, are distinct from those originating from India. This suggests that AYLCB in these two areas are evolving independently. Possibly FN432358, an "Indian" isolate of AYLCB occurring in Pakistan, indicates that infrequent exchanges between India and Pakistan take place. Also, the Pakistan AYLCB isolates characterized prior to 2000 are distinct from those characterized more recently, which may indicate temporal changes that have occurred in this betasatellite species.

Analysis of the Infectivity of AEV and Betasatellite Clones

Partial repeat constructs of the AEV and betasatellite clones isolated from *S. oleraceus* (isolate SOL) and *A. conyzoides* (isolate ACL) were produced for *Agrobacterium*-mediated inoculation of plants. The results of the infectivity studies are summarized in Table 2. Inoculation of *N. benthamiana* with the AEV and AYLCB clones isolated from *S. oleraceus* (AEV^{SOL} and AYLCB^{SOL}) led to the appearance of symptoms consisting of vein thickening and dimples (structures that, on the underside of the leaf, resembled enations but on the upper leaf surface were depressions; Figure 3, panels D and E) on veins at 12 days post inoculation (dpi) of leaves developing subsequent to inoculation. At approximately 18 to 20 dpi the dimples were more pronounced and were associated with severe leaf curling, followed by foliar chlorosis at 30 dpi with the veins being raised on the lower surface and depressed on the upper surface (Figure 3, panels F and G). Older plants were stunted, in comparison to healthy *N. benthamiana* plants, with leaves showing a distinct downward curling and chlorosis (panel H).

Table 2: Infectivity of AEV, isolates SOL and ACL, with their AYLCB by*Agrobacterium*-mediated inoculation

Inoculum	Plant Species	Infectivity (Plants Infected/Plants Inoculated)	Latent Period # (days)	PCR Detection		Symptoms *
				Virus	Betasatellite	
AEV^{SOL} and AYLCB^{SOL}	*N. benthamiana*	60/60	12–18	+	+	D, LC, C
	N. tabacum	20/20	12–18	+	+	VY, D
	S. lycopersicon	20/20	15–21	+	+	C, N
	A. conyzoides	2/2	15–18	+	+	CR, LC
	S. oleraceus	0/5	-	-	-	-
AEV^{SOL}	*N. benthamiana*	22/30	12–18	+	-	-
AYLCB^{SOL}	*N. benthamiana*	00/30	-	-	-	-
pGreen0029	*N. benthamiana*	00/20	-	-	-	-
AEV^{ACL} and AYLCB^{ACL}	*N. benthamiana*	10/10	12–18	+	+	R, LC, C
	S. lycopersicon	10/10	15–21	+	+	SG, C, R, N
	A. conyzoides	2/2	15–18	+	+	CR, LC
	S. oleraceus	0/5	-	-	-	-
none	*N. benthamiana*	00/20	-	-	-	-
pGreen0029	*N. benthamiana*	00/20	-	-	-	-

Days between inoculation and the first appearance of symptoms. * Symptoms exhibited by inoculated plants. Symptoms are denoted as vein yellowing (VY), leaf curl (LC), leaf crumpling (CR), stunted growth (SG), dimples on upper leaf surface (D), chlorosis (C), necrosis (N) reduced leaf area (R).

Figure 3: Symptoms of plants infected with AEV and AYLCB clones obtained from isolate SOL. *Nicotiana tabacum* plants infected with AEV^{SOL}/AYLCB^{SOL} initially (15 dpi) showed mild vein yellowing on leaves developing subsequent to inoculation (A) which developed into ...

For *N. tabacum*, inoculation with AEV^{SOL} and AYLCB^{SOL} initially resulted in mild vein yellowing (at approx. 15 dpi; Figure 3, panel A) followed by sunken veins (on the upper leaf surface; appearing as swollen vein on the lower surface)) and dimple structures, similar to those seen in *N. benthamiana*, at 20 dpi (Figure 3, panels B and C). However, *N. tabacum* did not exhibit the leaf curling and chlorosis evident in *N. benthamiana*.

A. conyzoides plants infected with AEV^{SOL}/AYLCB^{SOL} initially showed mild leaf crumpling at 12 dpi. At approx. 18 dpi the leaves developing subsequent to inoculation showed crumpling and some leaf curling (Figure 3, panel I).

For infected *S. lycopersicon* plants inoculated with AEV^{SOL} and AYLCB^{SOL} the leaves developing subsequent to inoculation initially showed vein yellowing at 15 dpi. By 21 dpi leaves were narrow, rolled at the edges and showed a patchy necrosis developing from the veins (Figure 3, panel J).

Inoculation of *N. benthamiana* with the clones obtained from isolate ACL (AEV^{ACL} and AYLCB^{ACL}) led to the first symptoms appearing within 12 dpi, consisting of narrow leaves with a pointed apex with some foliar chlorosis for leaves developing subsequent to inoculation. As plants developed, the leaves remained small and chlorotic with some twisting (Figure 4, panels B and C). Inoculated *A. conyzoides* plants similarly showed the first symptoms within 12 dpi consisting of crumpling in leaves developing after inoculation. Subsequently plants remained severely stunted, with severe crumpling and deformation of leaves (Figure 4, panel A). For tomato (*S. lycopersicon*) plants inoculated with AEV^{ACL} and AYLCB^{ACL} the symptoms were not very distinct with older plants showing stunting, reduced leaf size and a mild foliar chlorosis (Figure 4, panel D). On some older leaves a mild veinal necrosis developed, but this was not as pronounced as for tomato plants inoculated with AEV^{SOL}/AYLCB^{SOL}.

Figure 4: Symptoms of plants infected with AEV and AYLCB clones obtained from isolate ACL. *Ageratum conyzoides* plants infected with AEV^{ACL}/AYLCB^{ACL} with severely distorted leaves but no evidence of vein yellowing (A). The leaves of *N. benthamiana* plants developing...

The inoculations of AEV^{SOL} alone, AYLCB^{SOL} alone, and pGreen vector alone, in *N. benthamina* did not show any apparent symptom (Table 2). None of the symptoms was observed in healthy control *N. benthamiana* plants. The *N. benthamiana* plants, inoculated with AEV^{SOL} and AYLCB^{SOL}, produced dimples, leaf curl and chlorosis symptoms. A fragment of the expected size was obtained from *N. benthamiana* plants inoculated with AEV^{SOL} alone in a PCR. The PCR product was sequenced and showed 100% nucleotide sequence identity to the inoculated AEV. The results showed that AEV^{SOL} alone is infectious but does not induce symptoms.

All attempts to infect *S. oleraceus* with AEV and AYLCB were unsuccessful. This was likely due to the plants exuding a white latex following inoculation by the pin-prick method used here, meaning that the*Agrobacterium* inoculum was likely not retained by the plant long enough for T-DNA transfer and establishment of an infection.

DISCUSSION

The begomovirus species AEV was first established in 2003, when the presently applicable species demarcation criteria for the genus *Begomovirus* were established, based upon the sequence of the Nepal isolate (AJ437618), which is the type isolate of the species [30]. Since this time AEV has been shown to be widespread across northern India and to infect a number of distinct hosts [31,32,33,34,35].

The results obtained here, as well as the results of Kumar *et al.* [31] and Marwal *et al.* [36], suggest that the common partner of AEV is the betasatellite species AYLCB. For none of the other AEV sequences available in the databases has the associated betasatellite been identified. Of the other AYLCB isolates available in the databases, only for two have the helper begomovirus been identified –*Tobacco curly shoot virus* isolated from wild sunflower originating from India (for HQ407397) and *Alternanthera yellow vein virus* (AlYVV) in *Sonchus arvensis* originating from Pakistan (for FN432358) [37]. It is interesting to note that the AYLCB from *S. arvensis* was identified in co-infection with a second betasatellite, Cotton leaf curl Multan betasatellite (CLCuMB). The identification of AYLCB with several distinct begomoviruses suggests that, as has been shown for other betasatellites [9,10,35,38], this has the capacity to be trans-replicated and maintained by more than one begomovirus.

The results presented here demonstrated the infectivity of two of the AEV clones and their respective AYLCB clones. Both pairs of clones were highly infectious to a range of plants and, for AEV^{ACL}/AYLCB^{ACL}, infectivity was shown to the plant species from which the clones were isolated (*A. conyzoides*). However, the symptoms (crumpling and curling of leaves, with no evidence of yellowing) induced by the cloned virus/betasatellite in *A. conyzoides* did not resemble the symptoms exhibited by the plant from which the clones were obtained (vein yellowing). The reason for this is unclear. Possibly the plant contained additional viruses/betasatellites that were responsible for the yellow vein symptoms, although there was no evidence for this. Nevertheless, Koch's postulates for this isolate inducing yellow vein disease of*Ageratum* have not been satisfied. Possibly the *Ageratum* plant from which AYLCB^{ACL} was isolated contained more than one "type" of AYLCB. Certainly this betasatellite species is associated with, and thus likely can induce, a range of distinct symptoms. Kumar *et al.* [31] showed that an AEV isolate from carrot, with an AYLCB from *A. conyzoides*, induced typical yellow vein symptoms in *A. conyzoides.* AYLCB clones have also previously been shown to be infectious in the presence of *Ageratum yellow vein virus*, a virus that is not reported to be associated with this species of betasatellite, inducing typical yellow vein symptoms in *A. conyzoides* [35].

The "dimple" symptom, exhibited by the *A. conyzoides* plant from which AEV^{ACN} was isolated and induced by AEV^{SOL}/AYLCB^{SOL} infection of *N. benthamiana* and *N. tabacum* are unusual. This symptom, as far as we are aware, has not been previously reported for any begomovirus infection of plants and was not noted by Kumar *et al.* [31] for their inoculations of AEV and AYLCB. On the underside of the leaf these structures look much like the enations on veins typical of cotton leaf curl disease due to hypo- and hyperplasia of spongy parenchyma induced by the βC1 protein of CLCuMB [39]. However, whereas for CLCuMB there are no changes on the upper leaf surface, here there is a depression (dimple) of the vein, indicating that distinct anatomical changes are occurring. Microscopic analysis will be required to determine what histological changes are occurring. Amin *et al.* [40] have shown that expression of the AC4 (a homolog of the C4 proteins of monopartite begomoviruses) protein of *Cabbage leaf curl virus* (a bipartite begomovirus originating from the New World which is only distantly related to AEV) from a *Potato virus X* vector induced a very similar dimple structure on leaves of *N. benthamiana.* For AEV however, infection of *N. benthamiana* in the absence of the betasatellite did not induce symptoms, suggesting that in this case the unusual symptom is due betasatellite rather than the virus.

It is evident, both from the results presented here and from the work of others, that AEV has a large natural host range that includes species from the families *Brassicaceae* (turnip), *Asteraceae* (*A. conyzoides*, *Z. elegans*, *C. crepidioides*, *Tagetes patula* and *S. oleraceous*), *Amaranthaceae* (*A. cruentus*), *Cucurbitaceae* (*T. dioica*), *Apiaceae* (carrot) and *Cleomaceae* (*C. gynandra*). This indicates that AEV is predominantly a weed infecting begomovirus, although it apparently has the capacity to occasionally infect crop species and is reported to be causing losses of the minor grain crop *A. cruentus* [34] and carrot [31].

The availability of infectious AEV and AYLCB clones that induce distinct symptoms in common hosts opens up the possibility of investigating the molecular basis for these differences by, for example, exchanging betasatellite components. It will be particularly interesting to discover the basis for the unusual symptoms (dimples), which have not been reported for any other begomovirus. This will be the focus of our future studies.

ACKNOWLEDGEMENTS

RWB is supported by the Higher Education Commission (HEC), Government of Pakistan, under the "Foreign Faculty Hiring Program". Funding for the work came, in part, from a grant under the Pak-US Linkage Scheme. The authors are grateful for the support of NIBGE and the School of Biological Sciences, University of the Punjab, in conducting this study. R.W.B. is grateful to P.G. Markham for providing isolate ACN.

AUTHOR CONTRIBUTIONS

M.T. performed the molecular genetic studies, phylogenetic studies and infectivity. I.A. conducted genetic studies of isolate ABF (acc. no. AM701770), M.S.H, S.M. and R.W.B supervised and designed the study, and M.T. and R.W.B wrote the manuscript. All authors read and approved the final manuscript.

REFERENCES

1. Lefeuvre P., Martin D.P., Harkins G., Lemey P., Gray A.J.A., Meredith S., Lakay F., Monjane A., Lett J.M., Varsani A., et al. The spread of Tomato yellow leaf curl virus from the Middle East to the world. PLoS Pathog. 2010;6:e1001164. doi: 10.1371/journal.ppat.1001164.
2. Wren J.D., Roossinck M.J., Nelson R.S., Scheets K., Palmer M.W., Melcher U. Plant virus biodiversity and ecology. PLoS Biol. 2006;4:e80. doi: 10.1371/journal.pbio.0040080.
3. Mubin M., Akhtar S., Amin I., Briddon R.W., Mansoor S. *Xanthiumstrumarium*: A weed host of components of begomovirus-betasatellite complexes affecting crops. Virus Genes. 2012;44:112–119. doi: 10.1007/s11262-011-0662-0.

4. Ilyas M., Qazi J., Mansoor S., Briddon R.W. Molecular characterisation and infectivity of a "Legumovirus" (genus Begomovirus: Family Geminiviridae) infecting the leguminous weed Rhynchosia minima in Pakistan. Virus Res. 2009;145:279–284. doi: 10.1016/j.virusres.2009.07.018.

5. Brown J.K., Fauquet C.M., Briddon R.W., Zerbini M., Moriones E., Navas-Castillo J. In: Virus Taxonomy—Ninth Report of the International Committee on Taxonomy of Viruses. King A.M.Q., Lefkowitz E., Adams M.J., Carstens E.B., editors. Associated Press; London, UK: 2012. pp. 351–373.

6. Varsani A., Navas-Castillo J., Moriones E., Hernández-Zepeda C., Idris A., Brown J., Murilo Zerbini F., Martin D. Establishment of three new genera in the family *Geminiviridae*: *Becurtovirus, Eragrovirus and Turncurtovirus*. Arch. Virol. 2014;159:2193–2203. doi: 10.1007/s00705-014-2050-2.

7. Rojas M.R., Hagen C., Lucas W.J., Gilbertson R.L. Exploiting chinks in the plant's armor: Evolution and emergence of geminiviruses. Ann. Rev. Phytopathol. 2005;43:361–394. doi: 10.1146/annurev.phyto.43.040204.135939.

8. Briddon R.W., Brown J.K., Moriones E., Stanley J., Zerbini M., Zhou X., Fauquet C.M. Recommendations for the classification and nomenclature of the DNA-_ satellites of begomoviruses. Arch Virol. 2008;153:763–781. doi: 10.1007/s00705-007-0013-6.

9. Briddon R.W., Stanley J. Sub-viral agents associated with plant-infecting single-stranded DNA viruses.Virology. 2006;344:198–210. doi: 10.1016/j.virol.2005.09.042.

10. Saunders K., Briddon R.W., Stanley J. Replication promiscuity of DNA-β satellites associated with monopartite begomoviruses; deletion mutagenesis of the Ageratum yellow vein virus DNA- β satellite localises sequences involved in replication. J. Gen. Virol. 2008;89:3165–3172. doi: 10.1099/vir.0.2008/003848-0.

11. Sharma P., Ikegami M. RNA-silencing suppressors of geminiviruses. J. Gen. Plant Pathol. 2008;74:189–202. doi: 10.1007/s10327-008-0085-5.

12. Settlage S.B., See R.G., Hanley-Bowdoin L. Geminivirus C3 protein: Replication enhancement and protein interactions. J. Virol. 2005;79:9885–9895. doi: 10.1128/JVI.79.15.9885-9895.2005.

13. Vanitharani R., Chellappan P., Pita J.S., Fauquet C.M. Differential roles of AC2 and AC4 of cassava geminiviruses in mediating synergism and suppression of posttranscriptional gene silencing. J. Virol.2004;78:9487–9498. doi: 10.1128/JVI.78.17.9487-9498.2004.

14. Krake L.R., Rezaian M.A., Dry I.B. Expression of the tomato leaf curl geminivirus C4 gene produces viruslike symptoms in transgenic plants. Mol. Plant-Microbe Interact. 1998;11:413–417. doi: 10.1094/MPMI.1998.11.5.413.

15. Hanley-Bowdoin L., Settlage S.B., Robertson D. Reprogramming plant gene expression: A prerequisite to geminivirus DNA replication. Mol. Plant Pathol. 2004;5:149–156. doi: 10.1111/j.1364-3703.2004.00214.x.

16. Orozco B.M., Kong L.J., Batts L.A., Elledge S., Hanley-Bowdoin L. The multifunctional character of a geminivirus replication protein is reflected by its complex oligomerization properties. J. Biol. Chem.2000;275:6114–6122. doi: 10.1074/jbc.275.9.6114.

17. Argüello-Astorga G.R., Guevara-González L.R., Herrera-Estrella L.R., Rivera-Bustamante R.F. Geminivirus replication origins have a group-specific organization of iterative elements: A model for replication. Virology. 1994;203:90–100. doi: 10.1006/viro.1994.1458.

18. Argüello-Astorga G.R., Ruiz-Medrano R. An iteron-related domain is associated to motif 1 in the replication proteins of geminiviruses: Identification of potential interacting amino acid-base pairs by a comparative approach. Arch. Virol. 2001;146:1465–1485. doi: 10.1007/s007050170072.

19. Doyle J.J., Doyle J.L. Isolation of plant DNA from fresh tissue. Focus. 1990;12:13–15.

20. Mansoor S., Khan S.H., Hussain M., Zafar Y., Pinner M.S., Briddon R.W., Stanley J., Markham P.G. Association of a begomovirus and nanovirus-like molecule with *Ageratum* yellow vein disease in Pakistan.Plant Dis. 2000;84:101. doi: 10.1094/PDIS.2000.84.1.101A.

21. Tahir M., Haider M.S., Briddon R.W. Complete nucleotide sequences of a distinct bipartite begomovirus, Bitter gourd yellow vein virus, infecting *Momordica charantia*. Arch. Virol. 2010;155:1901–1905. doi: 10.1007/s00705-010-0819-5.

22. Shahid M.S., Mansoor S., Briddon R.W. Complete nucleotide sequences of cotton leaf curl Rajasthan virus and its associated DNA β molecule infecting tomato. Arch. Virol. 2007;152:2131–2134. doi: 10.1007/s00705-007-1043-9.

23. Briddon R.W., Markham P.G. Universal primers for the PCR amplification of dicot-infecting geminiviruses. Mol. Biotechnol. 1994;1:202–205. doi: 10.1007/BF02921559.

24. Briddon R.W., Bull S.E., Mansoor S., Amin I., Markham P.G. Universal primers for the PCR-mediated amplification of DNA β; a molecule associated with some monopartite begomoviruses. Mol. Biotechnol.2002;20:315–318. doi: 10.1385/MB:20:3:315.

25. Larkin M.A., Blackshields G., Brown N.P., Chenna R., McGettigan P.A., McWilliam H., Valentin F., Wallace I.M., Wilm A., Lopez R., et al. Clustal W and Clustal X version 2.0. Bioinformatics. 2007;23:2947–2948. doi: 10.1093/bioinformatics/btm404.

26. Page R.D.M. TREEVIEW: An application to display phylogenetic trees on personal computers. Comput. Appl. Biosci. 1996;12:357–358. [PubMed]

27. Fauquet C.M., Briddon R.W., Brown J.K., Moriones E., Stanley J., Zerbini M., Zhou X. Geminivirus strain demarcation and nomenclature. Arch. Virol. 2008;153:783–821. doi: 10.1007/s00705-008-0037-6.

28. Hellens R.P., Edwards E.A., Leyland N.R., Bean S.,

Mullineaux P.M. pGreen: A versatile and flexible binary Ti vector for *Agrobacterium*-mediated plant transformation. Plant Mol. Biol. 2000;42:819–832. doi: 10.1023/A:1006496308160.

29. Briddon R.W., Watts J., Markham P.G., Stanley J. The coat protein of beet curly top virus is essential for infectivity. Virology. 1989;172:628–633. doi: 10.1016/0042-6822(89)90205-5.
30. Fauquet C.M., Bisaro D.M., Briddon R.W., Brown J.K., Harrison B.D., Rybicki E.P., Stenger D.C., Stanley J. Revision of taxonomic criteria for species demarcation in the family *Geminiviridae*, and an updated list of begomovirus species. Arch. Virol. 2003;148:405–421. doi: 10.1007/s00705-002-0957-5.
31. Kumar J., Gunapati S., Singh S.P., Gadre R., Sharma N.C., Tuli R. Molecular characterization and pathogenicity of a carrot (*Daucus carota*) infecting begomovirus and associated betasatellite from India.Virus Res. 2013;178:478–485. doi: 10.1016/j.virusres.2013.10.010.
32. Kumar Y., Hallan V., Zaidi A.A. Chilli leaf curl Palampur virus is a distinct begomovirus species associated with a betasatellite. Plant Pathol. 2011;60:1040–1047. doi: 10.1111/j.1365-3059.2011.02475.x.
33. Kumar Y., Bhardwaj P., Hallan V., Zaidi A. Detection and characterization of *Ageratum enation virus*and a nanovirus-like satellite DNA1 from zinnia causing leaf curl symptoms in India. J. Gen. Plant Pathol.2010;76:395–398. doi: 10.1007/s10327-010-0270-1.
34. Raj S., Snehi S., Khan M., Tiwari A., Rao G. Detection of *Ageratum enation virus* from cat's whiskers (*Cleome gynandra* L.) with leaf curl symptoms in India. J. Gen. Plant. Pathol. 2010;76:292–294. doi: 10.1007/s10327-010-0243-4.
35. Briddon R.W., Bull S.E., Amin I., Idris A.M., Mansoor S., Bedford I.D., Dhawan P., Rishi N., Siwatch S.S., Abdel-Salam A.M., et al. Diversity of DNA β: A satellite molecule associated with some monopartite begomoviruses. Virology. 2003;312:106–121. doi: 10.1016/S0042-6822(03)00200-9.
36. Marwal A., Sahu A., Choudhary D., Gaur R.K. Complete nucleotide sequence of a begomovirus associated with satellites molecules infecting a new host *Tagetes patula* in India. Virus Genes. 2013;47:194–198. doi: 10.1007/s11262-013-0914-2.
37. Mubin M., Shahid M.S., Tahir M.N., Briddon R.W., Mansoor S. Characterization of begomovirus components from a weed suggests that begomoviruses may associate with multiple distinct DNA satellites. Virus Genes. 2010;40:452–457. doi: 10.1007/s11262-010-0470-y.
38. Nawaz-ul-Rehman M.S., Mansoor S., Briddon R.W., Fauquet C.M. Maintenance of an Old World betasatellite by a New World helper begomovirus and possible rapid adaptation of the betasatellite. J. Virol.2009;83:9347–9355. doi: 10.1128/JVI.00795-09.
39. Qazi J., Amin I., Mansoor S., Iqbal J., Briddon R.W. Contribution of the satellite encoded gene βC1 to cotton leaf curl disease symptoms. Virus Res. 2007;128:135–139. doi: 10.1016/j.virusres.2007.04.002.
40. Amin I., Patil B.L., Briddon R.W., Mansoor S., Fauquet C.M. A common set of developmental miRNAs are up regulated in *Nicotiana benthamiana* by diverse begomoviruses. Virol. J. 2011;8:e143. doi: 10.1186/1743-422X-8-143.

CHAPTER 6

BIOLOGICAL INVASIONS OF GEMINIVIRUSES: CASE STUDY OF TYLCV AND BEMISIA TABACIIN REUNION ISLAND

Frédéric Péréfarres[1, 2], Magali Thierry[1, 2], Nathalie Becker[3], Pierre Lefeuvre[1], Bernard Reynaud[1], Hélène Delatte[1], and Jean-Michel Lett[1]

[1]CIRAD, UMR PVBMT, Pôle de Protection des Plantes, 97410 Saint-Pierre, Ile de La Réunion, France

[2]Université de La Réunion, UMR PVBMT, Pôle de Protection des Plantes, 97410 Saint-Pierre, Ile de La Réunion, France

[3]Muséum National d'Histoire Naturelle, Département Systématique et Evolution, USM 601, CNRS UMR 5202 Origine, Structure et Evolution de la Biodiversité, 57 rue Cuvier, CP 50, 75005 Paris, France

ABSTRACT

In the last 20 years, molecular ecology approaches have proven to be extremely useful to identify and assess factors associated with viral emerging diseases, particularly in economically and socially important tropical crops such as maize (maize streak disease) and cassava (cassava mosaic disease). Molecular ecology approaches were applied in Reunion Island to analyze the epidemic of tomato yellow leaf curl disease, which has been affecting the island since the end of the 1990s. Before the invasive biotype B (currently known as Middle East-Asia Minor 1 cryptic species) of *Bemisia tabaci* spread across the world, Reunion Island (South West Indian Ocean) only hosted an indigenous biotype of *B. tabaci*, Ms (currently known as Indian Ocean cryptic species). Wild hybrids between invasive and indigenous species were subsequently characterized over multiple generations. Endosymbiont analysis of the hybrid population indicated that matings were non-random. Similarly, while no indigenous begomoviruses have ever been reported on Reunion Island, the two main strains of one of the most damaging and emerging plant viruses in the world, the Mild and Israel strains of the *Tomato yellow leaf curl virus* (TYLCV-Mld and TYLCV-IL), were introduced in 1997 and 2004 respectively. While these introductions extensively modified the agricultural landscape of Reunion Island, they also provided an invaluable opportunity to study the ecological and genetic mechanisms involved in biological invasion and competition.

INTRODUCTION

Biological invasions are a major agent of global change. They are often linked to emerging diseases [1] and often have negative effects on biodiversity and on the economy [2]. This is particularly true in insular ecosystems where small isolated populations allow rapid evolution and where the negative effects of invasion are more pronounced [3,4]. This makes the study of species introduced in insular ecosystems particularly appealing when it comes to dealing with agricultural issues. Humans have deliberately or accidentally introduced thousands of non-indigenous animal and plant species on such islands [2]. To become invasive in a new area, a species must successfully complete a series of fundamental steps: introduction, establishment, increase in number and geographical spread [5].

Citation: Frédéric Péréfarres, Magali Thierry, Nathalie Becker, Pierre Lefeuvre, Bernard Reynaud, Hélène Delatte, and Jean-Michel Lett1, Biological Invasions of Geminiviruses: Case Study of TYLCV and Bemisia tabaci in Reunion Island, doi: 10.3390/v4123665.

In this review, we focus on biological invasions in agricultural settings and on the worldwide emergence of the whitefly *Bemisia tabaci* and begomoviruses, most notably *Tomato yellow leaf curl virus* (TYLCV). Over the past 30 years, whiteflies and begomoviruses have become serious threats to the cultivation of a variety of vegetable crops of great importance in different parts of the world but especially in the tropics and sub-tropics. After a brief review of historically recent examples of successful biological invasions of geminiviruses in agricultural settings and their respective drivers, we trace the worldwide dissemination of invasive biotypes of *B. tabaci* and TYLCV. We also describe the fundamental role of endosymbionts in their whitefly host ecology and evolution through a wide range of effects, and their likely effect on virus acquisition and long-term retention. Finally, we describe what took place in Reunion Island, whose invasion by *B. tabaci* and TYLCV strains profoundly modified the agricultural landscape.

DISCUSSION

Emergence of Geminivirus Diseases in Agricultural Settings

Biological invasions are historically tightly linked with the spread of plants that human have domesticated for food, fibers, and medicinal and ornamental purposes [6,7]. Major migrations of modern man started over 8000 years ago and led to the propagation of cultivated plants and livestock. This process was so intensive that today only a small number cultivated plants (maize, rice, wheat) and animal species (domestic chickens, cattle) are found throughout the world and provide more than 98% of the world's food supply [8].

By moving cultivated plants away from their centers of domestication to distant regions or other continents, human populations have been responsible for new encounters between plants and pest species such as viruses, fungi, bacteria, phytoplasmas and nematodes [9,10]. While pest epidemics are generally considered to be rare in indigenous plant communities, modern agricultural practices, such as monocultures, favor frequent and damaging pest epidemics [11]. In addition, by decreasing the magnitude of the evolutionary response necessary for a pest to adapt to the conditions in new and remote territories, the worldwide homogenization of the agricultural landscape has improved the success of invasion. Among emerging plant diseases, viruses are the main pests and account for about the half of emerging diseases reviewed in Anderson *et al.* [12]. These epidemics may cause serious crop losses which have dramatic social and economic consequences when they affect staple crops (e.g., maize, rice, wheat and potato) or cash crops (e.g., citrus, cacao, coffee and banana) [1,12]. While viral emergences are generally multi-factorial and are still poorly understood, it is nevertheless possible to detect general trends driving plant viral emergences. The introduction of plant viruses to new regions is the most important driver of plant viral emergences followed by changes in the vector population, the effect of recombination and/or mutation, the weather, and finally changes in farming practices [1,12]. Because of the wide variety of factors involved, interdisciplinary approaches are needed to study the emergence of plant viruses. In this context, molecular ecology (a recent branch of biological sciences which combines biology, ecology, epidemiology, molecular evolution and genetics) has emerged and is now widely used to analyze the complex patterns of virus emergences. Among the economically and socially most important viral emerging plant diseases transmitted by arthropod vectors, maize streak disease (MSD) and cassava mosaic disease (CMD) have helped highlight the wide range of factors involved in a successful viral emergence at a continental scale. These two diseases will be used in this review to illustrate viral emerging diseases and their respective drivers.

Maize Streak Disease in Africa

Maize streak disease is a major constraint on maize production in all temperate and tropical regions in sub-Saharan Africa [13]. It has a complex, poorly understood epidemiology, with periodic outbreaks that devastate the maize yields of small- and medium-scale farmers [14]. Maize streak virus strain A (MSV-A, genus *Mastrevirus*, family *Geminiviridae*) is the causal agent of maize streak disease in Africa and surrounding islands. Whereas the majority of MSV strains (classified as MSV-A through -K) are apparently adapted to infecting wild grass species, only MSV-A isolates are adapted to infecting maize [14]. The contribution of recombination to the genetic diversification of MSV has, like for other geminiviruses, been widely demonstrated [15,16]. A recombination event between ancestral MSV-B and MSV-G/F variants is credited with having generated a maize-adapted MSV-A prototype, possibly within 20 years of the first credible reports of MSD in southern Africa in the 1870s [17]. Several observations suggest that this recombination event was the key to the emergence of MSV-A as a viral pathogen of maize. After its initial emergence and subsequent spread throughout the continent, five distinct MSV-A subtypes (MSV-A1, -A2, -A3, -A4, and -A6) evolved, and today, each has a different well-defined geographical range in sub-Saharan Africa [18]. Although it is evident that over the last century, the movement of certain MSV-A subtypes across Africa has been far less constrained than the movements of either other MSV strains or related African streak virus species, it is also evident that the different MSV-A subtypes are not all equally mobile. For example, whereas the MSV-A1 subtype is found throughout Africa, the MSV-A2, -A3, -A4, and -A6 subtypes have been found only in West Africa, East Africa, southern Africa, and Reunion Island, respectively. Using model-based phylogeographic analyses of hundreds of fully sequenced MSV-A isolates, a plausible history of MSV-A movements over the past 150 years was reconstructed [19]. Besides confirming southern Africa as the probable origin of MSV-A, an average rate of movement over the past century of 32.5 km per year was also inferred [19]. More than just the result of the evolutionary dynamics of the MSV-A strain, the epidemiology of MSD in Africa and its erratic nature is certainly the result of the complex interactions between multiple environmental and ecological factors [13]. The epidemic spread of MSD is usually attributed to the convergence of factors such as: (1) climatic conditions (temperature, rainfall, relative humidity), which increase the density of leafhopper populations, and environmental factors, which drive the long distance movement of leafhoppers; (2) the presence within leafhopper populations of a high percentage of

MSV transmitters; (3) the population density of both MSV-A and leafhopper populations in reservoirs of wild grasses; (4) cropping practices such as the staggered growing seasons of maize (for reviews see [13,14]). The complex interplay of all these factors mean MSD epidemics can cause insignificant to dramatic crop losses, which are extremely difficult to predict.

The African Cassava Mosaic Disease Pandemic

Cassava is a major staple food crop in tropical developing countries and is becoming a significant cash crop option. Cassava mosaic disease is the most important biotic constraint to production and is caused by a complex of African begomovirus species (family*Geminiviridae*) [20,21]. After its first description in Tanzania in 1894, CMD was reported in most of the cassava-growing regions in East and West Africa between the 1920s and the 1930s [22]. Plants affected by CMD are characterized not only by yellowing and curling of the leaves but also a serious reduction in yield. In the early 1990s, a severe epidemic began in Uganda and spread throughout East and Central Africa causing dramatic crop losses which required international intervention to prevent widespread famine [23]. The pandemic was also described as advancing along a 'front' which, in Uganda, was estimated to be moving 20 to 50 km per year [24]. Key characteristics of the pandemic were high incidences of severe symptoms of CMD, consistent mixed infection of the recombinant Uganda strain (also called Uganda variant) of East African mosaic virus (EACMV-UG) with African cassava mosaic virus (ACMV) [25], the synergism between these viruses [23,26], and rapid transmission by super-abundant *B. tabaci* populations [27]. The molecular mechanisms behind the synergism between ACMV and EACMV-like viruses include the combined action of two geminiviral suppressors of gene silencing [28]. While a clear association has been demonstrated between the CMD pandemic and the occurrence of EACMV-UG [23], the presence of EACMV-UG in countries not affected by the pandemic has also been reported in South Africa [29], Zimbabwe [29] or Burkina Faso [30]. From being a key driver of the CMD across Africa, these reports identify the Uganda variant as a contributory rather than a determining factor in the pandemic.

Although the causes of the CMD pandemic in Africa are not fully understood, it is clear that the synergism between viruses combined with the abundance of cassava whiteflies (newly adapted or introduced) has been crucial in the rapid development of CMD at a continental scale.

The Worldwide Emergence of a Whitefly Pest and Plant Virus Vector

The Bemisia Tabaci Species Complex

Rather than simply being a vector of CMD viruses, *Bemisia tabaci* (Gennadius, order Hemiptera, family Aleyrodidae) is the exclusive vector of the whole *Begomovirus*genus. This viral genus has recently been so successful that it has become the most diverse plant virus in terms of described species [31]. As a more exhaustive view of the *B. tabaci* species has been obtained, it has become difficult to determine whether begomoviruses are an intrinsically successful viral species or if this success is partially or mostly attributable to the success of *B. tabaci*.

B. tabaci belongs to a complex of genetic species formerly referred as biotypes [32]. Whereas, until recently, few biotypes had been described, it has become evident that a wide range of biotypes exists and that some are particularly invasive. Albeit morphologically indistinguishable, genetic and behavioural differences (isoenzyme profiling, barcoding based on conserved genes, life history traits) have been used to identify the biotypes, and up to 28 "putative" cryptic species are now suggested to belong to the *B. tabaci* species complex [32,33,34,35].

Although *B. tabaci* has a pan-tropical origin, these insects can be found in all warmer parts of the world as well as in greenhouses in temperate areas [33,36]. *B. tabaci* is now regarded as one of the world's top 100 invasive species (Global Invasive Species Database; [37]). In addition, it is one of the most damaging pests in open fields or protected cropping production worldwide [38]. As well as virus transmission (begomovirus but also criniviruses, carlaviruses, ipomoviruses and torradoviruses; [31]), *B. tabaci* also damages crops directly by feeding on phloem, excreting honeydew and causing phytotoxic disorders [39,40].

Shifts in Whitefly Populations

Two members of this complex, Middle East-Asia Minor 1 (MEAM1, commonly known as B biotype) and Mediterranean (MED, commonly known as Q biotype) have now progressed well beyond their home ranges mainly as a consequence of trade in ornamental plant species [41,42]. MEAM1 first invaded southwestern USA in the late 1980s, and subsequently became a serious outbreak pest. It probably originates from the Middle East-Asia Minor (including Iran, Israel, Jordan, Kuwait, Pakistan, Saudi Arabia, Syria, United Arab Republic, and Yemen) and has spread to about 50 other countries in the world [32].

MED has become a major threat to crops due to its recent worldwide expansion (around 10 years after MEAM1), its ability to reach high population densities [43] and its resistance to different insecticides [44,45]. MED, thought to be originally restricted to the Iberian Peninsula, has recently been observed in the Mediterranean Basin [45,46,47], sub-Saharan Africa [33,48,49] and in non-Mediterranean countries including China [50], Japan [51], Mexico and the USA [52]. Its probable origins are both eastern and western Mediterranean regions [32]. The eastern region is probably the source of the Mediterranean haplotypes found in the USA (Florida and to a lesser extent, California); the western region is probably the major source of the other invaded regions.

Following their introduction into new areas, displacement of native species of *B. tabaci* by the invasive MEAM1 or MED has been demonstrated throughout the world. For instance, the rapid displacement observed of Australia cryptic species (formerly AN biotype) and Asia II3 (formerly ZHJ1 biotype) by MEAM1 in Australia and China respectively occurred through competition and asymmetric reproductive interference. During these invasion and displacement processes, there appears to have been an increased frequency of copulation leading to increased production of female progeny among the invader (MEAM1 biotype), as well as a reduced copulation and female production in the native populations (Australia and Asia II3) [53]. Later, MEAM1 was also found to be completely displaced by MED in China [54]. However, in a similar invasion in Israel, where MEAM1 is native and MED is invasive, the two species coexist [47,55] as a result of niche partitioning and stochastic processes across landscapes [56].

Because biological and genetic characteristic of MED and MEAM1 are rather different, agricultural consequences and control strategies will depend on which of the two invasive species is present. MEAM1 appears to have higher fecundity and be more competitive than MED [57], and MED appears to present a high degree of resistance to certain insecticides [44,45]. They also differ in their hostrange [58,59] and mating behavior [60,61]. Studies on genetic divergence of this whitefly species complex traditionally used only a few genes (*Cytochrome Oxydase I*, *Intergenic transcribed Spacer I* and *16S* ribosomal DNA) but recent studies based notably on transcriptomic approaches have provided new insights into genetic differences between MEAM1 and MED, notably on specific sets of genes involved in biological invasion, host adaptation, and insecticide resistance [62,63]. Indeed, as areas invaded with pests are commonly treated with several pesticides, pesticide resistance of introduced whiteflies is probably the main prerequisite for their settlement and spread [64] and some "genetic combination" linked to insecticide resistance may be a key factor in invasion success [55].

Whitefly Population Differentiation Based on Endosymbionts

Another interesting feature of the arthropod lifestyle is their association with endosymbionts, which has greatly contributed to the evolutionary success of arthropods [65]. Heritable endosymbionts can be classified as being either obligate or facultative for the host. In contrast to obligate endosymbionts, facultative endosymbionts are not essential for host development or reproduction. *B. tabaci* harbours an obligate endosymbiont *Portiera aleyrodidarum* and can harbour up to seven facultative endosymbionts (*Cardinium*, *Arsenophonus*, *Rickettsia*, *Hamiltonella*,*Wolbachia*, *Fristchea*) [66,67,68,69]. Different endosymbiotic combinations have been described in the field, mainly according to former biotypes, suggesting that endosymbionts play a role in the biology of *B. tabaci*and in its differentiation into biotypes [48,70,71,72].

In the last decade, it has become clear that endosymbionts play a fundamental role in their host ecology and evolution through a wide range of effects [73,74,75], which include manipulating host reproduction [76,77,78], as well as increasing host survival or fecundity [79,80,81]. In *B. tabaci*, a role in manipulating reproduction between cryptic species has been suggested [72]. Recent studies have also pointed out that some endosymbionts can establish insecticide resistance in pest insects by detoxifying chemical agents [82]. In *B. tabaci* the link between endosymbionts and insecticides has been investigated: a high density of*Rickettsia* or *Wolbachia* was shown to be correlated with increased susceptibility, depending on the family of the insecticide compound concerned [83,84].

While these endosymbiont associations may form long-term evolutionary relationships with their arthropod hosts, they can also be dynamic, entering and leaving populations in spans of a few years to thousands of years [85,86]. For instance, Himler *et al.* [75] recently observed the rapid spread of the facultative endosymbiont *Rickettsia* sp. nr. *bellii* in the population of *B. tabaci* MEAM1 in the United States. Horizontal transmission routes, via parasitoids or plants, have recently been described for *Rickettsia* and MEAM1; these routes are possibly involved in this rapid spread [87,88]. Compared to uninfected populations, *Rickettsia*-infected populations performed better with an increase in offspring, better rates of survival until adulthood, faster development and an increased female: male sex ratio in offspring. The sex-ratio bias and performance benefits associated with *Rickettsia*-infections were sufficient to explain the sweep of *Rickettsia* in the fields across the southwestern United States [75]. Interestingly, similar differences in performance were observed between MEAM1 (harboring a greater than 90% prevalence of *Rickettsia*) and Indian Ocean species (harboring only 3% prevalence of *Rickettsia*) [72]. Moreover, the association of *Rickettsia* with increased tolerance to heat shock (40 °C) in MEAM1 [89] may explain the better performances of *Rickettsia*-harboring MEAM1 in tropical regions such as Reunion Island.

Involvement of Endosymbionts in the Survival of Begomoviruses in their Insect Vector

Endosymbionts are likely involved in the circulative and persistent transmission of begomoviruses [90]. A chaperonin GroEL homologue (hereafter named GroEL) produced by endosymbionts is required. Immunolocalization and functional studies suggest that GroEL interacts specifically with virus coat protein (CP, [91,92]), notably to protect them from degradation in the haemolymph [90,93,94,95,96,97,98].

The endosymbionts producing the GroEL responsible for CP binding were previously believed to be secondary [93]. In *B. tabaci*, GroELs have been described for a range of secondary endosymbionts:*Hamiltonella* [91], *Arsenophonus* [92], *Rickettsia* [91], and the primary symbiont *Portiera* [91]. Subsequent comparisons of TYLCV transmission by MEAM1 and MED species (harboring different symbiotypes) indicated that *Hamiltonella*, but not *Rickettsia* or *Arsenophonus*, may be the main actors during transmission [91]. However, in Reunion Island, previous comparisons between Indian Ocean (devoid of *Hamiltonella*, but harboring *Arsenophonus* and *Cardinium*) and MEAM1 (78% prevalence for *Hamiltonella*, 91% for*Rickettsia*) revealed no significant differences in TYLCV transmission [72,99]. Therefore, interaction of GroEL with the coat protein appears to be necessary but not sufficient to explain the differences observed in the efficiency of virus transmission by the insect vectors [94]. Last but not least, a rich microbial community was reported in a recent transcriptome analysis of *B. tabaci*: a total of 17,766 bacterial unigenes were classified into 322 genera [100]. Among the NCBI-nr best blast hits in the transcriptome of *B. tabaci*,GroELs closely related to that of other species were identified (*Rubrobacter xylanophilus; Candidatus Amoebophilus asiaticus; Stenotrophomonas maltophilia; Ishikawaella* symbiont of *Megacopta cribraria*); their possible role in the interaction with viral coat proteins is not yet known. Thus, the array of possible interactions between different GroELs and viruses remains to be deciphered.

The Worldwide Spread of TYLCV

Among the begomoviruses transmitted, some may cause plant diseases of considerable economic and social importance. Tomato yellow leaf curl disease (TYLCD) is one of the most devastating plant diseases in warm and temperate regions of the world [101,102]. TYLCD infections result in yellowing and upwards curling of leaflet margins, stunting of the plants and flower abortion. Infected plants are less vigorous and produce fruits with reduced market value. Moreover, infection during early growth stages can lead to total loss of the crop [103,104].

The first description of TYLCD-like symptoms dates back to the late 1920s in the Jordan Valley in Israel (cited in [105]). Severe outbreaks of the disease occurred in the early 1960s in Israel [105]. From the late 1980s on, a major and rapid geographic spread of TYLCD took place and its distribution area now stretches from Japan in the east to Spain in the west in the Old World, and to Reunion Island and Australia in the south [105]. In the 1990s, TYLCD reached the New World and rapidly spread in North America as well as in the Caribbean [106,107,108].

It was not until 1991 that the first molecular data on the viral agent associated with TYLCD became available and was named *Tomato yellow leaf curl virus*-Israel (TYLCV-IL; e.g. TYLCV-IL[IL:Reo:86]; [109,110]). Subsequently, another begomovirus strain associated with TYLCD infections was cloned and named *Tomato yellow leaf curl virus*-Mild (TYLCV-Mld; TYLCV-Mld[IL:93]) [111]. Ever since, an increasing number of tomato yellow leaf curl genomes have been sequenced and molecularly characterized revealing the wide viral diversity associated with TYLCD [20]. Based on phylogenetic relationships, there are currently six related species and 15 strains, referred to as TYLCV-like viruses, associated with TYLCD: TYLCV, *Tomato yellow leaf curl Sardinia virus* (TYLCSV), *Tomato yellow leaf curl Axiarqua virus* (TYLCAxV), *Tomato yellow leaf curl Malaga virus* (TYLCMalV), *Tomato yellow leaf curl Mali virus* (TYLCMV) and *Tomato leaf curl Sudan virus* (ToLCSDV) [112]. Among them, TYLCV, the best known and emergent species, have the broadest geographical range. Based on sequence identities, five strains of TYLCV are currently described including Gezira (e.g., TYLCV-Gez [SD:96]), Iran (e.g., TYLCV-IR [IR:Ira:98]), Israel (e.g., TYLCV-IL [IL:Reo:86]), Mild (e.g., TYLCV-Mld [IL:93]) and Oman strains (e.g., TYLCV-OM [OM:Alb:05]).

Because of the great TYLCV diversity encountered in Mediterranean basin and the Middle East, this region was suspected to be the center of origin of these viruses [113]. Using Bayesian phylogeopraphic inference and recombination analyses, Lefeuvre *et al.* [114] recently reconstructed a plausible history of the diversification and movement of TYLCV throughout the world (Figure 1). In accordance with historical records, this study suggests that the first TYLCV most probably arose in the Middle East between the 1930s and 1950s and that the global spread of TYLCVs only began in the 1980s after the emergence of TYLCV-Mld and TYLCV-IL. Evidence was also provided that the highly invasive TYLCV-IL has jumped at least twice to the New World, once from the Mediterranean basin in the 1990s, and once from Asia in the 2000s. This study also identified the Middle East and the regions surrounding Iran in particular as the most probable current and past centers of ongoing TYLCV diversification. Nevertheless, although this region displays intensive TYLCV diversification, it was also found to be epidemiologically isolated, suggesting that the novel variants found there may never disseminate outside the Middle East. In contrast, the Mediterranean basin was identified as the main launch-pad of global TYLCV movements and the direct source of the TYLCV variants currently spreading around the world.

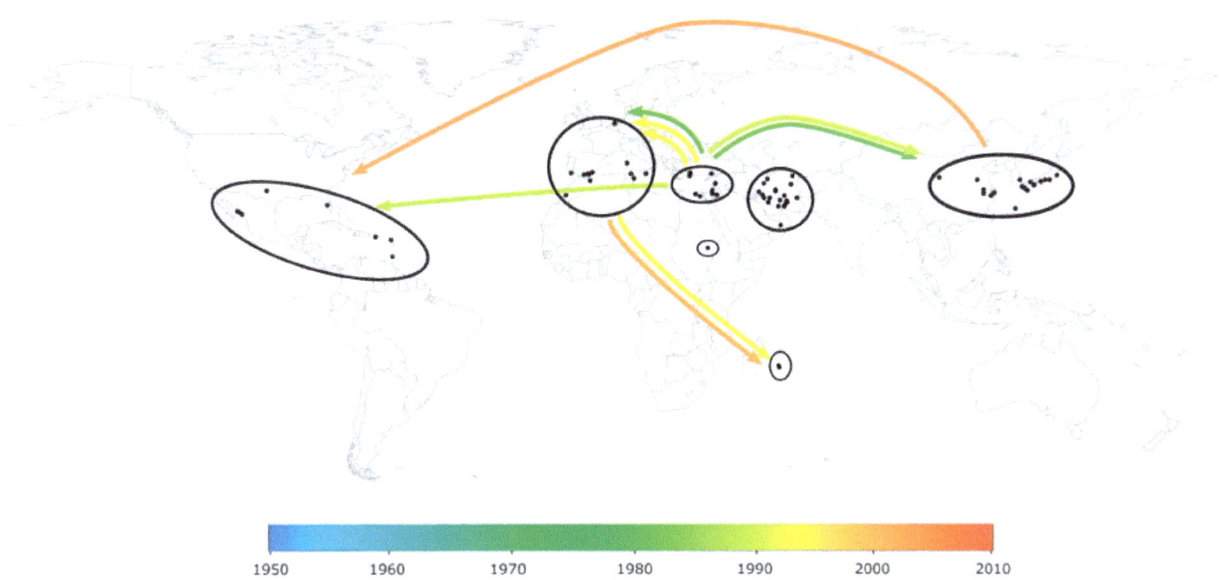

Figure 1: Worldwide spread of *Tomato yellow leaf curl virus* (TYLCV) inferred using phylogeography analyses on the coat protein and full genome datasets [114]. According to the color scale at the bottom of the figure, arrow colors depict the mean ages (in years) ...

It is important to note that this pattern of invasion resembles the recently described phenomenon of the invasive bridgehead effect [115]. In this scheme, widespread invasions do not result from isolates originating from across the native range of the invasive species but rather from individual particularly invasive sub-populations that serve as sources of colonists for the conquest of new territories. While the invasive bridgehead effect is evolutionarily parsimonious [115,116], data and examples remain too scarce to investigate whether this pattern of invasion is a common mechanism, and it would thus be of particular interest to explore genetic data sets from other worldwide invasive species.

Case Study in Reunion Island

Reunion Island is a French island situated in the Mascarenes archipelago, in the south-western part of the Indian Ocean (700 km east of Madagascar). As colonization and the development of agriculture in Reunion Island are recent, imprints of biological invasion are particularly easy to track and make the island an ideal venue to study biological invasions.

Invasion of the World Invasive MEAM1 Species of Bemisia Tabaci

B. tabaci was first described in Reunion Island in 1938 [117] and later by Luziau [118]. After its introduction in the late 1990s, the world invasive MEAM1 species of *B. tabaci* became a major threat to tomato crops as a vector of TYLCV [99,119,120,121]. RAPD-PCR and *COI* sequencing on *B. tabaci* populations of this island also revealed the presence of a new undescribed species, named Ms (Indian Ocean putative species, according to Dinsdale *et al.* [33]), indigenous to the south-western Indian Ocean [122,123]. The genetic diversity of these two putative species was later analyzed using microsatellite markers on field samples from 2001 and 2002, and wild hybrids of multiple generation between MEAM1 and Indian Ocean species were subsequently characterized [72,123]. In 2006, samples were collected at the same sites as in 2001 and 2002 and genetic analyses using principle component analyses (PCA, Figure 2) and a Bayesian clustering method (STRUCTURE) confirmed the presence of admixed individuals of multiple generations between MEAM1 and Indian Ocean putative species.

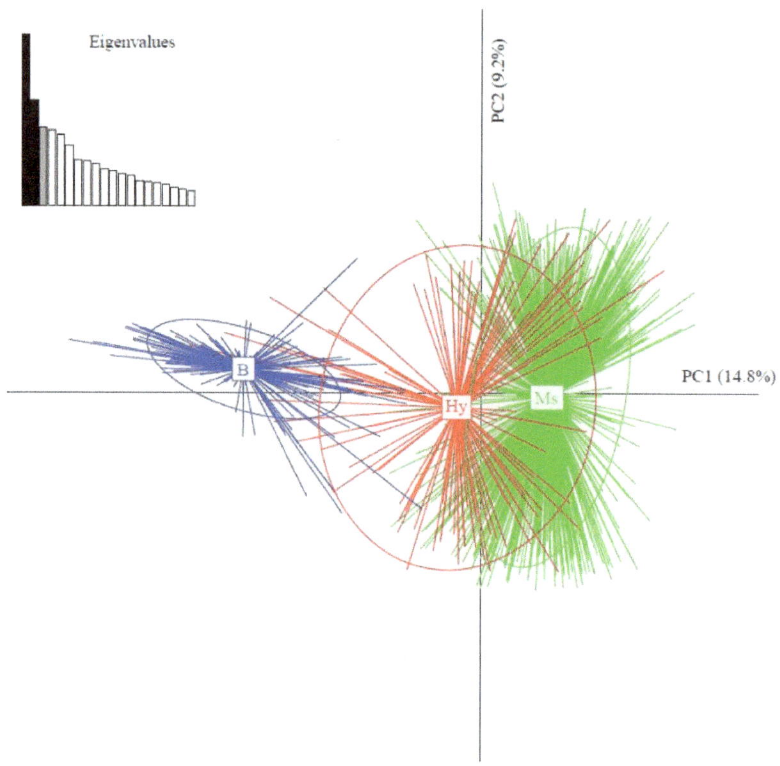

Figure 2: Multivariate analysis of *Bemisia tabaci* populations in Reunion Island [72]. First and second components of a principal component analysis of 10-loci microsatellites from *B. tabaci* females from the three populations MEAM1 (biotype B), Hybrids (Hy) and .

Co-Existence of Invasive and Indigenous Species and Hybrids of Bemisia Tabaci

More than 12 years of intensive field surveys after the introduction of MEAM1 species demonstrated that the indigenous species is still present in Reunion Island (unpublished data), although in other countries, MEAM1 or MED have displaced indigenous populations [32,124]. The limited success of MEAM1 in Reunion Island may be explained by a combination of several factors. Firstly, the endosymbiont populations of the different *B. tabaci*populations, which have been characterized in MEAM1 and Indian Ocean species (and the hybrids of these species; [72]) may hold some clues. MEAM1 harbors *Rickettsia* and *Hamiltonella* as in other world populations, and Indian Ocean harbours *Cardinium* and *Arsenophonus*. Endosymbiont analysis of the hybrid population indicated that matings involving Indian Ocean females and MEAM1 males were more successful than reciprocal matings (Figure 3). These results may partly explain why MEAM1 has not yet displaced the Indian Ocean species in Reunion Island. Secondly, coexistence of invasive and indigenous *B. tabaci* species might be facilitated by spatial and temporal niche partitioning [56]. In Reunion Island, Delatte *et al.* [123] demonstrated that MEAM1 and Indian Ocean coexist in sympatry throughout most of their geographical ranges, although they tend to segregate into different types of host plants with MEAM1 predominating in vegetable crops, and Indian Ocean being more frequently found on weeds. Endosymbionts could enhance postzygotic barriers involved in non-random hybridization between MEAM1 and Indian Ocean [72] and niche partitioning implicated in prezygotic barriers could explain the coexistence of an invasive and indigenous species.

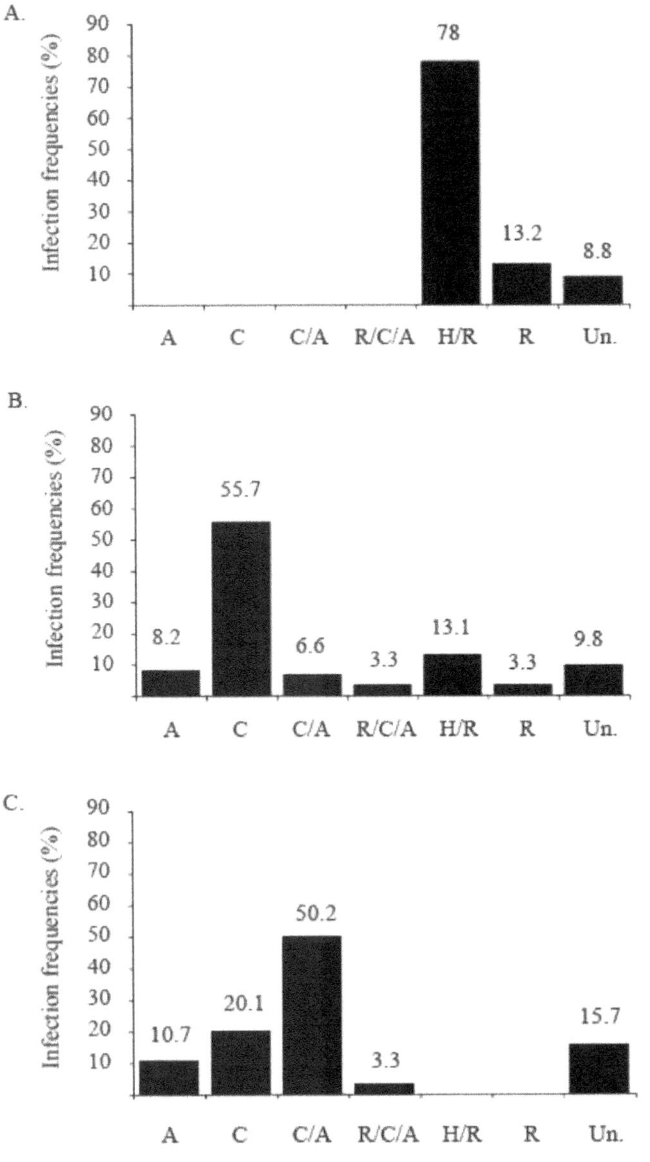

Figure 3: Infection frequencies of the different endosymbiont combinations observed in individuals of MEAM1 (A), Hybrids (B) and Indian Ocean (C) populations of *Bemisia tabaci* collected in the field in Reunion Island. A: *Arsenophonus*, C: *Cardinium*, H: *Hamiltonella* ...

Most interestingly, analyses of samples in Reunion Island in 2010 indicated the additional presence of MED [124]. The interactions between MEAM1 and Indian Ocean putative species, monitored in the field since 1997, will thus be complemented from 2010 onwards by MED as a new actor.

Successive Introduction of TYLCV Strains

Although indigenous populations of *B. tabaci* were detected in Reunion Island, no indigenous begomoviruses infecting tomato were described before the introduction of TYLCV-Mld in 1997. Following its first description in the southern part of the island, TYLCV-Mld was detected in almost all the tomato-growing area by the end of 1998. With severe economic losses observed (up to 85%) in outdoor and/or protected tomato crops, TYLCV became the main limiting factor to tomato production [119,120,125].

This accidental introduction in a remote agroecosystem, apparently free of any other tomato-infecting begomoviruses, provided an opportunity to study the molecular evolution of TYLCV-Mld from the initial inoculum. The genetic variation of TYLCV-Mld was monitored in a field survey conducted between 1997 and 2004 in the main tomato growing areas. The very low diversity of the isolates in 1997 did not provide any evidence of multiple TYLCV introductions. A quasi-linear increase in genetic diversity was observed in the following years. In addition, the population effective size of TYLCV-Mld underwent a sudden increase from 2001 to 2004, consistent with a founder effect due to the introduction of a small number of virus individuals [126].

In 2004, particularly severe symptoms of TYLCD were observed in the western part of the tomato cropping region. Molecular diagnosis revealed the introduction of the "Israel" strain also called "the severe" strain of TYLCV (TYLCV-IL) [127]. This introduction provided an ideal model to study the competition between two strains of one of the most emergent virus species. While studies describing epidemiological factors (*i.e.*, host range, vector transmission efficiencies) associated with TYLCD epidemics have been performed in temperate regions, extensive studies in tropical insular ecosystems are still rare. Different scenarios can be expected in such an epidemiological situation with competition: synergism, co-existence, displacement, or emergence of new variants. Studies describing epidemiological dynamics of TYLCV-like viruses in Spain or Italy revealed the displacement of one species by the other [128,129], the co-existence of the two species [129] or the emergence of a new severe recombinant variant [130,131,132]. After a seven-year period following its emergence in Reunion Island, TYLCV-IL has progressively displaced TYLCV-Mld, which is now predominantly detected in mixed infections [133]. The characterization of the key biological traits involved in this epidemiological pattern is underway [134] and should reveal clues to virus competition that may help predict the outcome of possible future introductions.

MAJOR CONCLUSIONS AND PERSPECTIVES

Invasions of arthropod vectors and viruses are the main factors associated with viral emerging diseases. This is particularly the case for begomoviruses transmitted by *B. tabaci*, which are responsible for the emergence of one of the most devastating plant diseases in the world (TYLCD). The first key factor identified in the worldwide dissemination of TYLCD is the worldwide expansion of invasive, polyphagous and in some cases pesticide resistant species (MEAM1 or MED) of *B. tabaci* [32,38,55]. It has also become clear that endosymbionts play a fundamental role in *B. tabaci* host ecology and evolution through a wide range of effects, such as increasing host survival or fecundity. Furthermore, the introduction of invasive species of *B. tabaci* into new ecosystems has provided opportunities for several begomoviruses to jump from natural flora to contaminate crops such as tomato. This phenomenon, recently named the weed-crop connection [31], is illustrated by the recent emergence and wide range of begomoviruses involved in viral epidemics on tomato in Brazil since the mid-1990s [31,135,136]. The second key factor is the introduction of plant viruses in a new area. This is the case for the worldwide spread of TYLCV invasive strains (TYLCV-IL and -Mld). In agreement with historical records, Lefeuvre *et al.* [114] recently suggested that the first TYLCVs most probably arose in the Middle East between the 1930s and 1950s, and that the global spread of TYLCVs only began in the 1980s after the emergence of TYLCV invasive strains. Moreover, the Mediterranean basin was identified as the main launch-pad of global TYLCV movements and the direct source of the TYLCV variants currently spreading around the world. This pattern of invasion is similar to the recently described phenomenon of the invasive bridgehead effect where a particularly invasive population serves as source of colonists for the conquest of new territories [115]. Recent studies have demonstrated that recombination plays a major role in the evolution and emergence of begomoviruses (for review see [1,137,138]). This is apparently the case of the invasive IL strain of TYLCV, which is a natural recombinant which arose by recombination between TYLCV-Mld and Tomato leaf curl virus (ToLCV)-like ancestors [139].

The introduction of plant viruses in a new area is mainly triggered by long distance exchange of plant material or vector insects due to the global trade in vegetables and ornamental plants [140]. A well-documented case of the importance of international trade causing long-distance dispersal of viruses was the original introduction of TYLCV-IL into the New World (Florida) from the Old World in 1992 via infected transplants that were probably purchased in Israel [141]. This example also illustrates the great capacity of the invasive TYLCV to very rapidly adapt to the new niches it invades; in some cases even causing the disappearance of indigenous begomovirus species. In the case of Reunion Island, the successive arrivals of TYLCV-Mld and MEAM1 (1997), TYLCV-IL (2004) and MED (2010) provide a wide range of possible interactions with the indigenous Indian Ocean species of *B. tabaci* in the field. The future of the interaction between the three species of *B. tabaci* and the competition between invasive strains of TYLCV in the fields of Reunion Island is a fortuitous experiment that is currently underway.

REFERENCES

1. Fargette D., Konaté G., Fauquet C., Muller E., Peterschmitt M., Thresh J.M. Molecular ecology and emergence of tropical plant viruses. Ann. Rev. Phytopathol. 2006;44:35–260.
2. Sax D.F., Brown J.H. The paradox of invasion. Global Ecol. Biogeogr. 2000;9:363–371. doi: 10.1046/j.1365-2699.2000.00217.x.
3. Gillespie R.G., Roderick G.K. Arthropods on islands: Colonization, speciation, and conservation. Ann. Rev. Entomol. 2002;47:595–632. doi: 10.1146/annurev.ento.47.091201.145244.
4. Gillespie R.G., Claridge E.M., Roderick G.K. Biodiversity dynamics in isolated island communities: Interaction between natural and human-mediated processes. Mol. Ecol. 2008;17:45–57.
5. Lockwood J.L., Cassey P., Blackburn T. The role of propagule pressure in explaining species invasions. Trends Ecol. Evol. 2005;20:223–228. doi: 10.1016/j.tree.2005.02.004.
6. Harlan J.R. Agricultural origins: Centers and noncenters. Science. 1971;174:468–474.
7. Balter M. Seeking agriculture's ancient roots. Science. 2007;316:1830–1835. doi: 10.1126/science.316.5833.1830.
8. Pimentel D., McNair S., Janecka J., Wightman J., Simmonds C., O'Connell C., Wong E., Russel L., Zern J., Aquino T., Tsomondo T. Economic and environmental threats of alien plant, animal, and microbe invasions. Agr. Ecosyst. Environ. 2001;84:1–20. doi: 10.1016/S0167-8809(00)00178-X.
9. Burdon J.J., Thrall P.H., Ericson L. The current and future dynamics of disease in plant communities.Ann. Rev. Phytopathol. 2006;44:19–39. doi: 10.1146/annurev.phyto.43.040204.140238.
10. Jones R.A.C., Salam M.U., Maling T.J., Diggle A.J., Thackray D.J. Principles or predicting plant virus disease epidemics. Ann. Rev. Phytopathol. 2010;48:179–203. doi: 10.1146/annurev-phyto-073009-114444.
11. Thrall P.H., Burdon J.J. The spatial scale of pathogen dispersal: Consequences for disease dynamics and persistence. Evol. Ecol. Res. 1999;1:681–701.
12. Anderson P.K., Cunningham A.A., Patel N.G., Morales F.J., Epstein P.R., Daszak P. Emerging infectious diseases of plants: Pathogen pollution, climate change and agrotechnology drivers. Trends Ecol. Evol. 2004;19:535–544. doi: 10.1016/j.tree.2004.07.021.
13. Martin D., Shepherd D. The epidemiology, economic impact and control of maize streak disease. Food Secur. 2009;1:305–315. doi: 10.1007/s12571-009-0023-1.
14. Shepherd D.N., Martin D.P., van der Walt E., Dent K., Varsani A., Rybicki E.P. Maize streak virus: An old and complex 'emerging' pathogen. Mol. Plant Pathol. 2010;11:1–12.
15. Lefeuvre P., Martin D.P., Hoareau M., Naze F., Delatte H., Thierry M., Varsani A., Becker N., Reynaud B., Lett J.M. Begomovirus 'melting pot' in the south-west Indian Ocean islands: Molecular diversity and evolution through recombination. J. Gen. Virol. 2007;88:3458–3468. doi: 10.1099/vir.0.83252-0.
16. Lefeuvre P., Lett J.M., Varsani A., Martin D.P. Widely conserved recombination patterns among single-stranded DNA viruses. J. Virol. 2009;83:2697–2707. doi: 10.1128/JVI.02152-08.
17. Harkins G.W., Martin D.P., Duffy S., Monjane A.L., Shepherd D.N., Windram O.P., Owor B.E., Donaldson L., van Antwerpen T., Sayed R.A., et al. Dating the origins of the maize-adapted strain of maize streak virus, MSV-A. J. Gen. Virol. 2009;90:3066–3074.
18. Varsani A., Shepherd D.N., Monjane A.L., Owor B.E.,

Erdmann J.B., Rybicki E.P., Peterschmitt M., Briddon R.W., Markham P.G., Oluwafemi S., et al. Recombination, decreased host specificity and increased mobility may have driven the emergence of maize streak virus as an agricultural pathogen. J. Gen. Virol.2008;89:2063–2074.

19. Monjane A.L., Harkins G.W., Martin D.P., Lemey P., Lefeuvre P., Shepherd D.N., Oluwafemi S., Simuyandi M., Zinga I., Komba E.K., et al. Reconstructing the history of maize streak virus strain a dispersal to reveal diversification hot spots and its origin in Southern Africa. J. Virol. 2011;85:9623–9636.

20. Fauquet C.M., Briddon R.W., Brown J.K., Moriones E., Stanley J., Zerbini M., Zhou X. Geminivirus strain demarcation and nomenclature. Arch. Virol. 2008;153:783–821.

21. Patil B.L., Fauquet C.M. Cassava mosaic geminiviruses: Actual knowledge and perspectives. Mol. Plant Pathol. 2009;10:685–701. doi: 10.1111/j.1364-3703.2009.00559.x.

22. Legg J.P., Thresh J.M. Cassava mosaic virus disease in East Africa: A dynamic disease in a changing environment. Virus Res. 2000;71:135–149. doi: 10.1016/S0168-1702(00)00194-5.

23. Legg J.P., Fauquet C.M. Cassava mosaic geminiviruses in Africa. Plant Mol. Biol. 2004;56:585–599. doi: 10.1007/s11103-004-1651-7.

24. Legg J.P., Ogwal S. Changes in the incidence of African cassava mosaic virus disease and the abundance of its whitefly vector along south-north transects in Uganda. J. Appl. Entomol. 1998;122:169–178. doi: 10.1111/j.1439-0418.1998.tb01480.x.

25. Zhou X., Liu Y., Calvert L., Munoz C., Otim-Nape G.W., Robinson D.J., Harrison B.D. Evidence that DNA-A of a geminivirus associated with severe cassava mosaic disease in Uganda has arisen by interspecific recombination. J. Gen. Virol. 1997;78:2101–2111.

26. Pita J.S., Fondong V.N., Sangaré A., Otim-Nape G.W., Ogwal S., Fauquet C.M. Recombination, pseudorecombination and synergism of geminiviruses are determinant keys to the epidemic of severe cassava mosaic disease in Uganda. J. Gen. Virol. 2001;82:655–665.

27. Legg J.P., French R., Rogan D., Okao-Okuja G., Brown J.K. A distinct Bemisia tabaci (Gennadius) (Hemiptera: Sternorrhyncha: Aleyrodidae) genotype cluster is associated with the epidemic of severe cassava mosaic virus disease in Uganda. Mol. Ecol. 2002;11:1219–1229. doi: 10.1046/j.1365-294X.2002.01514.x.

28. Vanitharani R., Chellappan P., Pita J.S., Fauquet C.M. Differential roles of AC2 and AC4 of cassava geminiviruses in mediating synergism and suppression of posttranscriptional gene silencing. J. Virol.2004;78:9487–9498. doi: 10.1128/JVI.78.17.9487-9498.2004.

29. Berry S., Rey M.E.C. Molecular evidence for diverse populations of cassava-infecting begomoviruses in southern Africa. Arch. Virol. 2001;146:1795–1802. doi: 10.1007/s007050170065.

30. Tiendrébéogo F., Lefeuvre P., Hoareau M., Traoré V.S.E., Barro N., Reynaud B., Traoré A.S., Konaté G., Traoré O., Lett J.-M. Occurrence of East African cassava mosaic virus-Uganda (EACMV-UG) in Burkina Faso. Plant Pathol. 2009;58:783–783.

31. Navas-Castillo J., Fiallo-Olivé E., Sánchez-Campos S. Emerging virus diseases transmitted by whiteflies. Ann. Rev. Phytopathol. 2011;49:219–248. doi: 10.1146/annurev-phyto-072910-095235.

32. De Barro J., Liu S., Boykin L., Dinsdale A. *Bemisia tabaci*: A statement of species status. Ann. Rev. Entomol. 2011;56:1–19. doi: 10.1146/annurev-ento-112408-085504.

33. Dinsdale A., Cook L., Riginos C., Buckley Y.M., Barro P.D. Refined global analysis of *Bemisia tabaci*(Hemiptera: Sternorrhyncha: Aleyrodoidea: Aleyrodidae) mitochondrial cytochrome oxidase 1 to identify species level genetic boundaries. Ann. Entomol. Soc. Am. 2010;103:196–208. doi: 10.1603/AN09061.

34. Liu S.S., Colvin J., De Barro P.J. Species concepts as applied to the whitefly *Bemisia tabaci* systematics: How many species are there? J. Integr. Agr. 2012;11:176–186.

35. Hu J., De Barro P., Zhao H., Wang J., Nardi F., Liu S.S. An extensive field survey combined with a phylogenetic analysis reveals rapid and widespread invasion of two alien whiteflies in China. PLoS One.2011;6:e16061.

36. Boykin L.M., Shatters R.G., Rosell R.C., McKenzie C.L., Bagnall R.A., De Barro P., Frohlich D.R. Global relationships of *Bemisia tabaci* (Hemipter: Aleyrodidae) revealed using Bayesian analysis of mitochondrial COI DNA sequences. Mol. Phylogenet. Evol. 2007;44:1306. doi: 10.1016/j.ympev.2007.04.020.

37. Global Invasive Species Database. [(accessed on 10 October 2012)]. Available online:www.issg.org/database/welcome.

38. Brown J.K. Phylogenetic Biology of the *Bemisia tabaci* Sibling Species Group. In: Stansly P.A., Naranjo S.E., editors. Bionomics and Management of a Global Pest. Springer; Amsterdam, The Netherlands: 2010. pp. 31–67.

39. Varma A., Malathi V.G. Emerging geminivirus problems: A serious threat to crop production. Ann. Appl. Biol. 2003;142:145–164. doi: 10.1111/j.1744-7348.2003.tb00240.x.

40. Jones D.R. Plant viruses transmitted by whiteflies. Eur. J. Plant Pathol. 2003;109:195. doi: 10.1023/A:1022846630513.

41. Cheek S., Macdonald O. Extended summaries sci pesticides group symposium management of *Bemisia tabaci*. Pestic. Sci. 1994;42:135–142. doi: 10.1002/ps.2780420211.

42. Dalton R. Whitefly infestations: The Christmas Invasion. Nature. 2006;443:898–900. doi: 10.1038/443898a.

43. Moya A., Guirao P., Cifuentes D., Beitia F., Cenis J.L. Genetic diversity of Iberian populations of*Bemisia tabaci* (Hemiptera: Aleyrodidae) based on random amplified polymorphic DNA-polymerase chain reaction. Mol. Ecol. 2001;10:891–897. doi: 10.1046/j.1365-294X.2001.01221.x.

44. Luo C., Jones C.M., Devine G., Zhang F., Denholm I., Gorman K. Insecticide resistance in *Bemisia tabaci* biotype Q (Hemiptera: Aleyrodidae) from China. Crop Prot. 2010;29:429–434. doi: 10.1016/j.cropro.2009.10.001.

45. Horowitz A.R., Kontsedalov S., Khasdan V., Ishaaya I. Biotypes B and Q of *Bemisia tabaci* and their relevance to neonicotinoid and pyriproxyfen resistance. Arch. Insect Biochem. 2005;58:216–225. doi: 10.1002/arch.20044.

46. Horowitz A., Denholm I., Gorman K., Cenis J.L., Kontsedalov S., Ishaaya I. Biotype Q of *Bemisia tabaci* identified in Israel. Phytoparasitica. 2003;31:94–98. doi: 10.1007/BF02979772.

47. Simon B., Cenis J.L., De La Rua P. Distribution patterns of the Q and B biotypes of *Bemisia tabaci* in the Mediterranean Basin based on microsatellite variation. Entomol. Exp. Appl. 2007;124:327–336. doi: 10.1111/j.1570-7458.2007.00586.x.

48. Gueguen G., Vavre F., Gnankine O., Peterschmitt M., Charif D., Chiel E., Gottlieb Y., Ghanim M., Zchori-Fein E., Fleury F. Endosymbiont metacommunities, mtDNA diversity and the evolution of the*Bemisia tabaci* (Hemiptera: Aleyrodidae) species complex. Mol. Ecol. 2010;19:4365–4376. doi: 10.1111/j.1365-294X.2010.04775.x.

49. Gnankiné O., Mouton L., Henri H., Terraz G., Houndeté T., Martin T., Vavre F., Fleury F. Distribution of *Bemisia tabaci* (Homoptera: Aleyrodidae) biotypes and their associated symbiotic bacteria on host plants in West Africa. Insect Conserv. Diver. 2012 doi: 10.1111/j.1752-4598.2012.00206.x.

50. Zhang L.P., Zhang Y.J., Zhang W.J., Wu Q.J., Xu B.Y., Chu D. Analysis of genetic diversity among different geographical populations and determination of biotypes of *Bemisia tabaci* in China. J. Appl. Entomol. 2005;129:121–128. doi: 10.1111/j.1439-0418.2005.00950.x.

51. Ueda S., Brown J.K. First report of the Q biotype of *Bemisia tabaci* in Japan by mitochondrial cytochrome oxidase I sequence analysis. Phytoparasitica. 2006;34:405–411. doi: 10.1007/BF02981027.

52. Brown J. The *Bemisia tabaci* Complex: Genetic and Phenotypic Variability Drives Begomovirus Spread and Virus Diversification. [(accessed on 10 October 2012)]. Available online:http://www.apsnet.org/online/feature/btabaci/

53. Liu S.-S., De Barro P.J., Xu J., Luan J.-B., Zang L.-S., Ruan Y.-M., Wan F.-H. Asymmetric mating interactions drive widespread invasion and displacement in a whitefly. Science. 2007;318:1769–1772. doi: 10.1126/science.1149887.

54. Chu D., Wan F.H., Zhang Y.J., Brown J.K. Change in the biotype composition of *Bemisia tabaci* in Shandong Province of China from 2005 to 2008. Environ. Entomol. 2010;39:1028–1036. doi: 10.1603/EN09161.

55. Kontsedalov S., Abu-Moch F., Lebedev G., Czosnek H., Horowitz A.R., Ghanim M. *Bemisia tabaci*biotype dynamics and resistance to insecticides in Israel during the years 2008–2010. J. Integr. Agr.2012;11:312–320.

56. Crowder D.W., Horowitz A.R., Breslauer H., Rippa M., Kontsedalov S., Ghanim M., Carriere Y. Niche partitioning and stochastic processes shape community structure following whitefly invasions. Basic Appl. Ecol. 2011;12:685–694. doi: 10.1016/j.baae.2011.09.007.

57. Pascual S. Mechanisms in competition, under laboratory conditions, between Spanish biotypes B and Q of *Bemisia tabaci* (Gennadius) Span. J. Agric. Res. 2006;4:351–354.

58. Muniz M. Host suitability of two biotypes of *Bemisia tabaci* on some common weeds. Entomol. Exp. Appl. 2000;95:63–70.

59. Muniz M., Nombela G. Differential variation in development of the B- and Q-biotypes of *Bemisia tabaci*(Homoptera: Aleyrodidae) on sweet pepper at constant temperatures. Environ. Entomol. 2001;30:720–727. doi: 10.1603/0046-225X-30.4.720.

60. Elbaz M., Lahav N., Morin S. Evidence for pre-zygotic reproductive barrier between the B and Q biotypes of Bemisia tabaci (Hemiptera: Aleyrodidae) Bull. Entomol. Res. 2010;100:581–590. doi: 10.1017/S0007485309990630.

61. Sun D.B., Xu J., Luan J.B., Liu S.S. Reproductive incompatibility between the B and Q biotypes of the whitefly Bemisia tabaci in China: Genetic and behavioural evidence. B. Entomol. Res. 2010;101:211–220.

62. Wang X.W., Luan J.B., Li J.M., Su Y.L., Xia J., Liu S.S. Transcriptome analysis and comparison reveal divergence between two invasive whitefly cryptic species. BMC Genom. 2011;12:458. doi: 10.1186/1471-2164-12-458.

63. Seal S., Patel M.V., Collins C., Colvin J., Bailey D. Next generation transcriptome sequencing and quantitative real-time PCR technologies for characterisation of the bemisia tabaci Asia 1 mtCOI Phylogenetic Clade. J. Integr. Agr. 2012;11:281–292.

64. Guillemaud T., Ciosi M., Lombaert É., Estoup A. Biological invasions in agricultural settings: Insights from evolutionary biology and population genetics. C. R. Biol. 2011;334:237–246. doi: 10.1016/j.crvi.2010.12.008.

65. Feldhaar H., Gross R. Insects as hosts for mutualistic bacteria. Int. J. Med. Microbiol. 2009;299:1–8. doi: 10.1016/j.ijmm.2008.05.010.

66. Chiel E., Gottlieb Y., Zchori-Fein E., Mozes-Daube N., Katzir N., Inbar M., Ghanim M. Biotype-dependent secondary symbiont communities in sympatric populations of *Bemisia tabaci*. Bull. Entomol. Res.2007;97:407. doi: 10.1017/S0007485307005159.

67. Everett K.D., Thao M., Horn M., Dyszynski G.E., Baumann P. Novel chlamydiae in whiteflies and scale insects: Endosymbionts 'Candidatus Fritschea bemisiae' strain Falk and 'Candidatus Fritschea eriococci' strain Elm. Int. J. Syst. Evol. Micr. 2005;55:1581–1587. doi: 10.1099/ijs.0.63454-0.

68. Gottlieb Y., Ghanim M., Chiel E., Gerling D., Portnoy V., Steinberg S., Tzuri G., Horowitz A.R., Belausov E.,

Mozes-Daube N., et al. Identification and localization of a Rickettsia sp in *Bemisia tabaci*(Homoptera: Aleyrodidae) Appl. Environ. Microb. 2006;72:3646–3652. doi: 10.1128/AEM.72.5.3646-3652.2006.

69. Zchori-Fein E., Brown J.K. Diversity of prokaryotes associated with *Bemisia tabaci* (Gennadius) (Hemiptera: Aleyrodidae) Ann. Entomol. Soc. Am. 2002;95:711–718. doi: 10.1603/0013-8746(2002)095[0711:DOPAWB]2.0.CO;2.

70. Chiel E., Gottlieb Y., Zchori-Fein E., Mozes-Daube N., Katzir N., Inbar M., Ghanim M. Biotype-dependent secondarysymbiontcommunitiesinsympatricpopulations of *Bemisia tabaci*. Bull. Entomol. Res.2007;97:407–413. doi: 10.1017/S0007485307005159.

71. Skaljac M., Zanic K., Ban S.G., Kontsedalov S., Ghanim M. Co-infection and localization of secondary symbionts in two whitefly species. BMC Microbiol. 2010;10:142. doi: 10.1186/1471-2180-10-142.

72. Thierry M., Becker N., Hajri A., Reynaud B., Lett J.M., Delatte H. Symbiont diversity and non-random hybridization among indigenous (Ms) and invasive (B) biotypes of *Bemisia tabaci*. Mol. Ecol.2011;20:2172–2187. doi: 10.1111/j.1365-294X.2011.05087.x.

73. Engelstadter J., Hurst G.D.D. The ecology and evolution of microbes that manipulate host reproduction.Ann. Rev. Ecol. Evol. Syst. 2009;40:127–149. doi: 10.1146/annurev.ecolsys.110308.120206.

74. Oliver K.M., Degnan P.H., Burke G.R., Moran N.A. Facultative symbionts in aphids and the horizontal transfer of ecologically important traits. Ann. Rev. Entomol. 2010;55:247–266. doi: 10.1146/annurev-ento-112408-085305.

75. Himler A.G., Adachi-Hagimori T., Bergen J.E., Kozuch A., Kelly S.E., Tabashnik B.E., Chiel E., Duckworth V.E., Dennehy T.J., Zchori-Fein E., Hunter M.S. Rapid spread of a bacterial symbiont in an invasive whitefly is driven by fitness benefits and female bias. Science. 2011;332:254–256.

76. Zchori-Fein E., Gottlieb Y., Kelly S.E., Brown J.K., Wilson J.M., Karr T.L., Hunter M.S. A newly discovered bacterium associated with parthenogenesis and a change in host selection behavior in parasitoid wasps. P. Natl. Acad. Sci. USA. 2001;98:12555–12560.

77. Zchori-Fein E., Perlman S.J., Kelly S.E., Katzir N., Hunter M.S. Characterization of a 'Bacteroidetes' symbiont in Encarsia wasps (Hymenoptera: Aphelinidae): proposal of 'Candidatus Cardinium hertigii'. Int. J. Syst. Evol. Microbiol. 2004;54:961–968. doi: 10.1099/ijs.0.02957-0.

78. Hunter M.S., Perlman S.J., Kelly S.E. A bacterial symbiont in the Bacteroidetes induces cytoplasmic incompatibility in the parasitoid wasp *Encarsia pergandiella*. Proc. R. Soc. B. 2003;270:2185–2190. doi: 10.1098/rspb.2003.2475.

79. Scarborough C.L., Ferrari J., Godfray H.C.J. Aphid protected from pathogen by endosymbiont. Science.2005;310:1781. doi: 10.1126/science.1120180.

80. Haine E.R. Symbiont-mediated protection. Proc. R. Soc. B. 2008;275:353–361. doi: 10.1098/rspb.2007.1211.

81. Vorburger C., Gehrer L., Rodriguez P. A strain of the bacterial symbiont *Regiella insecticola* protects aphids against parasitoids. Biol. Lett. 2010;6:109–111. doi: 10.1098/rsbl.2009.0642.

82. Kikuchi Y., Hayatsu M., Hosokawa T., Nagayama A., Tago K., Fukatsu T. Symbiont-mediated insecticide resistance. P. Natl. Acad. Sci. USA. 2012;109:8618–8622.

83. Ghanim M., Kontsedalov S. Susceptibility to insecticides in the Q biotype of *Bemisia tabaci* is correlated with bacterial symbiont densities. Pest Manag. Sci. 2009;65:939–942. doi: 10.1002/ps.1795.

84. Kontsedalov S., Zchori-Fein E., Chiel E., Gottlieb Y., Inbar M., Ghanim M. The presence of Rickettsia is associated with increased susceptibility of *Bemisia tabaci* (Homoptera: Aleyrodidae) to insecticides. Pest Manag. Sci. 2008;64:789–792. doi: 10.1002/ps.1595.

85. Hinrich J., Schulenburg G.V.D., Hurst G.D.D., Tetzlaff D., Booth G.E., Zakharov I.A., Majerus M.E.N. History of infection with different male-killing bacteria in the two-spot ladybird beetle *Adalia bipunctata*revealed through mitochondrial DNA sequence analysis. Genetics. 2002;160:1075–1086. [PMC free article]

86. Turelli M., Hoffmann A.A. Rapid spread of an inherited incompatibility factor in *California drosophila*.Nature. 1991;353:440–442. doi: 10.1038/353440a0.

87. Caspi-Fluger A., Inbar M., Mozes-Daube N., Mouton L., Hunter M.S., Zchori-Fein E. Rickettsia 'in' and 'out': Two different localization patterns of a bacterial symbiont in the same insect species. PLoS One.2011;6:e21096.

88. Caspi-Fluger A., Inbar M., Mozes-Daube N., Katzir N., Portnoy V., Belausov E., Hunter M.S., Zchori-Fein E. Horizontal transmission of the insect symbiont Rickettsia is plant-mediated. Proc Biol Sci.2012;279:1791–1796.

89. Brumin M., Kontsedalov S., Ghanim M. Rickettsia influences thermotolerance in the whitefly *Bemisia tabaci* B biotype. Insect Science. 2011;18:57–66. doi: 10.1111/j.1744-7917.2010.01396.x.

90. Czosnek H., Ghanim M. Back to basics: Are begomoviruses whitefly pathogens? J. Integr. Agr.2012;11:225–234.

91. Gottlieb Y., Zchori-Fein E., Mozes-Daube N., Kontsedalov S., Skaljac M., Brumin M., Sobol I., Czosnek H., Vavre F., Fleury F., Ghanim M. The transmission efficiency of To *mato yellow leaf curl virus* by the whitefly *Bemisia tabaci* is correlated with the presence of a specific symbiotic bacterium species. J. Virol.2010;84:9310–9317.

92. Rana V.S., Singh S.T., Priya N.G., Kumar J., Rajagopal R. Arsenophonus GroEL interacts with CLCuV and is localized in midgut and salivary gland of whitefly *B. tabaci*. PLoS One. 2012;7:e42168.

93. Morin S., Ghanim M., Zeidan M., Czosnek H., Verbeek M., van den Heuvel J. A GroEL homologue from endosymbiotic bacteria of the whitefly *Bemisia tabaci* is implicated in the circulative transmission of*Tomato yellow leaf curl virus*. Virology. 1999;256:75. doi: 10.1006/viro.1999.9631.

94. Morin S., Ghanim M., Sobol I., Czosnek H. The GroEL protein of the whitefly *Bemisia tabaci* interacts with the coat protein of transmissible and nontransmissible begomoviruses in the yeast two-hybrid system. Virology. 2000;276:404–416. doi: 10.1006/viro.2000.0549.

95. Azzam O., Frazer J., de la Rosa D., Beaver J.S., Ahlquist P., Maxwell D.P. Whitefly transmission and efficient ssDNA accumulation of bean golden mosaic geminivirus require functional coat protein. Virology.1994;204:289–296. doi: 10.1006/viro.1994.1533.

96. Briddon R.W., Pinner M.S., Stanley J., Markham P.G. Geminivirus coat protein gene replacement alters insect specificity. Virology. 1990;177:85–94. doi: 10.1016/0042-6822(90)90462-Z.

97. Hofer P., Bedford I.D., Markham P.G., Jeske H., Frischmuth T. Coat protein gene replacement results in whitefly transmission of an insect nontransmissible geminivirus isolate. Virology. 1997;236:288–295.

98. Noris E., Vaira A.M., Caciagli P., Masenga V., Gronenborn B., Accotto G.P. Amino acids in the capsid protein of *Tomato yellow leaf curl virus* that are crucial for systemic infection, particle formation, and insect transmission. J. Virol. 1998;72:10050–10057.

99. Delatte H., Martin D.P., Naze F., Goldbach R., Reynaud B., Peterschmitt M., Lett J.-M. South West Indian Ocean islands tomato begomovirus populations represent a new major monopartite begomovirus group. J. Gen. Virol. 2005;86:1533–1542. doi: 10.1099/vir.0.80805-0.

100. Xie W., Meng Q.S., Wu Q.J., Wang S.L., Yang X., Yang N.N., Li R.M., Jiao X.G., Pan H.P., Liu B.M., et al. Pyrosequencing the *Bemisia tabaci* transcriptome reveals a highly diverse bacterial community and a robust system for insecticide resistance. PLoS One. 2012;7:e35181.

101. Hanssen I.M., Lapidot M., Thomma B.P.H.J. Emerging viral diseases of tomato crops. Mol. Plant Microbe Int. 2010;23:539–548.

102. Moriones E., Navas-Castillo J., Díaz-Pendón J.A. Emergence of Begomoviruses Diseases. In: Caranta C., Aranda M.A., Tepfer M., Lopez-Moya J.J., editors. Recent Advances in Plant Virology. Caister Academic Press; Nortfolk, United Kingdom: 2011. pp. 301–320.

103. Nakhla M.K., Maxwell D.P., Martinez R.T., Carvalho M.G., Gilbertson R.L. Widespread occurrence of the eastern mediterranean strain of tomato yellow leaf curl geminivirus in tomatoes in the Dominican Republic. Plant Dis. 1994;78:926–926.

104. Picó B. Viral diseases causing the greatest economic losses to the tomato crop. II. The Tomato yellow leaf curl virus—A review. Sci. Hortic. 1996;67:151–196.

105. Cohen S., Lapidot M. Appearance and Expansion of TYLCV: A Historical Point of View. In: Czosnek H., editor. Tomato yellow leaf curl disease. Springer Netherlands; Dordrecht, The Netherlands: 2007. pp. 3–12.

106. Polston J.E., Anderson P.K. The emergence of whitefly-transmitted geminiviruses in tomato in the western hemisphere. Plant Dis. 1997;81:1358–1369. doi: 10.1094/PDIS.1997.81.12.1358.

107. Duffy S., Holmes E.C. Multiple introductions of the Old World begomovirus Tomato yellow leaf curl virus into the New World. Appl. Environ. Microb. 2007;73:7114–7117. doi: 10.1128/AEM.01150-07.

108. Lett J.M., Péréfarres F., Hoareau M., Lefeuvre P., De Bruyn A., Dottin M., Prior P., Wicker E., Umaharan P. Tomatoes showing yellow leaf curl symptoms in the island of Grenada exhibit an infection with Tomato yellow leaf curl virus either alone or in combination with Potato yellow mosaic virus. New Disease Reports. 2011;24:19–19. doi: 10.5197/j.2044-0588.2011.024.019.

109. Kheyr-Pour A., Bendahmane M., Matzeit V. Tomato yellow leaf curl virus from sardinia is a whitefly-transmitted monoparatite geminivirus. Nucleic Acids Res. 1991;19:6763–6769. doi: 10.1093/nar/19.24.6763.

110. Navot N., Pichersky E., Zeidan M., Zamir D., Czosnek H. Tomato yellow leaf curl virus: A whitefly-transmitted geminivirus with a single genomic component. Virology. 1991;185:151–161. doi: 10.1016/0042-6822(91)90763-2.

111. Antignus Y., Cohen S. Cloning of tomato yellow leaf curl virus (TYLCV) and the complete nucleotide sequence of a mild infectious clone. Phytopathology. 1994;84:707–712. doi: 10.1094/Phyto-84-707.

112. Abhary M., Patil B., Fauquet C. Molecular Biodiversity, Taxonomy, and Nomenclature of Tomato Yellow Leaf Curl-like Viruses. In: Czosnek H., editor. Tomato yellow leaf curl disease. Springer Netherlands; Dordrecht, The Netherlands: 2007. pp. 85–118.

113. Fauquet C.M., Sawyer S., Idris A.M., Brown J.K. Sequence analysis and classification of apparent recombinant begomoviruses infecting tomato in the Nile and Mediterranean Basins. Phytopathology.2005;95:549–555. doi: 10.1094/PHYTO-95-0549.

114. Lefeuvre P., Martin D.P., Harkins G., Lemey P., Gray A.J.A., Meredith S., Lakay F., Monjane A., Lett J.-M., Varsani A., Heydarnejad J. The spread of *Tomato eellow leaf curl virus* from the Middle East to the world. PLoS Pathogens. 2010;6:e1001164–e1001164. doi: 10.1371/journal.ppat.1001164.

115. Lombaert E., Guillemaud T., Cornuet J.-M., Malausa T., Facon B., Estoup A. Bridgehead effect in the worldwide invasion of the biocontrol harlequin ladybird. PLoS ONE. 2010;5:e9743–e9743.

116. Facon B., Genton B.J., Shykoff J., Jarne P., Estoup A., David P. A general eco-evolutionary framework for understanding bioinvasions. Trends Ecol. Evol. 2006;21:130–135. doi: 10.1016/j.tree.2005.10.012.

117. Bourriquet G. Note concernant les maladies des plantes cultivées à La Réunion. Revue Agricole Réunion. 1938;43:33–38.

118. Luziau R. Contribution à la prospection phytosanitaire de lʹîle de la Réunion. Phytoma. 1953;6:13–19.

119. Peterschmitt M., Granier M., Mekdoud R., Dalmon A., Gambin O., Vayssieres J.F., Reynaud B. First report of Tomato Yellow Leaf Curl virus in Réunion Island. Plant Dis. 1999;83:303–303.

120. Reynaud B., Wuster G., Delatte H., Soustrade I., Lett J.-M., Gambin O., Peterschmitt M. Les maladies à bégomovirus chez la tomate dans les départements français d›Outre-Mer—Le Tomato yellow leaf curl virus (TYLCV) à la Réunion. Phytoma, la défense des végétaux. 2003:13–17.

121. Delatte H., Lett J.-M., Lefeuvre P., Reynaud B., Peterschmitt M. An Insular Environment before and after TYLCV Introduction. In: Czosnek H., editor. Tomato yellow leaf curl disease. Springer Netherlands; Dordrecht, The Netherlands: 2007. pp. 13–23.

122. Delatte H., Holota H., Warren B.H., Becker N., Thierry M., Reynaud B. Genetic diversity, geographical range and origin of *Bemisia tabaci* biotype Ms. Bull. Ent. Res. 2011;101:487–497. doi: 10.1017/S0007485311000101.

123. Delatte H., David P., Granier M., Lett J.M., Goldbach R., Peterschmitt M., Reynaud B. Microsatellites reveal the coexistence and genetic relationships between invasive and indigenous whitefly biotypes in an insular environment. Genet. Res. 2006;87:109–124. doi: 10.1017/S0016672306008135.

124. Thierry M., Bile A., Grondin M., Reynaud B., Becker N., Delatte H. Evidence from mitochondrial and endosymbiotic data of the *Bemisia tabaci* Q biotype introduction in La Réunion. 2012 submitted.

125. Rimbaud L., Cottineau J.-S., Avril J.-P., Sorres V., Festin C., Minatchy J., Suzanne W., Leroux K., Tilma P., Maillary L., et al. La tomate face au TYLCV à la Réunion. Phytoma. 2012;650:48–52.

126. Delatte H., Holota H., Moury B., Reynaud B., Lett J-M., Peterschmitt M. Evidence for a founder effect after introduction of Tomato yellow leaf curl virus-mild in an insular environment. J. Mol. Evol.2007;65:112–118. doi: 10.1007/s00239-007-0005-x.

127. Delatte H., Holota H., Naze F., Peterschmitt M., Reynaud B., Lett J.M. The presence of both recombinant and non recombinant strains of *Tomato yellow leaf curl virus* on tomato in Réunion Island. Plant Pathol. 2005;54:262. doi: 10.1111/j.1365-3059.2004.01154.x.

128. Sánchez-Campos S., Navas-Castillo J., Camero R., Soria C., Díaz J.A., Moriones E. Displacement of*Tomato yellow leaf curl virus* (TYLCV)-Sr by TYLCV-Is in Tomato epidemics in Spain. Phytopathology. 1999;89:1038–1043. doi: 10.1094/PHYTO.1999.89.11.1038.

129. Davino S., Napoli C., Davino M., Accotto G.P. Spread of Tomato yellow leaf curl virus in Sicily: Partial displacement of another geminivirus originally present. Eur. J. Plant Pathol. 2006;114:293–299. doi: 10.1007/s10658-005-5805-5.

130. Monci F., Sanchez-Campos S., Navas-Castillo J., Moriones E. A natural recombinant between the geminiviruses *Tomato yellow leaf curl Sardinia virus* and *Tomato yellow leaf curl virus* exhibits a novel pathogenic phenotype and is becoming prevalent in Spanish populations. Virology. 2002;303:317–326. doi: 10.1006/viro.2002.1633.

131. García-Andrés S., Accotto G.P., Navas-Castillo J., Moriones E. Founder effect, plant host, and recombination shape the emergent population of begomoviruses that cause the tomato yellow leaf curl disease in the Mediterranean basin. Virology. 2007;359:302–312. doi: 10.1016/j.virol.2006.09.030.

132. Davino S., Napoli C., Dellacroce C., Miozzi L., Noris E., Davino M., Accotto G.P. Two new natural begomovirus recombinants associated with the tomato yellow leaf curl disease co-exist with parental viruses in tomato epidemics in Italy. Virus Res. 2009;143:15–23. doi: 10.1016/j.virusres.2009.03.001.

133. Péréfarres F., Lefeuvre P., Hoareau M., Thierry M., Becker N., Reynaud B., Dintinger J., Lett J.-M. Rapid displacement as a result of interaction between strains of TYLCV in Reunion Island. Acta Horticulturae. 2011;914:197–201.

134. Péréfarres F., Thébaud G., Lefeuvre P., Rimbaud L., Hoareau M., Chiroleu F., Reynaud B., Lett J.-M. Long-term viral competition monitoring: A case of epidemiological rescue. 2012. in preparation.

135. Castillo-Urquiza G.P., Beserra J.E.A., Bruckner F.P., Lima A.T.M., Varsani A., Alfenas-Zerbini P., Murilo Zerbini F. Six novel begomoviruses infecting tomato and associated weeds in southeastern Brazil.Arch. Virol. 2008;153:1985–1989. doi: 10.1007/s00705-008-0172-0.

136. Fernandes F.R., de Albuquerque L.C., de Britto Giordano L., Boiteux L.S., de Avila A.C., Inoue-Nagata A.K. Diversity and prevalence of Brazilian bipartite begomovirus species associated to tomatoes.Virus Genes. 2008;36:251–258. doi: 10.1007/s11262-007-0184-y.

137. Martin D.P., Biagini P., Lefeuvre P., Golden M., Roumagnac P., Varsani A. Recombination in eukaryotic single stranded DNA viruses. Viruses. 2011;3:1699–1738. doi: 10.3390/v3091699.

138. Rey M.E.C., Ndunguru J., Berrie L.C., Paximadis M., Berry S., Cossa N., Nuaila V.N., Mabasa K.G., Abraham N., Rybicki E.P., et al. Diversity of dicotyledenous-infecting geminiviruses and their associated DNA molecules in Southern Africa, including the south-west Indian Ocean islands. Viruses. 2012;4:1753–1791. doi: 10.3390/v4091753.

139. Navas-Castillo J., Sánchez-Campos S., Noris E., Louro D., Accotto G.P., Moriones E. Natural recombination between Tomato yellow leaf curl virus—Is and Tomato leaf curl virus. J. Gen. Virol.2000;81:2797–2801.

140. Jones R.A.C. Plant virus emergence and evolution: Origins, new encounter scenarios, factors driving emergence, effects of changing world conditions, and prospects for control. Virus Res. 2009;141:113–130. doi: 10.1016/j.virusres.2008.07.028.

141. Polston J.E., McGovern R.J., Brown L.G. Introduction of *Tomato yellow leaf curl virus* in Florida and implications for the spread of this and other geminiviruses of tomato. Plant Dis. 1999;83:984–988. doi: 10.1094/PDIS.1999.83.11.984.

CHAPTER 7

CHARACTERIZATION OF A PROPOSED DICHORHAVIRUS ASSOCIATED WITH THE CITRUS LEPROSIS DISEASE AND ANALYSIS OF THE HOST RESPONSE

José Luis Cruz-Jaramillo[1], Roberto Ruiz-Medrano[1], Lourdes Rojas-Morales[2], José Abel López-Buenfil[1,3], Oscar Morales-Galván[3], Claudio Chavarín-Palacio[3], José Abrahán Ramírez-Pool[1], and Beatriz Xoconostle-Cázares[1]

[1]Departamento de Biotecnología y Bioingeniería, Centro de Investigación y de Estudios Avanzados del Instituto Politécnico Nacional Av. IPN 2508, Zacatenco 07360, México D.F., Mexico

[2]LaNSE, Centro de Investigación y de Estudios Avanzados del IPN Av. IPN 2508, Zacatenco 07360, México D.F., Mexico

[3]Servicio Nacional de Sanidad Inocuidad y Calidad Agroalimentaria, Guillermo Pérez Valenzuela 127, Coyoacán 04100, México D.F., Mexico

ABSTRACT

The causal agents of Citrus leprosis are viruses; however, extant diagnostic methods to identify them have failed to detect known viruses in orange, mandarin, lime and bitter orange trees with severe leprosis symptoms in Mexico, an important citrus producer. Using high throughput sequencing, a virus associated with citrus leprosis was identified, belonging to the proposed *Dichorhavirus* genus. The virus was termed Citrus Necrotic Spot Virus (CNSV) and contains two negative-strand RNA components; virions accumulate in the cytoplasm and are associated with plasmodesmata—channels interconnecting neighboring cells—suggesting a mode of spread within the plant. The present study provides insights into the nature of this pathogen and the corresponding plant response, which is likely similar to other pathogens that do not spread systemically in plants.

INTRODUCTION

Citrus leprosis is a viral disease with important economic implications that is rapidly spreading in the New World. It has been reported throughout America, although not observed in the last four decades in the U.S., causing severe damage in different citrus cultivars [1,2,3,4]. Besides the appearance of symptoms, which result in lower fruit quality, leprosis negatively impacts productivity, which is affected by a decrease in foliar area, premature leaf abscission and branch death. This disease is widespread from Argentina to Costa Rica; recently, it has been detected in Chiapas, Southern Mexico, and different viruses appear to be associated with it [5]. However, the precise identity of the associated pathogen remains to be determined, at least in Mexico. More recently, *Citrus* spp. plants (sweet orange, mandarin and grapefruit) showing leprosis symptoms have been observed in the state of Jalisco, central-western Mexico, indicating the spread of the disease. Thus, the complete characterization of this pathogen is necessary to design plausible control strategies.

Leprosis symptoms include local, albeit severe, necrotic symptoms in infected leaves, as well as in fruits, in addition

Citation: José Luis Cruz-Jaramillo, Roberto Ruiz-Medrano, Lourdes Rojas-Morales, José Abel López-Buenfil, Oscar Morales-Galván, Claudio Chavarín-Palacio, José Abrahán Ramírez-Pool and Beatriz Xoconostle-Cázares, Characterization of a Proposed Dichorhavirus Associated with the Citrus Leprosis Disease and Analysis of the Host Response, doi:10.3390/v6072602.

to corked stems. The most notable cytopathic effects are the appearance of electron-dense inclusion bodies in the cytoplasm or in the nucleus [4]. The cytoplasmic leprosis is the more prevalent. Viruses that cause leprosis symptoms in various citrus cultivars have been identified [6,7,8,9,10,11]. The most widely distributed is the *Citrus leprosis virus* cytoplasmic type, CiLV-C, which harbors two single-stranded positive stranded RNA components, both of which are capped and polyadenylated [7]. This virus is the type member of the *Cilevirus* genus and is the etiological agent of cytoplasmic leprosis [7]. More viruses have been identified causing similar diseases in citrus, such as the newly discovered *Citrus leprosis virus* cytoplasmic type 2 (CiLV-C2) in Colombia, which is related to other Cileviruses; it consists of two single-stranded (ss) positive-stranded RNA components. While its genomic organization is similar to CiLV-C, it possesses an additional open reading frame in the RNA2 component [11]. Another related virus infecting *Citrus volkameriana*, *Hibiscus green spot virus* (HGSV), harbors three ssRNA (+) components, displaying a similar genomic organization to CiLV-C2 [9]. Phylogenetic analyses of these viruses indicate that they are type members of new genera. Furthermore, all these viruses have in common their mode of transmission by mites of the *Brevipalpus* genus (Acari: Tenuipalpidae) [6] and are not systemic in nature [6,7,8,9,10,11,12,13,14]. Seventy-one species of false spider mites representing five genera (*Pseudoleptus*, *Aegyptobia*, *Tenuipalpus*,*Brevipalpus* and *Priscapalpus*) in Mexico have been described [14]. Citrus leprosis can also be transmitted mechanically [15]. In all, this evidence suggests that different viruses could elicit similar responses from their hosts and, thus, symptomatology.

There are several members of the *Rhabdoviridae* and the *Cilevirus* genus among viruses transmitted by spider mites. These are related in terms of their genome organization, overall sequence similarity, virion morphology (enveloped bacilliform structures *vs.* bullet-shaped enveloped virions) and cytopathic effects on their hosts [7,13]. However, there are important differences, namely that the genome of rhabdoviruses consists of monopartite negative-ssRNA. Additional unrelated viruses causing leprosis symptoms belong to the *Mandarivirus* genus, which are related to *Potexvirus* and, thus, display a monopartite ssRNA genome of a positive polarity [8]. In all cases, the cytopathic effects in hosts are quite similar.

Rhabdovirus particles can accumulate either in the cytoplasm or in the nucleus, forming large electron-dense inclusion bodies, or viroplasms, where the replication of the virus occurs [16,17]. The genome assembly of a potential CiLV-N in Citrus trees with high similarity to *Orchid Fleck Virus* (OFV) was described [10].

As mentioned before, citrus leprosis has been detected in Mexico; given that different viruses may cause similar symptomatology in citrus, it was not clear whether the causal agent was CiLV (cytoplasmic or nuclear) or a hitherto unknown virus. ELISA and RT-PCR performed on symptomatic leaves from infected citrus in Jalisco and Chiapas failed to detect quarantine virus or other known pathogens in Mexico. In order to determine the identity of the pathogen causing the observed symptomatology, a different strategy was devised; RNA was obtained from bitter orange (*C.* × *aurantium*) leaves showing leprosis symptoms and employed for high throughput sequencing (HTS). Novel sequencing technologies have allowed the elucidation of large genomes, and RNA-seq has been used recently to analyze diverse transcriptomes [18]. Combined with bioinformatic analysis, this methodology can be used to compare sets of transcriptomes for two different biological conditions [19,20], as well as for the detection and analysis of low-abundance RNAs, which may be useful for the detection of novel pathogens or symbionts [18]. In the present work, we describe the characterization of a novel virus, termed Citrus necrotic spot virus (CNSV) belonging to a proposed new group, *Dichorhavirus*, associated with citrus leprosis in bitter oranges in Mexico. This virus was found to be similar to *Orchid fleck virus*, as well as to a recently sequenced Citrus Leprosis Virus-Nuclear (CiLV-N) [10] and Coffee ringspot virus (CRSV), the sequences of the potential open reading frames of CRSV, which have been deposited recently in GenBank (Accession Nos. AHH44825.1,AHH44826.1, AHH44827.1, AHH44828.1, AHH44829.1 and AHH44830.1). The cytoplasmic localization, as well as the non-coding regions of this virus suggest that it is different from other Dichorhaviruses.

MATERIALS AND METHODS

Plant Material

Leaves and fruits with necrotic and/or chlorotic spots were collected from 67 citrus cultivars *Citrus* ×*aurantium* (bitter orange), lime (*Citrus sinensis*) in temperate regions of Mexico, mostly in Guadalajara, Jalisco, (20.663626° N, 103.375854° W) and Tecpatán (16.99768° N, −93.46194° W) and Ocozocuautla (16°46'0" N, 93°22'0" W), Chiapas. These symptoms were similar to those described on Citrus leprosis-affected trees. The collection of plant material was performed by the Jalisco and Chiapas State Committees for Plant Health (Comité Estatal de Sanidad Vegetal de Jalisco and Comite Estatal de Sanidad Vegetal de Chiapas, respectively) [21,22], which are government agencies authorized by the Mexican Department of Agriculture for this purpose. False spider mites were observed in infected citrus tissue. All plant samples were shipped and stored at −80 °C until used. Field studies did not involve endangered or protected species.

RNA Isolation for High Throughput Sequencing

Total RNA from *Citrus* × *aurantium* asymptomatic and symptomatic leaves obtained from ten different trees were isolated using the RNeasy kit (Qiagen, Hilden, NRW, Germany); genomic DNA was treated with DNaseI (Invitrogen, Carlsbad, CA, USA) and sent for the Illumina Whole Transcriptome Shotgun Sequencing platform (RNA-seq) to Otogenetics© Corporation (Atlanta, GA, USA). DNA sequencing was performed from cDNA, synthesized from Citrus poly(A+) RNA, as requested from the sequencing service. Two replicates of RNA-seq sets for symptomatic and asymptomatic samples were delivered. Each RNA-seq set consisted of 7.5×10^7 reads and was 74 bases long, with a total of 15.0×10^7 for each biological condition.

Bioinformatic Analysis

A Bio-Linux v7.0 workstation was employed to analyze the raw data of RNA-seq [23,24,25,26], comprised of 15 million short reads of 74 bases in length, in files in the FastQ format [27]. Quality scores were verified with FastQC, which were acceptable. Raw data were retrieved from the sequencing service and compared against the reference genome of *Citrus* × *clementina* downloaded from the Phytozome V9.0 database [25]. The reference genome was indexed with the Bowtie-build indexer tool from the Bowtie package in order to align RNA-seq reads to the *C.* × *clementina* genome [27]. Modified parameters consisting of minor gap penalizations for TopHat 2 were used, which yielded two sets of files, the first set of reads of which had homology with the reference genome and the second set of which was not aligned with the *C.* × *clementina*genome. The homologous reads were analyzed to identify differential gene expression for both asymptomatic and symptomatic tissue [24,28,29,30,31]. The differential expression was performed with the Cufflinks, Cuffmerge and Cuffdiff default parameters [24]. A reference annotation file (GFF3 format) was downloaded, as well, for *C.* × *clementina*, from the Phytozome database. Unaccepted reads that had no homology with the reference genome were first analyzed with BLAST tools and, later, were overlapped with the assembly tool in the CLC Genomics Workbench version 6.0 (CLC bio, Aarhus, J, Denmark). Several contigs were retrieved, and a BLAST was performed against the GenBank database [32]. The BLAST provided information of which contigs could be part of viral genomic components. The further reassembly of previous contigs was performed using different assembly programs, including Mira v3.9.10 [33]. Potential open reading frames (ORF) were obtained using Artemis [34], and the annotation for the complete sequences was submitted to NCBI GenBank, Accession Nos. KF198064 for CNSV RNA1 and KF198065 for CNSV RNA2.

Phylogenetic Analysis

Amino acid sequences from the *Rhabdovirus* genus were retrieved from the NCBI non-redundant protein database. Alignments for sequences of Nucleocapsid (N) and RNA-dependent RNA polymerase (L) were made with Seaview [35] and ClustalX-2 [36]. Each alignment was optimized for L sequences (the ClustalW algorithm using the PAM350 substitution matrix, an iteration after each alignment) and N sequences (the multiple sequence comparison by log-expectation algorithm, known as MUSCLE with 6 maximum iterations). The best-fit substitution models for the alignments, as well as the amino acid frequencies (+F), the proportion of invariable sites (+I), the substitution rate categories and the gamma shape (+G) parameters were selected according to ProtTest 3.0 [37] for L protein sequence alignment (RtREV + I + G + F) and N protein sequence alignment (LG + I + G + F). Phylogenetic reconstruction was performed using the PHYML platform [38,39,40], and 100 bootstraps for branch support was selected. Rooted trees were plotted with FigTree v1.4.0 [41].

Differential Expression Analysis

Differential expression from both infected and asymptomatic samples was determined as described before following the Tuxedo Protocol [24,26]. Transcript abundance was quantified in fragments per kilobase of exon per million mapped (FPKM) using the Cufflinks software package. A combined annotation file for both conditions was merged with the reference annotation consisting of the *C.* × *clementina* GFF3 file (Cuffmerge). The log2 fold change for both conditions was used to cutoff for up- or down-regulated genes.*P*-values and E-values were tested to identify significant differences between the two conditions using Cufflinks [24,42]. Genes differentially expressed were categorized by known function or possible function. The gene annotation of differentially expressed transcripts was carried out first, by searching for their gene ID sequence and later by submitting to the automatic annotation server, Blast2GO [43,44]. Annotations consisted of automatic search for sequence homology, Gene Ontology annotation, as well as signature protein InterProScan collection. In order to graphically display differential gene expression, MapMan software was employed [45]. For this purpose, *Citrus* × *aurantium* up- and down-regulated genes were employed to identify orthologous *Arabidopsis thaliana* genes, considering the annotation provided by Phytozome V9.1.

RT-PCR Analysis

Total RNA purification was performed using the method reported by Logemann *et al.* with minor modifications [46]. The resulting RNA was then cleaned up using a commercial system, following the manufacturer's instructions (Qiagen, Santa Clarita, CA, USA). Once the identification of candidate viral contigs was completed, primers were designed to detect both RNA1 and RNA2 genomic components: for RNA1, CNSV1F (5›-GCTAATCCAAGTGAGATCGATTACATGAC-3›) and CNSV1R (5›-GCTGTCCTGCCTTGTCTTGATGTCCG-3›), the target sequence is located between 69 and 428 nt in RNA1, synthesizing a 360-bp fragment; for RNA2, CNSV2F (5›-TCCCGTCCGGACTTTCACTGTCCATAAGT-3›) and CNSV2R (5›-GATGTTTGGCG AAAGGTCCA TGTGT GGAT-3›), located between 836 and 1315 nt in RNA2, synthesizing a 480 bp fragment.

A One Step RT-PCR assay was performed for the molecular detection of CNSV using the SuperScript™ III One-Step RT-PCR System with the Platinum® Taq High Fidelity Kit (Invitrogen, Carlsbad, CA, USA). The total RNA extracted from asymptomatic *C.* × *aurantium* leaves was used as the negative control. As a template for the assay, 100 ng of Total RNA samples from infected tissue were used. The RT-PCR One-Step reaction was performed following the manufacturer's instructions. The primer mixtures used were CNSV1F/CNSV1R and CNSV2F/CNSV2R, expecting 360-bp and 480-bp amplicons, respectively. A 12.5-μL RT-PCR reaction was prepared for each RNA sample and each primer mixture with the following program: 1 cycle at 50 °C, 30 min for cDNA synthesis; 1 cycle at 94 °C, 5 min for Platinum® Taq activation; 40 cycles at 94 °C for 35 s, 62 °C for 20 s and 72 °C for 45 s; and 1 cycle at 72 °C for 5 min. Amplification products were resolved in 1.0% w/v agarose gel stained with ethidium bromide and visualized under UV light.

Validation of Differentially Gene Expression through Quantitative RT-PCR

Differential gene expression for both symptomatic and asymptomatic samples was validated through qRT-PCR for the following transcripts encoding for: germin-like protein, thaumatin-like protein, PR-3 class IV chitinase, proline-rich protein 4-like, aquaporin tip-2 like and housekeeping cytochrome c oxidase. Forward and reverse oligonucleotides were designed in order to amplify a segment of each transcript sequences as described in Table S2. The quantitative RT-PCR reaction was set up with the KAPA SYBR® FAST One-Step qRT-PCR Universal Kit (KAPA Biosystems, Boston, MA, USA) following the manufacturer's instruction and set up on a Rotor Gene 6000® real-time amplification system (Corbett Life Science, Hilden, Germany), using the following program: 1 cycle at 42 °C, 10 min for cDNA synthesis; 1 cycle at 95 °C, 5 min for Taq activation; 45 cycles at 95 °C for 5 s, 60 °C for 30 s and 72 °C for 5 s; and a melt profile from 50–99 °C, rising 1 °C each 5 s. Absolute expression levels were quantified by constructing a standard curve using cDNA dilutions of each gene, and statistical significance was tested using the F distribution.

Transmission Electron Microscopy

Leaf tissues with chlorotic and necrotic spots were collected from symptomatic trees and processed as follows: tissue was cut in 1-mm^2 squares and transferred to microtubes containing 2% glutaraldehyde in 1× PBS for 2 h. The fixative was removed, and the tissue was rinsed twice with 1× PBS. Postfixation was carried out with 1% osmium tetroxide in the same buffer for 12 h. Samples were dehydrated in ethanol series from 20% to 100% and incubated in propylene oxide, to finally be embedded in Araldite 502 resin (Electron Microscopy Sciences, Hatfield, PA, USA). Ultrathin sections (Ultracut E Reichert Jung, Reichert Technologies, Depew, NY, USA) were placed on 200-mesh Formvar-coated copper grids. Sections were then contrasted with 2% uranyl acetate, followed by lead citrate and then examined with a JEOL 2000EX transmission electron microscope (JEOL, Kyoto, Japan) at 80 KV.

RESULTS

Citrus Leprosis Symptomatology in Citrus Plants in Mexico

Leaves, fruits and stems from 67 symptomatic citrus trees with leprosis lesions were collected in conurbated zones of Guadalajara in the state of Jalisco and Tecpatán and Ocozocuautla, Chiapas, Mexico (Figure 1A). Guadalajara City is located at 1552 m above the sea level (masl) with temperate weather, while the cities of Tecpatán and Ocozocuautla are located at 310 masl with tropical weather. Fruits exhibited chlorotic and necrotic spots in both immature and mature fruits (Figure 1B), while infected stems are corked with brownish spots (Figure 1C, center and right stems), in contrast with healthy green stems (Figure 1C, left stem). Leaf lesions consisted of chlorotic areas, with necrotic borders, asymmetrically located on the leaves (Figure 1D,E). Defoliation was a recurrent symptom observed in trees affected by this disease.

Figure 1: Symptoms of leprosis in bitter orange in Mexico. (A) Plant tissue was collected in the urban zones of Guadalajara, Jalisco, and Tecpatán and Ocozocuautla, Chiapas; shown in blue in the map. (B) Typical symptoms found in bitter orange (*C.* × ...

A Virus Belonging to the Proposed *Dichorhavirus* Genus is Associated with Citrus Leprosis Symptoms in Mexico

The agent causing leprosis-like symptoms in citrus cultivars in Mexico had not been identified, and since extant diagnostic methods yielded negative results for bacteria and known viruses, it was hypothesized that an unidentified virus causes this disease. Samples from sweet orange, bitter orange and grapefruit showing such symptoms were tested for CiLV-C and HGSV, which yielded negative results (Figure S1). Thus, in order to gain insight into the etiology of this disease, high throughput sequencing of RNA from symptomatic leaves of sour orange (*Citrus* × *aurantium*) was carried out. This plant is a hybrid of *C. maxima* and *C. reticulata* and is widely grown in Mexico, and in Texas and Florida in the U.S. [47]. Since *Citrus* ×*clementina* (clementine, which is a *Citrus* × *aurantium* × *C. reticulata* hybrid) is the closest relative of *Citrus*× *aurantium*, its genome was used as a scaffold to determine the transcriptome of symptomatic leaf tissue. The 3 × 10^7 reads were obtained and compared to the *C.* x *clementina* genome database, as well as to another RNA sample from asymptomatic leaves from *C.* × *aurantium*. Approximately 1.2 × 10^4 reads did not match the *C.* x *clementina* nor *C. sinensis* genomes in symptomatic samples. The coverage of the viral genome by the contigs was redundant (Table 1). No other contigs absent from the *C.* x *aurantium*or *C. sinensis* genome databases were found, with the exception of a bacteriophage highly similar to φX174. Additionally, only 3 reads for CNSV RNA1 and 4 for CNSV RNA2 were found in the asymptomatic tissue used as a control; these reads showed close similarity to OFV and to a recently reported CiLV-N [10]. The reads were assembled into two genomic RNA components that are indeed similar to OFV and CiLV-N (90% overall homology).

Table 1: Reads of symptomatic and asymptomatic samples

Total no. of reads for symptomatic tissue	7,737,481
Total no. of reads for asymptomatic tissue	7,291,897
Total	15,029,378
Total no. of reads for RNA1 (6087 nt)	12,083
Average read	196.87 nt
Total no. of reads for RNA2 (6015 nt)	12,079
Average read	199.39 nt
Total length of coverage of cDNA reads for symptomatic tissue	2.01 Gb
Total length of coverage of cDNA reads for asymptomatic tissue	1.90 Gb

The assembled viral genome consists of two ssRNA components, RNA1, 6495 nt in length, and RNA2, 6018 nt in length. RNA1 harbors four open reading frames, potentially coding for the nucleocapsid protein (N) (ORF1, 450 aa in length, 60.5 kDa); a putative phosphoprotein (P) that also forms part of the nucleocapsid (ORF2, 237 aa, 19.8 kDa); a putative movement protein (MP) (ORF3, 370 aa, 41.6 kDa), which has been functionally characterized in other plant rhabdoviruses [48,49]; the putative matrix protein (M) involved in maintaining virion shape (ORF4, 183 aa, 26.4 kDa); and a potential glycoprotein (G) of unknown function (ORF5, 450 aa, 49.2 kDa) (Figure 2A). On the other hand, RNA2 consists of a single ORF (ORF6), potentially coding for the RNA-dependent RNA polymerase (L), which is an 853-aa protein with a theoretical MW of 212 kDa (Figure 2B). The putative function of each ORF was deduced on the basis of similarity to the OFV RNA1 and 2 sequences [50]. A comparison between OFV, CiLV-N and CNSV shows a highly conserved genome organization. All ORFs from these viruses are highly similar; however, both leader and trailer RNA sequences in the two components do differ between CNSV and OFV (Figure S3). The CNSV RNA1 leader displays a similarity of 89% with OFV, while CiLV-N shows 96% homology to OFV. Because of the striking similarity between CiLV-N and OFV sequences, they appear to be variants of the same strain. Both intergenic regions (IGR) and trailer RNA are highly conserved in the three compared sequences. A comparison of RNA1 and RNA2 components (Figure S3) shows differences in both leader and trailer sequences, *i.e.*, there are three deletions in both CILV-N and OFV in RNA1, when compared to CNSV, while intergenic region 1 (IGR1) also harbors a four-base deletion. Of note is the RNA1 trailer region, in which CNSV has the larger sequence: 34 bp more than OFV and 184 bp more than CiLV-N. As for the RNA2 leader, two deletions are again present in OFV and CILV-N (Figure S3), as well as two deletions of four and one bases, respectively, located toward the ORF6 5’ end. Based on these differences, we propose that this could be considered a different strain of OFV and CiLV-N, which was designated provisionally Citrus necrotic spot virus (CNSV).

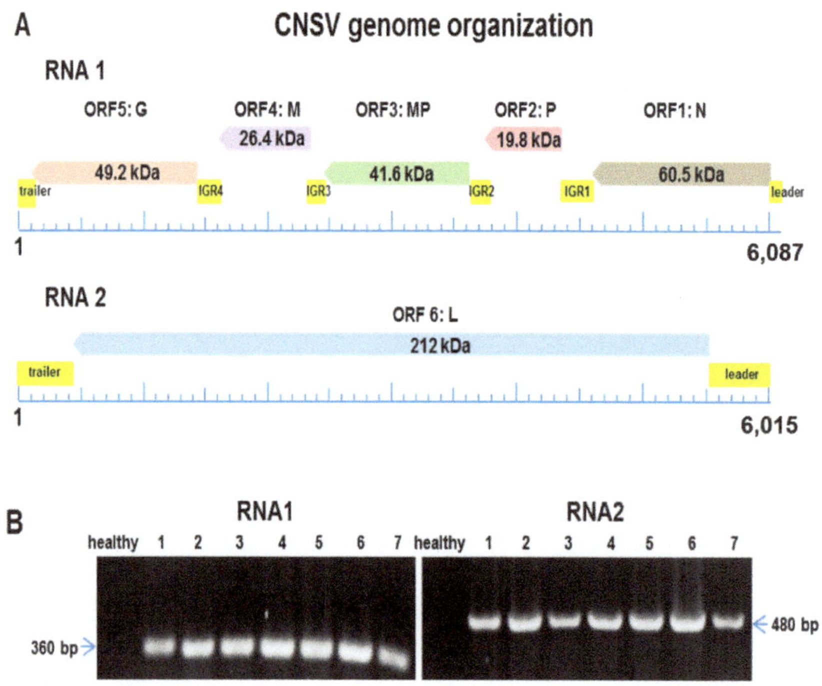

Figure 2: Map of Citrus Necrotic Spot Virus (CNSV) genome components and putative open reading frames in both RNA1 and two components. (A) Leader and trailer RNAs and intergenic regions (IGR) are also depicted. N, nucleocapsid protein; P, phosphoprotein; MP, movement...

It must be mentioned that while the assembled sequence of RNA2 displayed negative polarity, as do all rhabdoviruses hitherto described so far, the contigs assembled with Cufflinks corresponding to RNA1 and RNA2 displayed both polarities, suggesting the identification of possible RNA(+) replicative intermediates. Therefore, the ratio of sense *versus* anti-sense sequencing reads was also determined. As indicated by the log-odds (Lods) values of sense to antisense reads, RNA1 was marginally enriched for sense-mapping reads and RNA2 for antisense reads. Based on the genome organization and homology with the most closely related virus (OFV), it is likely that both RNA genome components display negative polarity.

The phylogenetic relationship of CNSV to other viruses was determined based on the ORF1 (nucleocapsid protein, N) and ORF6 (RNA-dependent RNA polymerase, L) amino acid sequences, as well as on the full-length genome. Rooted distance trees were constructed with 100 bootstraps. In all cases, CNSV forms a single clade with OFV and CiLV-N and, thus, can be considered a member of the proposed clade, Dichorhavirus ([50]: Figure 3 and Figure 4). This clade, based on different trees, is further located from other plant Rhabdovirus groups. A closer inspection of the tree based on the nucleocapsid protein sequence revealed that CNSV and OFV do not fall within the nucleorhabdovirus clade (Figure 3). Similar results were obtained with the RdRp and whole genome sequences (Figure 4). OFV, CiLV-N and CNSV likewise are grouped in a separate clade outside cytorhabdovirus and nucleorhabdovirus. This clade comprising CNSV, as well as OFV and CiLV-N, correspond to the newly proposed Dichorhavirus genus [6], which includes bipartite, negative-strand RNA plant viruses.

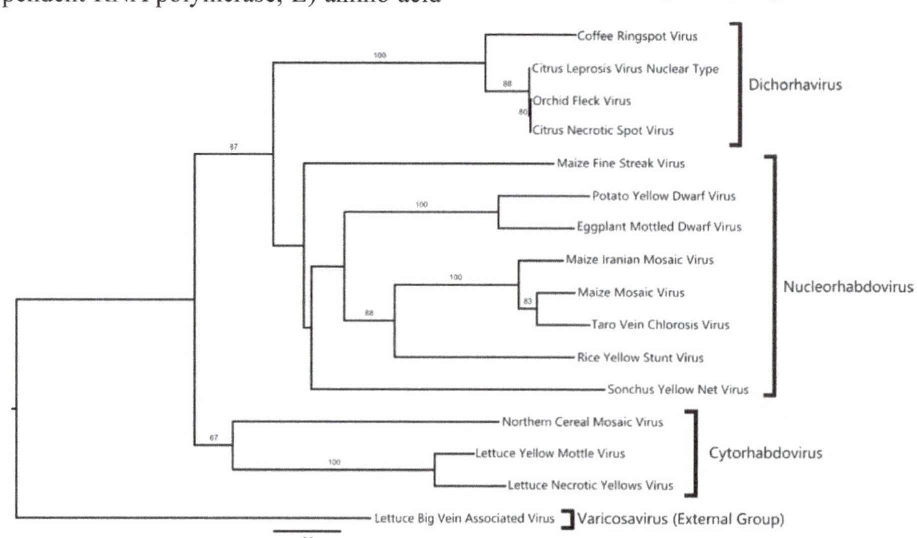

Figure 3: Phylogenetic reconstruction for nucleocapsid sequences. Tree construction is based on the maximum likelihood model with a bootstrap value indicated above supported branches. CNSV and*Orchid Fleck Virus* (OFV) and CiLV-N are grouped in a single branch,...

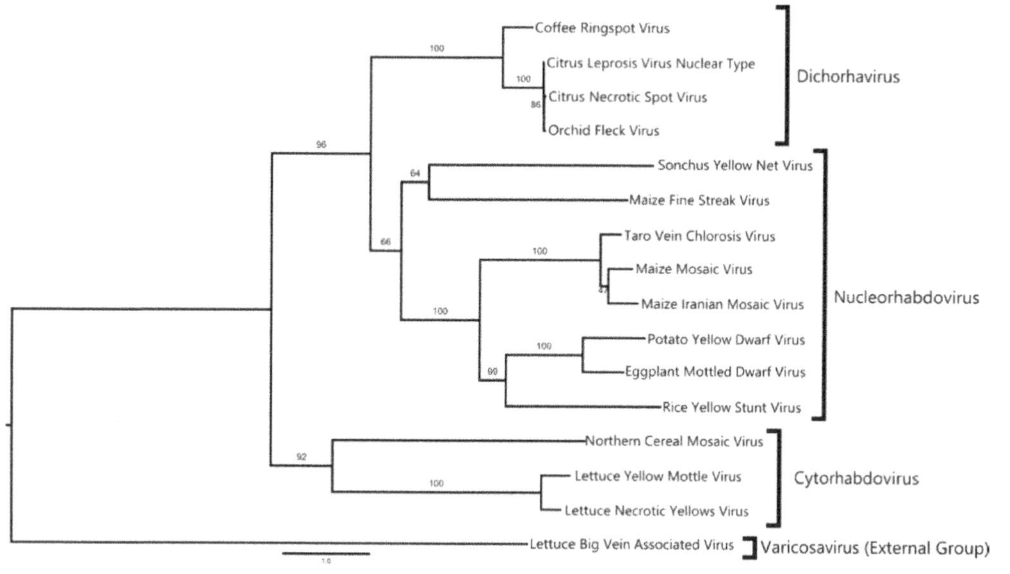

Figure 4: Phylogenetic reconstruction for RNA-dependent RNA polymerase sequences. Tree construction is based on the maximum likelihood model with a bootstrap value indicated above supported branches. The proposed Dichorhaviruses are grouped in a single branch.

RT-PCR analysis was carried out for 67 leaf samples showing leprosis symptoms from diverse citrus cultivars, including *C.* × *aurantium*, *C. sinensis*, *C. tangerine* and *C. paradisi*. Sixty five samples yielded positive results for both RNA1 and RNA2 CNSV components and failed to detect them in asymptomatic tissue. The analysis of eight samples is shown in Figure 2. Thus, there is a strong correlation between the presence of CNSV and leprosis symptoms, considering that no other extant virus was detected using diagnostic techniques, such as the PCR, RT-PCR and ELISA assays, as well as no additional RNA sequences were identified by deep-sequencing.

CNSV Particles Are Located in the Cytoplasm of Infected Cells

Considering CNSV belongs to a clade with similarity to cyto- and nucleo-rhabdovirus, it was necessary to determine the subcellular localization of CNSV in infected tissue. An analysis using transmission electron microscopy was carried out to determine the presence of virions in symptomatic sour orange (*Citrus* ×*aurantium*) samples showing leprosis symptoms. As a control, the leaves of asymptomatic plants were also processed for image comparison. Rounded, membranous, electron-dense structures were identified in the cytoplasm with an average size of 120 to 475 nm (Figure 5A–F). The structure with the largest diameter (Figure 5F) shows a cross-section of a symmetric viroplasm, surrounded by a host membrane, containing 16 rod-shaped viral structures, with a radial arrangement. These structures suggest that the viroplasm is a sphere containing rod-shaped virion particles with an average size of 67 × 27 nm. Unassembled structures are either linear or semicircular electron-dense filaments with an average length of 248 nm. The theoretical length of CNSV RNA components (6018 and 6495 bases) is circa 200 nm; the observed filaments of 248 nm may correspond to the viral genomes likely associated with N nucleoprotein. Interestingly, an electron-dense network was present in the chloroplast of infected leaves (Figure 5C,D,G–I); 20 × 18-nm crystals showed a geometrical arrangement; similar inclusion bodies are formed in plant cells infected with several other viruses, but not in chloroplasts [51,52]. However, similar electron-dense bodies have been detected in the chloroplasts of plants infected with the *Bamboo mosaic virus* [53]. Such protein arrangements, as well as vesicles (Figure 5C) in the chloroplast were present only in infected cells and absent in asymptomatic samples. This suggests that this protein crystal network is of a viral nature or at least induced by viral infection [54,55].

Regarding the presence of viral particles in the nucleus, no evidence of electron-lucent viroplasms were observed, characteristic of Nucleorhabdovirus; indeed, the cytopathic effects of Nucleorhabdovirus, *i.e.*, large amounts of granular or fibrous viroplasms of various sizes occupying most of the nucleus and that are frequently observed near the nucleolus [50], were not observed in the analyzed tissue; instead, rare, small translucent areas were detected in the nuclei (Figure 6K,L). However, the presence of membranous material covering the viroplasm and its "spoke wheel" structure suggest that virus replication and virion formation could occur in the cytoplasm, possibly in the endoplasmic reticulum and in the outer nuclear envelope, in a manner reminiscent of Rhabdoviruses [17]. No budding of viral particles from nuclei was observed in the analyzed samples, supporting the notion that CNSV accumulation and replication do not occur in the nucleus, although it cannot be discarded that some stages of replication and nucleocapsid assembly do occur in the nucleus.

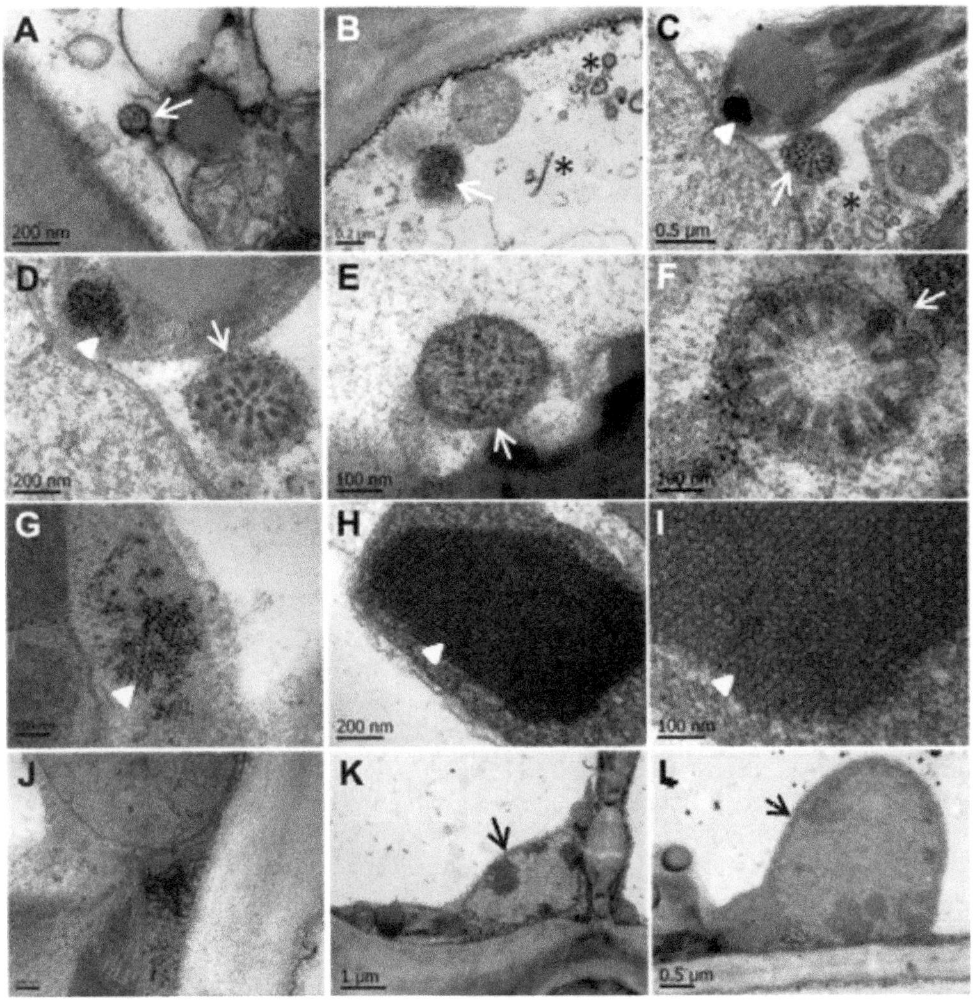

Figure 5: CNSV particles are located in the cytoplasm of infected cells. (A–F) Electron-dense viral particles are located in viroplasms within the cytoplasm of infected cells. Electron-dense structures with a membrane unit and average size of 120 to 475 ...

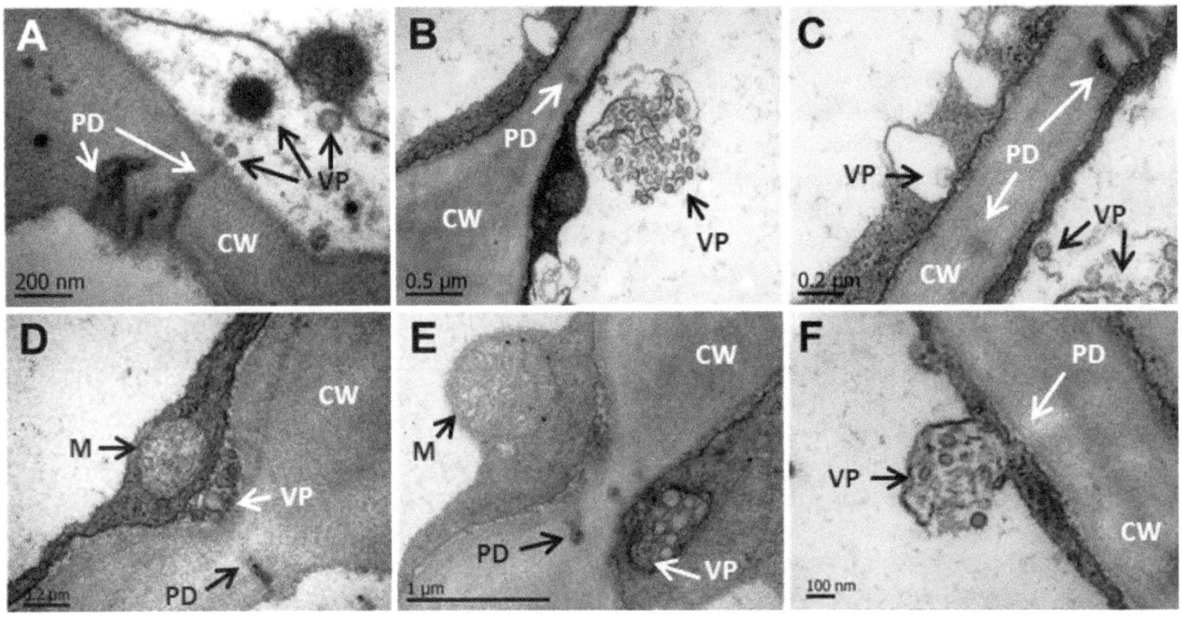

Figure 6: CNSV particles are associated with cell-to-cell interconnecting plasmodesmata. (A–F) Viroplasms (VP) and unassembled virions localize to secondary plasmodesmata (PD). Specialized membranous sacs show virus assembly-disassembly (C). Thinner cell ...

CNSV Particles Associate with Unmodified Plasmodesmata

TEM analysis showed numerous bacilliform virions and viroplasms associated with plasmodesmata (PD). Figure 6A shows two secondary PD associated with an electron-dense viroplasm; indeed, unassembled virions appearing as electron-dense filaments with an average length of 248 nm are in the close boundary of a PD, while a smaller and possibly immature viroplasm appears in the endoplasmic reticulum (ER) appressed onto the boundary of the neighboring cell (Figure 6B). In a magnified image, the transport of potential virion-associated macromolecules to the neighboring cell is suggested in (Figure 6C,F). On a speculative note, enveloped structures (probably virions) appear associated with the ER (Figure 6C–F). Interestingly, the thickness of the cell wall in the trafficking area (Figure 6D,E) is thinner than in other regions of the cell wall.

CNSV Elicits the Accumulation of Multiple Host Transcripts

Using mapped RNA-seq reads for both conditions, a total of 134 transcripts were found to differentially accumulate in symptomatic and asymptomatic leaf samples. One hundred eighteen of the corresponding transcripts displayed increased levels, while 16 were downregulated. Gene Ontology (GO) assignments were used to classify the functions of the up- and down-regulated genes. These fell into 12 categories of Level 2 GO terms (Tables S3 and S4; Figure S4), involved in the response to both biotic and abiotic stress, cell death, reproduction, development, metabolism, regulation, growth, signaling, localization and cellular organization and processes. Most of the genes with increased expression levels are involved in stress and defense responses, such as those encoding pathogenesis-related proteins 1 and 3, Mlo-like proteins, serine protease inhibitors, phenyl alanine ammonia lyase (PAL), BAHD acyl transferase-like proteins and chalcone synthase (CHS) [56,57]. Indeed, PAL and CHS are the central nodes for the synthesis of secondary metabolites, among them antimicrobial compounds [58,59]. Interestingly, some of the induced genes are involved in the defense response against mechanical damage and herbivory, such as a serine protease inhibitor (log2 fold-change = 4.8) and germin-like protein (log2 fold-change = 11.3). Germin-like proteins have been implicated in development and the defense response in various plant species [60]. The identity of other genes suggests a role in the response to viral pathogenesis, as in the case of a cysteine-rich receptor-like protein kinase (log2 fold-change = 6.0). Of note, several induced genes encode proteins involved in resistance against phytopathogenic fungi, such as *Cladosporium fulvum* (the Avr9 cf-9 rapidly elicited protein; log2 fold-change = 3.6) [61], which points to a non-specific response to viral infection. It must be mentioned that the control corresponds to an asymptomatic plant, and while extremely low viral levels were detected through HTS, it could still have elicited similar responses.

A graphic representation of the differentially expressed genes described above is shown in Figure S4, in which genes are classified according to cellular functions, metabolic processes, secondary metabolites and responses to biotic stress.

The deep sequencing of symptomatic and asymptomatic tissues revealed the unexpected presence of a phage with high similarity to ϕx174. Interestingly, there were approximately four-times more reads for this phage than for the total found for CNSV, although these were obtained in both symptomatic and asymptomatic leaves. It is likely that these RNAs correspond to an environmental sample from the phylloplane [62]. BLAST analysis of all the other assembled transcripts, in contrast, correspond to *Citrus* genes, disregarding that these may also be contaminants from the phylloplane.

Considering the high log2 cutoff, few downregulated genes were identified. These include a heavy metal transport detoxifying protein, aquaporin TIP2, and a number of genes associated with primary metabolism. An alternate validation of five differentially expressed genes was performed by quantitative real time RT-PCR (Figure 7). The fold change was statistically significant using the F distribution ($p = 0.05$), thus confirming the differential expression of genes in virus-infected Citrus plants.

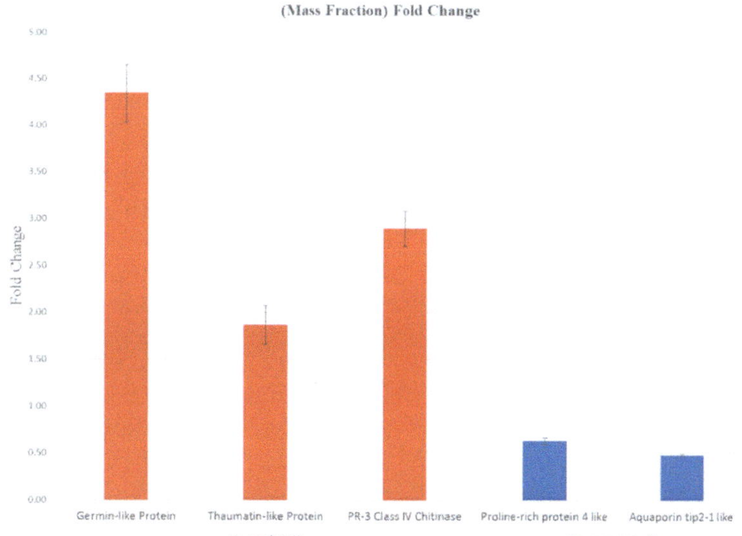

Figure 7: Validation of five differentially expressed genes through qRT-PCR. The fold change is statistically significant. Red, upregulated genes; blue, downregulated genes, according to RNA-seq analyzed data. The standard deviation is indicated with bars.

Considering the differential expression through bioinformatic analyses, the genes annotated as germin-like protein subfamily 1 member 14, thaumatin-like protein, PR-3 class IV chitinase were validated through qRT-PCR (Figure 7), indicating that these genes (associated principally with the plant response against pathogens) are overexpressed throughout infected samples. Two downregulated genes encoding for the proline-rich protein 4-like and aquaporin tip-2-like identified in the RNA seq analyses were also confirmed to reduce their expression in infected tissue (Figure 7 and Table S4).

DISCUSSION

Several diseases threaten the citrus industry, among them huanglongbing, citrus tristeza and citrus leprosis; the latter may be caused by different viruses, although eliciting common symptoms. Among the causal agents, HGSV [9] and, more recently, CiLV-C [7,11] and CiLV-N [10] have been already described. The etiological agent of nuclear leprosis, so-called because of the appearance of electron-lucent bodies in the nucleus, has been considered to be a species related to OFV [63].

CNSV Belongs to the New Proposed *Dichorhavirus* Genus

In this work, using deep sequencing of RNA from citrus plants showing leprosis symptoms, we identified a Dichorhavirus associated with leprosis symptoms, showing a high similarity to OFV, which was provisionally termed Citrus Necrotic Spot Virus (CNSV). The phylogeny of these viruses indicate that these fall within the same clade, though outside of the Nucleorhabdovirus and Cytorhabdovirus. Along with OFV and CiLV-N, they represent a unique clade of negative ssRNA virus, whose genome is bipartite. Since the global homology between CNSV and OFV/CiLV-N is close to 90%, these viruses can be considered strains or closely related species, although all belong to the dichorhavirus branch [6]. This clade is also very similar to a recently sequenced virus, Coffee ringspot virus (CRSV).

Rhabdoviruses are a diverse group of viruses infecting both plants and animals; aside from the negative polarity of their ssRNA genome, they all share the lipidic membranous envelope surrounding its capsid. The nuclei of infected cells harbor electron-lucent regions, as has been observed with OFV and nucleorhabdoviruses (OFV) [63,64]. The electron microscopy data presented here suggests that CNSV does not form viroplasms in the nuclei of infected cells (Figure 6K, L). Additionally, CNSV belongs to an unusual clade that includes OFV, which has been extensively described [6,63,64]. Bullet-shaped virus particles, a hallmark of rhabdovirus, were not observed; instead, long filaments that could consist of RNA and viral and/or host proteins were detected in large spoke-wheels structures, similar to those observed in OFV-infected *Odontoglossum* [50,55,64]. Interestingly, these structures reside in the cytoplasm, and no nucleus-budding viral structures could be identified in the analyzed nuclei; however, no CiLV-N ultrastructure is available for comparison. Furthermore, large electron-dense networks were observed within chloroplasts of CNSV-infected cells. It is possible that these large aggregates (perhaps consisting of viral and host proteins) may interfere with chloroplast function, causing some of the macroscopic symptoms, such as chlorosis. The close association of vesicles, possibly harboring viral RNA and filamentous particles, with plasmodesmata is suggestive of the mechanism for CNSV cell-to-cell transport, *i.e.*, through unmodified plasmodesmata. Indeed, no evidence was found for tubular structures formed during the intercellular transport of certain plant viruses. Experiments with protoplasts using labeled viral proteins, in particular the movement protein, which is probably involved in cell-to-cell transport, will help to clarify this question.

CNSV Induces a Characteristic Set of Genes during Infection

Mites of the genus, *Brevipalpus*, widely distributed in the studied areas, transmit leprosis disease in citrus. In general, viruses transmitted by these vectors do not move systemically in an extensive manner. Rather, they move cell-to cell, thriving in sectors of leaves and fruits, and it is likely that they do not reach the vasculature. Since most signals activating the defense response against pathogens are transported through the vascular tissue [65], this could result in a slower activation of a systemic response. Viral infection was found to induce strongly a germin-like protein mRNA. These proteins may have a role in defense to biotic and abiotic stress [58] and development [66], and the overexpression of a member of this large family in tobacco increases resistance to Geminivirus infection [67]. Interestingly, a beta glucanase encoding transcript was highly accumulated; this is a non-cell-autonomous pathway protein (NCAPP), described as increasing the gating capacity of plasmodesmata and, thus, facilitating viral movement [68]. While differentially regulated genes fell in several GO categories, most (54%) are classified within the stress response and defense categories. Indeed, tissue necrosis, if limited, is a plant response that results in pathogen limitation. Proteases and cell wall remodeling activities, among others, account for the slow progression of the cell-to-cell viral movement. Transcripts for proteins involved in cell signaling and innate plant immune response accumulate to considerably high levels; all of these data strongly suggest the notion of a coordinate plant defense against the pathogen. Secondary metabolism could also play an important role in response to viral infection, given that RNAs for two key enzymes, PAL and CHS, accumulate to high levels. PAL is the key enzyme in phenylpropanoid biosynthesis, and CHS is involved in the synthesis of flavonoids, among several other metabolites. However, it is not clear whether these responses have an essential role in slowing virus replication and/or spread. It must be mentioned that the relatively low number of differentially expressed genes is due to the high cutoff value set (a log2 fold value), and that asymptomatic plants are subliminally infected. Furthermore, several reads that did not match the *C.* × *clementina* or *C. sinensis* genomes could correspond to genes specific to *C.* × *aurantium* and, in particular, to untranslated regions or even small RNAs. The observed set of stress- and defense-related genes altering their expression profiles in response to viral infection are in agreement with those identified in general plant responses to viral infections [69,70].

Ongoing epidemiological studies will provide valuable information regarding the extent of Citrus Necrotic Spot Virus infections in Mexico, one of the main citrus producers in the world.

ACKNOWLEDGMENTS

This manuscript is dedicated to Javier Trujillo Arriaga, for his contributions to the Mexican Plant Health. The authors acknowledge help from the Federal Committee for Plant Health (Comités Estatales de Sanidad Vegetal) from Jalisco and Chiapas States, Mexico (CESAVEJAL and CESAVE Chiapas), for conducting the extensive search and collection of infected citrus samples. J.L.C.-J. and J.A.R.-P. are CONACyT (Consejo Nacional de Ciencia y Tecnología) fellows. The work was supported by CONACyT and SENASICA-SAGARPA grants.

AUTHOR CONTRIBUTIONS

B.X.-C. devised the experiments and the general strategy. J.L.C.-J. performed the molecular biology procedures and bioinformatic analysis. R.R.-M. helped with the experiment design. L.R.-M. and B.X.-C. carried out the electron microscopy analysis. J.L.C.-J., B.X.-C., R.R.-M., J.A.L.-B., O.M.-G., C.C.-P. and J.A.R.-P. analyzed the data, and B.X.-C., R.R.-M. and J.L.C.-J. wrote the paper, which was read and approved by all of the authors.

REFERENCES AND NOTES

1. Chagas C.M. Leprosis and zonate chlorosis. In: Timmer L.W., Garnsey S.M., Graham J.H., editors.Compendium of Citrus Diseases. 2nd ed. American Phytopathological Society Press; St. Paul, MN, USA: 2000. pp. 57–58.
2. Childers C.C., Rodrigues J.C.V., Derrick K.S., Achor D.S., French J.C., Welbourn W.C., Ochoa R., Kitajima E.W. Citrus leprosis and its status in Florida and Texas: Past and present. Exp. Appl. Acarol.2003;30:181–202. doi: 10.1023/B:APPA.0000006548.01625.72.
3. Bastianel M., Freitas-Astúa J., Nicolini F., Segatti N., Novelli V.M., Rodrigues V., Medina C.L., Machado M.A. Response of mandarin cultivars and hybrids to Citrus leprosis virus. J. Plant Pathol. 2008;90:305–310.
4. Bastianel M., Novelli V.M., Kitajima E.W., Kubo K.S., Bassanezi R.B., Machado M.A., Freitas-Astúa J. Citrus leprosis: Centennial of an unusual mite–virus pathosystem. Plant Dis. 2010;94:284–292. doi: 10.1094/PDIS-94-3-0284.
5. SENASICA (Servicio Nacional de Sanidad, Inocuidad y Calidad Agroalimentaria) Informe Sobre Situación de la leprosis de los cítricos. [(accessed on 1 May 2014)]. Available online:http://www.senasica.gob.mx/includes/asp/download.asp?IdDocumento=907&IdUrl=1519/
6. Dietzgen R.G., Kuhn J.H., Clawson A.N., Freitas-Astúa J., Goodin M.M., Kitajima E.W., Kondo H., Wetzel T., Whitfield A.E. Dichorhavirus: A proposed new genus for Brevipalpus mite-transmitted, nuclear, bacilliform, bipartite, negative-strand RNA plant viruses. Arch. Virol. 2014;159:607–619. doi: 10.1007/s00705-013-1834-0.
7. Locali-Fabris E.C., Freitas-Astúa J., Souza A.A., Takita M.A., Astúa-Monge G., Antonioli-Luizon R., Rodrigues V., Targon M.L., Machado M.A. Complete nucleotide sequence, genomic organization and phylogenetic analysis of Citrus leprosis virus cytoplasmic type. J. Gen. Virol. 2006;87:2721–2729. doi: 10.1099/vir.0.82038-0.
8. Prabha K., Baranwal V.K. The genome sequence of an isolate of Indian citrus ringspot virus infecting the sweet orange in India. J. Virol. 2012;86:12446–12447. doi: 10.1128/JVI.02084-12.
9. Melzer M.J., Sether D.M., Borth W.B., Hu J.S. Characterization of a virus infecting Citrus volkameriana with citrus leprosis-like symptoms. Phytopathology. 2012;102:122–127. doi: 10.1094/PHYTO-01-11-0013.
10. Roy A., Stone A., Otero-Colina G., Wei G., Choudhary N., Achor D., Shao J., Levy L., Nakhla M.K., Hollingsworth C.R., et al. Genome assembly of citrus leprosis virus nuclear type reveals a close association with orchid fleck virus. Genome Announc. 2013;1:e00519–e00513.
11. Roy A., Choudhary N., Guillermo L.M., Shao J., Govindarajulu A., Achor D., Wei G., Picton D.D., Levy L., Nakhla M.K., et al. A novel virus of the genus Cilevirus causing symptoms similar to citrus leprosis. Phytopathology. 2012;103:488–500.
12. Kitajima E.W., Chagas C.M., Rodrigues J.C. Brevipalpus-transmitted plant virus and virus-like diseases: Cytopathology and some recent cases. Exp. Appl. Acarol. 2003;30:135–160. doi: 10.1023/B:APPA.0000006546.55305.e3.
13. Rodrigues J.C., Childers C.C. Brevipalpus mites (Acari: Tenuipalpidae): Vectors of invasive, non-systemic cytoplasmic and nuclear viruses in plants. Exp. Appl. Acarol. 2013;59:165–175.
14. Baker E.W., Tuttle D.M., Abbatiello M.J. Smithsonian Contributions to Zoology. Smithsonian Institution Press; Washington, DC, USA: 1975. The False Spider Mites of Northwestern and North Central Mexico (Acarina: Tenuipalpidae) No. 194.
15. Colariccio A., Lovisolo O., Chagas C.M., Galletti S.R., Rossetti V.V., Kitajima E.W. Mechanical transmission and ultrastructural aspects of citrus leprosis disease. Fitopatol. Bras. 1995;20:208–213.
16. Jackson A.O., Dietzgen R.G., Goodin M.M., Bragg J.N., Deng M. Biology of plant rhabdoviruses.Annu. Rev. Phytopathol. 2005;43:623–660. doi: 10.1146/annurev.phyto.43.011205.141136.
17. Redinbaugh M.G., Hogenhout S.A. Plant rhabdoviruses. Curr. Top. Microbiol. Immunol. 2005;292:143–163.
18. Radford A.D., Chapman D., Dixon L., Chantrey J., Darby A.C., Hall N. Application of next-generation sequencing technologies in virology. J. Gen. Virol. 2012;93:1853–1868. doi: 10.1099/vir.0.043182-0.
19. Wang Z., Gerstein M., Snyder M. RNA-Seq: A revolutionary tool for transcriptomics. Nat. Rev. Genet.2009;10:57–63.
20. Marguerat S., Bähler J. RNA-seq: From technology to biology. Cell. Mol. Life Sci. 2010;67:569–579. doi: 10.1007/s00018-009-0180-6.
21. Comité Estatal de Sanidad Vegetal de Jalisco. [(accessed on 6 June 2013)]. Available online:http://www.cesavejal.org.mx/
22. Comité Estatal de Sanidad Vegetal de Chiapas. [(accessed on 6 June 2013)]. Available online:http://cesavechiapas.org.mx/

23. Nagalakshmi U., Waern K., Snyder M. Current Protocols in Molecular Biology. Suppl. 89. John Wiley & Sons, Inc.; Hoboken, NJ, USA: 2010. RNA-Seq: A method for comprehensive transcriptome analysis; pp. 4.11.1–4.11.13.

24. Trapnell C., Roberts A., Goff L., Pertea G., Kim D., Kelley D.R., Pimentel H., Salzberg S.L., Rinn J.L., Pachter L. Differential gene and transcript expression analysis of RNA-seq experiments with TopHat and Cufflinks. Nat. Protoc. 2012;17:562–578.

25. Goodstein D.M., Shu S., Howson R., Neupane R., Hayes R.D., Fazo J., Mitros T., Dirks W., Hellsten U., Putnam N., et al. Phytozome: A comparative platform for green plant genomics. Nucleic Acids Res.2012;40:1178–1186. doi: 10.1093/nar/gkr944.

26. Kim D., Pertea G., Trapnell C., Pimentel H., Kelley R., Salzberg S.L. TopHat2: Accurate alignment of transcriptomes in the presence of insertions, deletions and gene fusions. Genome Biol. 2013;14:R36. doi: 10.1186/gb-2013-14-4-r36.

27. Cock P.J., Fields C.J., Goto N., Heuer M.L., Rice P.M. The Sanger FASTQ file format for sequences with quality scores, and the Solexa/Illumina FASTQ variants. Nucleic Acids Res. 2010;38:1767–1771. doi: 10.1093/nar/gkp1137.

28. Langmead B. Current Protocols in Bioinformatics. Suppl. 32. John Wiley & Sons, Inc.; Hoboken, NJ, USA: 2010. Aligning short sequencing reads with Bowtie; pp. 11.7.1–11.7.14.

29. Langmead B., Trapnell C., Pop M., Salzberg S.L. Ultrafast and memory-efficient alignment of short DNA sequences to the human genome. Genome Biol. 2009;10:R25. doi: 10.1186/gb-2009-10-3-r25.

30. Mutz K.O., Heilkenbrinker A., Lönne M., Walter J.G., Stahl F. Transcriptome analysis using next-generation sequencing. Curr. Opin. Biotechnol. 2012;24:22–30.

31. Trapnell C., Williams B.A., Pertea G., Mortazavi A., Kwan G., van Baren M.J., Salzberg S.L., Wold B.J., Pachter L. Transcript assembly and quantification by RNA-Seq reveals unannotated transcripts and isoform switching during cell differentiation. Nat. Biotechnol. 2010;28:511–515. doi: 10.1038/nbt.1621.

32. Johnson M., Zaretskaya I., Raytselis Y., Merezhuk Y., McGinnis S., Madden T.L. NCBI BLAST: A better web interface. Nucleic Acids Res. 2008;36:W5–W9.

33. Chevreux B., Pfisterer T., Drescher B., Driesel A.J., Müller W.E.G., Wetter T., Suhai S. Using the miraEST assembler for reliable and automated mRNA transcript assembly and SNP detection in sequenced ESTs. Genome Res. 2004;14:1147–1159. doi: 10.1101/gr.1917404.

34. Rutherford K., Parkhill J., Crook J., Horsnell T., Rice P., Rajandream M.A., Barrell B. Artemis: Sequence visualization and annotation. Bioinformatics. 2000;10:944–945.

35. Galtier N., Gouy M., Gautier C. SEAVIEW and PHYLO_WIN: Two graphic tools for sequence alignment and molecular phylogeny. Comput. Appl. Biosci. 1996;12:543–548.

36. Larkin M.A., Blackshields G., Brown N.P., Chenna R., McGettigan P.A., McWilliam H., Valentin F., Wallace I.M., Wilm A., Lopez R., et al. lustal W and Clustal X version 2.0.0. Bioinformatics. 2007;23:2947–2948. doi: 10.1093/bioinformatics/btm404.

37. Abascal F., Zardoya R., Posada D. ProtTest: Selection of best-fit models of protein evolution.Bioinformatics. 2005;21:2104–2105. doi: 10.1093/bioinformatics/bti263.

38. Felsenstein J. PHYLIP-Phylogeny inference package (version 3.2) Cladistics. 1989;5:164–166.

39. Felsenstein J. PHYLIP (Phylogeny Inference Package) Department of Genome Sciences, University of Washington; Seattle, WA, USA: 2005. version 3.6; Distributed by the author.

40. Guindon S., Dufayard J.F., Lefort V., Anisimova M., Hordijk W., Gascuel O. New algorithms and methods to estimate maximum-likelihood phylogenies: Assessing the performance of PhyML 3.0. Syst. Biol.2010;59:307–321. doi: 10.1093/sysbio/syq010.

41. Molecular Evolution, Phylogenetics and Epidemiology. [(accessed on 5 December 2012)]. Available online: http://tree.bio.ed.ac.uk/software/figtree/

42. Oshlack A., Robinson M., Young M. From RNA-seq reads to differential expression results. Genome Biol. 2009;11:220. doi: 10.1186/gb-2010-11-12-220.

43. Conesa A., Götz S., García-Gómez J.M., Terol J., Talón M., Robles M. Blast2GO: A universal tool for annotation, visualization and analysis in functional genomics research. Bioinformatics. 2005;21:3674–3676. doi: 10.1093/bioinformatics/bti610.

44. Conesa A., Götz S. Blast2GO: A comprehensive suite for functional analysis in plant genomics. Int. J. Plant Genomics. 2008;2008:619832.

45. Thimm O., Blaesing O., Gibon Y., Nagel A., Meyer S., Krüger P., Selbig J., Müller L.A., Rhee S.Y., Stitt M. MAPMAN: A user-driven tool to display genomics data sets onto diagrams of metabolic pathways and other biological processes. Plant J. 2004;37:914–939. doi: 10.1111/j.1365-313X.2004.02016.x.

46. Logemann J., Schell J., Willmitzer L. Improved method for the isolation of RNA from plant tissues. Anal. Biochem. 1987;163:16–20. doi: 10.1016/0003-2697(87)90086-8.

47. Plants Profile for Citrus × aurantium [Maxima × Reticulata] (Sour Orange) [(accessed on 1 June 2013)];Available online: http://plants.usda.gov/java/profile?symbol=CIAU8/

48. Hiraguri A., Hibino H., Hayashi T., Netsu O., Shimizu T., Uehara-Ichiki T., Omura T., Sasaki N., Nyunoya H., Sasaya T. The movement protein encoded by gene 3 of rice transitory yellowing virus is associated with virus particles. J. Gen. Virol. 2012;93:2290–2298. doi: 10.1099/vir.0.044420-0.

49. Huang Y.W., Geng Y.F., Ying X.B., Chen X.Y., Fang R.X. Identification of a movement protein of rice yellow stunt rhabdovirus. J. Virol. 2005;79:2108–2114. doi: 10.1128/JVI.79.4.2108-2114.2005.

50. Peng W., Zheng G.H., Zheng Z.Z., Tong Q.X., Ming Y.L. Orchid fleck virus: An unclassified bipartite, negative-sense RNA plant virus. Arch. Virol. 2013;158:313–323. doi: 10.1007/s00705-012-1506-5.

51. Martelli G.P., Russo M. Plant virus inclusion bodies. Adv. Virus Res. 1977;21:175–266. doi: 10.1016/S0065-3527(08)60763-0.

52. Moshe A., Gorovits R. Virus-induced aggregates in infected cells. Viruses. 2012;4:2218–2232. doi: 10.3390/v4102218.

53. Lin N.S., Chen C.C. Association of Bamboo Mosaic Virus (BoMV) and BoMV-specific electron-dense crystalline bodies with chloroplasts. Phytopathology. 1991;81:1551–1555. doi: 10.1094/Phyto-81-1551.

54. Chang M.U., Arai K., Doi Y., Yora K. Morphology and intracellular appearance of orchid fleck virus.Ann. Phytopathol. Soc. Jpn. 1976;42:156–167. doi: 10.3186/jjphytopath.42.156.

55. Kitajima E.W., Kondo H., Mackenzie A., Rezende J.A.M., Gioria R., Gibbs A., Tamada T. Comparative cytopathology and immunocytochemistry of Japanese, Australian and Brazilian isolates of Orchid fleck virus.J. Gen. Plant Pathol. 2001;67:231–237. doi: 10.1007/PL00013018.

56. Jones J.D., Dangl J.L. The plant immune system. Nature. 2006;444:323–329. doi: 10.1038/nature05286.

57. Zheng Z., Qualley A., Fan B., Dudareva N., Chen Z. An important role of a BAHD acyl transferase-like protein in plant innate immunity. Plant J. 2009;57:1040–1053. doi: 10.1111/j.1365-313X.2008.03747.x.

58. Schreiber K., Desveaux D. Message in a bottle: Chemical biology of induced disease resistance in plants.Plant Pathol. J. 2008;24:245–268. doi: 10.5423/PPJ.2008.24.3.245.

59. Dao T.T., Linthorst H.J., Verpoorte R. Chalcone synthase and its functions in plant resistance.Phytochem. Rev. 2011;10:397–412. doi: 10.1007/s11101-011-9211-7.

60. Wang T., Chen X., Zhu F., Li H., Li L., Yang Q., Chi X., Yu S., Liang X. Characterization of peanut germin-like proteins, AhGLPs in plant development and defense. PLoS One. 2013;8:e61722.

61. Rowland O., Ludwig A.A., Merrick C.J., Baillieul F., Tracy F.E., Durrant W.E., Fritz-Laylin L., Nekrasov V., Sjölander K., Yoshioka H., et al. Functional analysis of Avr9/Cf-9 rapidly elicited genes identifies a protein kinase, ACIK1, that is essential for full Cf-9-dependent disease resistance in tomato. Plant Cell. 2005;17:295–310. doi: 10.1105/tpc.104.026013.

62. Jones J.B., Vallad G.E., Iriarte F.B., Obradovic A., Wernsing M., Jackson L.E., Balogh B., Hong J.C., Momol MT. Considerations for using bacteriophages for plant disease control. Bacteriophage. 2012;2:208–214. doi: 10.4161/bact.23857.

63. Kondo H., Maeda T., Shirako Y., Tamada T. Orchid fleck virus is a rhabdovirus with an unusual bipartite genome. J. Gen. Virol. 2006;87:2413–2421. doi: 10.1099/vir.0.81811-0.

64. Kondo H., Chiba S., Andika I.B., Maruyama K., Tamada T., Suzuki N. Orchid fleck virus structural proteins N and P form intranuclear viroplasm-like structures in the absence of viral infection. J. Virol.2013;87:7423–7434. doi: 10.1128/JVI.00270-13.

65. Ruiz-Medrano R., Xoconostle-Cázares B., Lucas W.J. The phloem as a conduit for inter-organ communication. Curr. Opin. Plant Biol. 2001;4:202–209. doi: 10.1016/S1369-5266(00)00162-X.

66. Ham B.K., Li G., Kang B.H., Zeng F., Lucas W.J. Overexpression of Arabidopsis plasmodesmata germin-like proteins disrupts root growth and development. Plant Cell. 2012;24:3630–3648. doi: 10.1105/tpc.112.101063.

67. Guevara-Olvera L., Ruíz-Nito M.L., Rangel-Cano R.M., Torres-Pacheco I., Rivera-Bustamante R.F., Muñoz-Sánchez C.I., González-Chavira M.M., Cruz-Hernandez A., Guevara-González R.G. Expression of a germin-like protein gene (CchGLP) from a geminivirus-resistant pepper (*Capsicum chinense* Jacq.) enhances tolerance to geminivirus infection in transgenic tobacco. Physiol. Mol. Plant Pathol. 2012;78:45–50. doi: 10.1016/j.pmpp.2012.01.005.

68. Whitham S.A., Yang C., Goodin M.M. Global impact: Elucidating plant responses to viral infection.Mol. PlantMicrobe Interact. 2006;19:1207–1215. doi: 10.1094/MPMI-19-1207.

69. Dorokhov Y.L., Komarova T.V., Petrunia I.V., Frolova O.Y., Pozdyshev D.V., Gleba Y.Y. Airborne signals from a wounded leaf facilitate viral spreading and induce antibacterial resistance in neighboring plants. PLoS Pathog. 2012;8:e1002640.

70. Postnikova O.A., Nemchinov L.G. Comparative analysis of microarray data in Arabidopsis transcriptome during compatible interactions with plant viruses. Virol. J. 2012;9:101.

CHAPTER 8

EMARAVIRUS: A NOVEL GENUS OF MULTIPARTITE, NEGATIVE STRAND RNA PLANT VIRUSES

Nicole Mielke-Ehret and Hans-Peter Mühlbach

Biocentre Klein Flottbek, University of Hamburg, Ohnhorststrasse 18, Hamburg 22609, Germany

ABSTRACT

Ringspot symptoms in European mountain ash (*Sorbus aucuparia* L.), fig mosaic, rose rosette, raspberry leaf blotch, pigeonpea sterility mosaic (*Cajanus cajan*) and High Plains disease of maize and wheat were found to be associated with viruses that share several characteristics. They all have single-stranded multipartite RNA genomes of negative orientation. In some cases, double membrane-bound virus-like particles of 80 to 200 nm in diameter were found in infected tissue. Furthermore, at least five of these viruses were shown to be vectored by eriophyid mites. Sequences of *European mountain ash ringspot-associated virus* (EMARaV),*Fig mosaic virus* (FMV), rose rosette virus (RRV), raspberry leaf blotch virus (RLBV), pigeonpea sterility mosaic virus and High Plains virus strongly support their potential phylogenetic relationship. Therefore, after characterization of EMARaV, the novel genus *Emaravirus* was established, and FMV was the second virus species assigned to this genus. The recently sequenced RRV and RLBV are supposed to be additional members of this new group of plant RNA viruses.

INTRODUCTION

In the taxonomy of viruses, genome organization as well as nucleic acid and deduced amino acid sequence data are the main attributes to describe a virus family or genus, in addition to particle morphology, transmission and serological typing. Today, a total number of 2,284 viruses are described, which are grouped in 18 families and 17 unassigned genera [1]. However, there are still several putative virus species which have been known for a long time but have not yet been classified. This might be due to inadequate characterization or dissimilarity to other known viruses. Among this group of putative novel viruses, agents associated with the ringspot disease of European mountain ash (*Sorbus aucuparia* L.), fig mosaic disease (FMD), rose rosette disease (RRD), High Plains disease (HPD) of wheat and maize, pigeonpea sterility mosaic disease (PPSMD) and raspberry leaf blotch disorder (RLBD) share similarities with respect to their genomic organization and some of their biological properties. They all have segmented genomes consisting of four or more negative sense RNAs, their putative virions are double membrane-bound particles, and most of the diseases were shown to be transmitted by eriophyid mites. After elucidation of the genome sequence of the *European mountain ash ringspot associated*

Citation: Nicole Mielke-Ehret and Hans-Peter Mühlbach, Emaravirus: A Novel Genus of Multipartite, Negative Strand RNA Plant Viruses, doi:10.3390/v4091515.

virus EMARaV [2], the unassigned genus *Emaravirus* was established [3]. Meanwhile, *Fig mosaic virus* FMV [4,5,6] has been assigned to the genus *Emaravirus* [7]; rose rosette virus (RRV) [8] as well as raspberry leaf blotch virus (RLBV) [9] might be the next additional members.

Although genomic sequence information is limited for High Plains virus (HPV, now referred to as maize red stripe virus—MRSV [10]) and pigeonpea sterility mosaic virus (PPSMV) [11], sequence similarities were found to EMARaV and to FMV, indicating a putative phylogenetic relationship [2,4].

This communication will resume the state of knowledge about these viruses and their associated diseases. It will give an idea about the virus characteristics as well as the relationship between each other, and to already described and classified plant viruses. Several aspects of the viruses in the group are summarized in Table 1.

Table 1: Structural and biological characteristics of *Emaravirus*-related viruses.

Virus Species	Particle Morphology	Vector (Putative)	Host Species
EMARaV	DMBs 80–120 nm	(*Phytoptus pyri*)	*Sorbus aucuparia*
FMV	DMBs 90–200 nm	*Aceria ficus*	*Ficus carica*
RRV	DMBs 120–150 nm	*Phyllocoptes fructiphilus*	*Rosa multiflora* cultivated hybrid roses
RLBV	indistinct filamentous bodies	*Phyllocoptes gracilis*	*Rubus spp.*
MRSV (HPV)	filamentous structures and enveloped particles 80–200 nm	*Aceria tosichella*	*Zea mays*, *Triticum aestivum*
PPSMV	filamentous structures and DMBs 100–150 nm	*Aceria cajani*	*Cajanus cajan*

VIRUSES AND ASSOCIATED DISEASES

EMARaV — Ringspot Disease of European Mountain Ash (*Sorbus aucuparia* L.)

Chlorotic ringspots and mottling have been described as typical symptoms of a disease affecting European mountain ash (*Sorbus aucuparia*), which was named 'ringfleck mosaic' or 'ringspot disease' [12]. Using nucleic acid-based techniques, EMARaV, the type member of the new genus *Emaravirus*, was found to be closely associated with the disease [2,13,14]. The two different symptoms, chlorotic mottling and ringspots, may occur separately or together, even on one single leaflet (Figure 1), and it was unclear for long time whether they are caused by the same pathogen. Recently it could be shown by quantitative reverse transcription (RT) PCR that in both areas of symptomatic leaves, the same amount of EMARaV-RNA was present [15]. Infected trees have often been found in clusters in many parts of Europe, the natural habitat of this species [16,17,18]. Close relatives of *S. aucuparia*, such as *S. aria* (whitebeam) and *S. torminalis*(service tree), have not been reported to be affected by 'ringspot disease' so far. Like many fruit tree species (e.g., apple, pear, plum) European mountain ash belongs to the family *Rosaceae*. Although the economical importance of *S. aucuparia* in fruit production is not that high, its ecological potential as a major pioneer tree species in reforestation projects, especially at erosion-endangered areas, is remarkable [19].

Figure 1: Typical symptoms of chlorotic ringspots and mottling on leaves of an EMARaV infected mountain ash tree.

FMV — Fig Mosaic Disease (*Ficus carica* L.)

Fig mosaic disease (FMD) was first described in the 1930s in California [20]. Since then, many fig cultivars have been found to be susceptible to the infection [21] and FMD has been reported all around the world. Leaves of affected trees show mosaic or chlorotic spots. Furthermore, mottling, ringspots and leaf deformations can be observed [5,20,22,23]. Fruit may also show mosaic and in some cultivars premature fruit drop and reduced growth were reported [24]. Based on sequencing data and EM studies, more than seven different viruses were discussed in conjunction with FMD, including viruses in the genera *Potyvirus,Umbravirus*, *Luteovirus* and *Closterovirus* [21,25,26]. More recently, a novel, multipartite negative strand RNA virus, named *Fig mosaic virus* (FMV), and closely related to EMARaV, was identified [4,5,6]. A much more detailed description of FMD and the pathogens associated with the disease is given by Martelli [27].

RRV — Rose Rosette Disease *(Rosa spp.)*

Rose rosette disease (RRD), first described in the 1940s [28], is an indigenous disease in North America. The disease, alternatively called Witches' broom of Rose, has been found in the central and eastern states of the USA on many cultivated hybrid roses as well as on a wide range of wild rose species [29]. Typical disease symptoms are described as rapid stem elongation, followed by breaking of axillary buds, leaflet deformation and wrinkling, bright red pigmentation, phyllody, and increased thorniness [29]. Among wild roses, *Rosa multiflora* is highly susceptible to the disease, being a putative source of infection of cultivated hybrid roses. During several decades in the last century *R. multiflora* was planted for hedges and wildlife improvement in rural areas as "living fences", but extensive natural seeding and development of the rose plant could not be controlled by farmers. Therefore, it was declared a noxious weed, and as a biocontrol means for its eradication grafting of shoots of RRD infected plants onto plants in established *R. multiflora* stands (augmentation) was suggested [30]. The etiology of the disease remained enigmatic for long time, but the discovery of virus-like particles of 120–150 nm in diameter [31], the identification of the eriophyid mite*Phyllocoptes fructiphilus* as a vector of the disease [32], and the isolation of double-stranded RNA from infected material [33] indicated that a virus is associated with the disease. The recent characterization of four genomic RNAs of the rose rosette virus (RRV), a putative member of the genus *Emaravirus*, suggested that it is probably causing rose rosette [8].

RLBV — Raspberry Leaf Blotch Disorder *(Rubus spp.)*

Infestation of raspberry and other *Rubus* species by the raspberry leaf and bud mite *Phyllocoptes gracilis*(*Eriophyidae*) causes a complex disease syndrome called raspberry leaf blotch disorder (RLBD). It is characterized by yellow blotching as well as twisting of the leaves and distortion of leaf margins, decrease of overall plant growth by killing the terminal growing tip and reduction of fruit quality [34,35]. Although these symptoms resemble to some extent viral infections, convincing evidence for the association of a plant virus with RLBD had not been reported until recently. However, using molecular techniques, a new multipartite negative-strand RNA virus was discovered in plants showing RLBD symptoms, which was tentatively named raspberry leaf blotch virus (RLBV) [9]. Its genome organization and sequence similarities suggest that it is another member of the genus *Emaravirus*.

MRSV (HPV) — High Plains Disease *(Zea mays, Triticum aestivum)*

High Plains disease (HPD) was first identified in 1993, where maize (*Zea mays*) and wheat (*Triticum aestivum*) with strong disease symptoms were observed in Idaho and Texas, and shortly after also in Kansas and Colorado (all USA) [36], but is probably the same disease as wheat spot mosaic, reported in the 1950s in the United States [37]. In the beginning, maize plants were tested positive for *Wheat streak mosaic virus*(WSMV) (*Potyvirus*) infection, but symptoms have been exceptionally strong, leading to the assumption that a second virus might be involved in the disease [38]. This second pathogen was tentatively named High Plains virus (HPV). Indeed, mixed infections of HPV and WSMV in wheat and maize are common and result in more severe symptoms [39]. Meanwhile, based on more detailed molecular studies, the name maize red stripe virus (MRSV) was suggested for the causative agent of HPD [10], which will be used invariably in the following paragraphs.

In maize, disease symptoms are multifold. First evidence of infection becomes visible in young plants at 30 to 45 cm height [38]. Diseased maize plants show stunting, chlorosis (mosaic or streaking) and reddening which may lead to necrosis from the tip down the leaf. Moreover, in 50%–60% of infected maize plants, no ears are produced or ears contain only rudimentary seed, while others show smaller ears with reduced seed. HPD affected wheat is characterized by chlorotic spots or mosaic on the leaves and even complete yellowing of the plant. The virus also occurs in several other monocots, such as oat (*Avena sativa*), rye (*Secale cereale*), yellow foxtail (*Setaria glauca*), green foxtail (*Setaria viridis*) and downy brome (*Bromus tectorum*) [40]. As maize and wheat are two of the three main crops worldwide and the USA is the main producer for maize and the third largest for wheat, HPD causes serious economic damage in this country. However, the disease also reached other wheat and maize growing areas. First reports from Australia, South America (Brazil and Chile), Israel and possibly China indicate a progressive distribution of this novel disease [41,42].

PPSMV — Sterility Mosaic Disease of Pigeon-pea *(Cajanus cajan)*

Sterility mosaic disease (SMD) of pigeonpea (*Cajanus cajan, Fabaceae*), also known as 'Green Plague', leads to reduced flowering, while vegetative growth is stimulated. Furthermore, SMD-affected plants show ringspots and mosaic. Leaf size may be reduced and, if existing, seeds are often small and deformed [43]. First reports of SMD in India and Myanmar were given as early as 1931 [44]. Today, the disease is distributed mainly on the Indian subcontinent (Nepal, India, Bangladesh) and in Myanmar. From the Eastern coast of Africa, where pigeonpea has been cultivated since the 1960s [45], and the Caribbean, another pigeonpea producer, no PPSMV-infection has been reported so far. Next to the crop *C. cajan*, its wild relatives, as e.g., *C. scarabaeoides*, can also be affected by SMD, supporting a spread of the 'Green Plague' [46].

Pigeonpea is one of the most important crop species in Southern Asia. It serves as main protein source for about 1.1 billion people. The seed contains 20%–30% of protein, including several essential amino acids, carbohydrates and vitamin A and C. Its stems are used for fuel and in handcraft, the leaves may serve as animal feed and the roots are able to nitrificate the soil and to release soil-bound phosphorus. In Nepal and India alone, yield losses of US$ 300 million due to SMD were already registered in 1993 and since then, infection has spread further [43]. PPSMV infection of plants younger than 45 days might result in a yield loss of 95%–100%; in the case of older pigeonpea plants, the losses vary between 26 and 97% [47]. Furthermore, SMD affected *C. cajan* is more susceptible to powdery mildew and spider mites [48,49], which increase the economical damage.

The detection of several RNA species and filamentous structures, as can be found in the genera *Tospovirus*and *Tenuivirus*, suggested that SMD is caused by a new RNA virus that was tentatively named pigeonpea sterility mosaic virus (PPSMV) [11]. Ultrathin leaf sections of infected *C. cajan* showed granular cell cytoplasm and deformation of the chloroplasts by large starch grains [50]. Today, several isolates of PPSMV differing in virulence are known. In addition, cultivars of *C. cajan* vary in susceptibility for PPSMV. Thus, PPSMV research is focused on identifying and producing resistant plants [46].

VIRUS MORPHOLOGY

Four of the six virus diseases, ringspot disease of mountain ash, FMD, RRD and HPD, were shown to be associated with the occurrence of electron-dense structures known as double-membrane-bound bodies (DMBs) [51] with diameters differing within the range of 80 to 200 nm. These particles are supposed to be the virions (Table 1).

In ultrathin sections of leaves from mountain ash trees affected by ringspot disease, DMBs of about 80–100 nm in diameter were detected in the cytoplasm [52], suggesting infection by tospoviruses, but further ELISA testing excluded the infection with *Tomato spotted wilt virus* (TSWV). Recently, we found similar structures in partially purified fractions from EMARaV-infected mountain ash leaves [3], obtained with a protocol developed for isolating TSWV virions [53] (Figure 2).

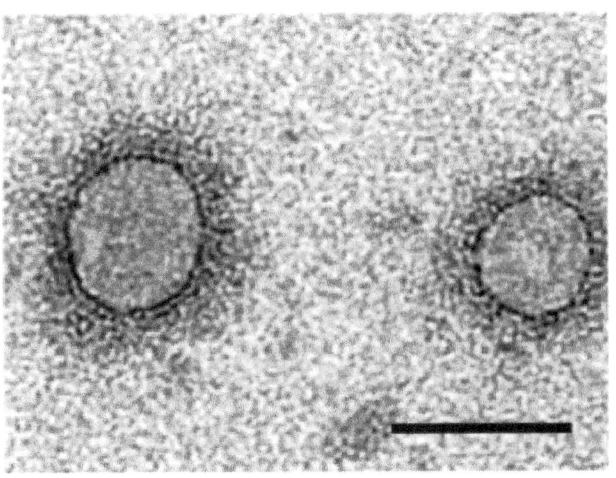

Figure 2: Double-membrane-bound bodies (DMBs) purified from EMARaV infected mountain ash leaves. Bar represents 100 nm.

DMBs resembling large tospoviral particles were also observed in infected fig tissues. The first DMBs in fig trees affected by fig mosaic were detected in 1970 [51]. Further studies supported the association of the disease with virus-like particles (VLPs), which vary in size, mainly between 120 and 160 nm in diameter [21,22,54,55].

Unusually large VLPs 120–150 nm in diameter were also found to be present in rose rosette affected tissue with typical disease symptoms [31]. In the same way as DMBs associated with infection of FMV and EMARaV [22,52], these putative RRV particles were surrounded by a double-membrane wall.

Sections of PPSMV-infected *C. cajan* and *Nicotiana benthamiana* plants indicated the presence of 100–150 nm DMBs and fibrous inclusions, various in size and adjacent to the nucleus [43]. Numerous DMBs could be found in all cell types of the leaf except in phloem and bundle sheath parenchyma cells, where no or just a few particles have been observed. These DMBs could be decorated with antibodies directed at a 32 kDa protein, the putative nucleocapsid (N) protein of PPSMV [50]. Higher decoration with this antiserum was observed for electron-dense material, which was also present in the cells. In contrast, fibrous inclusions adjacent to the nucleus, resembling that of tenuivirus-infected cells and which contain a viral non-capsid protein [56,57], did not react with the antiserum. No DMBs were detectable in leaf sap, which remains infectious for 15 minutes at most. Attempts at partial purification, based on a gradient centrifugation, did not result in an enrichment of these structures [11].

In electron microscopic studies of High Plains diseased wheat, maize and barley, DMBs of 80–200 nm in diameter have been observed [10,38,58,59]. They could be found in all cell types, also in vascular parenchyma cells. These VLPs have been observed in maize leaves as early as two days after inoculating with mites collected from infected plants [10] and, as described for PPSMV, they were associated with electron-dense, amorphous material in the cytoplasm. Here also, the VLPs as well as the associated amorphous inclusions could be immunogold-labeled with an antiserum directed against the putative N-protein [58]. The decoration was shown to be concentrated in the core of the DMBs, while the surrounding membranes remained unlabelled. Following a protocol for purification of tospoviral particles [60] it was possible to enrich these DMBs [10]. It was noticed that primarily damaged virions were decorated with gold particles, indicating that the N-protein might not be the main component of the double membrane, but rather a structural protein of enclosed nucleocapsids as shown for tospoviruses [61].

Often, DMBs of these novel viruses were found to be located near the ER and Golgi cisterns [50,58,62], indicating that particle morphogenesis might take place at these intracellular host membranes as described for tospoviruses [63]. In addition to the DMBs, partial purification of MRSV and PPSMV by gradient centrifugation delivered flexuous structures, 3–10 nm in diameter and of undefined length, resembling tospoviral or tenuiviral nucleocapsids [11,43,58]. MRSV associated structures were shown to be decorated by the N-protein specific antiserum [58]. Like DMBs, these putative nucleocapsids seemed to be essential for the virus infection cycle, as they have been observed in PPSMV purifications with plant material from different places in India, as well as from experimentally mite-infected *C. cajanus* and *N. benthamiana* [11].

For RLBV, no ultrastructural studies with infected raspberry leaf tissue were undertaken until now, but in extracts enriched for putative virus particles, indistinct filamentous bodies similar to those reported for PPSMV and MRSV were found by examination under the microscope [9].

GENOME ORGANIZATION AND VIRAL PROTEINS

The multipartite genomes of these novel viruses consist of single stranded (ss), negative-sense RNA. Double-strand RNA (dsRNA) preparations from infected plant tissue, Northern blot analyses as well as nucleic acid purification from enriched virus particle or nucleocapsid fractions indicated the presence of up to seven or eight RNA species, sized between 1.1 and 8 kb. Table 2 summarizes the number of presently known RNA species for each virus and their coding capacities.

Table 2: Genome organization and putatively encoded proteins of *Emaravirus*-related viruses

Virus Species	RNA 1	RNA 2	RNA 3	RNA 4	RNA 5	RNA 6
EMARaV	7040 nt P1: 266 kDa RdRp	2335 nt P2: 75 kDa Glycoprotein precursor	1559 nt P3: 35 kDa Nucleocapsid	1348 nt P4: 27 kDa unknown	-	-
FMV	7093 nt P1: 264 kDa RdRp	2252 nt P2: 73 kDa Glycoprotein precursor	1490 nt P3: 35 kDa Nucleocapsid	1472 nt P4: 40.5 kDa unknown	1752 nt P5: 59 kDa unknown	1212 nt P6: 22 kDa unknown
RRV	7026 nt P1: 265 kDa RdRp	2220 nt P2: 74 kDa Glycoprotein precursor	1544 nt P3: 36 kDa Nucleocapsid	1541 nt P4: 41 kDa MP	-	-
RLBV	7062 nt P1: 269 kDa RdRp	2135 nt P2: 75 kDa Glycoprotein precursor	1365 nt P3: 32 kDa Nucleocapsid	1675 nt P4: 42 kDa MP	1718 nt P5: 56 kDa unknown	-
MRSV [a]	RNA-L 7–8 kb RdRp?	RNA-M (double band) 2.5/2 kb	RNA-S 1.4 kb 32 kDa Nucleocapsid	?	?	?
PPSMV [b]	6.8 kb?	2.7 kb?	2.1 kb?	1.6 kb?	1.4 kb P5: 32 kDa Nucleocapsid	1.2 kb?

[a]: MRSV has probably more RNAs in the RNA-S complex [9]; [b]: PPSMV may contain an additional RNA species RNA 7 with 1.1 kb, respectively [11].

EMARaV

EMARaV was the first virus among this group, for which full-length sequences of genomic RNAs were established [2]. Four RNA fragments could be detected and despite extensive search by dsRNA analyses, Northern blotting and PCR using primer pairs derived from the conserved RNA ends, no additional RNAs were found [2]. The four RNAs sequenced to date are sized between 1.3 and 7.0 kb and contain in the complementary form a single open reading frame (ORF) (Figure 3). The largest RNA 1 (7,040 nt) encodes the 266 kDa viral RNA-dependent RNA-polymerase (RdRp), which shows similarity to RdRps of bunyaviruses and tenuiviruses [2,13]. All conserved motifs (premotif A and motifs A-E) as well as a putative endonucleolytic center could be identified. The latter may be involved in the mechanism of cap snatching, which is typical for ss(-)RNA virus families, such as *Arenaviridae*, *Orthomyxoviridae* and *Bunyaviridae*[64,65] and recently demonstrated for FMV [66]. The second largest RNA (RNA 2 with 2235 nt) carries the gene for a 75 kDa glycoprotein precursor [2]. Sequence analyses revealed a conserved domain of the glycoprotein precursor of the vertebrate infecting genus *Phlebovirus* (*Bunyaviridae*) [67]. Furthermore, proteomics software predicted all essential elements as N-glycosylation sites, transmembrane helices, an N-terminal signal peptide and a putative processing site, where the precursor might be cleaved into two single glycoproteins of 52 and 23 kDa. Structural comparison with other glycoprotein precursors of bunyaviruses revealed a similar configuration [2]. RNA 3 (1559 nt) most likely encodes the N-protein (35 kDa), since a protein of that size was specifically enriched in a virus particle purification procedure with leaf material of EMARaV-infected mountain ash. The smallest RNA 4, with a size of 1,348 nt in total, contains the ORF for a 27 kDa protein (P4) of yet undetermined function. Although typical motifs of protein function have not been detected in the P4 of EMARaV, the P4 sequence of RRV contains motifs suggesting a function in cell-to-cell movement [8] and RLBV P4 contains a signal peptide at its N-terminus and was localized to the plasma membrane in transgenic tobacco [9]. These features will be discussed in detail in chapters 4.3 and 4.4, but the data suggest that P4 of EMARaV might exhibit a similar function. EMARaV RNAs have complementary ends, typical for ss(-)RNA viruses. Depending on the RNA species, complementarity spans over a stretch of 19 to 23 nt with short interruptions at two or four positions. Furthermore, the terminal 13 nt is fully conserved and shows a high consensus with the RNA ends of orthobunyaviruses and hantaviruses. These terminal structures of viral nucleic acids are discussed to be of protective function by forming the so-called panhandle structure and have regulatory capacity during encapsidation, replication and transcription [68,69,70].

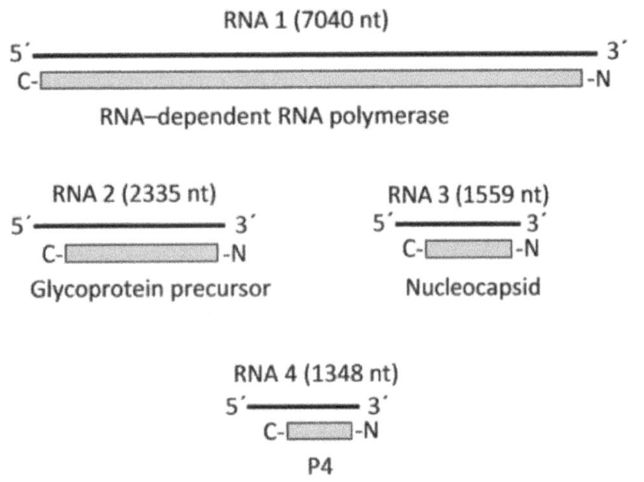

Figure 3: Genome organization of EMARaV. Virus genomic (minus-strand) RNAs are shown as black lines. Virus proteins encoded by the plus-strand mRNAs are shown as grey boxes.

FMV

Experimental strategies similar to the characterization of EMARaV allowed cDNA cloning and sequencing of six genomic RNAs of FMV [4,5,6,71,72]. Preparations of dsRNA from FMD-affected fig trees had revealed a changing number of molecules of up to 7 kb in length [6,24]. Initially, partial sequence information was obtained on two RNAs, indicating the relationship to viruses of the family *Bunyaviridae* and to EMARaV [6]. The genomic RNA molecules 1 to 4 of FMV exhibited some similarities to EMARaV with respect to the size and coding capacity of the individual RNAs. Each RNA species harbors in its complementary form (mRNA) one ORF and contains a highly conserved stretch of complementary nucleotides at the 3›- and 5›-end. Like EMARaV, RNA 1 encodes the viral RdRp, RNA 2 encodes a putative glycoprotein precursor, RNA 3 contains the gene of the nucleocapsid protein and RNA 4 encodes a protein of 40.5 kDa, which is larger than P4 of EMARaV, the function of which is also still unknown [4]. The newly-detected RNA-5 and RNA-6 contain two further proteins of unknown function [71,72]. FMV was assigned as new species *Fig mosaic virus* to the genus *Emaravirus* [7].

RRV

The strategy of dsRNA-based cDNA cloning and high throughput sequencing has recently allowed the characterizing of four genomic RNA segments of rose rosette virus (RRV) [8]. Genome organization and RNA sequences showed striking similarities to EMARaV and FMV, prompting the authors to conclude that it might be a further member of the genus *Emaravirus* [8]. As shown in Table 2, the four RNAs are quite similar in size to those of EMARaV and FMV and encode putative proteins with features similar to EMARaV and FMV orthologs. In addition, software-aided analyses predicted that P3, the putative nucleocapsid protein, can form homodimers and is bowl-shaped with interlocking protrusions and -sheets, properties which allow the protein to bind RNA [8]. P4 derived from RNA 4 is still of undetermined function, but it contains an ATPase motif and a dnaK motif, the latter being known from dnaK-containing heat shock proteins. Since closteroviral Hsp70h was proven to be involved in virus cell-to-cell movement [73]; the authors suggested that RRV P4 might have a similar function.

RLBV

The analysis of the genome organization of RLBV was started with random amplified cDNA derived from dsRNA [9]. A clone with some similarity to the nucleocapsid protein sequence of MRSV suggested that a new virus with similarities to the novel genus *Emaravirus* was present in raspberry plants exhibiting raspberry leaf blotch disorder. Therefore, RT-PCR with primers derived from the conserved terminal sequences of *Emaravirus* and stepwise using RNA-specific primers, followed by mass sequencing (Roche 454), led to characterization of five RNAs of the RLBV genome [9]. Four RNAs correspond to those found for other emaraviruses, but one was novel. RNA 1 and RNA 2 are very similar in size to the corresponding RNAs of the other emaraviruses (Table 2), and also encode the RdRp and a glycoprotein precursor, respectively. RNA 3, with 1,365 nt in size, is the shortest one of the five RNAs, but it encodes a putative nucleocapsid protein, similar to other emaraviruses, and is therefore called RNA 3, although RNA 4 and RNA 5 are larger. The similarity of the nucleocapsid sequence to other emaraviruses is varying; highest similarity (56%) was found to the nucleocapsid protein of MRSV, but only 44%–47% similarity was seen to EMARaV, RRV and FMV. The function of the putative 42 kDa protein P4, encoded by RNA 4, is still unknown, but the protein has some sequence similarities to the other unidentified proteins of emaraviruses. It contains a signal peptide at its N-terminus, and P4-GFP and P4-mRFP fusion proteins were localized to the plasma membrane and plasmodesmata in transgenic tobacco (*N. benthamiana and N. tabacum*) plants [9]. In addition, it co-localized with the tobacco mosaic virus 30K protein in plasmodesmata. Taken together, these findings led to the assumption that RLBV P4 could be involved in virus cell-to-cell movement. Up to now, P5 is unique to RLBV. It could be shown that the P5-GFP fusion protein was localized in aggregated structures in the cytoplasm, but its role in the life cycle of RLBV remains unknown [9].

MRSV

The RNA genome of MRSV is not fully characterized yet [10] (Table 2). Nucleic acid purifications from enriched virus preparations mainly delivered a large RNA-L of about 8 kb, a double band of 2.0 to 2.5 kb, called RNA-M, and a smaller fraction, called RNA-S, with an approximate size of 1.4 kb, which probably consists of multiple bands. By Northern blot-hybridization experiments with sense and antisense probes derived from RNA-S, viral genomic RNA was determined to be of negative polarity. Protein sequence predictions for RNA-L showed some similarities to the RdRp of viruses in the genus *Tospovirus.* While the predicted proteins encoded by the (probably) two different RNA species in the RNA-M complex showed no significant sequence similarities to proteins of other viruses, RNA-S has some similarities to EMARaV and PPSMV [2,3]. It is supposed to encode the 32 kDa nucleocapsid protein [10]. Partial virus purifications based on gradient centrifugation delivered predominantly a protein of 32 kDa, but sizes of 33 kDa have also been reported [10,38,74,75].

PPSMV

In contrast to the *Emaravirus* related virus species discussed above, the genome of PPSMV was reported to consist of seven or more RNAs [11]. Partial purification of virus-like particles, followed by nucleic acid preparation, revealed five to seven RNAs sized between 1.1 and 6.8 kb, although the largest RNA was not obtained in all purifications. The authors stated that the RNAs are of single stranded nature due to high sensitivity against RNase A even at high salt concentrations. Purified RNA was not infectious, indicating a possibly degradation during preparation or more likely the negative polarity of the viral genome. The main viral protein obtained during purification was supposed to be the N-protein and seems to be encoded by a 1.4 kb RNA named RNA 5. This protein differed in size, while Indian isolates from Andhra Pradesh (P) and Bangalore (B) possessed a 32 kDa N-protein, another isolate C from Tamil Nadu is characterized by a larger protein of about 35 kDa [43].

Sequence Similarities among All Emaravirus Related Viruses

Taken together, comparison of the so far published deduced amino acid (aa) sequences revealed remarkable similarities among EMARaV, FMV, RRV and RLBV as well as MRSV and PPSMV, but also to bunyaviruses and tenuiviruses. In particular, P1, the putative RdRp, shows significant conserved sections due to functional domains. Thus, the RdRps of EMARaV, FMV, RLBV and RRV were described as having aa identities of 49%–68% (56%–83% similarity) to their orthologs [2,4,5,6,8,9]. Unfortunately, no PPSMV or MRSV-specific RdRp sequences are available so far. The second protein P2 of these viruses, a putative glycoprotein precursor, also shows high similarities. The structural organization of EMARaV P2 was found to be very similar to the precursor molecules of bunyaviruses and tenuiviruses [2]. Nevertheless, their aa-sequences do not share any identity; only a short conserved motif of the phlebovirus glycoproteins (*Bunyaviridae*) has been identified, as was recently reported for RRV [8]. However, the putative glycoprotein precursors of EMARaV and FMV were reported to have aa-identities between 34% and 38% [4,5], while RRV P2 showed 51% aa identity with FMV and 40% with EMARaV [8] and RLBV has 46%–49% similarity to EMARaV, FMV and RRV [9]. Furthermore, for all six (putative) emaraviruses, major parts of the N-protein sequences (P3), are known and they show various similarities to each other [2,4,5,6,8,9,10,11]. Thus, FMV has highest similarity to EMARaV (49%) and PPSMV (42%), while identity with MRSV is only 27%. In contrast, RLBV shows 56% similarity to MRSV, but only 44%–47% to EMARaV, RRV and FMV. No similarity was found with members of the family *Bunyaviridae* or the genus *Tenuivirus*. The role of protein P4 could not be clarified for the *Emaravirus* related viruses under study, and there is no remarkable sequence similarity to proteins of other virus families or genera [2,4,8,9]. However, P4 of RRV shows 59% aa identity to the FMV ortholog [8], while P4 of RLBV has lower sequence similarities to EMARaV P4 (31%) and to FMV and RRV (41%) [9]. The size of P4 varies from 27 kDa (EMARaV) up to 42 kDa (RLBV). Since the predicted amino acid sequence for RRV P4 indicates an ATPase motif and a dnaK motif, a role of P4 in virus cell-to-cell movement is discussed [8]. This hypothesis is in agreement with the finding that P4 of RLBV contains an N-terminal signal peptide sequence and localized to the plasma membrane [9]. However, it is also conceivable that P4 proteins of the novel viruses have individual functions, which await further investigations.

Furthermore, recent studies showed that the complementary RNA termini, characterized so far, are more or less identical among EMARaV, FMV, RRV, and RLBV (5›-AGUAGUGUUCUCC … GGAGUUCACUACU-3›). The sequence of the only available MRSV RNA terminus is also similar. High similarity was also found with the RNA ends of the genera *Orthobunyavirus* and *Hantavirus* within the family *Bunyaviridae*, but not with that of the plant pathogenic tospo- and tenuiviruses [2,4]. For PPSMV, no terminus sequences are available in the databases up to now.

SEROLOGICAL RELATIONSHIP AND DIAGNOSTIC PROCEDURES

As particle and nucleocapsid morphology of EMARaV, PPSMV, MRSV and FMV resembled to some extent tospoviruses and tenuiviruses, efforts were made to detect these viruses with antisera against the two genera. But no reactions were obtained, indicating a low serological relationship, which can be easily explained by the recently published sequence comparison with viruses of these genera. Today, for the specific detection of identified proteins of this group of novel viruses, several antisera are available. However, the two antisera raised against the putative glycoproteins and the N-protein (P3) of EMARaV were not recommended for routine testing, due to weak reaction and background signals in Western blot with crude protein extracts [2,14]. But when the antiserum directed against P3 was used with partially purified virus fractions, a strong serological reaction was obtained with the 35 kDa nucleocapsid protein. The antisera for PPSMV and MRSV, directed against the 32 kDa N-proteins, are routinely used in ELISA testing and Western blot analyses [10,43,46,50,59,76,77,78]. The PPSMV-specific antiserum is able to detect different Indian isolates, although the reaction with the isolate C from Tamil Nadu is weaker than with the isolates P (Andhra Pradesh) and B (Bangalore) [43]. The MRSV antiserum was tested on different natural hosts of the virus and was found to be applicable in immunodetection with root material indicating a systemic infection [10]. None of these antisera was shown to cross-react with another virus of this group, reflecting the insufficient aa sequence similarities of their P3 proteins.

As most of the host plants infected with these novel plant RNA viruses are of high economical importance, specific and sensitive diagnostic tools are indispensable. Due to lack of sufficient sequence data for PPSMV and MRSV, primer pairs for RT-PCR, which are able to detect all RNA species and most of the existing isolates of these two putative viruses, are still rare. However, for the other four viruses, whose genomes are extensively sequenced, diagnostic primer pairs are available. Four diagnostic primer pairs were developed for EMARaV, which were shown to detect different isolates from Germany and Austria [14]. EMARaV RNA 3 was also detected by RT-PCR in single individuals of *Phytoptus pyri*, collected from galls on leaves of EMARaV infected mountain ash trees [79].

RT-PCR methods for FMV detection have been reported previously, amplifying parts of the RNAs carrying the N- and glycoprotein genes [5]. Meanwhile, primer pairs detecting all sequenced FMV RNAs were designed, which, for the first time, allowed a survey of the occurrence of FMV in countries from the African continent to Japan, just to mention a few [80,81,82,83,84]. Additionally, dot-spot-hybridization assays were successfully applied for FMV detection in fig seedlings, using probes synthesized from the positive strand of RNA 1 [6].

Primer pairs were developed for RRV that unambiguously detected RRV in 84 of 84 samples of cultivated roses and *R. multiflora*, showing typical symptoms of RRD, while RRV was not detected in 30 asymptomatic roses, used as negative controls [8]. RLBV-specific primer pairs detected all five RNAs of RLBV, not only in symptomatic field-grown raspberry plants, but also in experimentally infected *N. benthamiana* plants, in raspberry plants inoculated with viruliferous mites, and in RNA extracted from a bulked sample of mites collected from RLBD-affected plants [9].

Diagnostic primers are also available for the partially characterized PPSMV and MRSV [11,42]. Thus, in RT-PCR studies, PPSMV was sensitively detected in plants from different locations in India, as well as in pigeonpea experimentally infected by mites and grafting.

VIRUS TRANSMISSION

The diseases caused by RRV, RLBV FMV, PPSMV and MRSV were all shown to be transmitted by eriophyid mites, a group of arthropods which are very host specific and also serve as vector for several potyviruses [32,35,74,76,85,86]. This host specificity corresponds with the observation that just a few plant species are naturally infected by these viruses.

Only little information about the transmission mechanism of these eriophyid mites is available yet. Regarding EMARaV, vector transmission by an eriophyid mite species has not been shown convincingly so far. However, galls induced by the pear leaf blister mite *Phytoptus pyri* (*Eriophyidae*), are frequently found on EMARaV-infected *S. aucuparia* leaves (Figure 4). A detailed study using immune fluorescence microscopy and quantitative RT-PCR demonstrated that both the genomic and the complementary forms of RNA 3 of EMARaV and the putative nucleocapsid protein P3 are present inside the mite body [79], a finding which makes *Phytoptus pyri* a candidate vector for EMARaV.

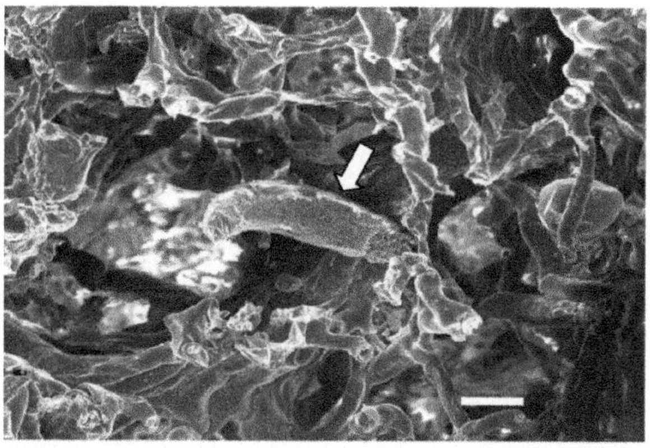

Figure 4: The eriophyid mite *Phytoptus pyri* (white arrow) within gall tissue from the undersurface of a leaf of an EMARaV infected mountain ash tree. Scanning electron microscopy. Bar represents 50 μm.

The vector for fig mosaic disease is the free living eriophyid mite *Aceria ficus* Cotte [85,87]. Feeding by a single mite was demonstrated to be sufficient to transmit the virus to healthy seedlings of *F. carica* [87]. First disease symptoms, such as chlorotic spotting, mosaic and yellowing of the leaves, as well as malformation of the laminae, appeared approximately 40 days post infection (dpi). Infection was confirmed by the observation of virus-like particles in parenchyma cells by EM studies. However, it cannot be excluded that *A. ficus* is transmitting more than one virus, as FMD symptoms are discussed as being caused by mixed virus infections [4,5,21,25,26].

The wheat curl mite (*Aceria tosichella* Keifer, *Eriophyidae*) is the vector for both MRSV and WSMV, explaining the fact that a mixed infection with these two viruses is common. An investigation in Nebraska (USA) in 1995/96 detected both viruses in 40% of the collected samples [39]. The authors stated that the shared transmission by the same mite population could indicate a putative synergistic interaction between MRSV and WSMV. Both viruses are not only found in the same plant, but also in the same cell. Electron microscopy showed the different virus particles next to each other within wheat leaf cells [58]. Furthermore, the number of MRSV-associated virus-like particles was usually higher in these doubly infected cells than in cells without WSMV inclusions. HPD transmission by *A. tosichella* is restricted to feeding; the mite was shown to be unable to transmit transovarially [10,74].

Aceria cajani lives free on the undersurface of the leaves of pigeonpea and transmits PPSMV probably in a semi-persistent matter [76]. It is assumed that the virus does not replicate within the mite, which stays infectious for about 6–13 h. Under experimental conditions, pigeonpea plants showed first symptoms within 1–2 weeks after inoculation with mites collected from infected plants. Inoculation access period for PPSMV was at least 90 min and could be reduced by prior starvation to 60 min. The acquisition access period is at least 15 min. In these experiments, mites lost their ability for transmission 2–10 h after feeding on healthy plants. Here also, no transovarial transmission could be observed [76].

Due to their short stylet, eriophyid mites, living on the undersurface of plant leaves, may reach predominantly epidermal cells or the adjacent layers of the mesophyll. Thus, for effective virus transmission, these cell types must be infected. Indeed, virus-like particles associated with an infection of FMV, PPSMV or MRSV were shown to be present in parenchyma cells as well as in subepidermal leaf cells [21,50,58].

As eriophyid mites are highly host specific, their use for virus transmission on typical experimental host plants, e.g., *Chenopodium quinoa*, *Solanaceae* and *Cucurbitaceae*, for an improved feasible virus characterization is restricted. Without feeding on their host, they die quickly. Indeed, *Aceria cajani* was shown to survive only up to 13 h [76].

Interestingly, the putative mite vector itself may cause disease symptoms similar to virus infection. This was the case for the raspberry leaf blotch disorder, which was primarily reported to be only caused by the infestation with the raspberry leaf and bud mite *Phyllocoptes gracilis*, because no indication of a virus infection was found using conventional techniques, such as mechanical inoculation and grafting [35].

As vegetative transmission is typical for plant virus infection, PPSMV, FMV and EMARaV were demonstrated to be graft transmissible or could be passed by cuttings [11,88,89,90]. However, mechanical transmission of these virus-induced diseases was found to be difficult; none of the novel viruses could be passed by abrasion on its host plant so far. The only exception is MRSV, which could be transmitted mechanically by vascular puncture inoculation on maize ([59,75,77,91]. Inhibiting secondary metabolites as polyphenols or instability of the virus particles were discussed as basic causes [11]. Nevertheless, with some difficulties PPSMV and FMV could be transmitted mechanically on tobacco plants. *N. clevelandii* and *N. benthamiana* were infected by PPSMV, inoculating a 2% nicotine solution homogenate with a success of 10%–40% [11]. Symptoms appeared not until 40 dpi on apical leaves as vein chlorosis and necrotic spots. However, symptoms were diverse and virus titer remained low, even after two passages, but virus infection could be confirmed by RT-PCR with specific primer pairs. FMV was also shown to be transmissible on*Nicotiana* species (*N. benthamiana* and *N. clevelandii*) as well as on *Chenopodium sp*., *Datura stramonium*and *Vigna unguiculata* [21]. For MRSV, no experimental host, except barley, is known.

Of the described viruses, no vertical transmission by seed or pollen has been reported so far. First efforts to detect EMARaV by highly specific and sensitive RT-PCR in plants grown from seed of infected mountain ash also gave no indication for seed transmissibility, although EMARaV is detectable in the seed itself (unpublished). FMV also repeatedly tested negative for seed transmission [22,89]. When pigeonpea seedlings derived from seed of PPSMV-infected plants were tested by ELISA, none of them contained the virus [92]. In the seed itself, PPSMV was only detectable in the coat, but not in the cotyledons.

However, MRSV showed a very low seed transmission rate of 0.008% [91]. Only three seedlings out of 38,482 and from different lots showed clearly visible mosaic symptoms. Furthermore, they were found positive for virus infection in ELISA testing and feeding experiments on barley. The authors remarked that a high greenhouse temperature was chosen for a better symptom development (34 °C day, 21 °C at night). Therefore, it remained unclear if seed transmission will occur under natural field conditions.

Transmission by soil has not been reported for any of these viruses so far.

CONCLUSIONS

The well characterized EMARaV, FMV, RRV and RLBV were shown to share many similarities, indicating a strong phylogenetic relationship, which suggests that also the latter two belong to the genus *Emaravirus*. Although extensive sequence information is still lacking for MRSV and PPSMV, their well-documented biological and structural properties also argue in favor of a relationship to emaraviruses. All six viruses have a segmented RNA genome of negative polarity, and recent sequence data analyses showed significant amino acid identities of some of the encoded proteins. Furthermore, they are all associated with the occurrence of DMBs, variable in size, which might represent the virions. Whereas vector transmission has not yet been conclusively shown for EMARaV, all the other viruses are vectored by eriophyid mites. Sequence comparison indicated that these viruses are related to the family *Bunyaviridae* and the genus *Tenuivirus*, but the collected data on virus morphology, genome organization, RNA and protein sequences, as well as biological properties, did not allow assigning them into existing families or genera. Therefore, the floating genus *Emaravirus* has been established with EMARaV and FMV as the first members, but RRV and RLBV are promising candidates for the next viruses to be assigned to this genus.

In addition to the six novel plant viruses discussed here, there remain still unassigned viruses or virus-like agents which might belong to the same group. Among them is a pathogen associated with Redbud yellow ringspot disease [93], which was shown to appear with DMBs and is also transmitted by eriophyid mites [94]. Leaves of affected redbud trees (*Cercis canadensis* L.) show bright yellow ringspots or blotches, and often necrotic spots. Furthermore, trees are characterized by many dead twigs and smaller branches. Very recently a virus with a genome organization highly similar to emaraviruses was shown to be associated with the disease, which was provisionally named Redbud yellow ringspot associated virus (RYRSaV) [95]. Furthermore, a disease with ringspot symptoms on the leaves of S*orbus scopulina* Greene (Greene's mountain ash) was found in Alaska. Virus-like membrane-bound spherical particles could be observed and an approximately 32 kDA protein was consistently present in extracts from only symptomatic leaves. Although antiserum directed against the 35 kDa N-protein of EMARaV did not react in Western blot analyses with the enriched 32 kDa protein, circumstantial evidence, such as disease symptoms, association of the disease with gall forming eriophyid mites and particle morphology of the putative viral agent, suggest some relationship to emaraviruses [96]. It is therefore very likely that more unidentified pathogens, which belong to the new plant RNA virus genus *Emaravirus*, might be detected.

Although in recent years we could collect new information about this novel genus, the genomic organization of emaraviruses and their biological interaction with host plants are far from understood. Finally, it has to be considered that in no case Koch's postulates were convincingly fulfilled. This is certainly due to the fact that no local lesion hosts have been identified yet and that the isolation of infectious virus particles was not possible so far. However, the close association of the novel viruses with typical disease patterns, the presence of virus-like particles only in symptomatic tissue, and the close relation of vector, virus and disease in almost all cases strongly support the hypothesis that the novel emaraviruses EMARaV and FMV and the*Emaravirus* related viruses RRV, RLBV and probably also MRSV and PPSMV are the causes of the corresponding plant diseases.

ACKNOWLEDGMENTS

The authors gratefully acknowledge the helpful comments and criticism of anonymous reviewers. Authors are grateful to the Deutsche Forschungsgemeinschaft (DFG) for the continuous support of their studies on EMARaV (grants Mu 559/8-1 to 8-7).

REFERENCES AND NOTES

1. Carstens E.B. Introduction to virus taxonomy. In: King A.M.Q., Adams M.J., Carstens E.B., Lefkovitz E.J., editors. Virus Taxonomy. Classification and Nomenclature of Viruses, Ninth Report of the International Committee on Taxonomy of Viruses. Elsevier; Oxford, UK: 2011. pp. 3–7.
2. Mielke N., Muehlbach H.-P. A novel, multipartite, negative-strand RNA virus is associated with the ringspot disease of European mountain ash (*Sorbus aucuparia* L.) J. Gen. Virol. 2007;88:1337–1346. doi: 10.1099/vir.0.82715-0.
3. Mühlbach H.-P., Mielke-Ehret N. Emaravirus. In: King A.M.Q., Adams M.J., Carstens E.B., Lefkovitz E.J., editors. Virus Taxonomy. Classification and Nomenclature of Viruses, Ninth Report of the International Committee on Taxonomy of Viruses. Elsevier; Oxford, UK: 2011. pp. 767–770.
4. Elbeaino T., Digiaro M., Martelli G.P. Complete nucleotide sequence of four RNA segments of fig mosaic virus. Arch. Virol. 2009;154:1719–1727. doi: 10.1007/s00705-009-0509-3.
5. Walia J.J., Salem N.M., Falk B.W. Partial sequence and survey analysis identify a multipartite, negative-sense RNA-virus associated with fig mosaic. Plant Dis. 2009;93:4–10. doi: 10.1094/PDIS-93-1-0004.[Cross Ref]
6. Elbeaino T., Digiaro M., Alabdullah A., de Stradis A., Minafra A., Mielke N., Castellano M.A., Martelli G.P. A multipartite single-stranded negative-sense RNA virus is the putative agent of fig mosaic disease. J. Gen. Virol. 2009;90:1281–1288. doi: 10.1099/vir.0.008649-0.
7. Adams M.J., Carstens E.B. Ratification vote on taxonomic proposals to the international committee on taxonomy of viruses (2012) Arch. Virol. 2012;157:1411–1422. doi: 10.1007/s00705-012-1299-6.
8. Laney A.G., Keller K.E., Martin R.R., Tzanetakis I.E. A discovery 70 years in the making: Characterization of the rose rosette virus. J. Gen. Virol. 2011;92:1727–1732. doi: 10.1099/vir.0.031146-0.
9. McGavin W.J., Mitchell C., Cock P.J.A., Wright K.M., MacFarlane S.A. Raspberry leaf blotch virus, a putative new member of the genus Emaravirus, encodes a novel genomic RNA. J. Gen. Virol. 2012;93:430–437. doi: 10.1099/vir.0.037937-0.
10. Skare J.M., Wijkamp I., Denham I., Rezende J.A.M., Kitajima E.W., Park J.W., Desvoyes B., Rush C.M., Michels G., Scholthof K.-B.G., et al. A new eriophyid mite-borne membrane-enveloped virus-like complex isolated from plants. Virology. 2006;347:343–353. doi: 10.1016/j.virol.2005.11.030.
11. Kumar P.L., Jones A.T., Reddy D.V.R. A novel mite-transmitted virus with a divided RNA genome closely associated with pigeonpea sterility mosaic disease. Phytopathology. 2003;93:71–81. doi: 10.1094/PHYTO.2003.93.1.71.
12. Kegler H. Das Ringfleckenmosaik der Eberesche (Sorbus aucuparia L.) (in German) Phytopathol. Z.1960;37:214–216. doi: 10.1111/j.1439-0434.1960.tb04035.x.
13. Benthack W., Mielke N., Büttner C., Mühlbach H.-P. Double-stranded RNA pattern and partial sequence data indicate plant virus infection associated with ringspot disease of European mountain ash (*Sorbus aucuparia* L.) Arch. Virol. 2005;150:37–52. doi: 10.1007/s00705-004-0397-5.
14. Mielke N., Weber M., Khan S., Muehlbach H.-P. Detection of European mountain ash ringspot associated virus (EMARAV) in *Sorbus aucuparia* L. by a specific antiserum and reverse transcription PCR. For. Pathol. 2008;38:371–380. doi: 10.1111/j.1439-0329.2008.00553.x.
15. Schlatermund N. Lokalisation und Quantifizierung der RNAs des European Mountain Ash Ringspot Associated Virus (EMARAV) in der Eberesche (*Sorbus aucuparia* L.) (in German) Ph.D. thesis, University of Hamburg, Hamburg, Germany, 2009.
16. Polak Z., Prochazkova Z., Braniaova H. Recent findings of viruses of forest trees on the territory of the Czech Republic. Arch. Phytopathol. Pflanzenschutz. 1990;26:389–393. doi: 10.1080/03235409009438995.
17. Polak Z., Zieglerova J. Towards ringspots and variegation in mountain ash leaves. Z. Pflanzenkr. Pflanzenschutz. 1996;103:432–435.
18. Kallinen A.K., Lindberg I.L., Tugume A.K., Valkonen J.P.T. Detection, distribution and genetic variability of European mountain ash ringspot-associated virus. Phytopathology. 2009;99:344–352. doi: 10.1094/PHYTO-99-4-0344.
19. Raspe O., Findlay C., Jaquemart A.-L. *Sorbus aucuparia* L. J. Ecol. 2000;88:910–930. doi: 10.1046/j.1365-2745.2000.00502.x.
20. Condit I.J., Horne W.T. A mosaic of the fig in California. Phytopathology. 1933;23:887–896.
21. Serrano L., Ramon J., Segarra J., Medina V., Achón M.A., López M. New approach in the identification of the causal agent of fig mosaic disease. Acta Hort. 2004;657:559–566.

22. Martelli G.P., Castellano M.A., Lafortezza R. An ultrastructural study of fig mosaic. Phytopathol. Mediterr. 1993;32:33–43.

23. Castellano M.A., Gattoni G., Minafra A., Conti M., Martelli G.P. Fig mosaic in Mexico and South Africa. J. Plant Pathol. 2007;89:441–444.

24. Acikgöz S., Döken T. The determination of sampling time for dsRNA isolation of the agent of fig mosaic disease prevalent in Aegean region: Turkey. Acta Hort. 2003;605:307–310.

25. Elbeaino T., Digiaro M., de Stradis A., Martelli G.P. Partial characterization of a closterovirus associated with a chlorotic mottling of fig. J. Plant Pathol. 2006;88:187–192.

26. Elbeaino T., Digiaro M., de Stradis A., Martelli G.P. Identification of a second member of the family closteroviridae in mosaic-diseased figs. J. Plant Pathol. 2007;89:119–124.

27. Martelli G.P. Fig mosaic disease and associated pathogens. In: Hadidi A., Barba M., Candresse T., Jelkmann W., editors. Virus and Virus-Like Diseases of Pome and Stone Fruits. APS Press; St. Paul, MN, USA: 2011. pp. 281–287.

28. Conners I.L. Twentieth Annual Report of the Canadian Plant Disease Survey 1940. Department of Agriculture; Ottawa, Canada: 1941. p. 98.

29. Crowe F.J. Witches' broom of rose: A new outbreak in several Central States. Plant Dis. 1983;67:544–546. doi: 10.1094/PD-67-544.

30. Epstein A.H., Hill J.H., Nutter F.W., Jr. Augmentation of rose rosette disease for biocontrol of multiflora rose (*Rosa multiflora*) Weed Sci. 1997;45:172–178.

31. Gergerich R.C., Kim K.S., Kitajima E.W. A particle of unique morphology associated with the disease of rose in northwest Arkansas. Phytopathology. 1983;73:500–501.

32. Amrine J.W., Jr., Hindal D.F., Stasny T.A., Williams R.L., Coffman C.C. Transmission of the rose rosette disease agent to *Rosa multiflora* by *Phyllocoptes fructiphylus* (Acari: Eriophyidae) Entomol. News.1988;99:239–252.

33. Di R., Hill J.H., Epstein A.H. Double-stranded RNA associated with the rose rosette disease of multiflora rose. Plant Dis. 1990;74:56–58. doi: 10.1094/PD-74-0056.

34. Gordon S.C., Taylor C.E. Some aspects of the biology of the raspberry leaf and bud mite (*Phyllocoptes*(*Eriophyes*) *gracilis* Nal.) Eriophyidae in Scotland. J. Hortic. Sci. 1976;51:501–508.

35. Jones A.T., Gordon S.C., Jennings D.L. A leaf-blotch disorder of tayberry associated with the leaf and bud mite (*Phyllocoptes gracilis*) and some effects of three aphid-borne viruses. J. Hortic. Sci. 1984;59:523–528.

36. Jensen S.G., Lane L.C. A new virus disease of corn and wheat in the High Plains. Phytopathology.1994;84:1158.

37. Slykhuis J.T. Wheat spot mosaic, caused by a mite-transmitted virus associated with *Wheat streak mosaic virus*. Phytopathology. 1956;46:682–687.

38. Jensen S.G., Lane L.C., Seifers D.L. A new disease of maize and wheat in the high plains. Plant Dis.1996;80:1387–1390. doi: 10.1094/PD-80-1387.

39. Mahmood T., Hein G.L., Jensen S.G. Mixed infection of hard red winter wheat with High Plains virus and Wheat streak mosaic virus from wheat curl mites in Nebraska. Plant Dis. 1998;82:311–315. doi: 10.1094/PDIS.1998.82.3.311.

40. Seifers D.L., Harvey T.L., Martin T.J., Jensen S.G. A partial host range of the High Plains virus of corn and wheat. Plant Dis. 1998;82:875–879. doi: 10.1094/PDIS.1998.82.8.875.

41. Jensen S.G. High plains virus: A new twist to an old story. Phytopathology. 1999;89:S84. doi: 10.1094/PHYTO.1999.89.1.84.

42. Lebas B.S.M., Ochoa-Corona F.M., Elliott D.R., Tang Z., Alexander B.J.R. Development of an RT-PCR for High Plains virus indexing scheme in New Zealand post-entry quarantine. Plant Dis. 2005;89:1103–1108. doi: 10.1094/PD-89-1103.

43. Jones A.T., Kumar P.L., Saxena K.B., Kulkarni N.K., Muniyappa V., Waliyar F. Sterility mosaic disease: The 'green plague' of pigeonpea. Plant Dis. 2004;88:436–445. doi: 10.1094/PDIS.2004.88.5.436.

44. Mitra M. Report of the imperial mycologist. Sci. Rep. Agric. Res. Inst. 1931;19:58–71.

45. Saxena K.B. Genetic improvement of pigeonpea. Trop. Plant Biol. 2008;1:159–178. doi: 10.1007/s12042-008-9014-1.

46. Kumar P.L., Latha T.K.S., Kulkarni N.K., Raghavendra N., Saxena K.B., Waliyar F., Rangaswamy K.T., Muniyappa V., Doriswamy S., Jones A.T. Broad-Based resistance to pigeonpea sterility mosaic disease in wild relatives of pigeonpea (Cajanus: Phaseolae) Ann. Appl. Biol. 2005;146:371–379. doi: 10.1111/j.1744-7348.2005.040091.x.

47. Kannaiyan J., Nene Y.L., Reddy M.V., Ryan J.G., Raju T.N. Prevalence of pigeonpea disease and associated crop losses in Asia, Africa and the Americas. Trop. Pest Manag. 1984;30:62–71. doi: 10.1080/09670878409370853.

48. Reddy M.V., Kannaiyan J., Nene Y.L. Increased susceptibility of sterility mosaic infected pigeonpea to powdery mildew. Int. J. Trop. Plant Dis. 1984;2:35–40.

49. Sithanantham S., Reddy M.V., Rameshwar R.V. Increased damage by the spider mite *Schizotetranychus cajani* in Pigeonpea Plants affected by sterility mosaic. In: Channabasavanna G.P., Viraktamath C.A., editors. Progress in Acarology. Vol. 2. Oxford & IBH Publishing Co; New Delhi, India: 1989. pp. 11–14.

50. Kumar P.L., Duncan G.H., Roberts A.M., Jones A.T., Reddy D.V.R. Cytopathology of pigeonpea sterility mosaic virus in pigeonpea and *Nicotiana benthamiana*: Similarities with those of eriophyid mite-borne agents of undefinded aetiology. Ann. Appl. Biol. 2002;140:87–96. doi: 10.1111/j.1744-7348.2002.tb00160.x.

51. Bradfute O.E., Whitmoyer R.E., Nault R.L. Ultrastructure of plant leaf tissue infected with mite-borne viral-like particles. Proc. Electron Microsc. Soc. Am. 1970;258:178–179.

52. Ebrahim-Nesbat F., Izadpanah K. Virus-Like particles

associated with ringfleck mosaic of mountain ash and a mosaic disease of raspberry in the Bavarian forest. Eur. J. For. Path. 1992;22:1–10. doi: 10.1111/j.1439-0329.1992.tb01426.x.

53. Kikkert M., van Poelwijk F., Storms M., Kassies W., Bloksma H., van Lent J., Kormelink R., Goldbach R. A protoplast system for studying Tomato spotted wilt virus infection. J. Gen. Virol. 1997;78:1755–1763.[PubMed]

54. Plavsic B., Milicic D. Intracellular changes in trees infected with fig mosaic. Acta Hort. 1980;110:281–286.

55. Appiano A., Conti M., Zini N. Cytopathological study of the double-membrane bodies in fig plants affected by fig mosaic disease. Acta Hort. 1995;386:585–592.

56. Ammar E.D., Gingery R.E., Nault L.R. Two types of inclusions in maize infected with maize stripe virus. Phytopathology. 1985;75:84–89. doi: 10.1094/Phyto-75-84.

57. Espinoza A.M., Pereira R., Macaya-Lizano A.V., Hernandez M., Goulden M., Rivera C. Comparative light and electron microscopic analyses of tenuivirus major noncapsid protein (NCP) inclusion bodies in infected plants, and of the NCP *in vitro*. Virology. 1991;195:156–166. [PubMed]

58. Ahn K.K., Kim K.S., Gergerich R.C., Jensen S.G. High Plains disease of corn and wheat: Ultrastructural and serological aspects. J. Submicrosc. Cytol. Pathol. 1998;30:563–571. [PubMed]

59. Louie R., Seifers D.L., Bradfute O.E. Isolation, transmission and purification of the High Plains virus. J. Virol. Meth. 2006;135:214–222. doi: 10.1016/j.jviromet.2006.03.023.

60. Black L.M., Brakke M.K., Vatter A.E. Purification and electron microscopy of tomato spotted-wilt virus. Virology. 1963;20:120–130. doi: 10.1016/0042-6822(63)90146-6.

61. Kitajima E.W., de Avila A.C., Resende R.D.O., Goldbach R.W., Peters D. Comparative cytological and immunogold labelling studies on different isolates of tomato spotted wilt virus. J. Submicrosc. Cytol. Pathol.1992;24:1–14.

62. Silvestro S.R., Chapman G.B. A transmission electron microscope study of "New Dawn" climber rose (*Rosa wichuraiana* x *safrano*) exhibiting rose rosette disease. Plant Cell Rep. 2004;23:345–351. doi: 10.1007/s00299-004-0838-3.

63. Kikkert M., van Lent J., Storms M., Bodegom P., Kormelink R., Goldbach R. Tomato spotted wilt virus particle morphogenesis in plant cells. J. Virol. 1999;73:2288–2297.

64. Duijsings D., Kormelink R., Goldbach R. *In vivo* analysis of the TSWV cap-snatching mechanism: Single base complementarity and primer length requirements. EMBO J. 2001;20:2535–2552.

65. Rao P., Yuan W., Krug R.M. Crucial role of CA cleavage sites in the cap-snatching mechanism for initiating viral mRNA synthesis. EMBO J. 2003;22:1188–1198. doi: 10.1093/emboj/cdg109.[PMC free article]

66. Walia J.J., Falk B.W. Fig mosaic virus mRNAs show generation by cap-snatching. Virology.2012;426:162–166. doi: 10.1016/j.virol.2012.01.035.

67. Liu D.Y., Tesh R.B., Travassos da Rosa A.P.A., Peters C.J., Yang Z., Guzman H., Xiao S.Y. Phylogenetic relationships among members of the genus Phlebovirus (Bunyaviridae) based on partial M segment sequence analyses. J. Gen. Virol. 2003;84:465–473. doi: 10.1099/vir.0.18765-0.

68. Fodor E., Pritlove D.C., Brownlee G.G. The influenza virus panhandle is involved in the initiation of transcription. J. Virol. 1994;68:4092–4096.

69. Barr J.N., Elliott R.M., Dunn E.F., Wertz G.W. Segment specific terminal sequences of Bunyamwera bunyavirus regulate genome replication. Virology. 2003;311:326–228. doi: 10.1016/S0042-6822(03)00130-2.

70. Kohl A., Lowen A.C., Leonard V.H.J., Elliott R.M. Genetic elements regulating packaging of the Bunyamwera orthobunyavirus genome. J. Gen. Virol. 2006;87:177–187. doi: 10.1099/vir.0.81227-0.

71. Elbeaino T., Digiaro M., Martelli G.P. RNA-5 and -6, two additional negative-sense RNA segments associated with fig mosaic virus. J. Plant Pathol. 2012;1

72. Ishikawa K., Maejima K., Komatsu K., Kitazawa Y., Hashimoto M., Takata D., Yamaji Y., Namba S. Identification and characterization of two novel genomic RNA segments of fig mosaic virus, RNA5 and RNA6. J. Gen. Virol. 2012;93:1612–1619. doi: 10.1099/vir.0.042663-0.

73. Dolja V.V., Kreuze J.F., Valkonen J.P. Comparative and functional genomics of closteroviruses. Virus Res. 2006;117:38–51.

74. Seifers D.L., Harvey T.L., Martin T.J., Jensen S.G. Identification of the wheat curl mite as the vector of the High Plains virus of corn and wheat. Plant Dis. 1997;81:1161–1166. doi: 10.1094/PDIS.1997.81.10.1161.

75. She Y.-M., Seifers D.L., Haber S., Ens W., Standing K.G. Characterization of the agent of 'High Plains disease' J. Biol. Chem. 2004;279:488–494.

76. Kulkarni N.K., Kumar P.L., Muniyappa V., Jones A.T., Reddy D.V.R. Transmission of pigeon pea sterility mosaic by the eriophyid mite, *Aceria cajani* (Acari: Arthropoda) Plant Dis. 2002;86:1297–1302. doi: 10.1094/PDIS.2002.86.12.1297.

77. Seifers D.L., Harvey T.L., Louie R., Gordon D.T., Martin T.J. Differential transmission of isolates of the High Plains virus by different sources of wheat curl mites. Plant Dis. 2002;86:138–142. doi: 10.1094/PDIS.2002.86.2.138.

78. Skare J.M., Wijkamp I., Rezende J.A.M., Michels G., Rush C., Scholthof K.-B.G., Scholthof H.B. Colony establishment and maintenance of the eriophyid wheat curl mite *Aceria tosichella* for controlled transmission studies on a new virus-like pathogen. J. Virol. Methods. 2003;108:133–137. doi: 10.1016/S0166-0934(02)00257-4.

79. Mielke-Ehret N., Thoma J., Schlatermund N., Mühlbach H.-P. Detection of European mountain ash ringspot-

associated virus-specific RNA and protein P3 in the pear leaf blister mite *Phytoptus pyri*(Eriophyidae) Arch. Virol. 2010;155:987–991. doi: 10.1007/s00705-010-0667-3.

80. Elbeshehy E.K., Elbeaino T. Viruses infecting figs in Egypt. Phytopathol. Mediterr. 2011;50:327–332.

81. El Air M., Mahfoudhi N., Elbeaino T., Dhouibi M.H., Digiaro M. Occurrence of *fig mosaic virus* in Tunisian fig orchards. J. Plant Pathol. 2012;1

82. Alhudaib K. Incidence of fig leaf mottle-associated virus and fig mosaic virus in Eastern Province of Saudi Arabia. Int. J. Virol. 2012;8:128–132. doi: 10.3923/ijv.2012.128.132.

83. Shahmirzaie M., Rakhshanderoo F., Zamanizadeh H.R., Elbeaino T. Current status of fig mosaic disease in Iran. J. Phytopathol. 2012;160:324–330. doi: 10.1111/j.1439-0434.2012.01908.x.

84. Ishikawa K., Maejima K., Nagashima S., Sawamura N., Takinami Y., Komatsu K., Hashimoto M., Yamaji Y., Yamamoto J., Namba S. First report of fig mosaic virus infecting common fig (*Ficus carica*) in Japan. J. Gen. Plant Pathol. 2012;78:136–139. doi: 10.1007/s10327-012-0359-9.

85. Flock R.A., Wallace J.M. Transmission of fig mosaic by the eriophyid mite *Aceria ficus*. Phytopathology.1955;45:52–54.

86. Seth M.L. Transmission of pigeonpea sterility by an eriophyid mite. Indian Phytopathol. 1962;15:225–227.

87. Proeseler G. Beziehungen zwischen Virus, Vektor und Wirtspflanze am Beispiel des Feigenmosaik-Virus und *Aceria ficus* Cotte (Eriophyoidea) (in German) Acta Phytopathol. Acad. Sci. Hung. 1972;7:179–186.

88. Capoor S.P. Observations on pigeonpea sterility disease in Bombay. Indian J. Agric. Sci. 1952;22:271–274.

89. Alfieri S.A., Jr. Plant Pathology Circular No. 64. Florida Department of Agriculture, Division of Plant Industry; Gainesville, FL, USA: 1967. Fig mosaic.

90. Führling M., Büttner C. Transmission experiments of viruses to woody seedlings (*Quercus robur* L. and*Sorbus aucuparia* L.) by grafting and mechanical inoculation. Eur. J. For. Path. 1995;25:129–135. doi: 10.1111/j.1439-0329.1995.tb00327.x.

91. Forster R.L., Seifers D.L., Strausbaugh C.A., Jensen S.G., Ball E.M., Harvey T.L. Seed transmission of the High Plains virus in sweet corn. Plant Dis. 2001;85:696–699. doi: 10.1094/PDIS.2001.85.7.696.

92. Divya P., Kumar L.L., Rangaswamy K.T., Muniyappa V. Detection of pigonpea sterility mosaic virus in floral parts and seeds. Indian J. Virol. 2005;16:36.

93. Kim K.S., Martin E.M. Virus-like particles associated with yellow ringspot of redbud. Phytopathol. News. 1978;12:119.

94. Ahn K.K., Kim K.S., Gergerich R.C., Jensen S.G., Anderson E.J. Comparative ultrastructure of double-membrane-bound particles and inclusions associated with eriophyid mite-borne plant diseases of unknown etiology: A potentially new group of plant viruses. J. Submicrosc. Cytol. Pathol. 1996;28:345–355.

95. Laney A.G., Gergerich R., Keller K., Martin R., Tzanetakis I. Rose rosette and redbud yellow ringspot are caused by two new emaraviruses. Phytopathology. 2010;100:S67.

96. Robertson N.L., Brown K.L. Ringspot leaf symptoms on *Sorbus scopulina* greene associated with virus-like particles. Phytopathology. 2009;99:S185–S186. doi: 10.1094/PHYTO-99-2-0185.

CHAPTER 9

DETECTION AND IDENTIFICATION OF THE FIRST VIRUSES IN CHIA (SALVIA HISPANICA)

Marcos G. Celli[1], Maria C. Perotto[1,2], Julia A. Martino[2], Ceferino R. Flores[3], Vilma C. Conci[1,2], and Patricia Rodriguez Pardina[1]

[1]Instituto de Patología Vegetal (IPAVE – CIAP – INTA) Camino. 60 cuadras Km 5,5, Córdoba, X5020ICA, Argentina

[2]Consejo Nacional de Investigaciones Científicas y Técnicas (CONICET), Camino. 60 cuadras Km 5,5, Córdoba, X5020ICA, Argentina

[3]Estación Experimental Agropecuaria Yuto (INTA), Ruta Nacional 34 Km 1286, Yuto, 4518, Jujuy, Argentina

ABSTRACT

Chia (Salvia hispanica), an herbaceous plant native to Latin America, has become important in the last 20 years due to its beneficial effects on health. Here, we present the first record and identification of two viruses in chia plants. The comparison of the complete nucleotide sequences showed the presence of two viral species with the typical genome organization of bipartite New World begomovirus, identified as Sida mosaic Bolivia virus 2 and Tomato yellow spot virus, according to the ICTV taxonomic criteria for begomovirus classification. DNA-A from Sida mosaic Bolivia virus 2 exhibited 96.1% nucleotide identity with a Bolivian isolate of Sida micrantha, and Tomato yellow spot virus showed 95.3% nucleotide identity with an Argentine bean isolate. This is the first report of begomoviruses infecting chia as well as of the occurrence of Sida mosaic Bolivia virus 2 in Argentina.

INTRODUCTION

Chia (*Salvia hispanica*) is an herbaceous plant of the *Lamiaceae* family, native to mountain areas of South America, from Mexico to Guatemala. Chia, along with corn and bean, was the basis of the diet of indigenous people and was third in economic importance [1]. Chia cultivation was drastically reduced by the Spanish colonizers; however, during the last two decades, the confirmation of its beneficial properties for human health has made chia one of the crops with the highest nutritional profile in Latin America, being commercially produced in Mexico, Bolivia and Argentina [1]. The cultivated area in Argentina was 70,000 ha in 2013, with a return that far exceeded that of soybean [2,3].

Chia seeds are rich in oil, dietary fibers, protein and mucilage. Its oil accounts for 33% of its seed weight, with 68% being α-linolenic acid, the highest percentage recorded in oilseeds [4]. Seed consumption has not shown any of the problems associated with other sources of omega-3 fatty acids, such as fishy taste, weight loss in animals or digestive problems [5,6,7]. Chia contains 25% dietary fiber (10% soluble fiber of very high molecular weight) and 20% content of gluten-free proteins, which makes it suitable for people suffering from celiac disease [4].

Citation: Marcos G. Celli, Maria C. Perotto, Julia A. Martino, Ceferino R. Flores, Vilma C. Conci, and Patricia Rodriguez Pardina, Detection and Identification of the First Viruses in Chia (Salvia hispanica), doi:10.3390/v6093450.

To date, several viruses naturally infecting the genus *Salvia* have been reported, such as *Cucumber mosaic virus* (CMV) in *S. uliginosa* [8] and *S. splendens* [9], *Broad bean wilt virus* 2 in *S. officinalis* [10], two begomoviruses (*Mung bean yellow mosaic virus* and *Tomato yellow leaf curl virus*) in *S. splendens* [11] and another putative begomovirus, *Clerodendron golden mosaic China virus* [12].

Up to the present, there has been no information of viruses affecting chia crop. In Argentina, a high percentage of plants with typical symptoms of viral infections, such as mosaic, deformed and stunted leaves and chlorosis, which notably affected crop production, have ben detected in recent crop seasons. The aim of this work was to detect and identify the infections produced by virus in chia plants collected from production areas.

MATERIALS AND METHODS

In 2013, samples of chia plants showing leaf deformation, chlorosis and dwarfing (Figure 1) were collected from production fields located in the north of Salta (23° 13'S, 64° 06'W).

Figure 1: Chia plants showing viral disease symptoms: dwarfism, chlorosis and deformation.

The symptomatic chia plants were screened for the presence of *Alfalfa mosaic virus, Cucumber mosaic virus,Potyvirus* genus, *Tobacco mosaic virus* and *Tospovirus* group (I, II and III) by double-antibody sandwich enzyme-linked immunosorbent assay (DAS-ELISA) [13] using commercial antisera (BIOREBA SRL Latin America, Mar del Plata, Argentina), according to the manufacturer's instructions. The samples were also tested by PCR using the oligonucleotides PAL1v1978/PAR1c496 [14], which amplify a 1,100-1,300 nt fragment corresponding to the 5'region of the *Rep* gene, the entire common region and the 5'end of the *CP*gene of begomoviruses.

Total DNA from chia plants was extracted using ZR-96 Plant/Seed DNA Kit™ (ZYMO, Irvine, CA, USA). PCR reactions were prepared in a 25 μL volume containing 5X buffer, 3.5 mM Mg_2Cl, 0.25 mM of dNTP mix, 0.75 μM of each primer, 1 unit of *Taq* DNA polymerase (Promega Corp., Madison, WI, USA) and 1 μL of DNA. PCR cycle parameters were as follows: 94 °C for 2 min, 30 cycles of 94 °C for 30 s, 55 °C for 90 s, 72 °C for 90 s, and a final extension of 72 °C for10 min. PCR products were analyzed by electrophoresis in 1.5% agarose gels and visualized with UV light after ethidium bromide staining.

Total DNA was subsequently used as template to amplify the putative full-length begomovirus genomes by rolling circle amplification (RCA) with Φ 29 DNA polymerase (Templiphi GE Healthcare, Piscataway, NJ, USA) as previously described [15]. To select an enzyme that could cut at a single site in the genome of begomovirus, generating unit-size molecules, the amplified DNA was digested independently with seven different restriction enzymes (*Eco*RI, *Eco*RV, *Xba*I, *Pst*I, *Kpn*I, *Hind*III and *Bam*HI). Linearized fragments were cloned in pBluescript SK+ [16] digested with the same restriction enzyme, transformed into*Escherichia coli* DH5α and sequenced at the Genomics Unit of the Biotechnology Institute-INTA (Argentina).

The nucleotide and deduced amino acid sequences were compared with those of other begomoviruses available in the GenBank [17]. Database searches were carried out using the Blastn algorithm [18]. The viral sequences showing highest identity were selected for the identity percentage analysis using Lasergene 8.0.2 software package (DNASTAR, Inc., Madison, WI, USA).

A phylogenetic analysis of the sequences was performed with Mega 5.2 software [19] using the Neighbor-Joining maximum likelihood method, GTR (general time reversible) model with G+I (invariant sites and distributed range). The bootstrap consensus tree was inferred from 1000 replicates.

RESULTS AND DISCUSSION

None of the tested samples reacted with any of the used antisera. PCR products of the expected size (1250 bp) were obtained for two of the five chia plants analyzed, confirming the presence of a begomovirus infection.

Circular DNA genomes amplified by RCA were linearized with the *Xba*I and *Kpn*I restriction enzymes for SM1 and SM2 clones, respectively, and with *BamHI* for TO1 and TO2 clones. The sequence analysis of clones SM1 (KJ742421) and TO1 (KJ742419) showed the typical genome organization of DNA-A of bipartite New World begomoviruses, with five open reading frames (ORFs), encoding replication-associated protein (Rep), transcriptional activator protein (TrAP), replication enhancer protein (REn), AC4, and coat protein (CP). Clones SM2 (KJ742422) and TO2 (KJ742420) encoded the movement proteins (MP) and nuclear shuttle proteins (NSP), included in the DNA-B of begomoviruses.

According to the ICTV criteria for differentiation of begomovirus species (89% nucleotide identity) [20], the comparison of the sequences obtained in this work and the geminivirus sequences from the database revealed a mixed infection by two different begomoviruses. The clones SM1 and SM2 showed 96.1% and 93.7% nt identity for DNA-A and DNA-B, respectively, with the sequences of *Sida mosaic Bolivia virus* 2 (SiMBoV2, HM585443/HM585444) previously detected infecting plants of *Sida micrantha* in Bolivia [21], therefore, considering it a new isolate of SiMBoV2.

The clone TO1 showed the highest nucleotide identity (95.3%) with DNA-A of an isolate of *Tomato yellow spot virus* (ToYSV, FJ538207) infecting bean (*Phaseolus vulgaris*) in Argentina [22], but also presented nucleotide identity above 89% with three other ToYSV isolates, four isolates of *Leonurus mosaic virus*(LeMV), and one of *Sida micrantha mosaic virus* (SimMV), whereas the isolates of *Okra mottle virus*(OMoV), *Sida yellow mosaic virus* (SiYMV), *Sida mosaic virus* (SiMoV) and *Sida yellow net virus*(SiYNV) showed identity between 80% and 89% (Table 1). The clone TO2 had higher identity with DNA-B of two Brazilian ToYSV isolates: 91.2% with the *Leonurus sibiricus* isolate (JX513953) [23] and 90.1% with the *Lycopersicon esculentum* isolate (DQ336351) [24]. Considering that *Leonurus mosaic virus* might be a strain of ToYSV [25] and that the two highest identity of DNA-A and DNA-B were obtained from the comparison with two ToYSV isolates, we propose that the isolate described here is ToYSV.

Table 1: Nucleotide identity of the DNA-A sequence of *Tomato yellow spot virus* isolated from chia (clone TO1)

Virus	Acronym	Origin	GenBank Accession Number	% of Identity
Tomato yellow spot virus	ToYSV	Argentina	FJ538207	95.3%
Tomato yellow spot virus	ToYSV	Brazil	JX513952	93.9%
Leonurus mosaic virus	LeMV	Brazil	JX863082	93.6%
Leonurus mosaic virus	LeMV	Brazil	JX863081	93.5%
Leonurus mosaic virus	LeMV	Brazil	JQ429791	93.4%
Leonurus mosaic virus	LeMV	Paraguay	KC683374	91.6%
Tomato yellow spot virus	ToYSV	Brazil	KC706628	89.9%
Tomato yellow spot virus	ToYSV	Brazil	DQ336350/NC_007726	89.8%
Sida micrantha mosaic virus	SimMV	Brazil	AJ557450	89.2%
Sida mottle virus	SiMoV	Brazil	AY090555/NC_004637	88.9%
Sida micrantha mosaic virus	SimMV	Brazil	FN557522	88.6%
Sida mottle virus	SiMoV	Brazil	JX871378	88.1%
Sida mottle virus	SiMoV	Brazil	JX871377	88.0%
Okra mottle virus	OMoV	Brazil	EU914817	87.1%
Okra mottle virus	OMoV	Brazil	EU914819	87.1%
Okra mottle virus	OMoV	Brazil	FJ686695	87.0%
Sida yellow net virus	SiYNV	Brazil	JX871376	86.6%
Sida yellow mosaic virus	SiYMV	Brazil	JX871369	83.1%
Sida yellow mosaic virus	SiYMV	Brazil	AY090558/NC_004639	83.1%

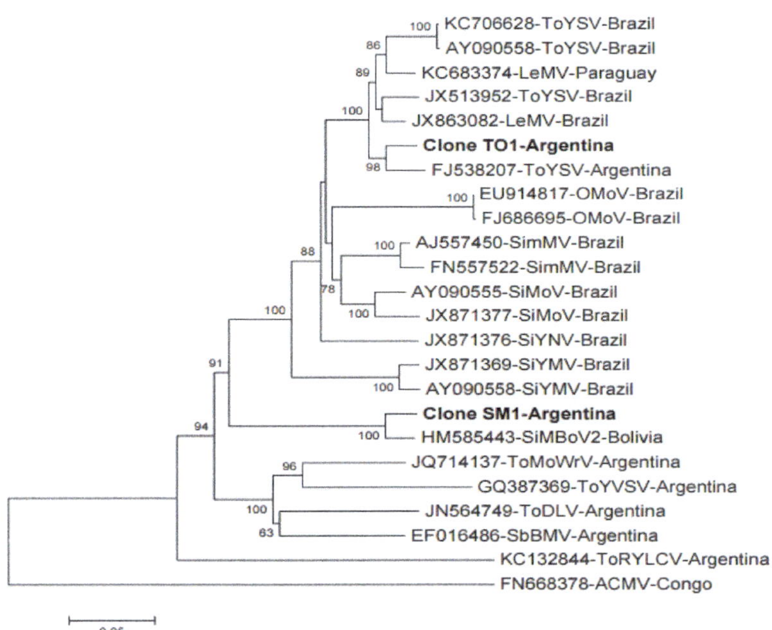

DNA-A

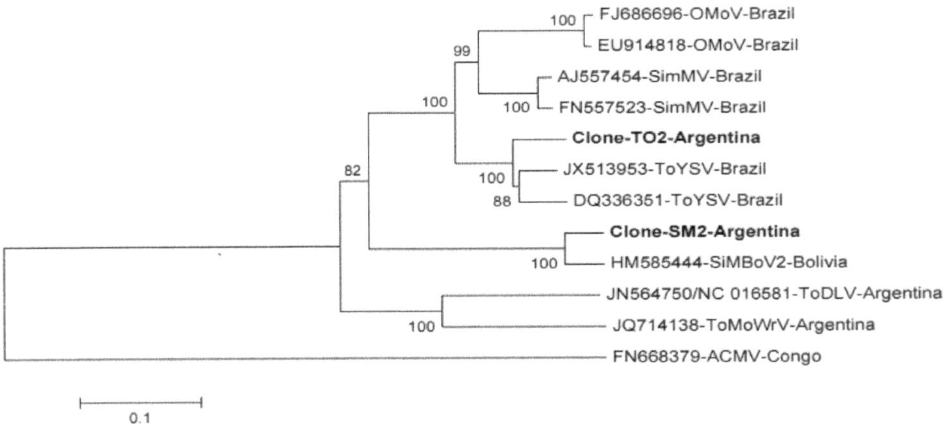

DNA-B

Figure 2: Neighbor-joining trees showing the phylogenetic relationship between DNA-A and DNA-B of the isolates detected in this study and other begomoviruses and from the international database GenBank [17] and using *African cassava mosaic virus* (ACMV) as an outgroup. ...

The phylogenetic tree (Figure 2) was constructed using complete nucleotide sequences of DNA-A and DNA-B of both isolates, sequences published in GenBank that showed more than 80% identity, and those of begomoviruses reported in Argentina. The genome sequences of DNA-A and DNA-B of SiMBoV2 obtained in this work grouped on a monophyletic branch, with 100% bootstrap confidence, with the SiMBoV2 sequence from Bolivia, confirming that these isolates are closely related. The phylogenetic relationship between the DNA-A sequence of ToYSV from Argentina and other begomoviruses (Figure 2) showed that the isolate obtained from the Argentine samples of chia clustered with other sequences of ToYSV and LeMV, but was more closely related to the Argentine isolate from ToYSV (FJ538207), with 99% bootstrap confidence value.

CONCLUSIONS

Here, we demonstrate the presence of two bipartite begomovirus infecting *Salvia hispanica* associated with severe disease symptoms in chia. Possibly, it was transmitted by whiteflies and the presence of these begomovirus in Argentina must be considered as a threat to the chia crop, since whitefly populations are present in the north-western Argentina [26]. This information is relevant for the implementation of control strategies to reduce disease damage. Given the increasing economic importance of this crop as agricultural product in the country [3], more studies are necessary to estimate the damage caused by SiMBo2 and ToYSV.

To the best of our knowledge, this is the first report of viral infection in chia plants, as well as of the presence of *Sida mosaic Bolivia virus* 2 in Argentina. The latter was previously reported to infect the weed *S. micrantha* [21], and the present results are the first references to the virus infecting a cultivated species.

ACKNOWLEDGMENTS

This study was carried out at IPAVE-CIAP-INTA and was partially supported by INTA and CONICET.

AUTHOR CONTRIBUTIONS

V.C.C. and P.R.P designed the study; M.G.C, M.C.P and J.A.M. conducted experiments; C.R.F. identified and collection of symptomatic plants. All authors contributed to the analysis of the results and writing the paper.

REFERENCES

1. Ayerza R., Wayne C. Chía, redescubriendo un olvidado alimento de los Aztecas. 1st ed. Del Nuevo Extremo; Buenos Aires, Argentina: 2006. p. 232.
2. Lobo Zavalía R., Alcocer M.G., Fuentes F.J., Rodriguez W.A., Morandini M., Devani M.R. Desarrollo del cultivo de chía en Tucumán. República Argentina. Av. Agroind. 2011;32:27–30.
3. La Gaceta Economía—*En el NOA se perdió entre el 70% y el 80% de la producción de chía, debido a la sequía.* 2013. [(accessed on 16 August 2013)]. Available online: http://www.lagaceta.com.ar.
4. Ayerza R. Seed composition of two chía (*Salvia hispanica* L.) genotypes which differ in seed color. Emir. J. Food Agric. 2013;25:495–500.
5. Ayerza R., Coates W. Dietary levels of chia: Influence on yolk cholesterol, lipid content and fatty acid composition, for two strains of hens. Poult. Sci. 2000;78:724–739. doi: 10.1093/ps/79.5.724.
6. Ayerza R., Coates W. The omega-3 enriched eggs: The influence of dietary linolenic fatty acid source combination on egg production and composition. Can. J. Anim. Sci. 2001;81:355–362. doi: 10.4141/A00-094.
7. Ayerza R., Coates W., Lauria M. Chia seed (*Salvia hispanica* L.) as an ω-3 fatty acid source for broilers: Influence on fatty acid composition, cholesterol and fat content of white and dark meat, growth performance and sensory characteristics. Poult. Sci. 2002;81:826–837.
8. Holcomb G.E., Valverde R.A. Natural Infection of *Salvia uliginosa* with *Cucumber mosaic cucumovirus*. HortScience. 1998;33:1215–1216.

9. Ali S., Khan A.A., Afreen B., Sharma R., Jahan T., Naqvi Q.A. Natural occurrence of *Cucumber mosaic virus* associated with mottling and mosaic disease on *Salvia splendens*, a new record from India. Australas. Plant Dis. Notes. 2008;3:118–120. doi: 10.1071/DN08047.

10. Mumford R.A., Jarvis B., Harju V., Boonham N., Skelton A. The first report of Broad bean wilt virus 2 in the UK: findings in foxglove and salvia. Plant Pathol. 2006;55:819.

11. Ara M.R., Masud M.M.H., Akanda A.M. Detection of plant viruses in some ornamental plants that act as alternate hosts. Agriculturists. 2012;10:46–54.

12. Valverde R.A., Singh R., Sabanadzovic S. Detection and identification of *Clerodendron golden mosaic China virus* in *Salvia splendens*. Eur. J. Plant Pathol. 2012;133:499–503. doi: 10.1007/s10658-011-9923-y.

13. Clark M.F., Adams A.N. Characteristics of the microplate method of enzyme-linked immunosorbent assay for the detection of plant viruses. J. Gen. Virol. 1977;34:475–483. doi: 10.1099/0022-1317-34-3-475.

14. Rojas M.R., Gilbertson R.L., Russell D.R., Maxwell D.P. Use of degenerate primers in the polymerase chain reaction to detect whitefly-transmitted geminiviruses. Plant Dis. 1993;77:340–347. doi: 10.1094/PD-77-0340.

15. Inoue-Nagata A.K., Albuquerque L.C., Rocha W.B., Nagata T. A simple method for cloning the complete begomovirus genome using the bacteriophage phi 29 DNA polymerase. J. Virol. Method.2004;116:209–211. doi: 10.1016/j.jviromet.2003.11.015.

16. Maniatis T., Fritsch E.F., Sambrook J. Molecular Cloning: A Laboratory Manual. Cold Spring Harbor Laboratory Press; Cold Spring Harbor, New York, NY, USA: 1982. p. 1626.

17. GenBank, National Center for Biotechnology Information – NCBI. [(accessed on 15 June 2014)];Available online: http://www.ncbi.nlm.nih.gov.

18. Altschul S.F., Gish W., Miller W., Myers E.W., Lipman D.J. Basic local alignment search tool. J. Mol. Biol. 1990;215:403–410.

19. Tamura K., Peterson D., Peterson N., Stecher G., Nei M., Kumar S. MEGA5: Molecular Evolutionary Genetics Analysis using Maximum Likelihood, Evolutionary Distance, and Maximum Parsimony Methods.Mol. Biol. Evol. 2011;28:2731–2739. doi: 10.1093/molbev/msr121.

20. Brown J.K., Fauquet C.M., Briddon R.W., Zerbini M., Moriones E., Navas Castillo J. Geminiviridae. In: King A.M.Q., Adams M.J., Carstens E.B., Lefkowitz E.J., editors. Virus taxonomy. Ninth Report of the International Committee on Taxonomy of Viruses. Elsevier Academic Press; London, UK: 2012. pp. 351–373.

21. Wyant P.S., Gotthardt D., Schäfer B., Krenz B., Jeske H. The genomes of four novel begomoviruses and a new *Sida micrantha mosaic virus* strain from Bolivian weeds. Arch. Virol. 2011;156:347–352. doi: 10.1007/s00705-010-0876-9.

22. Rodríguez Pardina P.E., Hanada K., Laguna I.G., Zerbini F.M., Ducasse D.A. Molecular characterization and relative incidence of bean- and soybean-infecting begomoviruses in northwestern Argentina. Ann. Appl. Biol. 2011;158:69–78. doi: 10.1111/j.1744-7348.2010.00441.x.

23. Barbosa J.C., Eckstein B., Bergamin Filho A., Rezende J.A.M., Dallagnol L.J. First Report of *Tomato yellow spot virus* Infecting *Leonurus sibiricus* in Brazil. Plant Dis. 2013;97:289. doi: 10.1094/PDIS-07-12-0692-PDN.

24. Calegario R.F., Ferreira S.S., Andrade E.C., Zerbini F.M. Characterization of *Tomato yellow spot virus*, a novel tomato-infecting begomovirus in Brazil. Pesquisa Agropecuária Brasileira. 2007;42:1335–1343.

25. Fernandes N.A.N., Boiteux L.S., Fonseca M.E.N., Segnana L.G., Kitajima E.W. Report of *Tomato yellow spot virus* infecting *Leonurus sibiricus* in Paraguay and within tomato fields in Brazil. Plant Dis.2013;97:289.

26. Alemandri V., Rodriguez Pardina P., Izaurralde J., García Medina S., Argüello Caro E., Mattio M.F., Dumón A., Rodriguez S.M., Truol G. Incidence of begomoviruses and climatic characterisation of *Bemisia tabaci*-geminivirus complex in soybean and bean in Argentina. Agriscientia. 2012;29:31–39.

CHAPTER 10

THE CELL BIOLOGY OF TOBACCO MOSAIC VIRUS REPLICATION AND MOVEMENT

Chengke Liu and Richard S. Nelson

Plant Biology Division, The Samuel Roberts Noble Foundation, Inc., Ardmore, OK, USA

ABSTRACT

Successful systemic infection of a plant by *Tobacco mosaic virus* (TMV) requires three processes that repeat over time: initial establishment and accumulation in invaded cells, intercellular movement, and systemic transport. Accumulation and intercellular movement of TMV necessarily involves intracellular transport by complexes containing virus and host proteins and virus RNA during a dynamic process that can be visualized. Multiple membranes appear to assist TMV accumulation, while membranes, microfilaments and microtubules appear to assist TMV movement. Here we review cell biological studies that describe TMV-membrane, -cytoskeleton, and -other host protein interactions which influence virus accumulation and movement in leaves and callus tissue. The importance of understanding the developmental phase of the infection in relationship to the observed virus-membrane or -host protein interaction is emphasized. Utilizing the latest observations of TMV-membrane and -host protein interactions within our evolving understanding of the infection ontogeny, a model for TMV accumulation and intracellular spread in a cell biological context is provided.

INTRODUCTION

Viruses, as obligate organisms, utilize host factors to accumulate and spread in their host. A successful infection by a plant virus includes entry and accumulation in the first cell, movement into neighboring uninfected cells, and systemic infection through the plant vascular tissue (Boevink and Oparka, 2005; Epel, 2009; Harries and Ding, 2011; Niehl and Heinlein, 2011; Schoelz et al., 2011; Tilsner et al., 2011). Plant viruses have varying strategies for infecting hosts which reflect their use of existing functionally redundant host developmental pathways. Therefore an understanding of virus infection processes also offers insight into normal host physiological processes. *Tobacco mosaic virus* (TMV) encodes four known functional proteins: the 126 and 183 kDa replication-associated proteins, the movement protein (MP), and the structural capsid or coat protein (CP). In order to have a successful infection, these four multifunctional proteins cooperate with many host

Citation: Chengke Liu and Richard S. Nelson, The cell biology ofTobacco mosaic virus replication and movement, doi: 10.3389/fpls.2013.00012.

components. The host membrane and cytoskeleton are subcellular structures important for TMV infection. TMV-induced granules or inclusion bodies that contain membranes also contain host proteins. In this review, we discuss the changing roles of host membranes, cytoskeleton, and inclusion body-associated proteins as infection progresses. Findings reported in the literature are first presented in the section(s) where the effect on virus physiology was observed rather than where it may additionally influence this activity. For example, the influence of synaptotagmin on TMV physiology (Lewis and Lazarowitz, 2010) was reported as an inhibition of intercellular spread of the TMV MP, although it likely influences the intracellular transport of this protein. This was done to clearly indicate what is in the published literature rather than what a reader may interpret the results to indicate. In some instances, however, the presumed influence of the observed outcome on the mechanism of virus movement is noted. As pertinent, findings from other tobamoviruses are mentioned to indicate the generality or specificity of a conclusion for the genus.

INITIAL INFECTION

Tobacco mosaic virus enters plant cells only through mechanical wounds which either transiently open the plasma membrane or allow pinocytosis (Palukaitis and Zaitlin, 1986; Shaw, 1999; Figure1). TMV begins to disassemble within 3 min after entry and disassembly of CP from the capsid is associated with translation of viral RNA (vRNA; Wu et al., 1994; reviewed in Shaw, 1999). TMV vRNA labeled with cyanine 3-UTP forms small granules in cytoplasm less than 5 min after entering the cell (Christensen et al., 2009). The vRNA-containing granules form where CP and vRNA co-localize as well as in the absence of CP, suggesting that although CP was not needed for granule formation the disassembly of TMV capsids occurred at the site of granule formation. Removal or mutation of *cis* and *trans* elements necessary for virus replication (i.e., the vRNA 3′ untranslated region and replicase) did not prevent granule formation, although they were smaller and less stable. The granules were shown to associate with fluorescently labeled ER. The 5′ methylguanosine cap on the vRNA was necessary to anchor vRNA to the ER/actin complex: absence of the cap leading to vRNA degradation and no granules (Christensen et al., 2009). Considering that uncoating of vRNA may make it accessible to the silencing surveillance system (reviewed in Niehl and Heinlein, 2011), it will be important to determine to which host factors viral proteins are attached during granule formation and transport to cortical and perinuclear replication sites. Identifying host factors in the granules will be difficult due to their presumed low quantities, but will be necessary to understand the steps prior to virus replication.

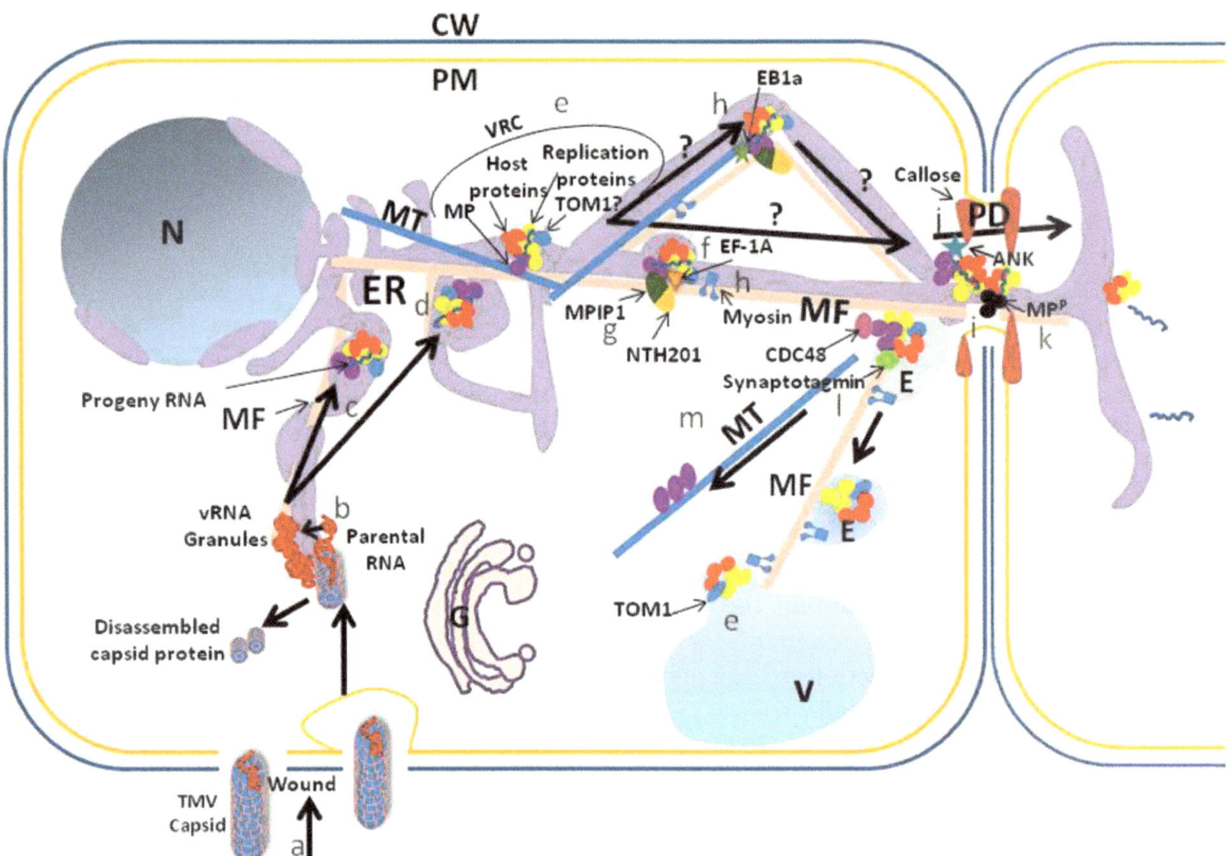

Figure 1: Schematic of a proposed accumulation and movement pathway for TMV within cells. To simplify the model we do not address the possibility that the MP or any other viral protein moves within the cell, with or without viral RNA, independently of the virus ...

REPLICATION

Tobacco mosaic virus and the very closely related *Tomato mosaic virus* (ToMV) use their parental genomes to synthesize complementary negative strands which serve as templates for the synthesis of progeny full-length positive strands and subgenomic mRNAs containing MP and CP open reading frames (ORFs;Ishikawa and Okada, 2004; Ishibashi et al., 2010). Although the 183 kDa protein encoded by the 5′ ORF of these viruses can replicate the genome, the 126 kDa protein, produced by termination of translation at an amber stop codon within the 183 kDa protein ORF, is necessary for maximum progeny RNA production (Ishikawa et al., 1986; Ishikawa et al., 1991; Lewandowski and Dawson, 2000). The 126 and 183 kDa proteins contain methyltransferase and helicase domains, while the 183 kDa protein alone contains the C-terminal domain encoding an RNA-dependent RNA polymerase. The 126 and 183 kDa proteins together will be referred to as the replication proteins in this review. TMV/ToMV replication is believed to occur in a membrane-associated complex containing the replication proteins, MP, vRNA, and host proteins (Hagiwara et al., 2003; Heinlein et al., 1998; Más and Beachy, 1999; reviewed in Ishibashi et al., 2010; Laliberté and Sanfaçon, 2010; de Castro et al., 2012; Figure1). The ER was implicated as a site for virus replication complex (VRC) formation through co-localization of an ER marker, BiP, with MP-GFP and in turn, MP-GFP co-localization with replication proteins during immunofluorescence studies (Heinlein et al., 1998).Reichel and Beachy (1998), using transgenic plants expressing an ER-targeted GFP, determined that the ER formed large cortical aggregates at reticulate vertices and fewer membrane tubules during TMV accumulation, but returned to a normal structure after replication ended. However, later studies with ToMV using both fluorescence microscopy and biochemical fractionation methods determined that the replication proteins and replicase activity were associated predominantly with the vacuolar membrane, although they also showed some localization and activity with other less defined membrane fractions which included the ER (Hagiwara et al., 2003). Interestingly, ToMV can replicate in cells that are vacuole-diminished (Nishikiori et al., 2006). This finding supports the notion that although the tonoplast may function to support ToMV/TMV accumulation, other membranes such as the ER are important for this activity. Clearly, additional work is necessary to determine which membranes are essential for tobamovirus accumulation (Figure1).

Regardless of the membrane used for tobamovirus accumulation, it is clear that a characteristic VRC is not uniformly induced by tobamoviruses. TMV, ToMV, and *Tobacco mild green mosaic virus* (TMGMV), form different subcellular structures containing replication proteins late in the infection cycle: TMV forming X-bodies and ToMV and TMGMV forming virus bundles when analyzed through immunocytochemical EM studies (Das and Hari, 1992). For a fourth tobamovirus, *Turnip vein clearing virus* (TVCV), X-bodies are rare (Resconich, 1961). Early fluorescence localization studies determined that for strains of TMV the size of the VRCs varied and was positively correlated with the level of disease observed (Liu et al., 2005, 2006). Recently, however, it was determined that silencing expression of the gamma subunit of ATP synthase, a nuclear-encoded chloroplast protein, resulted in smaller but more numerous VRCs and severe disease symptoms (Bhat et al., 2012). Thus, the size of the VRC is not a perfect indicator of disease intensity and the number of VRCs may influence this phenotype.

The form of inclusions induced by tobamoviruses is correlated with differences in the replication protein sequences (Liu et al., 2005; Harries et al., 2009). While ectopically expressed 126 kDa protein from TMV fused with GFP forms intracellular inclusions, the 125 kDa protein homolog from TVCV does not form inclusions (Harries et al., 2009). Interestingly, the intracellular inclusions formed by the TMV 126 kDa protein co-localized with microfilaments, as observed for the TMV VRC (Liu et al., 2005). Domain(s) within the 126 kDa protein necessary for inclusion body formation are not identified, however, it is known that the helicase domain when expressed alone is able to form octomers *in vitro* (Goregaoker and Culver, 2003) and thus may be a domain important for this activity. In addition, a 126 kDa protein-GFP construct expressing only the N terminal 781 amino acids of the 126 kDa protein associated with the ER and formed inclusions (dos Reis Figueira et al., 2002). The 781 amino acid protein includes the methyltransferase and non-conserved bridge domain that previously was determined to influence the RNA silencing suppression function of this protein. There is unpublished data indicating that the methyltransferase domain alone can form inclusions (Knapp et al., 2007). More work is needed to further identify the domains responsible for inclusion formation and the relevance of inclusion formation to tobamovirus physiology.

Ectopically expressed TMV MP fused with fluorescent reporter proteins also can form cytoplasmic inclusion bodies (Heinlein et al., 1998; Reichel and Beachy, 1998; Sambade et al., 2008). These small inclusions are similar to those visualized in the cortical periphery during infection with tobamoviruses expressing an MP-GFP fusion (Padgett et al., 1996; Heinlein et al., 1998; Reichel and Beachy, 1998; Boyko et al., 2007). The cortical MP-GFP inclusions that appear during virus infection likely represent the inclusions containing replication proteins and MP observed by Szécsi et al. (1999), but this requires confirmation. Inclusions containing MP-GFP associate with microtubules both early and late in the infection cycle (Heinlein et al., 1998; Boyko et al., 2007; Sambade et al., 2008). Studies with cellular markers and an MP-mRFP construct containing a downstream non-viral stemloop-forming RNA aptamer that can be fluorescently labeled determined that MP-mRFP associates with ER and its own RNA (Sambade et al., 2008). In this regard, it will be important to determine if multiple types of inclusions are formed independently during infection or are always part of a continuum, with progeny inclusions appearing from parental inclusions.

The host proteins within VRCs or inclusions that contain them are not fully identified. Late in infection, TMV-induced X-bodies have been shown to contain the microtubule component, β-tubulin, through immunocytochemical EM studies (X. S. Ding and R. S. Nelson, personal communication). The function of this protein in X-bodies is unknown, but perhaps it could be to aid the degradation of body components analogous to the suggested function of microtubules during TMV MP turnover (Kragler et al., 2003). The host translation factor, elongation factor 1A (EF-1A), is present in the membrane-associated fraction where viral replicase activity was observed (Osman and Buck, 1996; Watanabe et al., 1999) and in X-bodies produced by TMV (Ding et al., 1998). It also interacts with the 3′-UTR of genomic vRNA and with the methyltransferase domain of the replication proteins (Zeenko et al., 2002; Yamaji et al., 2006). EF-1A has additional activities beyond supporting translation including forming complexes with tubulin and actin, the actin interaction possibly linking the cytoskeleton to protein synthesis, and ubiquitin-mediated degradation (Durso and Cyr, 1994; Gonen et al., 1994; Kim and Coulombe, 2010). The function of EF-1A during TMV accumulation was hypothesized to aid minus strand synthesis (Yamaji et al., 2006). However, down-regulation of EF-1A through virus-induced gene silencing resulted in the reduced size of green fluorescent lesions induced by TMV-expressing GFP, but no reduction in lesion numbers or translation activity in the silenced leaves (Yamaji et al., 2010). This result suggests that the function of EF-1A is not for translation or virus accumulation, but for virus movement that may, in some manner, be linked to the cytoskeleton (Figure1).

Tobamovirus multiplication 1 (TOM1) is a predicted multi-pass transmembrane protein required for tobamovirus accumulation (Ishikawa et al., 1993; Yamanaka et al., 2000). Surprisingly, over-expression of TOM1 decreases ToMV multiplication (Hagiwara-Komoda et al., 2008). This observation is associated with the finding that accumulation of the ToMV replication proteins in membrane-free (soluble) fractions was lower for plants over-expressing TOM1 compared with those not over-expressing this protein (Hagiwara-Komoda et al., 2008). This result indicates that the level of the soluble form or the ratio of soluble and membrane-bound forms of the replication proteins is critical for normal virus accumulation. It was hypothesized that the soluble form is important for RNA silencing suppressor activity and it was shown that the loss of suppressor activity is correlated with diminished accumulation of the virus (Kubota et al., 2003;Hagiwara-Komoda et al., 2008). TOM1 interacts with the helicase domain of the 130 kDa protein from the related tobamovirus, *Tobacco mosaic virus-Cg* (crucifer-infecting virus), in a yeast two-hybrid screen (Yamanaka et al., 2000). This interaction was shown to be with the helicase core region based on predictions from the crystal structure of the helicase domain (Nishikiori et al., 2012). The replication proteins from ToMV and TOM1 share similar subcellular fractionation pattern in extracts from infected BY-2 cells, residing mostly in the tonoplast-containing fractions, but also in fractions with other membranes, including the ER (Hagiwara et al., 2003). It is hypothesized that TOM1 forms a link between the host membrane in which it resides and the tobamovirus replication proteins (Figure1). This interaction is likely important for VRC formation, but the co-localization of TOM1 and tobamovirus replication proteins in live cells has not been reported.

INTRACELLULAR MOVEMENT

For TMV to establish a systemic infection, the virus or its components must move within a cell to establish an infection site, multiply and finally position for movement to the next cell. The granules of vRNA that form on initial infection, the VRCs that form during infection and the 126 kDa protein- and MP/vRNA-containing inclusions observed during ectopic expression all move within the cell (e.g., Kawakami et al., 2004; Liu et al., 2005; Sambade et al., 2008; Christensen et al., 2009). These complexes may change their form and constituents with time.

During initial infection, granules containing vRNA anchor to cortical ER and move to cortical ER vertices and the perinuclear ER where virus replication and translation occurs (Reichel and Beachy, 1998;Christensen et al., 2009; Figure1). Indeed, vRNA has been visualized in the perinuclear bodies by bimolecular fluorescence complementation using a modified sequence-specific RNA-binding protein, Pumilio1, or by classical *in situ* RNA labeling (Tilsner et al., 2008). The vRNA must contain a sequence that targets the ER membrane directly or through a protein that targets the ER. Why the vRNA moves to particular cortical ER vertices or the perinuclear ER for replication is unknown. However, considering that most of the ribosome-containing, or rough, ER is present in the perinuclear region (Carrasco and Meyer, 2011) it is likely that this location, or a cortical ER vertex also containing ribosomes, is best suited for virus protein synthesis. Neither cytochalasin D nor latrunculin B (LatB) treatment, both microfilament antagonists, affect granule formation suggesting that microfilaments are not involved in this initial activity (Christensen et al., 2009). However, disruption of microfilaments results in granules hovering in the cortical ER, suggesting microfilaments help transport the granules in the cell. In contrast, depolymerizing microtubules does not stop vRNA granule movement along the tubular cortical ER (Christensen et al., 2009).

Membranes may be involved in intracellular trafficking of TMV components and the virus itself, as a VRC or vRNP, during virus accumulation and later spread. TMV replication occurs in association with ER and other membranes and both the MP and the replication proteins associate intrinsically or through a protein linker with membranes (Brill et al., 2000; Hagiwara et al., 2003; Fujiki et al., 2006). Interestingly, however, interruption of ER to Golgi secretory transport, mediated by the host coat complex II (COP-II) with brefeldin A (BFA) or through over-expression of a dominant-negative GTPase, Sar1p, did not alter targeting of ectopically expressed, fluorescently labeled MP to plasmodesmata (PD; Boutant et al., 2009; Genovés et al., 2010). This result, as well as BFA studies with infectious virus (see below), indicates that the COP-II-mediated transport system is not utilized by TMV MP or TMV to target viral products to the cell periphery. The actual pathway used by the virus, however, likely includes ER since that membrane is present in early- and late-formed virus inclusions (by fluorescence microscopy and EM), with early forming inclusions paired at the cell wall (e.g., Esau and Cronshaw, 1967; Heinlein et al., 1998; Szécsi et al., 1999; see below).

Much information is available on the movement of inclusion bodies containing the viral replication proteins. Several laboratories pursued EM-based immunocytolocalization studies of TMV infections with antibody against the replication proteins (Hills et al., 1987; Saito et al., 1987). They noted that the structure of the inclusions likely changed during development, going from smoothly granular to containing electron-dense rope-like structures, composed at least partly of 126 kDa protein, in a ribosome-rich matrix. Saito et al. (1987) referred to the former as viroplasms and the latter as X-bodies. Szécsi et al. (1999) showed through light and EM immunocytolocalization studies that TMV-induced inclusion bodies with viroplasm and X-body characteristics changed position and content as infections within cells developed.

Near the infection front the inclusion bodies were paired on either side of the cell wall and contained both the replication proteins and MP, while four to six cells back from the front the bodies were not paired, had moved away from the cell wall and only contained the replication proteins. Through fluorescence microscopy of cells near the infection front, Tilsner et al. (2008) observed vRNA in small cortical bodies in the peripheral cytoplasm. They suggested these cortical bodies may represent the inclusions at the cell periphery observed by Szécsi et al. (1999). Motionless small fluorescent bodies in the cell were detected at 12 h post-inoculation and these bodies were moving by 14 h post-inoculation when tracking TMV expressing an MP-GFP fusion: a period when both MP and the replication proteins would co-localize (Kawakami et al., 2004). Movement of the fluorescent inclusions early in infection, when both replication proteins and MP would be present, was aligned with microfilaments, and through pharmacological and gene-silencing studies, the inclusions (also referred to as VRCs) were shown to require these intact microfilaments for their intracellular movement (Kawakami et al., 2004; Liu et al., 2005). The degradation of the microfilaments was not sufficient to decrease virus accumulation to levels that would prevent virus movement or VRC formation and thus the influence of microfilaments on TMV intracellular movement was not confounded by a significant inhibition of virus replication (Liu et al., 2005). Treatment with a general myosin motor inhibitor, 2,3-butanedione monoxime (BDM), impaired the intracellular movement of VRCs (Kawakami et al., 2004).

Boyko et al. (2007), using a chimeric TMV expressing the MP from the Ob tobamovirus (Padgett et al., 1996), determined that MP-GFP inclusions at the infection front likely representing VRCs were associated with and typically trafficked along cortical microtubules for short distances in a stop-and-go manner (microtubules labeled with a microtubule-associated protein fused with GFP). Later studies obtained similar findings using the native MP of TMV fused with mRFP (Sambade et al., 2008). It was hypothesized that microtubules serve to anchor and then, through polymerization, release the VRC for movement (Sambade and Heinlein, 2009; reviewed in Peña and Heinlein, 2012). A second hypothesis, to be discussed further in the section on TMV intercellular movement, was that microtubule polymerization pushes the VRC along the ER (Peña and Heinlein, 2012). Other research, however, found that disruption of microtubules through pharmacological treatment with aminoprophos-methyl, colchicine, or oryzalin, or by silencing α-tubulin, had no significant effect on the transport of the MP-GFP within cells or to the PD area during infection (Gillespie et al., 2002; Kawakami et al., 2004; Wright et al., 2007). Regarding microfilaments, early studies indicated an association of TMV MP with rhodamine-conjugated phalloidin microfilaments after probing cells with polyclonal antibody against the MP and fluorescein-conjugated secondary antibody (McLean et al., 1995).Wright et al. (2007) utilized fluorescence recovery after photobleaching (FRAP) to observe MP-GFP movement and found that microfilament antagonists, LatB and cytochalasin B, inhibited MP targeting to the PD. However, a later study determined that the MP-GFP expressed during virus infection was not observed to co-align with fluorescently labeled microfilaments (Hofmann et al., 2009). Lastly, in studying membrane-mediated transport of the MP, inhibition of ER–Golgi membrane trafficking with BFA at low concentration (10 μg/ml) did not inhibit MP targeting (Tagami and Watanabe, 2007; Wright et al., 2007), but did influence the structure of the MP-GFP inclusion bodies (Tagami and Watanabe, 2007). At high concentration of BFA (100 μg/ml), which disrupts cortical ER structure, there was a significant effect on MP targeting to the PD (Wright et al., 2007). Thus, during virus infection the trafficking of the MP within the cell likely requires intact ER and may require microfilaments and microtubules (Figure1), although evidence against a direct action for microfilaments exists and microtubules may not be in an intact form or the required microtubule array is unusual in that it is impervious to certain pharmacological agents (Seemanpillai et al., 2006).

Although the TMV VRC moves within cells, the replication proteins and the MP, independent of one another and in the absence of vRNA, also can transport within the cell. Their independent transport may have physiological relevance. Expression of a fusion of the TMV 126 kDa protein with GFP yields an intracellular inclusion body that, like the VRC, co-aligns with and traffics along microfilaments (Liu et al., 2005). As observed for the VRC, disruption of microfilaments ends intracellular transport of the 126 kDa protein-GFP inclusion body. Furthermore, normal intracellular movement of the 126 kDa-GFP inclusion body, like TMV sustained intercellular movement, requires myosin XI-2 (see section on "INTERCELLULAR MOVEMENT" for discussion of myosin influence on TMV physiology; Harries et al., 2009; C. Liu, and R. S. Nelson, personal communication). Whether the 126 kDa protein directly interacts with microfilaments or myosin XI-2 requires further study. If there is no direct interaction between these proteins, trafficking of virus proteins may be through interaction of myosin XI-2 with host components associated with a virus-host protein complex or through the creation of a bulk flow network of cytoplasmic constituents (i.e., cytoplasmic streaming). Considering that the viral replication proteins may transport independently of the MP or other viral components to complete their functions, additional studies of their ectopic transport in relationship to their transport during virus infection are needed. The difficulty in pursuing studies on the viral replication proteins is that no fusion between the replication proteins and a fluorescent marker has been developed that yields a viable virus. This needs to be addressed for further progress to occur.

Transient expression of the MP fused with fluorescent markers results in the formation of inclusions that associate with RNA and associate with and traffic along the ER, perhaps interacting with a microtubule scaffold necessary for movement (Sambade and Heinlein, 2009). Both microfilament and microtubule antagonists inhibited intracellular transport of the MP-fluorescent marker fusion, the results from the latter treatment (using aminoprophos-methyl) being a different finding from many during virus infection where MP-GFP fusion movement was not impeded by microtubule antagonists. To complicate this situation further, additional research determined that neither oryzalin or aminoprophos-methyl, both antagonists of microtubules, nor LatB, a microfilament antagonist, inhibited the formation of MP-GFP inclusions or their localization to the cell periphery or PD in cultured cells or leaves (Prokhnevsky et al., 2005; Boutant et al., 2009). These apparently conflicting results highlight the difficulty interpreting findings from pharmacological studies. Clearly, additional work is required to determine, in real time and through methods using non-pharmacological techniques, the influence of the cytoskeleton on transiently expressed MP intracellular trafficking. Assuming MP trafficking independent of the VRC has physiological relevance, findings from these additional studies would provide further insight into the mechanism of TMV intracellular movement. Some work investigating the influence of MP transport in the absence of pharmacological treatment has been published. For example, Kotlizky et al. (2001) determined that an ectopically expressed MP mutant-GFP fusion, MP^{NT-1}-GFP, which inhibits TMV spread when expressed in transgenic plants, was able to associate with microtubules but did not target PD or move between cells. This suggests that the microtubule binding domain resides in a different location from the region important for PD localization and supports the pharmacological studies indicating that MP PD localization and initial virus spread requires more than microtubule association by the MP.

INTERCELLULAR MOVEMENT

As for intracellular movement, our understanding of TMV intercellular movement is fragmented. It is certain, however, that TMV utilizes PD to move between cells. PD are bounded by the plasma membrane and contain a cytoplasmic sleeve between this membrane and intact ER, the ER referred to as the desmotubule in this tissue (Lucas et al., 2009; Benitez-Alfonso et al., 2010; White and Barton, 2011; Figure1). Callose is present in the neck region of the PD (Northcote et al., 1989; see Benitez-Alfonso et al., 2010). Actin and myosin are among multiple protein components in the cytoplasmic sleeve (Fernandez-Calvino et al., 2011;White and Barton, 2011). The size exclusion limit (SEL) of PD allows the passive diffusion of small molecules ~1 kDa in size. This presents an impediment to virus movement because virus structures that are hypothesized to move between cells require a far larger SEL. The MP of TMV modifies the SEL of the PD to facilitate virus movement through PD and fluorescently labeled MP is observed in PD after virus spread has occurred (reviewed in Benitez-Alfonso et al., 2010; Niehl and Heinlein, 2011). The MP itself also can move between cells when ectopically expressed (reviewed in Niehl and Heinlein, 2011). The increase in PD SEL during TMV spread is transient, returning to a restricted size after passage of the infection front as measured by fluorescence expressed from the TMV-GFP genome (Oparka et al., 1997). Microfilament antagonists lead to a significant increase in the PD SEL while those that stabilize microfilaments prevent the MP from increasing the PD SEL (White et al., 1994; Ding et al., 1996; Su et al., 2010). The TMV MP exhibits microfilament severing activity (Su et al., 2010). These findings, in total, suggest a mechanism for virus spread where TMV MP opens PD through its microfilament severing activity, mimicking the phenotype induced with microfilament antagonists (Figure1). Su et al. (2010) hypothesized that the severing activity of the TMV MP was limited to the PD area by analogy with findings from a mutant MP of *Cucumber mosaic virus* altered in its ability to target the PD. This mutant MP, which also has severing activity, fragments microfilaments in the cytoplasm (Su et al., 2010). Besides the MP, evidence for the VRC moving between cells has been published (Kawakami et al., 2004), but there have been no subsequent reports supporting this finding. The 126 kDa protein fused with GFP has not been reported to move between cells during ectopic expression, suggesting that VRC intercellular movement requires expression of additional viral proteins (likely the MP) or the presence of the vRNA.

In addition to innate actin severing activity by the TMV MP that may enlarge PD, the TMV MP interacts with ankyrin repeat-containing protein, ANK, at the PD and this association is correlated with an increase the PD SEL (Ueki et al., 2010). ANK has multiple activities, including binding to and delivering chloroplasts to their destination, supporting disease resistance against bacteria and virus challenge and participating in reactive oxygen scavenging, but it does not have callose degrading activity. Over-expressing ANK in transgenic plants resulted in more extensive MP-YFP and TMV-DsRed spread between cells. During transient expression of ANK and MP, the level of callose in the neck region of the PD decreased. Previously it was shown that enhanced TMV intercellular spread was correlated with enhanced expression of the callose-degrading enzyme, β -1, 3-glucanase (Bucher et al., 2001). Whether predominantly cytoplasmic ANK, through its MP interaction and the MP interaction with membrane, targets β -1, 3-glucanase in ER-derived vesicles to the cell wall or if ANK functions directly to inhibit callose synthase activity, remains to be determined (Ueki et al., 2010; Figure1). Interactions of the TMV MP with other host proteins that influence TMV spread have been described (e.g., pectin methyl-esterase; Chen et al., 2000). However, in many instances cell biological studies to observe the interaction between the MP and host proteins within a live cell have not been conducted to further determine the location where the interaction may influence intercellular spread.

Regarding microtubules and TMV intercellular movement, there is evidence that MP interaction with the microtubule end-binding protein 1a (EB1a) is important for TMV spread. Brandner et al. (2008) determined that EB1a-GFP and MP-GFP form complexes both *in vitro* and *in vivo*. EB1a-GFP sub-cellular localization during TMV infection was altered from end labeling comet-like structures representing growing microtubules to labeling the length of microtubules. The length-wise decoration of microtubules was associated with a co-localization of MP-RFP. This unexpected re-localization of the EB1a protein at the infection front was correlated with an inhibition of virus intercellular spread. Ouko et al. (2010) determined that a mutant tobacco expressing a detyrosinated α-tubulin had a slower moving GFP-EB1 and inhibited intercellular spread by TMV. These studies suggest that modified microtubule dynamics inhibits TMV intercellular spread, perhaps by inhibiting microtubule polymerization that normally would push the VRC along the ER during intracellular transport (reviewed in Peña and Heinlein, 2012; Figure1). These findings support those using a mutant strain of TMV whose temperature sensitive intercellular movement is correlated with the temperature sensitive localization of MP with microtubules (Boyko et al., 2007). However, other studies using pharmacological agents or silencing of the α-tubulin gene found no effect of these treatments on TMV intercellular spread or on the presence of MP in PD (Gillespie et al., 2002). Additionally, a virus expressing a modified MP with limited affinity for microtubules displayed enhanced intercellular spread (Gillespie et al., 2002). These and other findings from studies of a microtubule-binding protein, MPB2C, which binds to the MP and when overexpressed has a negative effect on intercellular movement of a related tobamovirus (Ruggenthaler et al., 2009), led to the suggestion that the function of microtubules in TMV spread is for degradation of the MP. Recently, it was determined that a CELL-DIVISION CYCLE protein48 (CDC48) from *Arabidopsis*, which localizes in the cytoplasm near the cortical ER network, interacts with the TMV MP to extract it from ER for its subsequent accumulation on microtubules (Niehl et al., 2012). These authors also determined that overexpression of CDC48 in infected tissue inhibited virus spread. Thus, it appears that removal of the MP from the ER at the infection front and its movement to the microtubules through CDC48 activity is directed toward processing and possibly, degrading, the MP (Figure1).

Microfilaments have been shown to be important for TMV intercellular movement, but the interpretation of their involvement in this activity is evolving. Findings from early studies using a GFP-labeled virus showed that disruption of microfilaments, through pharmacological methods or by silencing actin, inhibited sustained (2 days and beyond post-inoculation) virus intercellular movement (Kawakami et al., 2004; Liu et al., 2005). This inhibition in sustained intercellular spread was not associated with a decrease in virus accumulation per cell that would affect virus movement or prevent VRC formation (Liu et al., 2005). In addition, the sustained virus intercellular movement was not correlated with VRC size (Liu et al., 2005). This movement required myosin motor activity and specifically myosin XI-2 motor activity (Kawakami et al., 2004; Harries et al., 2009). Surprisingly, intercellular movement of the related tobamovirus, TVCV, was unaffected by disruption of microfilaments or silencing of any myosin studied to date (Harries et al., 2009). During this time, it also was shown that TMV was not inhibited in spread early after LatB treatment (i.e., movement up to 24 h after treatment; Hofmann et al., 2009). In addition, these researchers determined that actin binding domain 2 (ABD2)-GFP, expressed in transgenic plants, inhibited TMV movement. They concluded that this disruption in virus movement was primarily due to a loss of membrane fluidity caused by the ABD2-GFP marker and that TMV intercellular movement was predominantly influenced by membrane diffusion characteristics. Thus, results from studies of both TVCV and early TMV intercellular movement suggest membranes as the predominant vehicle controlling tobamovirus intercellular movement. These findings also support those from a previous study showing that both the viral replication proteins and MP are necessary to allow maximum diffusion through PD of GFP-fused probes representing soluble ER membrane-bound proteins (Guenoune-Gelbart et al., 2008). It was concluded that the results best support a model in which the virus complex, perhaps consisting of viral RNA, MP and other proteins, diffuses on the ER membrane within the PD from infected to uninfected cells driven by a concentration gradient (Guenoune-Gelbart et al., 2008). The involvement of the ER directly in virus movement, without prior transport through Golgi, is supported by the finding that inhibition of the COP II transport system through expression of a dominant negative GTPase mutant protein, Sar1, did not inhibit sustained TMV intercellular movement (Genovés et al., 2010).

Irrespective of the mechanism of early movement between cells by TMV, an explanation of this virus's requirement for microfilaments for sustained movement is required. Harries et al. (2010) suggested that the influence of microfilaments on this activity may be in preventing a stress response from occurring at the PD that later signaled to cells in advance of virus spread. These cells would then modify their metabolism in preparation for the arrival of the virus. Here we hypothesize that it is the transport of the VRC from the cell wall/PD area to an internal subcellular location on the actomyosin array that is necessary to prevent the stress signal from moving to the next cell (Figure1). On treatment with LatB or silencing of myosin XI-2, the TMV VRCs would remain at the cell wall/PD interface, inhibiting normal cell–cell communications. This stress would be signaled to the cells in advance of the virus spread and modify these cells to inhibit subsequent virus spread. This hypothesis is supported by the finding that the VRCs of TMV move away from the cell wall as the infection passes (Szécsi et al., 1999). This interpretation would also accommodate the finding that early TMV movement is unaffected by actomyosin inhibitors since it would take some time to signal in advance of the infection front to stop virus movement. In addition, it would explain the lack of effect of LatB on TVCV movement since the 125 kDa protein, the homolog of the 126 kDa protein of TMV, does not form intracellular inclusions that associate with microfilaments (Harries et al., 2009) and the virus itself produces few visible VRCs (Resconich, 1961). The transport of the TMV VRC or its remnants from the wall may require membranes and vesicles since the TMV MP binds with a plant synaptotagmin *in vitro* (Lewis and Lazarowitz, 2010). Synaptogamins are a family of Ca^{2+}- and lipid-binding proteins that modulate, through interaction with SNARE proteins, the fusion of vesicles with membranes (Chapman, 2008). A dominant negative synaptotagmin mutant caused depletion of endosomes and inhibited intercellular trafficking of the MP-GFP fusion (Lewis and Lazarowitz, 2010). The direct influence of synaptotagmin on virus physiology appears to be on the endocytic pathway, implying that the effect of the dominant negative synaptotagmin on virus movement is through blocking the recycling of membrane used by the virus to reach the cell wall/PD area, thus backing up the system (Figure1). This hypothesis, suggesting a requirement for actomyosin-mediated vesicle trafficking of the VRC or VRC components from the wall membranes for sustained virus movement, can be evaluated through cell biological studies.

Regarding non-cytoskeletal or membrane-associated proteins and TMV intercellular movement, a gene encoding a class II KNOTTED1-like protein, NTH201, was cloned from *Nicotiana tabacum* and its expression levels were positively correlated with MP accumulation, VRC numbers and virus spread (Yoshii et al., 2008). NTH201 has no ability to traffic between cells or traffic GFP, unlike class I KNOX-like plant proteins, and the authors speculated that NTH201 may function as a transcription factor that helps to stabilize or fold MP and VRCs through its regulation of other genes. This mystery was partially explained when a second host factor, MPIP1, a DnaJ-like protein with potential chaperone activity, was determined to interact with TMV MP and NTH201 in a yeast-three hybrid screen (Shimizu et al., 2009). Silencing MPIP1 inhibits TMV spread, as determined by GFP fluorescence, and thus its silencing phenocopies observations of TMV spread when NTH201 expression was silenced. It is possible that an interaction between two host proteins and the TMV MP aid in transport of virus between cells (Figure1).

CONCLUSION

Cell biological studies over the last 20 years have tremendously aided our understanding of TMV accumulation and spread. Without advanced molecular and biochemical technologies allowing virus and virus component labeling and advanced imaging hardware our understanding of the individual processes during virus spread would be diminished. For example, if virus intercellular movement were studied by genetics alone the importance of the transport of TMV vRNA granules to the perinuclear region of the ER versus the transport of TMV VRCs and MPs to the PD could go unrecognized. This said, some cell biological studies have yielded conflicting or controversial results. This is especially true of pharmacological studies and researchers must carefully control as many variables in these studies as possible. In addition, conclusions from pharmacological studies should be verified using other methods. Use of novel virus labeling techniques and advanced microscopes will allow further advances in this area. For example, the identification of small fluorescent tags that do not influence the function of the viral protein to which it is fused will be helpful. The iLOV protein, derived from the blue light receptor phototropin and much smaller than GFP, has been available for some time and was a first step toward utilizing smaller fluorescent tags (Chapman et al., 2008). More recently a MYB-related transcription factor, Rosea 1, which is also smaller than GFP, has been placed in TMV as a marker for virus location (Bedoya et al., 2012). Advanced microscopes with super high resolution will allow us to more easily determine whether proteins are interacting or simply co-localizing (Tilsner and Oparka, 2010). Access to the *N. benthamiana* genome (Bombarely et al., 2012) will allow better identification of gene family members projected to influence virus transport and the ability to target individual members for knockdown, over-expression and labeling. Lastly, recent findings suggest that while transport of TMV to the PD is important, it is also important to understand what happens to the virus inclusions left in the cell after virus movement, since their proper degradation or storage may influence sustained intercellular movement by the virus. As an analogy, although humans can function well for a time in their home (i.e., cell) without working plumbing, a plugged drain, like a plugged PD, will eventually be noticed. With our improving technologies, resources, and knowledge the future is bright, literally, for cell biological studies on TMV accumulation and spread.

ACKNOWLEDGMENTS

The authors thank Elison Blancaflor and Liangfa Ge for critically reading the manuscript and Kim Cooper for reference preparation. This work was supported by the Agriculture and Food Research Initiative Competitive Grants Program Grant no. 2010-65108-20525 from the USDA National Institute of Food and Agriculture.

REFERENCES

1. Bedoya L. C., Martínez F., Orzáez D, Daròs J. A. (2012). Visual tracking of plant virus infection and movement using a reporter MYB transcription factor that activates anthocyanin biosynthesis. *Plant Physiol.* 158 1130–1138
2. Benitez-Alfonso Y., Faulkner C., Ritzenthaler C., Maule A. (2010). Plasmodesmata: gateways to local and systemic virus infection. *Mol. Plant Microbe Interact.* 23 1403–1412
3. Bhat S., Folimonova S. Y., Cole A. B., Ballard K. D., Lei Z., Watson B. S., et al. (2012). Influence of host chloroplast proteins on Tobacco mosaic virus accumulation and intercellular movement. *Plant Physiol.* 161 134–147
4. Boevink P., Oparka K. J. (2005). Virus-host interactions during movement processes. *Plant Physiol.*138 1815–1821
5. Bombarely A., Rosli H. G., Vrebalov J., Moffett P., Mueller L. A., Martin G. B. (2012). A draft genome sequence of *Nicotiana benthamiana* to enhance molecular plant-microbe biology research.*Mol. Plant Microbe Interact.* 25 1523–1530
6. Boutant E., Fitterer C., Ritzenthaler C., Heinlein M. (2009). Interaction of the Tobacco mosaic virus movement protein with microtubules during the cell cycle in tobacco BY-2 cells. *Protoplasma* 237 3–12
7. Boyko V., Hu Q., Seemanpillai M., Ashby J., Heinlein M. (2007). Validation of microtubule-associated Tobacco mosaic virus RNA movement and involvement of microtubule-aligned particle trafficking. *Plant J.* 51 589–603
8. Brandner K., Sambade A., Boutant E., Didier P., Mély Y., Ritzenthaler C., et al. (2008). Tobacco mosaic virus movement protein interacts with green fluorescent protein-tagged microtubule end-binding protein 1. *Plant Physiol.* 147 611–623
9. Brill L. M., Nunn R. S., Kahn T. W., Yeager M., Beachy R. N. (2000). Recombinant Tobacco mosaic virus movement protein is an RNA-binding, alpha-helical membrane protein. *Proc. Natl. Acad. Sci. U.S.A.* 97 7112–7117
10. Bucher G. L., Tarina C., Heinlein M., Serio F. D., Meins F. J., Iglesias V. A. (2001). Local expression of enzymatically active class I β -1, 3-glucanase enhances symptoms of TMV infection in tobacco.*Plant J.* 28 361–369
11. Carrasco S., Meyer T. (2011). STIM proteins and the endoplasmic reticulum-plasma membrane junctions. *Annu. Rev. Biochem.* 80 973–1000
12. Chapman E. R. (2008). How does synaptotagmin trigger neurotransmitter release? *Annu. Rev. Biochem.* 77 615–641
13. Chapman S., Faulkner C., Kaiserli E., Garcia-Mata C., Savenkov E. I., Roberts A. G., et al. (2008).The photoreversible fluorescent protein iLOV outperforms GFP as a reporter of plant virus infection.*Proc. Natl. Acad. Sci. U.S.A.* 105 20038–20043
14. Chen M. H., Sheng J., Hind G., Handa A. K., Citovsky V. (2000). Interaction between the Tobacco mosaic virus movement protein and host cell pectin methylesterases is required for viral cell-to-cell movement. *EMBO J.* 19 913–920

15. Christensen N., Tilsner J., Bell K., Hammann P., Parton R., Lacomme C., et al. (2009). The 5′ cap of Tobacco mosaic virus (TMV) is required for virion attachment to the actin/endoplasmic reticulum network during early infection. *Traffic* 10 536–551

16. Das P., Hari V. (1992). Intracellular distribution of the 126K/183K and capsid proteins in cells infected by some tobamoviruses. *J. Gen. Virol.* 73 3039–3043

17. de Castro I. F., Volonté L., Risco C. (2012). Virus factories: biogenesis and structural design. *Cell. Microbiol.* 15 24–34

18. Ding B., Kwon M. O., Warnberg L. (1996). Evidence that actin filaments are involved in controlling the permeability of plasmodesmata in tobacco mesophyll. *Plant J.* 10 157–164

19. Ding X. S., Bao Y., Carter S. A., Nelson R. S. (1998). "Host factors involved in viurs replication and cell-to-cell movement," in *1997 Annual Report of the Samuel Rorberts Noble Foundation* (Ardmore, OK: Noble Foundation Communiations;), 94–96

20. dos Reis Figueira A., Golem S., Goregaoker S. P., Culver J. N. (2002). A nuclear localization signal and a membrane association domain contribute to the cellular localization of the Tobacco mosaic virus 126-kDa replicase protein. *Virology* 301 81–89

21. Durso N. A., Cyr R. J. (1994). A calmodulin-sensitive interaction between microtubules and a higher plant homolog of elongation factor-1α . *Plant Cell* 6 893–905

22. Epel B. L. (2009). Plant viruses spread by diffusion on ER-associated movement-protein-rafts through plasmodesmata gated by viral induced host beta-1, 3-glucanases. *Semin. Cell Dev. Biol.* 20 1074–1081

23. Esau K., Cronshaw J. (1967). Relation of Tobacco mosaic virus to the host cells. *J. Cell Biol.* 33 665–678

24. Fernandez-Calvino L., Faulkner C., Walshaw J., Saalbach G., Bayer E., Benitez-Alfonso Y., et al. (2011). *Arabidopsis* plasmodesmal proteome. *PLoS ONE* 6:e18880 10.1371/journal.pone.0018880

25. Fujiki M., Kawakami S., Kim R. W., Beachy R. N. (2006). Domains of Tobacco mosaic virus movement protein essential for its membrane association. *J. Gen. Virol.* 87 2699–2707

26. Genovés A., Navarro J. A, Pallás V. (2010). The intra- and intercellular movement of Melon necrotic spot virus (MNSV) depends on an active secretory pathway. *Mol. Plant Microbe Interact.* 23 263–272

27. Gillespie T., Boevink P., Haupt S., Roberts A. G., Toth R., Valentine T., et al. (2002). Functional analysis of a DNA-shuffled movement protein reveals that microtubules are dispensable for the cell-to-cell movement of Tobacco mosaic virus. *Plant Cell* 14 1207–1222

28. Gonen H., Smith C. E., Siegel N. R., Kahana C., Merrick W. C., Chakarburtty K., et al. (1994). Protein synthesis elongation factor EF-1α is essential for ubiquitindependent degradation of certain Nα -acetylated proteins and may be substituted for by the bacterial elongation factor EF-Tu. *Proc. Natl. Acad. Sci. U.S.A.* 91 7648–7652

29. Goregaoker S. P., Culver J. N. (2003). Oligomerization and activity of the helicase domain of the Tobacco mosaic virus 126- and 183-kilodalton replicase proteins. *J. Virol.* 77 3549–3556

30. Guenoune-Gelbart D., Elbaum M., Sagi G., Levy A., Epel B. L. (2008). Tobacco mosaic virus (TMV) replicase and movement protein function synergistically in facilitating TMV spread by lateral diffusion in the plasmodesmal desmotubule of *Nicotiana benthamiana*. *Mol. Plant Microbe Interact.*21 335–345

31. Hagiwara-Komoda Y., Hirai K., Mochizuki A., Nishiguchi M., Meshi T., Ishikawa M. (2008). Overexpression of a host factor TOM1 inhibits Tomato mosaic virus propagation and suppression of RNA silencing. *Virology* 376 132–139

32. Hagiwara Y., Komoda K., Yamanaka T., Tamai A., Meshi T., Funada R., et al. (2003). Subcellular localization of host and viral proteins associated with tobamovirus RNA replication. *EMBO J.* 22 344–353

33. Harries P., Ding B. (2011). Cellular factors in plant virus movement: at the leading edge of macromolecular trafficking in plants. *Virology* 411 237–243

34. Harries P. A., Park J.-W., Sasaki N., Ballard K. D., Maule A., Nelson R. S. (2009). Differing requirements for actin and myosin by plant viruses for sustained intercellular movement. *Proc. Natl. Acad. Sci. U.S.A.* 106 17594–17599

35. Harries P. A., Schoelz J. E., Nelson R. S. (2010). Intracellular transport of viruses and their components: utilizing the cytoskeleton and membrane highways. *Mol. Plant Microbe Interact.* 231381–1393

36. Heinlein M., Padgett H. S., Gens J. S., Pickard B. G., Casper S. J., Epel B. L., et al. (1998). Changing patterns of localization of the Tobacco mosaic virus movement protein and replicase to the endoplasmic reticulum and microtubules during infection. *Plant Cell* 10 1107–1120 [PMC free article]

37. Hills G. J., Plaskitt K. A., Young N. D., Dunigan D. D., Watts J. W., Wilson T. M. A., et al. (1987). Immunogold localization of the intracellular sites of structural and nonstructural Tobacco mosaic virus proteins. *Virology* 161 488–496

38. Hofmann C., Niehl A., Sambade A., Steinmetz A., Heinlein M. (2009). Inhibition of Tobacco mosaic virus movement by expression of an actin-binding protein. *Plant Physiol.* 149 1810–1823

39. Ishibashi K., Nishikiori M., Ishikawa M. (2010). Interactions between tobamovirus replication proteins and cellular factors: their impacts on virus multiplication. *Mol. Plant Microbe Interact.* 23 1413–1419

40. Ishikawa M., Naito S., Ohno T. (1993). Effects of the toml mutation of *Arabidopsis thaliana* on the multiplication of Tobacco mosaic virus RNA in protoplasts. *J. Virol.* 67 5328–5338 [PMC free article]

41. Ishikawa M., Okada Y. (2004). Replication of tobamovirus RNA. *Proc. Jpn. Acad. Ser. B* 80 215–224

42. Ishikawa M., Meshi T., Motoyoshi F., Takamatsu N.,

Okada Y. (1986). *In vitro* mutagenesis of the putative replicase genes of Tobacco mosaic virus. *Nucleic Acids Res.* 14 8291–8305 [PMC free article]

43. Ishikawa M., Meshi T., Ohno T., Okada Y. (1991). Specific cessation of minus-strand RNA accumulation at an early stage of tobacco mosaic virus infection. *J. Virol.* 65 861–868

44. Karpova O. V., Ivanov K. I., Rodionova N. P., Dorokhov Y. L., Atabekov J. G. (1997).Nontranslatability and dissimilar behavior in plants and protoplasts of viral RNA and movement protein complexes formed *in vitro*. *Virology* 230 11–21

45. Kawakami S., Watanabe Y., Beachy R. N. (2004). Tobacco mosaic virus infection spreads cell to cell as intact replication complexes. *Proc. Natl. Acad. Sci. U.S.A.* 101 6291–6296 [PMC free article]

46. Kim S., Coulombe P. A. (2010). Emerging role for the cytoskeleton as an organizer and regulator of translation. *Nat. Rev. Mol. Cell Biol.* 11 75–81

47. Knapp E., Achor D., Lewandowski D. J. (2007). Tobacco mosaic virus defective RNAs expressing C-terminal methyltransferase domain sequences are severely impaired in long-distance movement in*Nicotiana benthamiana*. *Virology* 367 82–91

48. Kotlizky G., Katz A., Van Der Laak J., Boyko V., Lapidot M., Beachy R. N., et al. (2001). A dysfunctional movement protein of Tobacco mosaic virus interferes with targeting of wild-type movement protein to microtubules. *Mol. Plant Microbe Interact.* 14 895–904

49. Kubota K., Tsuda S., Tamai A., Meshi T. (2003). Tomato mosaic virus replication protein suppresses virus-targeted posttranscriptional gene silencing. *J. Virol.* 77 11016–11026 [PMC free article]

50. Kragler F., Curin M., Trutnyeva K., Gansch A., Waigmann E. (2003). MPB2C, a microtubule-associated plant protein binds to and interferes with cell-to-cell transport of Tobacco mosaic virus movement protein. *Plant Physiol.* 132 1870–1883

51. Laliberté J. F., Sanfaçon H. (2010). Cellular remodeling during plant virus infection. *Annu. Rev. Phytopathol.* 48 69–91

52. Lee J.-Y., Taoka K.-I., Yoo B.-C., Ben-Nissan G., Kim D.-J., Lucas D. J. (2005). Plasmodesmal-associated protein kinase in tobacco and *Arabidopsis* recognizes a subset of non-cell-autonomous proteins. *Plant Cell* 17 2817–2831

53. Lewandowski D. J., Dawson W. O. (2000). Functions of the 126- and 183-kDa proteins of TMV.*Virology* 271 90–98

54. Lewis J. D., Lazarowitz S. G. (2010). Arabidopsis synaptotagmin SYTA regulates endocytosis and virus movement protein cell-to-cell transport. *Proc. Natl. Acad. Sci. U.S.A.* 107 2491–2496

55. Liu J. Z., Blancaflor E. B., Nelson R. S. (2005). The Tobacco mosaic virus 126-kilodalton protein, a constituent of the virus replication complex, alone or within the complex aligns with and traffics along microfilaments. *Plant Physiol.* 138 1853–1865

56. Liu J. Z., Blancaflor E. B., Nelson R. S. (2006). "The structure of the Tobacco mosaic virus replication complex is modulated by the 126-kDa protein and complex and protein traffic along microfilaments," in *Biology of Plant–Microbe Interactions* Vol. 5, eds Sánchez F., Quinto C., Lopez-Lara I. M., Geiger O., editors. (St Paul: IS-MPMI Press;) 410–415

57. Lucas W. J., Ham B. K., Kim J. Y. (2009). Plasmodesmata - bridging the gap between neighboring plant cells. *Trends Cell Biol.* 19 495–503

58. Más P., Beachy R. N. (1999). Replication of Tobacco mosaic virus on endoplasmic reticulum and role of the cytoskeleton and virus movement protein in intracellular distribution of viral RNA. *J. Cell Biol.*147 945–958

59. McLean B. G., Zupan J., Zambryski P. C. (1995). Tobacco mosaic virus movement protein associates with the cytoskeleton in tobacco cells. *Plant Cell* 7 2101–2114

60. Niehl A., Heinlein M. (2011). Cellular pathways for viral transport through plasmodesmata.*Protoplasma* 248 75–99

61. Niehl A., Khalid A., Dalya G., Brandner K., Mély Y., Heinlein M. (2012). Control of *Tobacco mosaic virus* movement protein fate by CELL-DIVISION-CYCLE protein48. *Plant Physiol.* 160 2093–2108

62. Nishikiori M., Dohi K., Mori M., Meshi T., Naito S., Ishikawa M. (2006). Membrane-bound tomato mosaic virus replication proteins participate in RNA synthesis and are associated with host proteins in a pattern distinct from those that are not membrane bound. *J. Virol.* 80 8459–8468 [PMC free article]

63. Nishikiori M., Sugiyama S., Xiang H., Niiyama M., Ishibashi K., Inoue T., et al. (2012). Crystal structure of the superfamily 1 helicase from Tomato mosaic virus. *J. Virol.* 86 7565–7576

64. Northcote D. H., Davey R., Lay J. (1989). Use of antisera to localize callose, xylan and arabinogalactan in the cell-plate, primary and secondary walls of plant cells. *Planta* 178 353–366

65. Oparka K. J., Prior D. A., Cruz S. S., Padgett H. S., Beachy R. N. (1997). Gating of epidermal plasmodesmata is restricted to the leading edge of expanding infection sites of Tobacco mosaic virus (TMV). *Plant J.* 14 781–789

66. Osman T. A., Buck K. W. (1996). Complete replication in vitro of Tobacco mosaic virus RNA by a template-dependent, membrane-bound RNA polymerase. *J. Virol.* 70 6227–6234 [PMC free article]

67. Ouko M. O., Sambade A., Brandner K., Niehl A., Peña E., Ahad A., et al. (2010). Tobacco mutants with reduced microtubule dynamics are less susceptible to TMV. *Plant J.* 62 829–839

68. Padgett H. S., Epel B. L., Kahn T. W., Heinlein M., Watanabe Y., Beachy R. N. (1996). Distribution of tobamovirus movement protein in infected cells and implications for cell-to-cell spread of infection.*Plant J.* 10 1079–1088

69. Palukaitis P., Zaitlin M. (1986). "Tobacco mosaic virus: infectivity and replication," in *The Plant Viruses*, Vol. 2, eds Van Regenmortel M. H. V., Fraenkel-Conrat H., editors. (New York: Plenum;) 105–131

70. Peña E. J., Heinlein M. (2012). RNA transport during TMV cell-to-cell movement. *Front. Plant Sci.*3:193 10.3389/fpls.2012.00193

71. Prokhnevsky A. I., Peremyslov V. V., Dolja V. V. (2005). Actin cytoskeleton is involved in targeting of a viral Hsp70 homolog to the cell periphery. *J. Virol.* 79 14421–14428

72. Reichel C., Beachy R. N. (1998). Tobacco mosaic virus infection induces severe morphological changes of the endoplasmic reticulum. *Proc. Natl. Acad. Sci. U.S.A.* 95 11169–11174

73. Resconich E. C. (1961). Interpretation of the forms of inclusions in bean systemically infected with Tobacco mosaic virus. *Virology* 15 16–21

74. Ruggenthaler P., Fichtenbauer D., Krasensky J., Jonak C., Waigmann E. (2009). Microtubule-associated protein AtMPB2C plays a role in organization of cortical microtubules, stomata patterning, and tobamovirus infectivity. *Plant Physiol.* 149 1354–1365

75. Saito T., Hosokawa D., Meshi T., Okada Y. (1987). Immunocytochemical localization of the 130K and 180K proteins (putative replicase components) of Tobacco mosaic virus. *Virology* 160 477–481

76. Sambade A., Brandner K., Hofmann C., Seemanpillai M., Mutterer J., Heinlein M. (2008). Transport of TMV movement protein particles associated with the targeting of RNA to plasmodesmata. *Traffic* 92073–2088

77. Sambade A., Heinlein M. (2009). Approaching the cellular mechanism that supports the intercellular spread of Tobacco mosaic virus. *Plant Signal. Behav.* 4 35–38

78. Schoelz J. E., Harries P. A., Nelson R. S. (2011). Intracellular transport of plant viruses: finding the door out of the cell. *Mol. Plant* 4 813–831

79. Seemanpillai M., Elamawi R., Ritzenthaler C., Heinlein M. (2006). Challenging the role of microtubules in Tobacco mosaic virus movement by drug treatments is disputable. *J. Virol.* 80 6712–6715

80. Shaw J. G. (1999). Tobacco mosaic virus and the study of early events in virus infections. *Philos. Trans. R. Soc. Lond. B Biol. Sci.* 354 603–611

81. Shimizu T., Yoshii A., Sakurai K., Hamada K., Yamaji Y., Suzuki M., et al. (2009). Identification of a novel tobacco DnaJ-like protein that interacts with the movement protein of Tobacco mosaic virus.*Arch. Virol.* 154 959–967

82. Su S., Liu Z., Chen C., Zhang Y., Wang X., Zhu L., et al. (2010). Cucumber mosaic virus movement protein severs actin filaments to increase the plasmodesmal size exclusion limit in tobacco. *Plant Cell*22 1373–1387

83. Szécsi J., Ding X. S., Lim C. O., Bendahmane M., Cho M. J., Nelson R. S., et al. (1999).Development of Tobacco mosaic virus infection sites in *Nicotiana benthamiana*. *Mol. Plant Microbe Interact.* 12 143–152

84. Tagami Y., Watanabe Y. (2007). Effects of brefeldin A on the localization of tobamovirus movement protein and cell-to-cell movement of the virus. *Virology* 361 133–140

85. Tilsner J., Amari K., Torrance L. (2011). Plasmodesmata viewed as specialised membrane adhesion sites. *Protoplasma* 248 39–60

86. Tilsner J., Linnik O., Christensen N. M., Bell K., Roberts I. M., Lacomme C., et al. (2008). Live-cell imaging of viral RNA genomes using a Pumilio-based reporter. *Plant J.* 57 758–770

87. Tilsner J., Oparka K. J. (2010). Tracking the green invaders: advances in imaging virus infection in plants. *Biochem. J.* 430 21–37

88. Ueki S., Spektor R., Natale D. M., Citovsky V. (2010). ANK, a host cytoplasmic receptor for the Tobacco mosaic virus cell-to-cell movement protein, facilitates intercellular transport through plasmodesmata. *PLoS Pathog.* 6:e1001201 10.1371/journal.ppat.1001201 [PMC free article]

89. Watanabe T., Honda A., Iwata A., Ueda S., Hibi T., Ishihama A. (1999). Isolation from Tobacco mosaic virus-infected tobacco of a solubilized template-specific RNA-dependent RNA polymerase containing a 126K/183K protein heterodimer. *J. Virol.* 73 2633–2640

90. White R. G., Badelt K., Overall R. L., Vesk M. (1994). Actin associated with plasmodesmata. *Protoplasma* 180 169–184

91. White R. G., Barton D. A. (2011). The cytoskeleton in plasmodesmata: a role in intercellular transport?*J. Exp. Bot.* 62 5249–5266

92. Wright K. M., Wood N. T., Roberts A. G., Chapman S., Boevink P., Mackenzie K. M., et al. (2007).Targeting of TMV movement protein to plasmodesmata requires the actin/ER network: evidence from FRAP. *Traffic* 8 21–31

93. Wu X. J., Xu Z., Shaw J. G. (1994). Uncoating of Tobacco mosaic virus RNA in protoplasts. *Virology*200 256–262

94. Yamaji Y., Hamada K., Yoshinuma T., Sakurai K., Yoshii A., Shimizu T., et al. (2010). Inhibitory effect on the Tobacco mosaic virus infection by a plant RING finger protein. *Virus Res.* 153 50–57

95. Yamaji Y., Kobayashi T., Hamada K., Sakurai K., Yoshii A., Suzuki M., et al. (2006). *In vivo*interaction between Tobacco mosaic virus RNA-dependent RNA polymerase and host translation elongation factor 1A. *Virology* 347 100–108

96. Yamanaka T., Ohta T., Takahashi M., Meshi T., Schmidt R., Dean C., et al. (2000). TOM1, an*Arabidopsis* gene required for efficient multiplication of a tobamovirus, encodes a putative transmembrane protein. *Proc. Natl. Acad. Sci. U.S.A.* 97 10107–10112

97. Yoshii A., Shimizu T., Yoshida A., Hamada K., Sakurai K., Yamaji Y., et al. (2008). NTH201, a novel class II KNOTTED1-like protein, facilitates the cell-to-cell movement of Tobacco mosaic virus in tobacco. *Mol. Plant Microbe Interact.* 21 586–596

98. Zeenko V. V., Ryabova L. A., Spirin A. S., Rothnie H. M., Hess D., Browning K. S., et al. (2002).Eukaryotic elongation factor 1A interacts with the upstream pseudoknot domain in the 3′ untranslated region of Tobacco mosaic virus RNA. *J. Virol.* 76 5678–5691

CHAPTER 11

GRAPEVINE PATHOGENIC MICROORGANISMS: UNDERSTANDING INFECTION STRATEGIES AND HOST RESPONSE SCENARIOS

Grace Armijo, Rudolf Schlechter, Mario Agurto, Daniela Muñoz, Constanza Nuñez, and Patricio Arce-Johnson

Laboratorio de Biología Molecular y Biotecnología Vegetal, Departamento de Genética Molecular y Microbiología, Facultad de Ciencias Biológicas, Pontificia Universidad Católica de Chile, Santiago, Chile

ABSTRACT

Grapevine (*Vitis vinifera* L.) is one of the most important fruit crop worldwide. Commercial cultivars are greatly affected by a large number of pathogenic microorganisms that cause diseases during pre- and/or post-harvest periods, affecting production, processing and export, along with fruit quality. Among the potential threats, we can find bacteria, fungi, oomycete, or viruses with different life cycles, infection mechanisms and evasion strategies. While plant–pathogen interactions are cycles of resistance and susceptibility, resistance traits from natural resources are selected and may be used for breeding purposes and for a sustainable agriculture. In this context, here we summarize some of the most important diseases affecting *V. vinifera*together with their causal agents. The aim of this work is to bring a comprehensive review of the infection strategies deployed by significant types of pathogens while understanding the host response in both resistance and susceptibility scenarios. New approaches being used to uncover grapevine status during biotic stresses and scientific-based procedures needed to control plant diseases and crop protection are also addressed.

INTRODUCTION

During their lifetime, plants are exposed to a wide variety of pathogens, such as bacteria, viruses, fungi, and nematodes. According to their lifecycle and infection strategies, pathogenic microorganisms can be classified as necrotrophics, biotrophics and hemibiotrophics: necrotrophic pathogens feed on dead tissue, secreting lytic enzymes and phytotoxins to promote cell death into the host plant. Biotrophic pathogens on the other hand, feed on living tissue, developing structures in order to invade the cell and obtain metabolism products. Finally, hemibiotrophic pathogens start with a biotrophic infection phase and then turn to a final necrotrophic phase, killing its host at the end of the infection cycle (Glazebrook, 2005).

Plant defense mechanisms are tightly regulated by hormone-mediated signaling pathways, mainly jasmonic acid (JA) and salicylic acid (SA). It is generally considered that JA and ethylene (JA/Et) mediate necrotrophic pathogens defense, while SA is involved against the biotrophic and hemibiotrophic ones (Pieterse et al., 2009).

Plants respond to necrotrophic pathogens through the induction of JA/Et biosynthesis, which is increased locally

Citation: Grace Armijo, Rudolf Schlechter, Mario Agurto, Daniela Muñoz, Constanza Nuñez and Patricio Arce-Johnson, Grapevine Pathogenic Microorganisms: Understanding Infection Strategies and Host Response Scenarios, doi: 10.3389/fpls.2016.00382.

and systemically when microorganisms secrete cell wall-degrading lytic enzymes (Robert-Seilaniantz et al., 2011). In addition to wall components, some phospholipids released by the degradation of the plasma membrane directly activate JA biosynthesis (Turner et al., 2002). This increase in JA levels induces the expression of defense related genes coding for glucanases, chitinases, protease inhibitors, and enzymes involved in the biosynthesis of secondary metabolites such as phytoalexins (Glazebrook, 2005).

On the other hand, SA mediates the response to biotrophic and hemibiotrophic pathogens, inducing an increase of reactive oxygen species (ROS) and, therefore, a localized programmed cell death (PCD) in the infected tissue (Pieterse et al., 2009). This defense mechanism, called hypersensitive response (HR), restricts pathogen growth by limiting their access to nutrients and water (Glazebrook, 2005).

For these events to occur, plants must be able to recognize these pathogens. Three types of plant–pathogen interactions have been described to date. The first, PTI or PAMPs-Triggered Immunity (recently called MTI or MAMPs-Triggered Immunity), is a basal immune response activated after recognition of non-adapted pathogens. This corresponds to a first line of defense, common in plants of the same species facing a potentially pathogenic microorganism (Jones and Dangl, 2006; Pieterse et al., 2009). PTI is mediated by plasma membrane localized pattern recognition receptors (PRRs), which are composed of an extracellular domain able to detect PAMPs or MAMPs (i.e., generally structural components of the pathogen) and a intracellular domain that amplifies the signal inside the cell (Monaghan and Zipfel, 2012). A second type of interaction is the effector-triggered susceptibility (ETS), named after the ability of certain microorganisms to overcome the basal plant response through the secretion of virulence factors (effectors) which inhibit PTI, thus promoting the disease (Pieterse et al., 2009). Lastly, there is a third type of interaction known as effector-triggered immunity (ETI). In this case, plants of a particular genotype can identify pathogen effectors through a second class of receptors, called resistance proteins (R). If an effector is recognized by an R protein, either directly or indirectly, it is considered an avirulence factor (AVR) and then the pathogen is avirulent to that plant, since this interaction leads to the activation of HR (Jones and Dangl, 2006; Hayward et al., 2009).

The grapevine (*Vitis vinifera* L.) is one of the most important fruit crops worldwide. This specie is greatly affected by a large number of pathogens that cause diseases in pre- and post-harvest periods, affecting production, processing and export, along with fruit quality. Some of the most important diseases in *V. vinifera*are the gray mold, powdery mildew, and downy mildew (DM), caused by *Botrytis cinerea*, *Erysiphe necator*and *Plasmopara viticola*, respectively, among others (Figure1). The overall purpose of this review is to synthetize the most important grapevine pathogens and their infection strategies along with the host responses in resistance and susceptibility scenarios, taking into consideration the understanding of *Vitis*-pathogens interactions due to its agronomic importance.

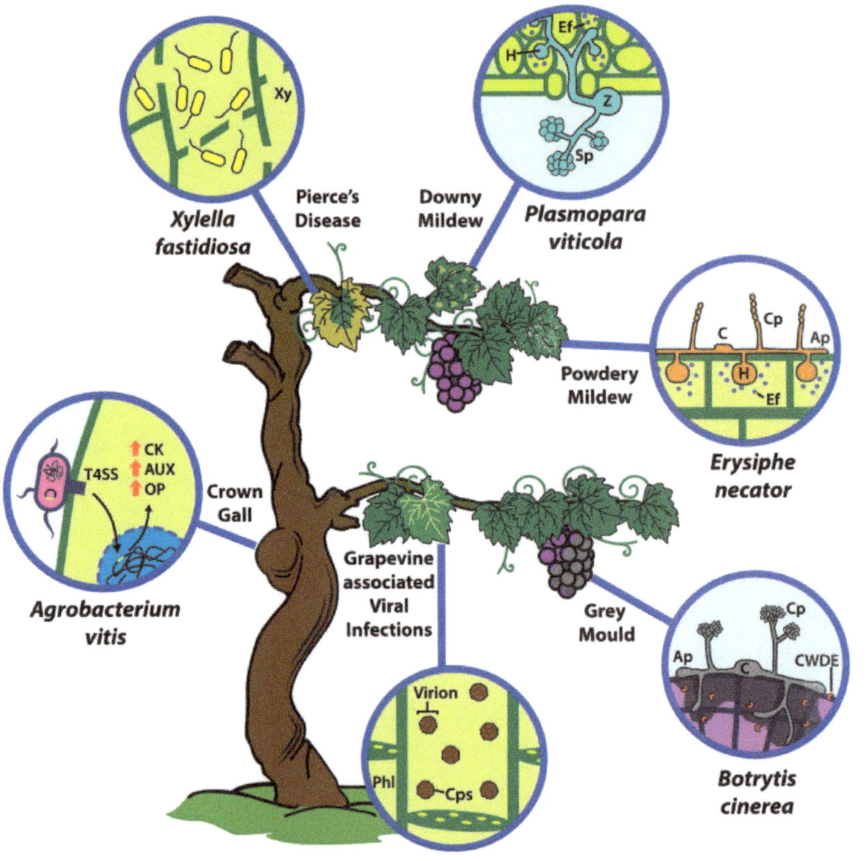

Figure 1: Infection strategy of main grapevine pathogenic microorganisms and disease-associated symptoms. Grapevine is affected by different types of microorganisms, i.e., bacteria, fungi, oomycetes and viruses.*Agrobacterium vitis* causes the grapevine crown gall ...

NECROTROPHIC PATHOGENS

Botrytis Cinerea

The fungus *B. cinerea* has the ability to live as a parasite in green tissue and as a saprophyte in dead or decaying ones. This is the reason for its wide distribution in nature and host unspecificity. Particularly in *V. vinifera*, *B. cinerea* infection causes a disease known as "gray mold" (Figure1) in post-harvest periods, affecting complete berry clusters during packing, transport and commercialization and thus becoming one of the most important pathogens affecting export wine and table grapes (Dean et al., 2012).

Infection Strategy

Botrytis cinerea infects grapevine by two main mechanisms: (1) direct mycelium penetration through skin pores or injury, and (2) early invasion, where conidia infect mainly the flower receptacle, and to a lesser extent the stigma and styles, remaining latent within the berry until maturity (Viret et al., 2004). This fungus presents high variability in several biological traits, which can be explained by the great diversity among isolates. However, attachment to a solid surface is a common requisite for conidial germination, though nutritional requirements may differ (Cotoras et al., 2009). When the conidium perceives characteristics and nutrients from the host surface, it develops an infective structure called appressorium that breaches the cuticle by means of a penetration peg (Rolke et al., 2004). It has been described that appressoria-mediated penetration of *B. cinerea* requires a membrane-associated tetraspanin-encoding gene termed *BcPLS1*, since *Bcpls1*-deficient mutants fail to infect intact host cells (Gourgues et al., 2004). The newly formed appressorium secretes lytic enzymes to cross cuticle and outer epithelial wall, like cutinases and lipases, to breach these layers and penetrate the plant cell.

Primary lesion formation triggers an oxidative burst that helps the fungus to kill and degrade the host tissue (Rolke et al., 2004), along with secretion of cell wall degrading enzymes (CWDE) such as endopolygalacturonases (Kars et al., 2005a), pectin methylesterases (Kars et al., 2005b), cellulases and hemicellulases, causing decomposition and consumption of plant biomass. In fact, it has been described that pectin-degrading enzymes are the most abundant plant cell wall modifying proteins expressed during *B. cinerea* infection in grapevines, such as xyloglucan transglucosylase/hydrolases and glucanases (Blanco-Ulate et al., 2014). Pectin breakdown during infection increases the plant cell wall porosity and may facilitate further polysaccharides degradation. Along with this class of enzymes, *B. cinerea* secretes toxins and oxalic acid during infection; the latter acidifies locally the infected region, allowing the activation of pectinases and laccases secreted by the fungus, favoring hyphal growth and inducing morphogenic signaling of infectious structures (van Kan, 2006). Furthermore, oxalic acid contributes to the virulence of the pathogen, promoting both host tissue cell death and sporulation in adjacent cells.

The initial step of infection appears to be similar in berries at different developmental stages, since a recent study of Kelloniemi et al. (2015) revealed that germinated conidia developed appressoria on both veraison and berry maturity, and in the two developmental stages expression of genes related to fungal virulence as endopolygalacturonase (*BcPG2*), pectin methyl esterase (*BcPME2*), superoxide dismutase (*BcSOD1*), glutatione-*S*-transferase (*BcGST1*), among others, were induced during the first 16–24 h. Nevertheless, *B. cinerea* failed to colonize berries at veraison and the infection was only successful in mature berries. In the latter, up-regulated genes were enriched in functional categories related to degradation of plant cell wall (CWDE), proteolysis, membrane transport, ROS generation and detoxification, phytotoxic proteins and secondary metabolites such as botrydial and botcinic acid, which suggest that the initial process of infection and penetration by the fungus may be common in these stages of development, but degradation of plant tissue is a critical step for successful infection.

Physiologically, infection of *B. cinerea* causes significant metabolic changes in berries. High levels of proline, glutamate, arginine, and alanine accumulated in these tissues, accompanied with the production of glycerol, gluconic acid, succinate and largely degraded phenylpropanoids, flavonoid compounds and sucrose (Hong et al., 2012). Moreover, it has been shown that hexokinase activity is required for the development of*B. cinerea* in the presence of hexoses (Rui and Hahn, 2007), suggesting that the fungus is able to manipulate plant metabolism to use carbon sources required for its growth, in addition to products of tissue degradation.

Host Response and Resistance against *B. cinerea*

Studies conducted in the *Vitis-B. cinerea* pathosystem showed that grapevines use natural and pre-existing skin qualities as defense structures, such as number of cell layers, cuticle, and wax content of the fruit (Gabler et al., 2003), along with the quantity and density of stomata and leaf trichomes. Also, activation of inducible defense mechanisms during the infection has been reported, such as the expression of polygalacturonase-inhibiting proteins (PGIP) to reduce the extensive pectin degradation caused by fungal attack (De Lorenzo et al., 2001) or an oxidative burst and phytoalexin accumulation in grape cells and primed leaves (i.e., trans-resveratrol and α-viniferin) necessary for defense activation (Aziz et al., 2003; Verhagen et al., 2011). Structural defenses are related to fungal primary infection processes (i.e., appressoria formation and plant tissue penetration), while inducible responses are associated with subsequent infection processes.

Grapevines are also able to secrete a set of chitinases, e.g., *Vvchit1a* (acidic class I chitinase), *Vvchit1b* (basic class I chitinase), and other hydrolytic enzymes to degrade the fungus cell wall (Robert et al., 2002; Chong et al., 2008). However, *B. cinerea* has the capacity to catalyze the conversion of chitin to chitosan, bypassing the degradation of its cell wall (El Gueddari et al., 2002). Thus, a fine-tuning regulation of different layers of plant defense can be deployed to cope at some extend with the pathogen infection.

As described in other plants, necrotrophic-type infection of *V. vinifera* activates the JA/ET pathway along with the induction of genes related to phytoalexin biosynthesis, such as phenylalanine ammonia lyase (*PAL*) and stilbene synthase (*STS*; Chong et al., 2008; Girault et al., 2008; Thaler et al., 2012; Wang et al., 2015). It has been described that methyl jasmonate (MeJA) treatments in harvested grape berries could induce resistance against *B. cinerea* and reduce disease incidence. This is associated with an increase of H_2O_2, enhanced expression of the defense-related gene *VvNPR1.1* and accumulation of stilbene phytoalexins, such as tran-resveratrol and its oligomer form, trans-ε-viniferin (Wang et al., 2015). MeJA treatments also induce the biosynthesis of defensive compounds, such as alkaloids, phenolics, flavonoids, and terpenoids along with accumulation of peroxidase, chitinase, β-1,3-glucanase and *PAL* (Jia et al., 2016).

Wan et al. (2015) demonstrated that a minimal production of ROS and a timely elevation of antioxidative capacity were correlated with a high level of resistance in Chinese wild *Vitis*, while highly susceptible cultivar 'Red Globe' suffered massive infection and sustained ROS production due to relatively unchanged antioxidative activities, suggesting the importance of ROS response for the timely recognition and defense to*B. cinerea*.

Synthesis and secretion of pathogenesis-related proteins (PR) are also part of the inducible defense mechanism deployed by *V. vinifera*. For instance, up-regulation of *PR-2* in less susceptible plants and accumulation of beta-1,3-glucanases have been described during infection (Derckel et al., 1999). A recent proteomic analysis on *V. vinifera* cell suspension inoculated with *B. cinerea* identified proteins involved in defense (e.g., PR10), response to oxidative stress, cell wall modification and protein folding (Dadakova et al., 2015). Besides, the transcripts of five defensin-like genes (*DEFL*) were identified as significantly up-regulated in grape tissues infected with *B. cinerea* and three novel defensins were proven to inhibit conidia germination (Giacomelli et al., 2012), suggesting a role of these genes in the defense against this pathogen.

A number of transcription factors have also been identified in *V. vinifera* with a significant role in the defense response against *B. cinerea*, whose functional orthologs were previously described in the plant model*Arabidopsis thaliana*. For example, induction of *VvWRKY33* (Group I from WRKY protein family) correlates with the expression of *VvPR10.1*. Complementation of resistance-compromised *Arabidopsis wrky33-1* mutant lines by the constitutive expression of *VvWRKY33* restores the resistance against *B. cinerea*(Merz et al., 2015). Also, transcription factors from the NAC family are known to be involved in the response to environmental stresses, as *VvNAC1* expression is induced in *B. cinerea* infected berries and leaves. Moreover, *VvNAC1*-overexpressing *Arabidopsis* plants exhibit enhanced resistance to this fungus. These plants present a modified expression pattern of defense-associated gene markers (*AtPR-1*, *AtPDF1.2*), suggesting that *VvNAC1* could be a regulatory component of the plant signaling defense cascade (Le Henanff et al., 2013).

The defenses induced in grapevine berries against *B. cinerea* differ during berry ripening. Agudelo-Romero et al. (2015) described the putative involvement of JA/Et, polyamines and auxins, and evidence of a reprogramming of carbohydrate and lipid metabolisms toward increased synthesis of secondary metabolites involved in plant defense, such as trans-resveratrol and gallic acid, and that, contrary to healthy berries, infected green berries did not activate the SA pathway, suggesting that the pathogen is able to shut down some defenses.

On the other hand, inoculated veraison berries accumulate ROS, activate the salicylate-dependent defense pathway, synthesis of the resveratrol phytoalexin, cell wall reinforcement and formation of papillae underneath the appressoria, thus stopping further infection. As was mentioned above, mature grapevine berries are more susceptible than veraison berries. Infected mature berries activate the jasmonate-dependent pathway, which is unable to stop the fungal necrotrophic infection (Kelloniemi et al., 2015).

Grape genotypes vary in their resistance to infection, degree of fungal colonization and severity of disease. Low or no resistant phenotypes in most common table grape *V. vinifera* cultivars have been described, whereas high level of resistance has only been found in the species *Muscadinia rotundifolia* (*V. rotundifolia*),*V. labrusca*, and other grape hybrids. This resistance appears to be related to the number and thickness of epidermal and hypodermal cell layers and cuticle and wax contents (Gabler et al., 2003). However, no genetic studies have been conducted to date in these resistant plants.

To date, pre-existing or basal defenses seem to be an important part of the defense against *B. cinerea*, along with activation of inducible defense mechanisms mediated by SA or JA/Et pathways, depending on the development state, together with an appropriate kinetics between ROS production and generation of antioxidant compounds.

BIOTROPHIC AND HEMIBIOTROPHIC PATHOGENS

Erysiphe Necator

The biotrophic fungus *E. necator* is the etiologic agent of the grapevine powdery mildew (PM; Figure1), affecting species of the genus *Vitis*. This disease appears as a white-grayish powder on the surface of the infected tissue, mostly leaves and stems. This generates large production losses, reduced yields and fruit quality, mainly by a declining in the sugar content and acidity of the berries (Gadoury et al., 2001; Calonnec et al., 2004).

Infection Strategy

Erysiphe necator depends on its host for growth and development. For this, the conidium attaches to the tissue cells of the plant, allowing the formation of a primary germ tube that differentiates into a specialized infectious structure (i.e., appressorium), which generates a mechanical pressure in order to penetrate and invade the host cell. The successful invasion results in the haustorium formation, by which the fungus absorbs nutrients necessary to complete their life cycle. The haustorium facilitates the dynamic exchange of molecules: the fungus retrieves hexoses, amino acids, vitamins, and other nutrients from host cells, while at the same time secretes proteins to suppress host defenses (Qiu et al., 2015). Once this structure is established, secondary hyphae spread along the infected tissue and finally asexual reproductive bodies (i.e., conidiophores and conidia) emerge from them. When environmental or nutritional conditions become unfavorable, *E. necator* develops cleistothecia, structures of sexual reproduction, that contain four to six asci at maturity, each of which usually contains four ascospores. However, physiological maturity may not be reached for several months, particularly in colder climates. Like conidia, ascospores germinate with a single germ tube, which terminates in appressorium formation (Gadoury et al., 2012).

As in other obligate biotrophic pathogens, *E. necator* genome shows a reduced number of genes related to secondary metabolism (i.e., polyketide synthase, non-ribosomal peptide synthase, dimethylallyl tryptophan synthase and terpene synthase), nitrate and sulfate metabolism, and other core ascomycete genes, such as those related to amino acid metabolism, fermentation, channels/transporters and stress response, among others (Jones et al., 2014). Due to these gene losses, nutrient acquisition from the host is essential for *E. necator*development. Nitrate transporter *VvNPF3.2* has shown to be up regulated in the susceptible grapevine cultivar 'Cabernet sauvignon' during PM infection, probably to increase nitrate or nitrite transport, whereas this was not observed in the resistant cultivar 'Norton' (Pike et al., 2014). It has also been described in PM-infected *V. vinifera* cv. 'Cabernet sauvignon' an increase in the abundance of proteins related to amino acid metabolism (i.e., Alanine aminotransferase and alanine glyoxylate aminotransferases), vitamin biosynthesis and lipid/sterol biosynthesis (Marsh et al., 2010).

In addition, PM exploits as well plant components for its successful penetration and establishment in the host cell. The *MLO*-like class of *R*-genes (Mildew Locus O, MLO, belonging to a family of seven-transmembrane domain proteins) is known to be required for successful host-cell invasion. Loss-of-function *mlo* alleles in barley and *A. thaliana* lead to enhanced resistance to adapted PM species (Freialdenhoven et al., 1996;Consonni et al., 2006). This family is composed by six members in grapevine (*VvMLO3*, *VvMLO4*,*VvMLO6*, *VvMLO9*, *VvMLO13*, and *VvMLO17*), but early induction of *VvMLO3, VvMLO4*, and *VvMLO17*coincided with the establishment of fungal penetration in grape leaves (Feechan et al., 2009). Recently, the locus Sen1 (susceptibility to *E. necator* 1) mapped in *V. vinifera* cv. 'Chardonnay' was described as a novel source of PM susceptibility (Barba et al., 2014). Although the biological functions of these genetic components remains elusive, host genes products are likely to be modulated by PM for its successful infection.

Host Response and Resistance against E. necator

Wine and table grape *V. vinifera* cultivars are very susceptible to *E. necator*. Marsh et al. (2010) showed that susceptible plants (*V. vinifera* cv. 'Cabernet sauvignon') are able to initiate a basal defense response but unable to restrict fungal growth or reduce the progression of the disease. Differentially expressed proteins during the infection of *E. necator* are involved in photosynthesis, metabolism, protein destination, and protein synthesis; suggesting that the pathogen manipulates plant energy processes, according to its needs. Furthermore, defense proteins, such as PRs, defensin-like proteins (DEFLs) and proteins involved in ROS detoxification are induced, resembling a basal defense response, however, these mechanisms are inadequate in timing and/or intensity in order to control disease progression (Marsh et al., 2010).

Many species of the *Vitis* genus, mostly from North America, have varying levels of resistance to PM.Feechan et al. (2011) classified this resistance into four types: susceptibility, partial resistance, penetration resistance, and resistance associated with PCD, based on the degree of development of *E. necator* in host tissues and the triggered response. Among these, and as described above, *V. vinifera* is classified as a susceptible species, *V. riparia* as partially resistant and *M. rotundifolia* is among the most resistant species mainly due to the induction of PCD.

At the first level of resistance, *PEN1*- and *PEN2/PEN3*-like pathways [PENETRATION (PEN) genes] are important components of PTI in grapevine, suggesting a role against non-adapted PM. These genes were described in *A. thaliana* and their grapevines orthologs may have similar functions. AtPEN1 is a member of the SNARE (soluble *N*-ethylmaleimide-sensitive factor attachment protein receptor) family which includes proteins that mediate membrane fusion events; AtPEN2 is a myrosinase involved in the biosynthesis of antimicrobial molecules that are delivered to the site of PM penetration via AtPEN3, which is an ATP-binding cassette transporter (Qiu et al., 2015). Their role in PTI response remains elusive, however, it has been shown their importance and functionality in grapevines.

In a second level of resistance, different loci have been identified in several species of the *Vitaceae* family that confer resistance to *E. necator*: REN1 (Hoffmann et al., 2008); REN2 (Dalbó et al., 2001); REN3 (Welter et al., 2007); REN4 (Ramming et al., 2011); REN5 (Blanc et al., 2012); REN6, REN7 (Qiu et al., 2015); RUN1 (Barker et al., 2005); RUN2.1, and RUN2.2 (Riaz et al., 2011), although molecular defense mechanisms triggered by these resistance loci are yet poorly understood. Among them, the RUN1 locus of*M. rotundifolia* was mapped in chromosome 12 and the functional characterization of resistance gene analogs (RGA) within this locus identified a single gene (*RGA10* or *MrRUN1*) that is associated to the complete resistance to PM (Feechan et al., 2013). This gene codes for a resistance protein of the Toll/interleukin-1 receptor-like (TIR)-NB-leucine-rich repeat (LRR) class of immune receptors, composed by a Toll/interleukin-1 receptor-like (TIR), a nucleotide-binding site (NBS) and a leucine-rich repeat (LRR) domains, acting on pathogen effector recognition and signal amplification (McHale et al., 2006). The other resistance loci have also been associated with this type of resistance gene functions; though this has not been experimentally verified to date. Amrine et al. (2015) sequenced and analyzed the transcriptomes of Central Asian grape accessions that were previously shown to carry a REN1-like local haplotype. They confirmed the partial resistance phenotype of these accessions, which is associated with a late post-penetration response to *E. necator* as it was previously described (Hoffmann et al., 2008; Coleman et al., 2009). In all Central Asian accessions, *E. necator* germinated and established primary hyphae, but it failed to colonize extensively the leaf tissues. Necrotic spots that colocalize with appressoria of secondary hyphae suggest that late HR and PCD may be key in slowing down fungal spread in the Central Asian accessions (Amrine et al., 2015).

Effective resistance responses of grapevine against *E. necator* include enhancement of both JA-mediated and systemic acquired resistance (SAR) responses and accumulation of phytoalexins. For example, transcriptomic changes in resistant Chinese wild grapes (*V. pseudoreticulata*) during PM-infection are related to SA and JA responses, SAR, HR, plant–pathogen interaction, flavonoid biosynthesis and plant hormone signal transduction(Wengetal.,2014).Upregulationofdefense-related genes such as glycosyl hydrolases, lipases,*PR-5* thaumatin-like proteins and proteinases, among others, has been described in successful response to *E. necator*. For example *VpPR-10.1*, isolated from *V. pseudoreticulata*, has nuclease activity (both RNase and DNase activities) and can lead to PCD and DNA degradation. A correlation between *VpPR-10.1* levels and its antifungal property suggest that the nuclease activity is the biochemical basis for the resistance. Thus,*VpPR-10.1* could potentially play a dual role degrading pathogen RNA and inducing programmed death of host cells (Xu et al., 2014). Also, the transcription factor *VpWRKY1* was rapidly induced in *V. pseudoreticulata* after *E. necator* inoculation, and the expression level was found to be correlated with the resistance magnitude (Li et al., 2010). A number of other genes have also been implicated in PM resistance in certain wild *Vitis* species, showing differential expression levels between PM-resistant wild *Vitis* species and susceptible *V. vinifera* cultivars and conferring increased levels of resistance to PM when overexpressed transiently in grapevine leaves or stably transformed into wild-type or mutant lines of *A. thaliana* (Qiu et al., 2015). The key positive regulator of SA pathway, ENHANCED DISEASE SUSCEPTIBILITY 1 (*EDS1*), was differentially expressed in 'Norton' (*V. aestivalis* cv. 'Norton') and 'Cabernet sauvignon,' where a constitutive expression in the first and an induced in the latter were observed after inoculation with PM (Fung et al., 2008). This also correlates with the high and constitutive SA levels in 'Norton' resistant cultivar. Moreover, analysis of the *EDS1* family in grapevine identified both *EDS1* and *EDL2* (EDS1-LIKE 2) in the resistant and susceptible cultivars as necessary components of the regulatory node EDS-PAD4 in a SA-mediated pathway (Gao et al., 2014).

Lignin, flavonoid phytoalexins, and phenolic compounds also play important roles in the defense response of grapes. The rapid production of resveratrol, major compound of the stilbene family and its transformation into viniferins appear to enhance PM resistance in grapevine cultivars. For example, in the resistant cultivar 'Norton,' *STS* are induced 24 h after inoculation with PM, accumulate higher levels of *STS* transcripts in mature leaves than the susceptible cultivar 'Cabernet sauvignon' and has a differential expression pattern during the berry formation compared to the susceptible cultivar, with a markedly increase at veraison (Dai et al., 2012).

As described above, the presence of resistance loci associated to *R* gene clusters would be crucial to define the susceptibility or resistance of grapevine plants against *E. necator*. This suggests that SA pathway is the main mechanism to lead a successful defense response in *V. vinifera*, since R – AVR signaling is mediated by this hormone. However, other hormones such as JA and the production of secondary metabolites in different cultivars may have a significant role in the defense response against this pathogen.

Plasmopara Viticola

Plasmopara viticola (Berk. and Curt.) Berl. and De Toni is the causal agent of grapevine DM (Figure1) and one of the most important pathogens affecting grapevine production worldwide. It was first collected in North East of the USA in 1834. Since then, it has been classified sequentially as the fungus *B. cana*, *B. viticola*,*Peronospora viticola* and finally as the obligate biotrophic oomycete *P. viticola* (Burruano, 2000; Gessler et al., 2011).

Plasmopara viticola spread to Europa by the year 1878, probably due to the use of cuttings from American grapes. All cultivated European *V. vinifera* cultivars are susceptible to *P. viticola*. Consequently, the production of *V. vinifera* has been affected by DM to the present. However, several North American *Vitis*species show resistance to this disease at different levels. This characteristic has also been found in*Muscadinia* species and some Asian *Vitis* species. Indeed, the use of natural sources of resistance to pathogens is still a promising tool in breeding programs nowadays (Gessler et al., 2011).

Infection Strategy

Plasmopara viticola infection occurs specifically through stomata. It starts early in the season, when oospores in fallen leaves or mycelium in dormant twigs are activated by adequate climate conditions to produce sporangia. In presence of water, mature sporangium releases self-motile biflagellate zoospores that infect plants tissues. Zoospores are able to place on the abaxial surface of leaves close to stomata, then germinate and penetrate through the stomatal cavity, where they form a substomatal vesicle. This vesicle gives rise to the primary hyphae and mycelium, which grows through intercellular spaces, enclosed by the veins of the leaf and enters to the cell of the mesophyll by its cell-wall-penetrating and feeding haustoria, which invaginates the plasma membrane of the parenchyma cells. As a result, the adaxial surface of the leaf exhibits a typical oil-spot lesion visible in plants affected by this pathogen. The mycelium also develops to form sporangiophores emerging from the stoma and releasing sporangia to the surrounding susceptible tissues (leaves, twigs, or grape clusters) or plants (Burruano, 2000; Perfect and Green, 2001;Gessler et al., 2011).

Stomata remain abnormally open and unresponsive to abscisic acid in grapevine leaves infected by DM. Recently, Guillier et al. (2015) identified two *V. vinifera* glycoproteins: a phototropin and a lysophospholipase that are induced by pathogen infection, accumulated in apoplastic fluids and considered as candidates for stomatal deregulation.

Little is known about genes involved in pathogenesis in *P. viticola*. During leaf infection, a gene encoding NADH-ubiquinone oxidoreductase has been indicated to be involved in the generation of the proton motive force to allow ATP biosynthesis and active nutrient transport before acquiring them from the host, mainly during the development of the first hyphal structures and haustoria. Analysis of ESTs from germinated zoospores has shown that the most represented unigene encodes a protein similar to fungal laccases, possibly contributing to pathogenesis by playing a role in stilbene detoxification. This could correspond to one of the mechanisms used by the pathogen to counterbalance the production of plant chemical arsenal. Also, an invertase-coding gene would be involved in pathogen proliferation in plant tissues by allowing *P. viticola* to uptake carbohydrates from the host (Mestre et al., 2012; Luis et al., 2013).

In cDNA-AFLP analysis carried out on infected grapevine leaves at oil spot stage, Polesani et al. (2008)identified nine transcripts derived fragments uniquely induced by infection, absent in sporangia, which represent putative virulence factors. In order to identify candidate genes encoding effectors from *P. viticola*several predicted secreted proteins have been identified. Genes corresponding to hydrolytic enzymes, protein inhibitors, elicitor-like proteins and members of the RXLR family of effectors are expressed upon infection. Also, the expression of genes coding for an INL11B-like elicitin, a protein with Kazal-like protease inhibitor fold and a RXLR protein, in infected tissues and germinated zoospores, has been described (Mestre et al., 2012).

Host Response and Resistance against P. viticola

Yu et al. (2012) established five immunity levels to DM in grapevine: (i) if accumulation of callose deposits around the stomata and inhibition of zoospores germination early in the infection process are observed; (ii) emergence of callose deposits around and in the stomata but unable to inhibit the formation of hyphae; (iii) with callose near the stomata and around haustoria, incapable to stop infection; (iv) with more development of hyphae than in (iii); and (v) with hyphae in all mesophyll tissue intercellular spaces. According to this classification, *M. rotundifolia* was classified as level (i), several Chinese species as level (iii), except for *V. amurensis* as level (iv) and some Chinese species and *V. vinifera*as level (v).

Liu et al. (2015) classified three Chinese wild *Vitis* genotypes according to the histological response to infection. *V. piasezkii* Liuba-8 was classified as 'highly resistant,' showing no sporulation, callose deposition and early production of H_2O_2, which corresponds to level (i), according to Yu et al. (2012) classification. On the other hand, *V. pseudoreticulata* Baihe-35-1 and *V. davidii* var. cyanocarpa Langao-5 showed decreased levels of sporulation and callose deposits in infected tissues, which could be related to level (ii) in Yu classification (Yu et al., 2012).

In grapevine, some QTLs with major DM-resistance effects have been identified, named Rpv (Resistance to*P. viticola*). Examples of these are Rpv2, derived from *M. rotundifolia* and located on chromosome 18 (Merdinoglu et al., 2003; Peressotti et al., 2010); Rpv3 found in 'Bianca' probably from a *V. rupestris*background and mapped on the reference chromosome 18 (Bellin et al., 2009; Casagrande et al., 2011; Di Gaspero et al., 2012); Rpv8, and Rpv12 located on chromosome 14 and Rpv10 located on chromosome 9, from *V. amurensis* (Blasi et al., 2011; Schwander et al., 2012; Venuti et al., 2013).

In the Rpv1 locus derived from *M. rotundifolia* and located on chromosome 12 (Merdinoglu et al., 2003), one of the first grapevine resistance genes to be cloned and functionally characterized was identified, designated *MrRPV1* and conferring strong resistance to *P. viticola* with a broad isolate specificity described by the severe restriction of hyphal growth and sporangiophore development. *MrRPV1* produces four alternatively spliced variants, encoding only one full-length functional TIR-NB-LRR protein (Feechan et al., 2013).

In compatible interactions between grapevine and *P. viticola*, analysis of differentially expressed grapevine genes during infection have demonstrated that most of these genes are down regulated and only 30% up regulated. Genes related to photosynthesis and primary carbon metabolism were the most negatively affected. Other down-regulated genes are related to protein metabolism, transport and signal transduction. Genes involved in secondary metabolism, defense and response to external stimuli were up regulated (Polesani et al., 2008). Among the genes related to defense response, expression of the *PR-2*, *PR-4*, *OSM-1* (osmotin),*GLP-2* (germin-like protein), *GLP-7*, *TLP-4* (thaumatin-like proteins), *CHI* (chalcone isomerase), and *NAC*genes are induced in microdissected stomata and surrounding cells at early times of infection. Thus, there would be a site-specific regulation of grapevine response to *P. viticola*, with short distance signals released from stomata to adjacent cells (Lenzi et al., 2016).

Stress and defense-related genes have shown higher levels of expression in resistant grapevines, so it has been suggested to be linked to constitutive resistance against *P. viticola*. Also, high levels of inositol, caffeic acid, and other antimicrobial substances have been related to constitutive response in uninfected resistant leaves. Up regulation of genes coding for PR proteins occurs as soon as resistant grapevines are inoculated with *P. viticola*. Incompatible interactions are also associated to modulation of genes involved in defense response, photosynthesis, primary and secondary metabolism, signal transduction and transport. This is also followed by an accumulation of metabolites such as resveratrol and viniferins (Gessler et al., 2011; Malacarne et al., 2011).

In *Vitis* progeny analysis carried out by Malacarne et al. (2011), higher stilbenoid accumulation is associated to individuals with least severe DM symptoms. These compounds were trans-resveratrol, trans-piceid, trans-pterostilbene, and several viniferins. Among the genes induced in the incompatible interaction there are two receptor-like protein kinases, one TIR-NBS receptor, a calcium-dependent protein kinase, as well as transcripts concerning ethylene biosynthesis and responsive factors, *PR1*, *PR2*, *PR5*, *PR10*, two isoforms of*STS*, genes encoding a caffeoyl-CoA *O*-methyltransferase, a flavonoid 3′ 5′-hydroxylase and a dihydroflavonol reductase, among others (Malacarne et al., 2011).

The resistance level of grapevines has been correlated to the quantity of the callose deposition. In this context, strong callose deposition in the *P. viticola* resistant *M. rotundifolia* was observed, along with an increased expression of callose synthases genes *CalS1* and *CalS10* at early times of infection (Yu et al., 2012).

VvWRKY33 has been suggested to play an important role by inducing early defense-related pathways in grapevine leaves against fungal pathogens. VvWRKY33, a class I WRKY transcription factor, activates the promoter of the *VvPR10.1* gene. Both are strongly induced when incompatible interaction occurs; in the case of *VvWRKY33* just 2 h after *P. viticola* infection. Ectopic expression of *VvWRKY33* in grapevine leaves of a susceptible grapevine cultivar has proved to strongly increase resistance to *P. viticola*, reducing sporulation between 50 and 70% (Merz et al., 2015).

Thus, as described in *E. necator*-grapevine pathosystem, some QTLs with major DM-resistance have been identified and would be crucial to define the susceptibility or resistance of grapevine plants against this pathogen. SA pathway, expression of related genes and also production of secondary metabolites may have a significant role in the defense response against this pathogen.

Agrobacterium Vitis

Agrobacterium vitis is a specific pathogen of *V. vinifera* and is the causal agent of the grapevine crown gall disease (Figure1). Virulent strains of this bacterium induce the formation of tumorigenic structures at the site of infection for nutrient uptake, while necrosis and a HR-like response has been reported in grapevine roots and in non-host plants infected with this bacterium, respectively. *A. vitis* can be defined as a biotrophic pathogen since it maintains a parasitic relationship with living tissues of their host to complete its life cycle.

Infection Strategy

Similar to *Rhizobium radiobacter* (formerly known as *Agrobacterium tumefaciens*), *A. vitis*contains Ti plasmids carrying *vir* genes and T-DNA sequences that allows it to transfer stable genetic material into the genome of its host, particularly genes involved in the biosynthesis of opines, auxins, cytokinins, and the utilization of tartrate (Burr and Otten, 1999). The infection starts commonly through plant injuries, particularly by freezing and/or mechanical wounds, to favor the release of phenolic compounds that act as chemoattractants to the attached bacteria. These compounds activate the transcription of *vir* genes, whose products induce the transfer of the T-DNA into the host genome. VirA is a protein that can detect and interact with the chemical signal and phosphorylate the VirG protein, which in turn transcriptionally activate several other *vir* genes. For instance, VirG activates the expression of *virD2*, whose product has the necessary endonuclease activity to release the T-DNA from the Ti plasmid, and *virE2*, involved in the protection and transport of the bacterial genetic material into the plant (Burr et al., 1998; Burr and Otten, 1999).

The genetic diversity of *A. vitis* is defined by the nature of their Ti plasmids and T-DNA sequences. Ti plasmids are commonly defined according to the type of opine they produce. In grape crown gall, three Ti plasmids have been defined as the octopine/cucumopine (O/C), the nopaline and vitopine plasmids (Szegedi et al., 1988). These three plasmids contain T-DNA sequences that can be categorized into two functional groups, whether are involved in opine synthesis or in gall induction. Opines are low molecular weight compounds derived from aminoacids, keto-acids, and sugar phosphates that only can be catabolized by*Agrobacterium* strains (Escobar and Dandekar, 2003). T-DNAs contain genes for the synthesis of these molecules, such as *acs* (agropine synthase), *cus* (cucumopine synthase), *nos* (nopaline synthase), and/or *vis*(vitopine synthase) depending on the nature of the T-DNA. Moreover, T-DNA sequences also contain genes involved in tumor formation, where genes involved in the biosynthesis of auxins (*iaaM* and *iaaH*) and citokinins (*ipt* genes) can be found (Burr et al., 1998). Oncogenes are also present in these regions, such as the gene *6b* that encodes a protein that stimulates cell growth in an auxin-independent manner. For example, the overexpression of T-6b in transgenic tobacco plants leads to an increased cell expansion in leaf disks but with no apparent changes in auxin content (Clement et al., 2006).

A hallmark of the genetic structure of *A. vitis* is the presence of plasmids with genes encoding proteins involved in tartrate utilization (pTr), an abundant compound in grapevines. Three types of these plasmids have been found in *A. vitis* and were defined as TAR-I and TAR-II from the AB3 strain, and TAR-III from the AB4 strain (Burr and Otten, 1999). Although there are differences in the size of these plasmids, the genes associated with tartrate utilization (*ttuA-E* genes) are conserved among the strains of *A. vitis*. It is suggested that TtuA act as a transcription factor that induce the expression of other *ttu* genes such as *ttuB*, which is involved in the tartrate entry into the bacteria; *ttuC* and *ttuD*, whose product catalyze the degradation of tartrate; and *ttuE*, which codes a tartrate-inducible pyruvate kinase that seems not essential for the utilization of tartrate (Salomone et al., 1996). It is important to notice that the particular characteristic of *A. vitis* to utilize a grape-derived compound can partially explain the high host specificity of this pathogen.

Host Response and Resistance against A. vitis

Natural resistance sources against the grape crown gall are important for the study of the interaction between tumor-forming pathogenic bacteria and its host, and for the introduction of these sources into further breeding programs. Some examples of both cultivated and wild *Vitis*species are *V. riparia* cv. 'Glorie de Montpellier' and *V. amurensis*, respectively, who are resistant to the infection of *A. vitis*. In the case of *V. riparia*, it has been suggested that the T-DNA from *A. vitis* is not stably integrated into the host genome given by the low frequency of T-DNA integration and the absence of opine content in infected tissues (Süle et al., 1994). In *V. amurensis*, a dominant locus has been identified in the chromosome 15 of this specie (*Rcg1* locus), where no signs of crown gall formation were observed on populations carrying this resistant locus (Kuczmog et al., 2012). Even though this resistance seems to affect a wider range of *Agrobacterium* species, the molecular basis underlying the *Rcg1* locus remains unclear.

The Korean cultivar *V. vinifera* cv. 'Tamnara,' originated from a grape breeding program between 'Campbell Early' (*V. labruscana*) and 'Himrod Seedless' (*Vitis* sp.; Park et al., 2004), also showed resistance to the grapevine crown gall. Interestingly, genes related to defense responses in Tamnara were expressed in both inoculations with *A. vitis* and SA-treatments, categorized in groups of ESTs whose genes encode proteins involved in defense (i.e., glucanase, chitinase, thaumatin-like protein, PR-10), signal transduction (NBS-LRR type protein), oxidative burst (i.e., GST, CAT, Glutathione peroxidase) and cell wall fortification (i.e., hydroxyproline-rich protein, extensin-like protein, cinnamyl alcohol dehydrogenase), among others (Choi et al., 2010). It might not be unexpected to find a group of defense related genes that correlate *A. vitis*inoculations and SA treatments in resistant cultivars, since it has been described that SA and SAR have a role in the defense against *R. radiobacter* in *N. benthamiana*, where pre-treatments with SA interferes with the growth and virulence of this bacteria on its host and might affect the expression of several *vir* genes (Anand et al., 2008).

Thus, it seems that the SA signaling pathway may be operating in the interaction between *A. vitis* and resistant cultivars of *V. vinifera*, inducing a defense response similar to those against other biotrophic pathogens mentioned above.

Xylella Fastidiosa

Xylella fastidiosa is a gram-negative, xylem-limited bacterium and the causal agent of Pierce's disease (PD) in *V. vinifera* (Figure1). This bacterium can be classified as a biotrophic pathogen, because it does not kill the host tissue until later stages of its life cycle. The growth of this bacterium depends mostly on climate. The optimum growth conditions for *X. fastidiosa* are warmer environments close to 28°C, this is the main reason that this pathogen is not prevalent in areas where winter temperature drops under 0°C (Lieth et al., 2011).

Infection Strategy

Xylella fastidiosa is transmitted to new host plants by insect vectors such as sharpshooters (*Homalodisca coagulate*), leafhoppers (*Cicadellidae* family) and spittlebugs (superfamily *Cercopoidea*) during xylem sap feeding. This is the major cause for this pathogen to spread and infect several agriculturally important plants, such as citrus, almond, coffee, and grapevine (Newman et al., 2003). In *V. vinifera*, *X. fastidiosa* causes the Pierce's disease (PD), an economically important disease that affects wine, table and raisin grape production. The symptoms of this disease can be severe, including leaf scorching, desiccated fruit, cordon die back, and finally vine death (Roper et al., 2007).

Xylella fastidiosa infects the vine creating a biofilm in xylem vessels that disrupt water and nutrients flow throughout it. This occlusion is composed by host gums, bacterial exopolysaccharide or degradation products from the host cell wall, or a combination of the three. The xylem is a water transport network of vessels composed of lignified dead cells. The vessels are interconnected by bordered pits (channels) and the pit membranes are composed of pectin, cellulose, hemicellulose and proteins. Pectin polymers determine the pore size in the pit membrane, being sizes between 5 nm in grapevine, allowing the passage of sap through xylem but blocking the passage of larger objects, like *X. fastidiosa*, whose diameter range from 0.3 to 0.5 μm (Davis et al., 1978; Newman et al., 2003; Roper et al., 2007).

Pierce's disease biofilm formation can be divided in three stages: early biofilm formation, growth, and adaptation. The first stage starts with the transmission of *X. fastidiosa* directly by insect vectors into the xylem vessel. After entry, *X. fastidiosa* needs to move through the vascular system. PilT, a type IV fimbriae, is responsible for retracting and extending fimbriae IV in a twitching motility, allowing the bacteria to move through the vascular system. These processes depend on cell–cell aggregation, which involves membrane attachment proteins, such as chaperonin GroEL, and two-component system regulatory proteins (PopP, FeuP, or PhoP) implicated in the correct adhesion of bacterium-bacterium and biofilm formation (Silva et al., 2011). Following this, the bacteria start to break down xylem pit membranes. In order to do this, they secrete CWDs enzymes, such as endo-1,4-β-glucanases, endoxylanases, β-xylosidases, cellobiohydrolase, and a polygalacturonase (PG). Bacterium needs PG to successfully infect and is a critical virulence factor to promote pathogenesis. This enzyme digests cell wall polymers that form the xylem pit membrane. The digestion of pectin component exposes the cell-wall polysaccharides that are target by the other CWDs enzymes. Thus, the pectin digestion allows *X. fastidiosa* both to use as a carbon source and to move from vessels to vessels (Roper et al., 2007).

The second stage involves a group of metabolic processes to favor biofilm growth, such as amino acid and carbon metabolism and cofactors biosynthesis. Furthermore, at this stage the phenomenon called quorum-sensing (QS) occurs; a bacterial mechanism to sense the population and react to their changing environment. This mechanism is regulated by acylated homoserine lactones (AHLs), which are molecules that trigger the expression of QS regulated genes. QS is involved in the production of *cis*-11-methyl-2-dodecenoic acid, which participates in virulence, motility, toxin production, aerobic respiration, biofilm dispersal and extracellular polysaccharide (EPS). Another important enzyme for this stage is a polynucleotide phosphorylase that degrades plant chloroplasts to provide carbon sources and facilitate the migration through xylem vessels (Roper et al., 2007).

The third and last stage corresponds to adaptation and protection, involving proteins which confer advantages to the bacterial population like resistance to environmental factors. Within this context, toxin production and heat shock protein (sHsp) expression have been observed. sHsp are a small stress induced proteins that prevent aggregation and are important in protein refolding. A predicted membrane localized GTPase was also found, which may be involved in a signal transduction mechanism, but their specific role is to date unclear. Finally, the two-component system regulatory protein (popP, feuP, or phoP) is required for virulence at the last and long stage of infection (Silva et al., 2011).

Host Response and Resistance against Xylella fastidiosa

The immune response and molecular mechanisms that *V. vinifera* has developed against infection of *X. fastidiosa* remain unclear. However, there are some species, such as *V. labrusca* or *M. rotundifolia* that when infected with *X. fastidiosa*, their symptoms are not as severe as in *V. vinifera* (Lieth et al., 2011). It has been shown that vines employ physical or chemical barriers post infection, such as tylose and gums. Tylose is an outgrowth of parenchyma cells through vessel pit pairs into the lumen of tracheary elements (Sun et al., 2006) and gums are originated from the cell wall or middle lamella (Fry and Milholland, 1990). Some authors suggest that the development of tylose and gums around the infection site is a resistance mechanism. Sun et al. (2006) found that tylose initiation occurred too slowly to restrict spread of mobile pathogens such as *X. fastidiosa*. However, since this bacterium has a slow basipetal movement in systemic infection, tylose could represent a significant obstacle to this type of infection. Fry and Milholland (1990) studied different species of the *Vitis*: *V. vinifera* genus, which is very susceptible to *X. fastidiosa*, and muscadine grapevine (*V. rotundfolia*), whose cultivars are tolerant to PD. The authors demonstrated that the susceptible *V. vinifera* cv. 'French Colombard' generates predominantly occlusion pectin, whereas the tolerant *V.*

rotundfolia cv. 'Carlos' and 'Noble' generate occlusion tylose and gums. Thus, the synthesis of tylose and gums could serve as a defense response to contain the infection of *X. fastidiosa*.

Mapping for resistance traits against PD has been reported in hybrids of Mexican cultivars of *V. arizonica*. Within these populations a major QTL has been identified, called PdR1 (Pierce's disease Resistance 1) in the linkage group 14, which explains 72% of the phenotypic variation (Krivanek et al., 2006; Riaz et al., 2006). Plants carrying this locus have lower levels of bacterial titer in the stems. Moreover, a fine-scale mapping identified two allelic forms in this locus, denominated PdR1a and PdR1b, derived from full siblings of a PD resistant population (*V. rupestris* 'A. de Serres' × *V. amazonica/candicans* 'b43-17'). Within this region, 14 putative genes have been identified and 5 of them were annotated with a molecular function (Riaz et al., 2008). Thus, it is yet to be proved whether PdR1a and PdR1b have different genes within this locus or two alleles of the same gene. A subtractive suppression hybridization (SSH) analysis was performed in leaf, stem, and shoot tissues between resistant and susceptible sibling genotypes (*V. rupestris* × *V. amazonica*) after inoculation with *X. fastidiosa* (Lin et al., 2007). Among these tissues, ESTs derived from resistance genotypes where associated with primary cell wall modifying and metabolic enzymes and PR proteins, such as endo-xyloglucan transglycosylase (EXT), xyloglucan endotransglycosylase (XET), PR-2 (β-1,3-glucanase) and PR-10, among others.

To date, identification of resistance locus to PD, called PdR1, and physical or chemical barriers such as tylose and gums are the most relevant known defense mechanism of grapevine to *X. fastidiosa*.

VIRUSES

Nearly 70 virus species have been identified to date that are able to infect the *Vitis* genus, accounting for at least 25 different diseases in grapevine (Martelli, 2014). As other viruses, they are classified by the International Committee on Taxonomy of Viruses (ICTV), classification that is based on several parameters such as: size particle, genome structure (ORFs), nucleotide or amino acid sequence identity of different virus proteins, like heat shock protein 70 (HSP70), polymerases, coat proteins (CP), type of transmission vector and serological information. It should be noted that in many cases they are found associated as a multiplex virus infection complex (Prosser et al., 2007; Martelli, 2014).

From an economic point of view, the most important grapevine viruses are those who cause the leafroll diseases, known as Grapevine Leaf Roll associated Viruses (GLRaV -1, -2, -3, -4, and -7). Moreover, several genetic variants of GLRaV-4 and GLRaV-6 have been described, as well as GLRaV-Pr, GLRaV-Car, and GLRaV-De, based on phylogenetic relationships with the *HSP70h* gene (Maliogka et al., 2009; Ghanem-Sabanadzovic et al., 2010, 2012). All of the above belong to the *Ampelovirus* genus, which harbor a monopartite genome composed of genomic and subgenomics positive (+) single stranded (ss) RNAs with around 12 open reading frames. Two exceptions to the rule have been described: GLRaV-2, classified as a Closterovirus, and GLRaV-7, which has not been assigned to any genus to date (Al Rwahnih et al., 2012a;Naidu et al., 2015).

Other important grape viruses are the Grapevine Virus A and B (GVA and GVB, respectively), belonging to the *Vitivirus* genus, a monopartite genome composed of genomic and subgenomics (+)ssRNAs with five open reading frames (Adams et al., 2004). GVA is related to the Kober stem grooving symptom (Goszczynski et al., 2008; Alabi et al., 2014) and can cause reddening of leaf margins and petioles, poor vigor and leafroll; while GVB is associated with the corky bark syndrome consisting of soft, rubbery, and abnormal swelling of the basal internodes of the canes, longitudinal cracks and cork forming, typical of the rugose wood complex (Bonavia et al., 1996). *Vitivirus* genus has other less ubiquitous species: GVD, GVE, and most the recently discovered GVF, causing similar corky rugose wood-like symptoms (Al Rwahnih et al., 2012b).

On the other hand, the most representative species of grape viruses in the *Nepovirus* genus are the grapevine fanleaf virus (GFLV) and the arabis mosaic virus (ArMV), both causing typical fan-leaf degeneration and leaf decline, reduced fruit quality, short internodes, and abnormal bifurcations. In addition, GFLV is considered to be the major threat to grape industry due to its ability to reduce crop yield up to 80% depending on the isolate, the susceptibility of the grapevine variety and environmental factors (Martelli, 2014).

Not only ssRNA viruses can infect grapes. Al Rwahnih et al. (2013) described a monopartite circular ssDNA virus which belongs to an evolutionarily distinct lineage of the *Geminiviridae* family, the grapevine red blotch-associated virus (GRBaV). This virus causes symptoms such as red blotch, marginal reddening, and red vein in leaves and reduced sugar accumulation in fruits of *V. vinifera* cv. 'Californian Cabernet Franc,' 'Cabernet sauvignon' and 'Zinfandel.' Similarly, a new ssDNA virus from Brazil was described which is able to infect pears, apples and different grapes varieties causing symptoms such as shrinkage, reddening or red blistering of leaves (Basso et al., 2015).

Infection Strategy

Virus infection starts with the entry of the virus into the plant cell through wounds or graftings, since they cannot cross the cell wall by themselves. In several cases, the spread occurs by transmission vectors, like mealybugs (*Pseudococcidae* family) and dagger nematodes (*Xiphinema index*, *X. italiae*, and *X. diversicaudatum* among others), which feed on sap phloem or roots, allowing movement of these pathogens into the plant. Also, due to grafting techniques, viruses can flow from an infected bottom (rootstock) to top (scion) through the phloem (Martelli, 2014). Once inside the cell, they must disassemble in order to release their own genome and replicate. Near 70% of plant viruses genomes, including grape-associated ones, are composed of a positive single stranded RNA molecule, or ss(+) RNA. Thus, replication process starts with the expression of a viral replicase, generating negative RNA strands and then subsequently, new positive strands through the activity of a viral RNA dependent RNA polymerase (RdRp; Zabel et al., 1976; Dorssers et al., 1984).

Early translated proteins are involved in the replication process, such as polymerases, helicases, and proteinases. Then, structural proteins such as capsid (CP) and movement proteins (MP) are transcribed, with different functions during viral lifecycle. For example, MP and/or CP are necessary for viral translocation, while CP proteins are also required for virus assembly and cell-to-cell movement (Lucas, 2006). MP proteins can target plasmodesmatas and actively use the host actin/endoplasmic reticulum network to increase the size exclusion limit (SEL) of this structure by either destroying actin filaments and allowing virions to move between cells (Wolf et al., 1989; Waigmann et al., 1994; Harries et al., 2010; Niehl and Heinlein, 2011) or removing desmotubules and expanding membrane pores converting them to tubules (Kawakami et al., 2004;Liu et al., 2005; Schoelz et al., 2011). GFLV replication process occurs in viral compartments from endoplasmic reticulum-derived membranes, and it cannot only can be transported inside Golgi-derived vesicles through microtubule- or microfilament-dependent pathways (Laporte et al., 2003), but they can interact with type-I plasmodesmata located proteins (PDLP) too; these proteins are located on the tubule base of modified plasmodesmatas and genetic disruptions of this interaction can reduce tubule formation, delay infection, and attenuate symptoms (Amari et al., 2010). Nevertheless, not only PDLPs are involved in grape virus movement; class XI myosins (XI-K and XI-2) inactivation produced a mislocalization of PDLP and MP, suppressing GFLV movement, suggesting that these proteins could facilitate the necessary tubule formation for virus translocation (Amari et al., 2011). However, virus translocation is not dependent on virion assembly. This type of cell-to-cell movement can be divided into three steps: first, genome transfer from replication sites to intracellular transport systems; then, assisted intracellular transport of the viral genome and finally, the intercellular transport of genomes through plasmodesmata-TMV-like mechanisms (Carrington et al., 1996).

Finally, in phloem-restricted viruses (Figure1), particles must translocate throughout the plant tissues, moving from mesophyll cells to sieve elements and thus achieve a systemic infection. Viruses take advantage of the phloem vasculature and sink-to-source flux in order to passively move through the plant at a very fast rate (Leisner and Turgeon, 1993). Although many viral determinants have been proposed depending on the viral genus, CP and MP seem to play a major role in this long-distance movement (Hipper et al., 2013). In the end, they are released into systemic tissues, thus starting new infection sites.

Host Response and Resistance against Viral Infection

The common plant cell response to viral infection is the activation of HR, including cell wall fortification and induction of PCD to block the systemic propagation of the virus (Stange et al., 2008; Komatsu et al., 2010). This response has been associated to an R gene-mediated immunity (i.e., ETI), through recognition of AVRs by this class of host immune receptors (Jones and Dangl, 2006). Nevertheless, this kind of response has not been reported within *Vitis* genus and no direct resistance gene has been described (Martelli, 2014).

During compatible grapevine–virus interactions, host transcriptional reprogramming in both leaves and berries has been reported. For example, presence of GRLaV-3 in *V. vinifera* cv. 'Carmenere' was correlated with leaf red coloration and chlorosis. Transcriptomic and developmental changes were studied during fruit ripening in *V. vinifera* cv. 'Cabernet sauvignon' infected with GRLaV-3. Reduction in sugar and total anthocyanins content, followed by differentially expressed genes, were observed during grape–virus compatible interaction. Among repressed genes, those associated with sugar metabolism, transport, and the phenylpropanoid pathway were severely inhibited at the ripening stage (E-L38), while genes associated to cell rescue, defense, death and aging were up-regulated (Vega et al., 2011). Similar results were found byEspinoza et al. (2007a,b) were same cellular responses plus proteasome-ubiquitin pathway and other biological processes as transport, transcription and RNA processing were induced during viral infections. Moreover, in *V. vinifera* cv. 'Merlot' infected with GRLaV-3, there is an induction of genes related to the flavonoid biosynthetic pathway, specifically the anthocyanin branch, which contributes with the reddish-purple leaf coloration phenotype in GRLaV-3-infected grapevines (Gutha et al., 2010). The same author showed that accumulation of anthocyanins, flavonols and proanthocyanidins are correlated with the induction of chalcone synthase (CHS), flavonoid-3′-hydroxylase (F3′H), flavonoid-3′, 5′-hydroxylase (F3′5′H), dihydroflavonol reductase (DFR), leucoanthocyanidin dioxygenase (LDOX), UDP-glucose:flavonoid 3-*O*-glycosyltransferase (UFGT) and the transcription factor MYBA1.

A physiological, agronomical and transcriptomic analysis of the infection of *V. vinifera* cv. 'Bosco' by GRSPaV (Grapevine rupestris stem pitting associated-virus) has been conducted by Gambino et al. (2012). At veraison (E-L35), infected grape leaves showed a decrease in their chlorophyll content and net photosynthesis, although up-regulation in photosynthesis, hormones, and secondary metabolism-related genes was observed. However, defense, signal transduction and primary metabolism (TCA, glycolysis and pentose phosphate pathway) related genes were down regulated in this infected tissue. In the same study, GRSPaV-infected berries at veraison showed that 233 genes were differentially expressed, showing down regulation of main functional categories (i.e., stress, disease, senescence, ROS detoxification, ethylene and JA signaling). It is believed that GRSPaV and grapevine have co-existed for a long period of time and these changes account as a form of adaptation and co-evolution between these two species, particularly in the inactivation of stress and defense responses and the favoring of the induction of photosynthesis associated genes.

Infected hosts are able to trigger molecular changes during compatible interactions in order to restrain virus infections. Plants deploy two main endogenous mechanisms to regulate gene expression: transcriptional gene silencing (TGS), which involves decreased RNA synthesis because of promoter methylation (Meyer et al., 1993; Sijen et al., 2001) and post-transcriptional gene silencing (PTGS; de Carvalho et al., 1992; Van Blokland et al., 1994) that recognizes dsRNA as a target (Hammond et al., 2001), which are then trimmed by type-III RNAses (i.e., DICER), producing small RNA fragments (i.e., siRNA). These short fragments act as a signal for a complementary sequence RNA that can be degraded at the end of the cycle. TGS and PTGS were considered as separate pathways. TGS was thought to regulate mainly transposons and accidental transgenes that mimic transposons. By contrast, PTGS was thought to regulate viral infection and accidental transgenes that encode some types of aberrant RNA that mimic viral RNA. The major breakthrough in the distinction came with the discovery that viruses and transgenes encoding dsRNA induce either TGS or PTGS (Vaucheret and Fagard, 2001).

As such, host PTGS machinery might be able to recognize viral dsRNA as a surveillance mechanism to cope with viral infections. For this reason, efforts have been made in order to obtain viral induced silencing transgenic lines expressing sense, antisense or both viral sequences in grapevines. For instance, *V. rupestris*and 110 Richter rootstock were successfully transformed with a short CP sequence using embryogenic cultures (Krastanova et al., 1995). Another transformations involved sense and antisense-oriented CP gene sequences from GFLV and GLRaV-3, a truncated HSP90-related gene of GLRaV-3 in rootstocks Couderc 3309, *V. riparia*, Teleki 5C, Millardet et De Grasset 101-14, and 110 Richter (Xue et al., 1999), and the GVA MP sense or antisense sequence expressed in *V. rupestris* plantlets (Martinelli et al., 2002).

However, virus can counterdefense plant response against infections in at least two known ways: RNAi viral suppressor proteins (VSRs) and RNA silencing suppressors (RSSs; Li and Ding, 2006; Ding and Voinnet, 2007; Murray et al., 2013). In either case, viruses can block the RNA silencing pathway at different stages (Burgyan and Havelda, 2011), affecting dsRNA processing and silencing signal amplification, decreasing siRNA stabilization, suppressing RISC activity or acting as suppressors with unspecified functions (Alvarado and Scholthof, 2009). For example, GLRaV-2 p21 protein suppresses PTGS through direct binding to siRNAs (Ye and Patel, 2005). Also, GLRaV-2 p24 protein and GVA p10 protein have been described to possess silencing suppressor activity (Chiba et al., 2006; Zhou et al., 2006). Another way to counterdefense plant response are protein domains encoded by RNA viruses which decrease alkylation damage. Van den Born et al. (2008) demonstrated that an Alk-B domain encoded by five viruses from the Flexiviridae family, is able to demethylate RNA with a robust repair activity due to its N-terminal extensions beyond the Alk-B core. Several other enzymes are also involved in viral RNAi suppression. Proteinase1/Helper component-proteinase (P1/HC-Pro) acts as a suppressor of both transgene induced and virus-induced gene silencing in tobacco plants (Anandalakshmi et al., 1998). Also, a tandem GLRaV-2 papain-like leader proteases, L1 and L2, are essential to establish the infection process of GLRaV-2 in initially inoculated grapevine (Liu et al., 2009).

Because viruses can affect plant post-transcriptional gene silencing mechanisms, some authors worked on a way to attack this viral counterdefense response. Soler et al. (2012) found that when three VSRs: p25, p20, and p23 from citrus tristeza virus (CTV) are silenced, complete resistance to viral infection in some transgenic lines were obtained. Thus, this strategy could be a promising approach to induce virus resistance in grapevine.

To date, identification of resistance genes has not been reported within *Vitis* genus, however, virus defense on plants is a common mechanism involving different steps on regulatory pathways. Future research for a deeper understanding of these mechanisms in grapes and other species is required.

DISEASE MANAGEMENT OUTLOOK

Disease diagnosis is a time-consuming laborious task, especially regarding the identification of a particular pathogen and selection of the best control strategy to prevent or reduce economic losses. However, concerning grapevine disease management, some strategies available are very expensive and/or not tolerated by the importing markets, mainly due to their potentially negative environmental impact.

Since the vast majority of *V. vinifera* cultivars are susceptible to the pathogens abovementioned in this review, wine and fresh-grape production depends greatly on the use of effective control methods. Cultural control techniques, including thinning and pruning, "cold curing" (i.e., exposing plants to low temperatures to prevent or eliminate the pathogen), chemical control including contact and systemic compounds, biological control and natural products have been included in management systems. Also, in some grapevine disease models applied to decision support systems, other variables have been correlated with the pathogen life cycle, as meteorological conditions and phenological stage of the plant to reduce the use of pesticides applications and increase management efficiency (Gessler et al., 2011; Lieth et al., 2011; Kamoun et al., 2015).

Among the most effective methods, chemical control is one of the most used strategies. Based on the type of chemical (active ingredient), "pesticides" can be classified as copper compounds, sulfur compounds, dithiocarbamates, benzimidazoles, antibiotics, among others. Although these chemicals are effective against fungal and bacterial pathogens, they are not effective against viral diseases. Despite this, certain highly volatile chemicals compounds are toxic for organisms including insects, common viral vectors. Nevertheless, this technique is not highly effective and cannot avoid future infections. For this reason, management control programs nowadays use healthy mother plants insuring virus-free next generation plants (Almeida et al., 2013). Another strategy is to use silencing sequences of suppressor viral proteins that have the potential to avoid future viral infections. For example, the first attempts to artificially induce RNA silencing pathways in grapevines has been done in transgenic *V. vinifera* cv. "Chardonnay" expressing artificial miRNA targeting GFLV CP (Jelly et al., 2012).

Currently, main concerns in disease management include the environmental and health impact of chemical compounds. Biological control of grapevine and plant diseases in general has become more significant, especially against fungi as *B. cinerea*, by using antagonistic microorganisms before or after infection (Calvo-Garrido et al., 2014; Parafati et al., 2015). However, the use of disease-resistant cultivars of *V. vinifera* is the most cost-effective, safe, and environmentally desirable option, since resistant cultivars sometimes offer the only practical control option available. The identification and inclusion of natural sources of resistance in grapevine breeding programs by pyramiding two or more resistance genes or loci, looking for a sustainable viticulture in the future is one of the most commonly used practices. Some of these plants are used as rootstocks, since some plants have substantial levels of resistance, but are not of commercial interest. However, a few *Vitis* species have been found to be resistant to these pathogens, nevertheless, the source of this resistance has not been completely identified.

The use of biotechnology, for example in the generation of genetically modified plants carrying these resistance elements, could be an effective strategy accepted in the future in agriculture. New molecular approaches are being developed, suggesting that the editing of *cis*-regulatory elements (CRE) in defense related genes could be a new target in order to increase pathogen resistance (Swinnen et al., 2016). All the aforementioned are excellent candidates to use with novel genome editing tools, such as CRISPR/Cas9being introduced for crop plants with promising results. Big data approaches are also being develop for plant-pathogen interactions in order to facilitate information accessibility for crop developers. Examples as the PhytoPath (Pedro et al., 2016), which provides access to all plant pathogen genomes submitted to the International Nucleotide Sequence Database Nucleotide Consortium are of great significance. It would be an advantage to include this type of approaches for the development of resistant grape cultivars with minimal genome modifications.

At present, integrated disease management, using combined methods to eliminate or reduce pathogen threats and negative impact on the environment and health is a partially effective, economical, and sustainable way to cope with these problems.

CONCLUDING REMARKS

Understanding plant responses to different kind of pathogens may show some light on the co-evolution of plant and pathogen strategies, and their impact on resistance or susceptibility to infections.

The study of plant–pathogen interactions in crop plants, like grapevine, is essential to understand how pathogens infect the plant and how plant defenses are activated and enhanced. The understanding of these processes allows looking for biotechnological applications to reduce the economic losses associated to these microorganisms. It is crucial to reduce the handling and treatment with chemical agents, but it is necessary to go deeply into the molecular processes that underlie these interactions in order to achieve this.

AUTHOR CONTRIBUTIONS

GA, RS, MA, DM, CN, and PA-J designed the work; made the work drafting, did data collection and reviewed it critically for the final version. RS designed the figure.

ACKNOWLEDGMENTS

This work was partially supported by the National Commission for Science and Technology CONICYT (FONDECYT grant number 1150220) and the Millennium Nucleus for Plant Synthetic Biology and Systems Biology NC130030. GA was supported by FONDECYT postdoctoral grant number 3140324. We thank Dr. F. Restovic (Departamento de Genética Molecular y Microbiología, Facultad de Ciencias Biológicas, Pontificia Universidad Católica de Chile) for helping with the final manuscript revision.

REFERENCES

1. Adams M. J., Antoniw J. F., Bar-Joseph M., Brunt A. A., Candresse T., Foster G. D., et al. (2004).The new plant virus family Flexiviridae and assessment of molecular criteria for species demarcation.*Arch. Virol.* 149 1045–1060. 10.1007/s00705-004-0304-0
2. Agudelo-Romero P., Erban A., Rego C., Carbonell-Bejerano P., Nascimento T., Sousa L., et al. (2015). Transcriptome and metabolome reprogramming in *Vitis vinifera* cv. Trincadeira berries upon infection with *Botrytis cinerea*. *J. Exp. Bot.* 66 1769–1785. 10.1093/jxb/eru517
3. Al Rwahnih M., Dave A., Anderson M. M., Rowhani A., Uyemoto J. K., Sudarshana M. R. (2013).Association of a DNA virus with grapevines affected by red blotch disease in California.*Phytopathology* 103 1069–1076. 10.1094/PHYTO-10-12-0253-R
4. Al Rwahnih M., Dolja V. V., Daubert S., Koonin E. V., Rowhani A. (2012a). Genomic and biological analysis of *Grapevine leafroll-associated virus* 7 reveals a possible new genus within the family Closteroviridae. *Virus Res.* 163 302–309. 10.1016/j.virusres.2011.10.018
5. Al Rwahnih M., Sudarshana M. R., Uyemoto J. K., Rowhani A. (2012b). Complete genome sequence of a novel vitivirus isolated from grapevine. *J. Virol.* 86 9545 10.1128/JVI.01444-12
6. Alabi O. J., Al Rwahnih M., Mekuria T. A., Naidu R. A. (2014). Genetic diversity of Grapevine virus A in Washington and California vineyards. *Phytopathology* 104 548–560. 10.1094/PHYTO-06-13-0179-R
7. Almeida R. P., Daane K. M., Bell V. A., Blaisdell G. K., Cooper M. L., Herrbach E., et al. (2013).Ecology and management of grapevine leafroll disease. *Front. Microbiol.* 4:9410.3389/fmicb.2013.00094
8. Alvarado V., Scholthof H. B. (2009). Plant responses against invasive nucleic acids: RNA silencing and its suppression by plant viral pathogens. *Semin. Cell Dev. Biol.* 20 1032–1040. 10.1016/j.semcdb.2009.06.001
9. Amari K., Boutant E., Hofmann C., Schmitt-Keichinger C., Fernandez-Calvino L., Didier P., et al. (2010). A family of plasmodesmal proteins with receptor-like properties for plant viral movement proteins. *PLoS Pathog.* 6:e1001119 10.1371/journal.ppat.1001119
10. Amari K., Lerich A., Schmitt-Keichinger C., Dolja V. V., Ritzenthaler C. (2011). Tubule-guided cell-to-cell movement of a plant virus requires class XI myosin motors. *PLoS Pathog.* 7:e100232710.1371/journal.ppat.1002327
11. Amrine K. C., Blanco-Ulate B., Riaz S., Pap D., Jones L., Figueroa-Balderas R., et al. (2015).Comparative transcriptomics of Central Asian *Vitis vinifera* accessions reveals distinct defense strategies against powdery mildew. *Hortic. Res.* 2 15037 10.1038/hortres.2015.37
12. Anand A., Uppalapati S. R., Ryu C. M., Allen S. N., Kang L., Tang Y., et al. (2008). Salicylic acid and systemic acquired resistance play a role in attenuating crown gall disease caused by*Agrobacterium tumefaciens*. *Plant Physiol.* 146 703–715. 10.1104/pp.107.111302
13. Anandalakshmi R., Pruss G. J., Ge X., Marathe R., Mallory A. C., Smith T. H., et al. (1998). A viral suppressor of gene silencing in plants. *Proc. Natl. Acad.*

Sci. U.S.A. 95 13079–13084. 10.1073/pnas.95.22.13079

14. Aziz A., Poinssot B., Daire X., Adrian M., Bezier A., Lambert B., et al. (2003). Laminarin elicits defense responses in grapevine and induces protection against *Botrytis cinerea* and *Plasmopara viticola*. *Mol. Plant Microbe Interact.* 16 1118–1128. 10.1094/MPMI.2003.16.12.1118
15. Barba P., Cadle-Davidson L., Harriman J., Glaubitz J. C., Brooks S., Hyma K., et al. (2014).Grapevine powdery mildew resistance and susceptibility loci identified on a high-resolution SNP map.*Theor. Appl. Genet.* 127 73–84. 10.1007/s00122-013-2202-x
16. Barker C. L., Donald T., Pauquet J., Ratnaparkhe M. B., Bouquet A., Adam-Blondon A. F., et al. (2005). Genetic and physical mapping of the grapevine powdery mildew resistance gene, Run1, using a bacterial artificial chromosome library. *Theor. Appl. Genet.* 111 370–377. 10.1007/s00122-005-2030-8
17. Basso M. F., da Silva J. C., Fajardo T. V., Fontes E. P., Zerbini F. M. (2015). A novel, highly divergent ssDNA virus identified in Brazil infecting apple, pear and grapevine. *Virus Res.* 210 27–33. 10.1016/j.virusres.2015.07.005
18. Bellin D., Peressotti E., Merdinoglu D., Wiedemann-Merdinoglu S., Adam-Blondon A. F., Cipriani G., et al. (2009). Resistance to *Plasmopara viticola* in grapevine 'Bianca' is controlled by a major dominant gene causing localised necrosis at the infection site. *Theor. Appl. Genet.* 120 163–176. 10.1007/s00122-009-1167-2
19. Blanc S., Wiedemann-Merdinoglu S., Dumas V., Mestre P., Merdinoglu D. (2012). A reference genetic map of *Muscadinia rotundifolia* and identification of Ren5, a new major locus for resistance to grapevine powdery mildew. *Theor. Appl. Genet.* 125 1663–1675. 10.1007/s00122-012-1942-3
20. Blanco-Ulate B., Morales-Cruz A., Amrine K. C., Labavitch J. M., Powell A. L., Cantu D. (2014).Genome-wide transcriptional profiling of *Botrytis cinerea* genes targeting plant cell walls during infections of different hosts. *Front. Plant Sci.* 5:435 10.3389/fpls.2014.00435
21. Blasi P., Blanc S., Wiedemann-Merdinoglu S., Prado E., Ruhl E. H., Mestre P., et al. (2011).Construction of a reference linkage map of *Vitis amurensis* and genetic mapping of Rpv8, a locus conferring resistance to grapevine downy mildew. *Theor. Appl. Genet.* 123 43–53. 10.1007/s00122-011-1565-0
22. Bonavia M., Digiaro M., Boscia D., Boari A., Bottalico G., Savino V., et al. (1996). Studies on "corky rugose wood" of grapevine and on the diagnosis of grapevine virus B. *Vitis* 35 53–58.
23. Burgyan J., Havelda Z. (2011). Viral suppressors of RNA silencing. *Trends Plant Sci.* 16 265–272. 10.1016/j.tplants.2011.02.010
24. Burr T. J., Bazzi C., Süle S., Otten L. (1998). Crown gall of grape: biology of *Agrobacterium vitis* and the development of disease control strategies. *Plant Dis.* 82 1288–1297. 10.1094/PDIS.1998.82.12.1288
25. Burr T. J., Otten L. (1999). CROWN GALL OF GRAPE: biology and disease management. *Annu. Rev. Phytopathol.* 37 53–80. 10.1146/annurev.phyto.37.1.53
26. Burruano S. (2000). The life-cycle of *Plasmopara viticola*, cause of downy mildew of vine. *Micologist*14 179–182. 10.1016/S0269-915X(00)80040-3
27. Calonnec A., Cartolaro P., Poupot C., Dubourdieu D., Darriet P. (2004). Effects of *Uncinula necator*on the yield and quality of grapes (*Vitis vinifera*) and wine. *Plant Pathol.* 53 434–445. 10.1111/j.0032-0862.2004.01016.x
28. Calvo-Garrido C., Vinas I., Elmer P. A., Usall J., Teixido N. (2014). Suppression of *Botrytis cinerea*on necrotic grapevine tissues by early-season applications of natural products and biological control agents. *Pest Manag. Sci.* 70 595–602. 10.1002/ps.3587
29. Carrington J. C., Kasschau K. D., Mahajan S. K., Schaad M. C. (1996). Cell-to-cell and long-distance transport of viruses in plants. *Plant Cell* 8 1669–1681. 10.1105/tpc.8.10.1669
30. Casagrande K., Falginella L., Castellarin S. D., Testolin R., Di Gaspero G. (2011). Defence responses in Rpv3-dependent resistance to grapevine downy mildew. *Planta* 234 1097–1109. 10.1007/s00425-011-1461-5
31. Chiba M., Reed J. C., Prokhnevsky A. I., Chapman E. J., Mawassi M., Koonin E. V., et al. (2006).Diverse suppressors of RNA silencing enhance agroinfection by a viral replicon. *Virology* 346 7–14. 10.1016/j.virol.2005.09.068
32. Choi Y. J., Yun H. K., Park K. S., Noh J. H., Heo Y. Y., Kim S. H., et al. (2010). Transcriptional profiling of ESTs responsive to *Rhizobium vitis* from 'Tamnara' grapevines (*Vitis* sp.). *J. Plant Physiol.*167 1084–1092. 10.1016/j.jplph.2010.02.005
33. Chong J., Le Henanff G., Bertsch C., Walter B. (2008). Identification, expression analysis and characterization of defense and signaling genes in *Vitis vinifera*. *Plant Physiol. Biochem.* 46 469–481. 10.1016/j.plaphy.2007.09.010
34. Clement B., Pollmann S., Weiler E., Urbanczyk-Wochniak E., Otten L. (2006). The *Agrobacterium vitis* T-6b oncoprotein induces auxin-independent cell expansion in tobacco. *Plant J.* 45 1017–1027. 10.1111/j.1365-313X.2006.02663.x
35. Coleman C., Copetti D., Cipriani G., Hoffmann S., Kozma P., Kovacs L., et al. (2009). The powdery mildew resistance gene REN1 co-segregates with an NBS-LRR gene cluster in two Central Asian grapevines. *BMC Genet.* 10:89 10.1186/1471-2156-10-89
36. Consonni C., Humphry M. E., Hartmann H. A., Livaja M., Durner J., Westphal L., et al. (2006).Conserved requirement for a plant host cell protein in powdery mildew pathogenesis. *Nat. Genet.* 38716–720. 10.1038/ng1806
37. Cotoras M., Garcia C., Mendoza L. (2009). *Botrytis cinerea* isolates collected from grapes present different requirements for conidia germination. *Mycologia* 101 287–295. 10.3852/08-012
38. Dadakova K., Havelkova M., Kurkova B., Tlolkova I., Kasparovsky T., Zdrahal Z., et al. (2015).Proteome and

transcript analysis of *Vitis vinifera* cell cultures subjected to *Botrytis cinerea* infection. *J. Proteomics* 119 143–153. 10.1016/j.jprot.2015.02.001

39. Dai R., Ge H., Howard S., Qiu W. (2012). Transcriptional expression of Stilbene synthase genes are regulated developmentally and differentially in response to powdery mildew in Norton and Cabernet Sauvignon grapevine. *Plant Sci.* 197 70–76. 10.1016/j.plantsci.2012.09.004
40. Dalbó M., Ye G., Weeden N., Wilcox W., Reisch B. (2001). Marker-assisted selection for powdery mildew resistance in grapes. *J. Am. Soc. Hortic. Sci.* 126 183–189.
41. Davis M. J., Purcell A. H., Thomson S. V. (1978). Pierce's disease of grapevines: isolation of the causal bacterium. *Science* 199 75–77. 10.1126/science.199.4324.75
42. de Carvalho F., Gheysen G., Kushnir S., Van Montagu M., Inze D., Castresana C. (1992).Suppression of beta-1,3-glucanase transgene expression in homozygous plants. *EMBO J.* 11 2595–2602.
43. De Lorenzo G., D'Ovidio R., Cervone F. (2001). The role of polygalacturonase-inhibiting proteins (PGIPs) in defense against pathogenic fungi. *Annu. Rev. Phytopathol.* 39 313–335. 10.1146/annurev.phyto.39.1.313
44. Dean R., Van Kan J. A., Pretorius Z. A., Hammond-Kosack K. E., Di Pietro A., Spanu P. D., et al. (2012). The Top 10 fungal pathogens in molecular plant pathology. *Mol. Plant Pathol.* 13 414–430. 10.1111/j.1364-3703.2011.00783.x
45. Derckel J. P., Baillieul F., Manteau S., Audran J. C., Haye B., Lambert B., et al. (1999). Differential induction of grapevine defenses by two strains of *Botrytis cinerea*. *Phytopathology* 89 197–203. 10.1094/PHYTO.1999.89.3.197
46. Di Gaspero G., Copetti D., Coleman C., Castellarin S. D., Eibach R., Kozma P., et al. (2012).Selective sweep at the Rpv3 locus during grapevine breeding for downy mildew resistance. *Theor. Appl. Genet.* 124 277–286. 10.1007/s00122-011-1703-8
47. Ding S. W., Voinnet O. (2007). Antiviral immunity directed by small RNAs. *Cell* 130 413–426. 10.1016/j.cell.2007.07.039
48. Dorssers L., van der Krol S., van der Meer J., van Kammen A., Zabel P. (1984). Purification of cowpea mosaic virus RNA replication complex: identification of a virus-encoded 110,000-dalton polypeptide responsible for RNA chain elongation. *Proc. Natl. Acad. Sci. U.S.A.* 81 1951–1955. 10.1073/pnas.81.7.1951
49. El Gueddari N., Rauchhaus U., Moerschbacher B., Deising H. (2002). Developmentally regulated conversion of surface-exposed chitin to chitosan in cell walls of plant pathogenic fungi. *New Phytol.*156 103–112. 10.1046/j.1469-8137.2002.00487.x
50. Escobar M. A., Dandekar A. M. (2003). *Agrobacterium tumefaciens* as an agent of disease. *Trends Plant Sci.* 8 380–386. 10.1016/S1360-1385(03)00162-6
51. Espinoza C., Medina C., Somerville S., Arce-Johnson P. (2007a). Senescence-associated genes induced during compatible viral interactions with grapevine and *Arabidopsis*. *J. Exp. Bot.* 58 3197–3212. 10.1093/jxb/erm165
52. Espinoza C., Vega A., Medina C., Schlauch K., Cramer G., Arce-Johnson P. (2007b). Gene expression associated with compatible viral diseases in grapevine cultivars. *Funct. Integr. Genomics* 795–110. 10.1007/s10142-006-0031-6
53. Feechan A., Anderson C., Torregrosa L., Jermakow A., Mestre P., Wiedemann-Merdinoglu S., et al. (2013). Genetic dissection of a TIR-NB-LRR locus from the wild North American grapevine species*Muscadinia rotundifolia* identifies paralogous genes conferring resistance to major fungal and oomycete pathogens in cultivated grapevine. *Plant J.* 76 661–674. 10.1111/tpj.12327
54. Feechan A., Jermakow A. M., Dry I. B. (2009). Grapevine MLO candidates required for powdery mildew pathogenicity? *Plant Signal. Behav.* 4 522–523. 10.1071/FP08173
55. Feechan A., Kabbara S., Dry I. B. (2011). Mechanisms of powdery mildew resistance in the Vitaceae family. *Mol. Plant Pathol.* 12 263–274. 10.1111/j.1364-3703.2010.00668.x
56. Freialdenhoven A., Peterhansel C., Kurth J., Kreuzaler F., Schulze-Lefert P. (1996). Identification of genes required for the function of non-race-specific mlo resistance to powdery mildew in barley. *Plant Cell* 8 5–14. 10.1105/tpc.8.1.5
57. Fry S. M., Milholland D. (1990). Response of resistant, tolerant, and susceptible grapevine tissue to invasion by the Pierce's disease bacterium *Xylella fastidiosa*. *Phytopathology* 80 66–69. 10.1094/Phyto-80-66
58. Fung R. W., Gonzalo M., Fekete C., Kovacs L. G., He Y., Marsh E., et al. (2008). Powdery mildew induces defense-oriented reprogramming of the transcriptome in a susceptible but not in a resistant grapevine. *Plant Physiol.* 146 236–249. 10.1104/pp.107.108712
59. Gabler F. M., Smilanick J. L., Mansour M., Ramming D. W., Mackey B. E. (2003). Correlations of morphological, anatomical, and chemical features of grape berries with resistance to *Botrytis cinerea*.*Phytopathology* 93 1263–1273. 10.1094/PHYTO.2003.93.10.1263
60. Gadoury D. M., Cadle-Davidson L., Wilcox W. F., Dry I. B., Seem R. C., Milgroom M. G. (2012).Grapevine powdery mildew (*Erysiphe necator*): a fascinating system for the study of the biology, ecology and epidemiology of an obligate biotroph. *Mol. Plant Pathol.* 13 1–16. 10.1111/j.1364-3703.2011.00728.x
61. Gadoury D. M., Seem R. C., Ficke A., Wilcox W. F. (2001). The epidemiology of powdery mildew on concord grapes. *Phytopathology* 91 948–955. 10.1094/PHYTO.2001.91.10.948
62. Gambino G., Cuozzo D., Fasoli M., Pagliarani C., Vitali M., Boccacci P., et al. (2012). Co-evolution between *Grapevine rupestris* stem pitting-associated

virus and *Vitis vinifera* L. leads to decreased defence responses and increased transcription of genes related to photosynthesis. *J. Exp. Bot.* 63 5919–5933. 10.1093/jxb/ers244

63. Gao F., Dai R., Pike S. M., Qiu W., Gassmann W. (2014). Functions of EDS1-like and PAD4 genes in grapevine defenses against powdery mildew. *Plant Mol. Biol.* 86 381–393. 10.1007/s11103-014-0235-4

64. Gessler C., Pertot I., Perazzolli M. (2011). *Plasmopara viticola*: a review of knowledge on downy mildew of grapevine and effective disease management. *Phytopathol. Mediterr.* 50 3–44.

65. Ghanem-Sabanadzovic N. A., Sabanadzovic S., Gugerli P., Rowhani A. (2012). Genome organization, serology and phylogeny of Grapevine leafroll-associated viruses 4 and 6: taxonomic implications. *Virus Res.* 163 120–128. 10.1016/j.virusres.2011.09.001

66. Ghanem-Sabanadzovic N. A., Sabanadzovic S., Uyemoto J. K., Golino D., Rowhani A. (2010). A putative new *Ampelovirus* associated with grapevine leafroll disease. *Arch. Virol.* 155 1871–1876. 10.1007/s00705-010-0773-2

67. Giacomelli L., Nanni V., Lenzi L., Zhuang J., Dalla Serra M., Banfield M. J., et al. (2012).Identification and characterization of the defensin-like gene family of grapevine. *Mol. Plant Microbe Interact.* 25 1118–1131. 10.1094/MPMI-12-11-0323

68. Girault T., Francois J., Rogniaux H., Pascal S., Delrot S., Coutos-Thevenot P., et al. (2008).Exogenous application of a lipid transfer protein-jasmonic acid complex induces protection of grapevine towards infection by *Botrytis cinerea*. *Plant Physiol. Biochem.* 46 140–149. 10.1016/j.plaphy.2007.10.005

69. Glazebrook J. (2005). Contrasting mechanisms of defense against biotrophic and necrotrophic pathogens. *Annu. Rev. Phytopathol.* 43 205–227. 10.1146/annurev.phyto.43.040204.135923

70. Goszczynski D. E., du Preez J., Burger J. T. (2008). Molecular divergence of *Grapevine* virus A (GVA) variants associated with Shiraz disease in South Africa. *Virus Res.* 138 105–110. 10.1016/j.virusres.2008.08.014

71. Gourgues M., Brunet-Simon A., Lebrun M. H., Levis C. (2004). The tetraspanin BcPls1 is required for appressorium-mediated penetration of *Botrytis cinerea* into host plant leaves. *Mol. Microbiol.* 51619–629. 10.1046/j.1365-2958.2003.03866.x

72. Guillier C., Gamm M., Lucchi G., Truntzer C., Pecqueur D., Ducoroy P., et al. (2015). Toward the identification of two glycoproteins involved in the stomatal deregulation of downy mildew-infected grapevine leaves. *Mol. Plant Microbe Interact.* 28 1227–1236. 10.1094/MPMI-05-15-0115-R

73. Gutha L. R., Casassa L. F., Harbertson J. F., Naidu R. A. (2010). Modulation of flavonoid biosynthetic pathway genes and anthocyanins due to virus infection in grapevine (*Vitis vinifera* L.) leaves. *BMC Plant Biol.* 10:187 10.1186/1471-2229-10-187

74. Hammond S. M., Caudy A. A., Hannon G. J. (2001). Post-transcriptional gene silencing by double-stranded RNA. *Nat. Rev. Genet.* 2 110–119. 10.1038/35052556

75. Harries P. A., Schoelz J. E., Nelson R. S. (2010). Intracellular transport of viruses and their components: utilizing the cytoskeleton and membrane highways. *Mol. Plant Microbe Interact.* 231381–1393. 10.1094/MPMI-05-10-0121

76. Hayward A. P., Tsao J., Dinesh-Kumar S. P. (2009). Autophagy and plant innate immunity: defense through degradation. *Semin. Cell Dev. Biol.* 20 1041–1047. 10.1016/j.semcdb.2009.04.012

77. Hipper C., Brault V., Ziegler-Graff V., Revers F. (2013). Viral and cellular factors involved in Phloem transport of plant viruses. *Front. Plant Sci.* 4:154 10.3389/fpls.2013.00154

78. Hoffmann S., Di Gaspero G., Kovacs L., Howard S., Kiss E., Galbacs Z., et al. (2008). Resistance to*Erysiphe necator* in the grapevine 'Kishmish vatkana' is controlled by a single locus through restriction of hyphal growth. *Theor. Appl. Genet.* 116 427–438. 10.1007/s00122-007-0680-4

79. Hong Y. S., Martinez A., Liger-Belair G., Jeandet P., Nuzillard J. M., Cilindre C. (2012).Metabolomics reveals simultaneous influences of plant defence system and fungal growth in *Botrytis cinerea*-infected *Vitis vinifera* cv. Chardonnay berries. *J. Exp. Bot.* 63 5773–5785. 10.1093/jxb/ers228

80. Jelly N. S., Schellenbaum P., Walter B., Maillot P. (2012). Transient expression of artificial microRNAs targeting *Grapevine fanleaf virus* and evidence for RNA silencing in grapevine somatic embryos. *Transgenic Res.* 21 1319–1327. 10.1007/s11248-012-9611-5

81. Jia H., Zhang C., Pervaiz T., Zhao P., Liu Z., Wang B., et al. (2016). Jasmonic acid involves in grape fruit ripening and resistant against *Botrytis cinerea*. *Funct. Integr. Genomics* 16 79–94. 10.1007/s10142-015-0468-6

82. Jones J. D., Dangl J. L. (2006). The plant immune system. *Nature* 444 323–329. 10.1038/nature05286

83. Jones L., Riaz S., Morales-Cruz A., Amrine K. C., McGuire B., Gubler W. D., et al. (2014). Adaptive genomic structural variation in the grape powdery mildew pathogen, *Erysiphe necator*. *BMC Genomics* 15:1081 10.1186/1471-2164-15-1081

84. Kamoun S., Furzer O., Jones J. D., Judelson H. S., Ali G. S., Dalio R. J., et al. (2015). The Top 10 oomycete pathogens in molecular plant pathology. *Mol. Plant Pathol.* 16 413–434. 10.1111/mpp.12190

85. Kars I., Krooshof G. H., Wagemakers L., Joosten R., Benen J. A., van Kan J. A. (2005a). Necrotizing activity of five *Botrytis cinerea* endopolygalacturonases produced in *Pichia pastoris*. *Plant J.* 43 213–225. 10.1111/j.1365-313X.2005.02436.x

86. Kars I., McCalman M., Wagemakers L., Ja V. A. N. K. (2005b). Functional analysis of *Botrytis cinerea* pectin methylesterase genes by PCR-based targeted mutagenesis: Bcpme1 and Bcpme2 are dispensable for virulence of strain B05.10. *Mol. Plant Pathol.* 6 641–652. 10.1111/j.1364-3703.2005.00312.x

87. Kawakami S., Watanabe Y., Beachy R. N. (2004). *Tobacco mosaic virus* infection spreads cell to cell as intact replication complexes. *Proc. Natl. Acad. Sci. U.S.A.* 101 6291–6296. 10.1073/pnas.0401221101

88. Kelloniemi J., Trouvelot S., Heloir M. C., Simon A., Dalmais B., Frettinger P., et al. (2015). Analysis of the molecular dialogue between gray mold (*Botrytis cinerea*) and grapevine (*Vitis vinifera*) reveals a clear shift in defense mechanisms during berry ripening. *Mol. Plant Microbe Interact.* 28 1167–1180. 10.1094/MPMI-02-15-0039-R

89. Komatsu K., Hashimoto M., Ozeki J., Yamaji Y., Maejima K., Senshu H., et al. (2010). Viral-induced systemic necrosis in plants involves both programmed cell death and the inhibition of viral multiplication, which are regulated by independent pathways. *Mol. Plant Microbe Interact.* 23 283–293. 10.1094/MPMI-23-3-0283

90. Krastanova S., Perrin M., Barbier P., Demangeat G., Cornuet P., Bardonnet N., et al. (1995).Transformation of grapevine rootstocks with the coat protein gene of grapevine fanleaf nepovirus.*Plant Cell Rep.* 14 550–554. 10.1007/BF00231936

91. Krivanek A. F., Riaz S., Walker M. A. (2006). Identification and molecular mapping of PdR1, a primary resistance gene to Pierce's disease in *Vitis*. *Theor. Appl. Genet.* 112 1125–1131. 10.1007/s00122-006-0214-5

92. Kuczmog A., Galambos A., Horvath S., Matai A., Kozma P., Szegedi E., et al. (2012). Mapping of crown gall resistance locus Rcg1 in grapevine. *Theor. Appl. Genet.* 125 1565–1574. 10.1007/s00122-012-1935-2

93. Laporte C., Vetter G., Loudes A. M., Robinson D. G., Hillmer S., Stussi-Garaud C., et al. (2003).Involvement of the secretory pathway and the cytoskeleton in intracellular targeting and tubule assembly of *Grapevine fanleaf virus* movement protein in tobacco BY-2 cells. *Plant Cell* 15 2058–2075. 10.1105/tpc.013896

94. Le Henanff G., Profizi C., Courteaux B., Rabenoelina F., Gerard C., Clement C., et al. (2013).Grapevine NAC1 transcription factor as a convergent node in developmental processes, abiotic stresses, and necrotrophic/biotrophic pathogen tolerance. *J. Exp. Bot.* 64 4877–4893. 10.1093/jxb/ert277

95. Leisner S. M., Turgeon R. (1993). Movement of virus and photoassimilate in the phloem: a comparative analysis.*Bioessays* 15 741–748. 10.1002/bies.950151107

96. Lenzi L., Caruso C., Bianchedi P. L., Pertot I., Perazzolli M. (2016). Laser microdissection of grapevine leaves reveals site-specific regulation of transcriptional response to *Plasmopara viticola.Plant Cell Physiol.* 57 69–81. 10.1093/pcp/pcv166

97. Li F., Ding S. W. (2006). Virus counterdefense: diverse strategies for evading the RNA-silencing immunity. *Annu. Rev. Microbiol.* 60 503–531. 10.1146/annurev.micro.60.080805.142205

98. Li H., Xu Y., Xiao Y., Zhu Z., Xie X., Zhao H., et al. (2010). Expression and functional analysis of two genes encoding transcription factors, VpWRKY1 and VpWRKY2, isolated from Chinese wild*Vitis pseudoreticulata*. *Planta* 232 1325–1337. 10.1007/s00425-010-1258-y

99. Lieth J. H., Meyer M. M., Yeo K. H., Kirkpatrick B. C. (2011). Modeling cold curing of Pierce's disease in *Vitis vinifera* 'Pinot Noir' and 'Cabernet Sauvignon' grapevines in California.*Phytopathology* 101 1492–1500. 10.1094/PHYTO-08-10-0207

100. Lin H., Doddapaneni H., Takahashi Y., Walker M. A. (2007). Comparative analysis of ESTs involved in grape responses to *Xylella fastidiosa* infection. *BMC Plant Biol.* 7:8 10.1186/1471-2229-7-8

101. Liu J. Z., Blancaflor E. B., Nelson R. S. (2005). The *Tobacco mosaic virus* 126-kilodalton protein, a constituent of the virus replication complex, alone or within the complex aligns with and traffics along microfilaments. *Plant Physiol.* 138 1853–1865. 10.1104/pp.105.065722

102. Liu R., Wang L., Zhu J., Chen T., Wang Y., Xu Y. (2015).Histological responses to downy mildew in resistant and susceptible grapevines. *Protoplasma* 252 259–270. 10.1007/s00709-014-0677-1

103. Liu Y. P., Peremyslov V. V., Medina V., Dolja V. V. (2009). Tandem leader proteases of *Grapevine leafroll-associated virus*-2: host-specific functions in the infection cycle. *Virology* 383 291–299. 10.1016/j.virol.2008.09.035

104. Lucas W. J. (2006). Plant viral movement proteins: agents for cell-to-cell trafficking of viral genomes. *Virology* 344 169–184. 10.1016/j.virol.2005.09.026

105. Luis P., Gauthier A., Trouvelot S., Poinssot B., Frettinger P. (2013). Identification of *Plasmopara viticola* genes potentially involved in pathogenesis on grapevine suggests new similarities between oomycetes and true fungi. *Phytopathology* 103 1035–1044. 10.1094/PHYTO-06-12-0121-R

106. Malacarne G., Vrhovsek U., Zulini L., Cestaro A., Stefanini M., Mattivi F., et al. (2011). Resistance to*Plasmopara viticola* in a grapevine segregating population is associated with stilbenoid accumulation and with specific host transcriptional responses. *BMC Plant Biol.* 11:114 10.1186/1471-2229-11-114

107. Maliogka V. I., Dovas C. I., Lotos L., Efthimiou K., Katis N. I. (2009). Complete genome analysis and immunodetection of a member of a novel virus species belonging to the genus *Ampelovirus*. *Arch. Virol.* 154 209–218. 10.1007/s00705-008-0290-8

108. Marsh E., Alvarez S., Hicks L. M., Barbazuk W. B., Qiu W., Kovacs L., et al. (2010). Changes in protein abundance during powdery mildew infection of leaf tissues of Cabernet Sauvignon grapevine (*Vitis vinifera* L.). *Proteomics* 10 2057–2064. 10.1002/pmic.200900712

109. Martelli G. P. (2014). "Virus diseases of grapevine," in *eLS* (Chichester: John Wiley & Sons, Ltd;). 10.1002/9780470015902.a0000766.pub3

110. Martinelli L., Candioli E., Costa D., Minafra A. (2002). Stable insertion and expression of the movement protein gene of Grapevine Virus A (GVA) in grape (*Vitis rupestris* S.). *Vitis* 41 189–193.

111. McHale L., Tan X., Koehl P., Michelmore R. W. (2006). Plant NBS-LRR proteins: adaptable guards. *Genome Biol.* 7 212 10.1186/gb-2006-7-4-212

112. Merdinoglu D., Wiedeman-Merdinoglu S., Coste P., Dumas V., Haetty S., Butterlin G., et al. (2003). "Genetic analysis of Downy Mildew resistance derived from *Muscadinia rotundifolia*," in *Proceedings of the Eighth International Conference on Grape Genetics and Breeding*, Leuven, 451–456.

113. Merz P. R., Moser T., Holl J., Kortekamp A., Buchholz G., Zyprian E., et al. (2015). The transcription factor VvWRKY33 is involved in the regulation of grapevine (*Vitis vinifera*) defense against the oomycete pathogen *Plasmopara viticola*. *Physiol. Plant.* 153 365–380. 10.1111/ppl.12251

114. Mestre P., Piron M. C., Merdinoglu D. (2012). Identification of effector genes from the phytopathogenic Oomycete *Plasmopara viticola* through the analysis of gene expression in germinated zoospores. *Fungal Biol.* 116 825–835. 10.1016/j.funbio.2012.04.016

115. Meyer P., Heidmann I., Niedenhof I. (1993). Differences in DNA-methylation are associated with a paramutation phenomenon in transgenic petunia. *Plant J.* 4 89–100. 10.1046/j.1365-313X.1993.04010089.x

116. Monaghan J., Zipfel C. (2012). Plant pattern recognition receptor complexes at the plasma membrane.*Curr. Opin. Plant Biol.* 15 349–357. 10.1016/j.pbi.2012.05.006

117. Murray G. G., Kosakovsky Pond S. L., Obbard D. J. (2013). Suppressors of RNAi from plant viruses are subject to episodic positive selection. *Proc. Biol. Sci.* 280:20130965 10.1098/rspb.2013.0965

118. Naidu R. A., Maree H. J., Burger J. T. (2015). Grapevine leafroll disease and associated viruses: a unique pathosystem. *Annu. Rev. Phytopathol.* 53 613–634. 10.1146/annurev-phyto-102313-045946

119. Newman K. L., Almeida R. P., Purcell A. H., Lindow S. E. (2003). Use of a green fluorescent strain for analysis of *Xylella fastidiosa* colonization of *Vitis vinifera*. *Appl. Environ. Microbiol.* 69 7319–7327. 10.1128/AEM.69.12.7319-7327.2003

120. Niehl A., Heinlein M. (2011). Cellular pathways for viral transport through plasmodesmata.*Protoplasma* 248 75–99. 10.1007/s00709-010-0246-1

121. Parafati L., Vitale A., Restuccia C., Cirvilleri G. (2015). Biocontrol ability and action mechanism of food-isolated yeast strains against *Botrytis cinerea* causing post-harvest bunch rot of table grape. *Food Microbiol.* 47 85–92. 10.1016/j.fm.2014.11.013

122. Park K. S., Yun H. K., Suh H. S., Jeong S. B., Cho H. M. (2004). Breeding of early season grape cultivar 'Tamnara' grapevines (*Vitis hybrid*) with high quality and disease resistance. *Korean J. Hortic. Sci. Technol.* 22 458–461.

123. Pedro H., Maheswari U., Urban M., Irvine A. G., Cuzick A., McDowall M. D., et al. (2016).PhytoPath: an integrative resource for plant pathogen genomics.*Nucleic Acids Res.* 44 D688–D693. 10.1093/nar/gkv1052

124. Peressotti E., Wiedemann-Merdinoglu S., Delmotte F., Bellin D., Di Gaspero G., Testolin R., et al. (2010). Breakdown of resistance to grapevine downy mildew upon limited deployment of a resistant variety. *BMC Plant Biol.* 10:147 10.1186/1471-2229-10-147

125. Perfect S. E., Green J. R. (2001). Infection structures of biotrophic and hemibiotrophic fungal plant pathogens.*Mol. Plant Pathol.* 2 101–108. 10.1046/j.1364-3703.2001.00055.x

126. Pieterse C. M., Leon-Reyes A., Van der Ent S., Van Wees S. C. (2009). Networking by small-molecule hormones in plant immunity. *Nat. Chem. Biol.* 5 308–316. 10.1038/nchembio.164

127. Pike S., Gao F., Kim M. J., Kim S. H., Schachtman D. P., Gassmann W. (2014). Members of the NPF3 transporter subfamily encode pathogen-inducible nitrate/nitrite transporters in grapevine and*Arabidopsis*. *Plant Cell Physiol.* 55 162–170. 10.1093/pcp/pct167

128. Polesani M., Desario F., Ferrarini A., Zamboni A., Pezzotti M., Kortekamp A., et al. (2008). cDNA-AFLP analysis of plant and pathogen genes expressed in grapevine infected with *Plasmopara viticola.BMC Genomics* 9:142 10.1186/1471-2164-9-142

129. Prosser S. W., Goszczynski D. E., Meng B. (2007). Molecular analysis of double-stranded RNAs reveals complex infection of grapevines with multiple viruses. *Virus Res.* 124 151–159. 10.1016/j.virusres.2006.10.014

130. Qiu W., Feechan A., Dry I. (2015). Current understanding of grapevine defense mechanisms against the biotrophic fungus (*Erysiphe necator*), the causal agent of powdery mildew disease. *Hortic. Res.*2:15020 10.1038/hortres.2015.20

131. Ramming D. W., Gabler F., Smilanick J., Cadle-Davidson M., Barba P., Mahanil S., et al. (2011). A single dominant locus, ren4, confers rapid non-race-specific resistance to grapevine powdery mildew.*Phytopathology* 101 502–508. 10.1094/PHYTO-09-10-0237

132. Riaz S., Krivanek A. F., Xu K., Walker M. A. (2006). Refined mapping of the Pierce's disease resistance locus, PdR1, and Sex on an extended genetic map of *Vitis rupestris* x *V. arizonica*. *Theor. Appl. Genet.* 113 1317–1329. 10.1007/s00122-006-0385-0

133. Riaz S., Tenscher A. C., Ramming D. W., Walker M. A. (2011). Using a limited mapping strategy to identify major QTLs for resistance to grapevine powdery mildew (*Erysiphe necator*) and their use in marker-assisted breeding. *Theor. Appl. Genet.* 122 1059–1073. 10.1007/s00122-010-1511-6

134. Riaz S., Tenscher A. C., Rubin J., Graziani R., Pao S. S., Walker M. A. (2008). Fine-scale genetic mapping of two Pierce's disease resistance loci and a major segregation distortion region on chromosome 14 of grape. *Theor. Appl. Genet.* 117 671–681. 10.1007/s00122-008-0802-7

135. Robert N., Roche K., Lebeau Y., Breda C., Boulay M., Esnault R., et al. (2002). Expression of grapevine chitinase genes in berries and leaves infected by fungal or bacterial pathogens. *Plant Sci.*162 389–400. 10.1016/S0168-9452(01)00576-3

136. Robert-Seilaniantz A., Grant M., Jones J. D. (2011). Hormone crosstalk in plant disease and defense: more than just jasmonate-salicylate antagonism. *Annu. Rev. Phytopathol.* 49 317–343. 10.1146/annurev-phyto-073009-114447

137. Rolke Y., Liu S., Quidde T., Williamson B., Schouten A., Weltring K. M., et al. (2004). Functional analysis of H(2) O(2)-generating systems in *Botrytis cinerea*: the major Cu-Zn-superoxide dismutase (BCSOD1) contributes to virulence on French bean, whereas a glucose oxidase (BCGOD1) is dispensable. *Mol. Plant Pathol.* 5 17–27. 10.1111/j.1364-3703.2004.00201.x

138. Roper M. C., Greve L. C., Warren J. G., Labavitch J. M., Kirkpatrick B. C. (2007). *Xylella fastidiosa*requires polygalacturonase for colonization and pathogenicity in *Vitis vinifera* grapevines. *Mol. Plant Microbe Interact.* 20 411–419. 10.1094/MPMI-20-4-0411

139. Rui O., Hahn M. (2007). The *Botrytis cinerea* hexokinase, Hxk1, but not the glucokinase, Glk1, is required for normal growth and sugar metabolism, and for pathogenicity on fruits. *Microbiology* 1532791–2802. 10.1099/mic.0.2007/006338-0

140. Salomone J. Y., Crouzet P., De Ruffray P., Otten L. (1996). Characterization and distribution of tartrate utilization genes in the grapevine pathogen *Agrobacterium vitis*. *Mol. Plant Microbe Interact.* 9 401–408. 10.1094/MPMI-9-0401

141. Schoelz J. E., Harries P. A., Nelson R. S. (2011). Intracellular transport of plant viruses: finding the door out of the cell. *Mol. Plant* 4 813–831. 10.1093/mp/ssr070

142. Schwander F., Eibach R., Fechter I., Hausmann L., Zyprian E., Topfer R. (2012). Rpv10: a new locus from the Asian *Vitis* gene pool for pyramiding downy mildew resistance loci in grapevine. *Theor. Appl. Genet.* 124 163–176. 10.1007/s00122-011-1695-4

143. Sijen T., Vijn I., Rebocho A., van Blokland R., Roelofs D., Mol J. N., et al. (2001). Transcriptional and posttranscriptional gene silencing are mechanistically related. *Curr. Biol.* 11 436–440. 10.1016/S0960-9822(01)00116-6

144. Silva M. S., De Souza A. A., Takita M. A., Labate C. A., Machado M. A. (2011). Analysis of the biofilm proteome of *Xylella fastidiosa*. *Proteome Sci.* 9 58 10.1186/1477-5956-9-58

145. Soler N., Plomer M., Fagoaga C., Moreno P., Navarro L., Flores R., et al. (2012). Transformation of Mexican lime with an intron-hairpin construct expressing untranslatable versions of the genes coding for the three silencing suppressors of *Citrus tristeza virus* confers complete resistance to the virus.*Plant Biotechnol. J.* 10 597–608. 10.1111/j.1467-7652.2012.00691.x

146. Stange C., Matus J. T., Dominguez C., Perez-Acle T., Arce-Johnson P. (2008). The N-homologue LRR domain adopts a folding which explains the TMV-Cg-induced HR-like response in sensitive tobacco plants. *J. Mol. Graph. Model.* 26 850–860. 10.1016/j.jmgm.2007.05.006

147. Süle S., Mozsar J., Burr T. J. (1994). Crown gall resistance of *Vitis* spp. and grapevine rootstocks. *Phytopathology* 84 607–611. 10.1094/Phyto-84-607

148. Sun Q., Rost T. L., Matthews M. A. (2006). Pruning-induced tylose development in stems of current-year shoots of *Vitis vinifera* (Vitaceae). *Am. J. Bot.* 93 1567–1576. 10.3732/ajb.93.11.1567

149. Swinnen G., Goossens A., Pauwels L. (2016). Lessons from domestication: targeting *cis*-regulatory elements for crop improvement. *Trends Plant Sci.* [Epub ahead of print].10.1016/j.tplants.2016.01.014

150. Szegedi E., Czakó M., Otten L., Koncz C. (1988). Opines in crown gall tumours induced by biotype 3 isolates of *Agrobacterium tumefaciens*. *Physiol. Mol. Plant Pathol.* 32 237–247. 10.1016/S0885-5765(88)80020-1

151. Thaler J. S., Humphrey P. T., Whiteman N. K. (2012). Evolution of jasmonate and salicylate signal crosstalk. *Trends Plant Sci.* 17 260–270. 10.1016/j.tplants.2012.02.010

152. Turner J. G., Ellis C., Devoto A. (2002). The jasmonate signal pathway. *Plant Cell* 14(Suppl.), S153–S164.

153. Van Blokland R., Van der Geest N., Mol J. N., Kooter J. M. (1994). Transgene-mediated suppression of chalcone synthase expression in *Petunia hybrida* results from an increase in RNA turnover. *Plant J.*6 861–877. 10.1046/j.1365-313X.1994.6060861.x

154. Van den Born E., Omelchenko M. V., Bekkelund A., Leihne V., Koonin E. V., Dolja V. V., et al. (2008). Viral AlkB proteins repair RNA damage by oxidative demethylation. *Nucleic Acids Res.* 365451–5461. 10.1093/nar/gkn519

155. van Kan J. A. (2006). Licensed to kill: the lifestyle of a necrotrophic plant pathogen. *Trends Plant Sci.*11 247–253. 10.1016/j.tplants.2006.03.005

156. Vaucheret H., Fagard M. (2001). Transcriptional gene silencing in plants: targets, inducers and regulators. *Trends Genet.* 17 29–35. 10.1016/S0168-9525(00)02166-1

157. Vega A., Gutierrez R. A., Pena-Neira A., Cramer G. R., Arce-Johnson P. (2011). Compatible GLRaV-3 viral infections affect berry ripening decreasing sugar accumulation and anthocyanin biosynthesis in*Vitis vinifera*. *Plant Mol. Biol.* 77 261–274. 10.1007/s11103-011-9807-8

158. Venuti S., Copetti D., Foria S., Falginella L., Hoffmann S., Bellin D., et al. (2013). Historical introgression of the downy mildew resistance gene Rpv12 from the Asian species *Vitis amurensis* into grapevine varieties. *PLoS ONE* 8:e61228 10.1371/journal.pone.0061228

159. Verhagen B., Trotel-Aziz P., Jeandet P., Baillieul F., Aziz A. (2011). Improved resistance against*Botrytis cinerea* by grapevine-associated bacteria that induce a prime oxidative burst and phytoalexin production. *Phytopathology* 101 768–777. 10.1094/PHYTO-09-10-0242

160. Viret O., Keller M., Jaudzems V. G., Cole F. M. (2004). *Botrytis cinerea* infection of grape flowers: light and electron microscopical studies of infection

sites. *Phytopathology* 94 850–857. 10.1094/PHYTO.2004.94.8.850

161. Waigmann E., Lucas W. J., Citovsky V., Zambryski P. (1994). Direct functional assay for *Tobacco mosaic virus* cell-to-cell movement protein and identification of a domain involved in increasing plasmodesmal permeability. *Proc. Natl. Acad. Sci. U.S.A.* 91 1433–1437. 10.1073/pnas.91.4.1433
162. Wan R., Hou X., Wang X., Qu J., Singer S. D., Wang Y., et al. (2015). Resistance evaluation of Chinese wild *Vitis* genotypes against *Botrytis cinerea* and different responses of resistant and susceptible hosts to the infection. *Front. Plant Sci.* 6:854 10.3389/fpls.2015.00854
163. WangK.,LiaoY.,KanJ.,HanL.,ZhengY.(2015).Response of direct or priming defense against*Botrytis cinerea* to methyl jasmonate treatment at different concentrations in grape berries. *Int. J. Food Microbiol.* 194 32–39. 10.1016/j.ijfoodmicro.2014.11.006
164. Welter L., Göktürk-Baydar N., Akkurt M., Maul E., Eibach R., Töpfer R., et al. (2007). Genetic mapping and localization of quantitative trait loci affecting fungal disease resistance and leaf morphology in grapevine (*Vitis vinifera* L). *Mol. Breed.* 20 359–374. 10.1007/s11032-007-9097-7
165. Weng K., Li Z. Q., Liu R. Q., Wang L., Wang Y. J., Xu Y. (2014). Transcriptome of *Erysiphe necator*-infected *Vitis pseudoreticulata* leaves provides insight into grapevine resistance to powdery mildew. *Hortic. Res.* 1 14049 10.1038/hortres.2014.49
166. Wolf S., Deom C. M., Beachy R. N., Lucas W. J. (1989). Movement protein of *Tobacco mosaic virus*modifies plasmodesmatal size exclusion limit. *Science* 246 377–379. 10.1126/science.246.4928.377
167. Xu T. F., Zhao X. C., Jiao Y. T., Wei J. Y., Wang L., Xu Y. (2014). A pathogenesis related protein, VpPR-10.1, from *Vitis pseudoreticulata*: an insight of its mode of antifungal activity. *PLoS ONE*9:e95102 10.1371/journal.pone.0095102
168. Xue B., Ling K. S., Reid C. L., Krastanova S., Sekiya M., Momol E. A., et al. (1999). Transformation of five grape rootstocks with plant virus genes and a virE2 gene from *Agrobacterium tumefaciens*. *In Vitro Cell. Dev. Biol. Plant* 35 226–231. 10.1007/s11627-999-0083-0
169. Ye K., Patel D. J. (2005). RNA silencing suppressor p21 of Beet yellows virus forms an RNA binding octameric ring structure. *Structure* 13 1375–1384. 10.1016/j.str.2005.06.017
170. Yu Y., Zhang Y., Yin L., Lu J. (2012). The mode of host resistance to *Plasmopara viticola* infection of grapevines. *Phytopathology* 102 1094–1101. 10.1094/PHYTO-02-12-0028-R
171. Zabel P., Jongen-Neven I., Van Krammen A. (1976). In vitro replication of cowpea mosaic virus RNA. II. Solubilization of membrane-bound replicase and the partial purification of the solubilized enzyme. *J. Virol.* 17 679–685.
172. Zhou Z., Dell'Orco M., Saldarelli P., Turturo C., Minafra A., Martelli G. P. (2006). Identification of an RNA-silencing suppressor in the genome of Grapevine virus A. *J. Gen. Virol.* 87 2387–2395. 10.1099/vir.0.81893-0

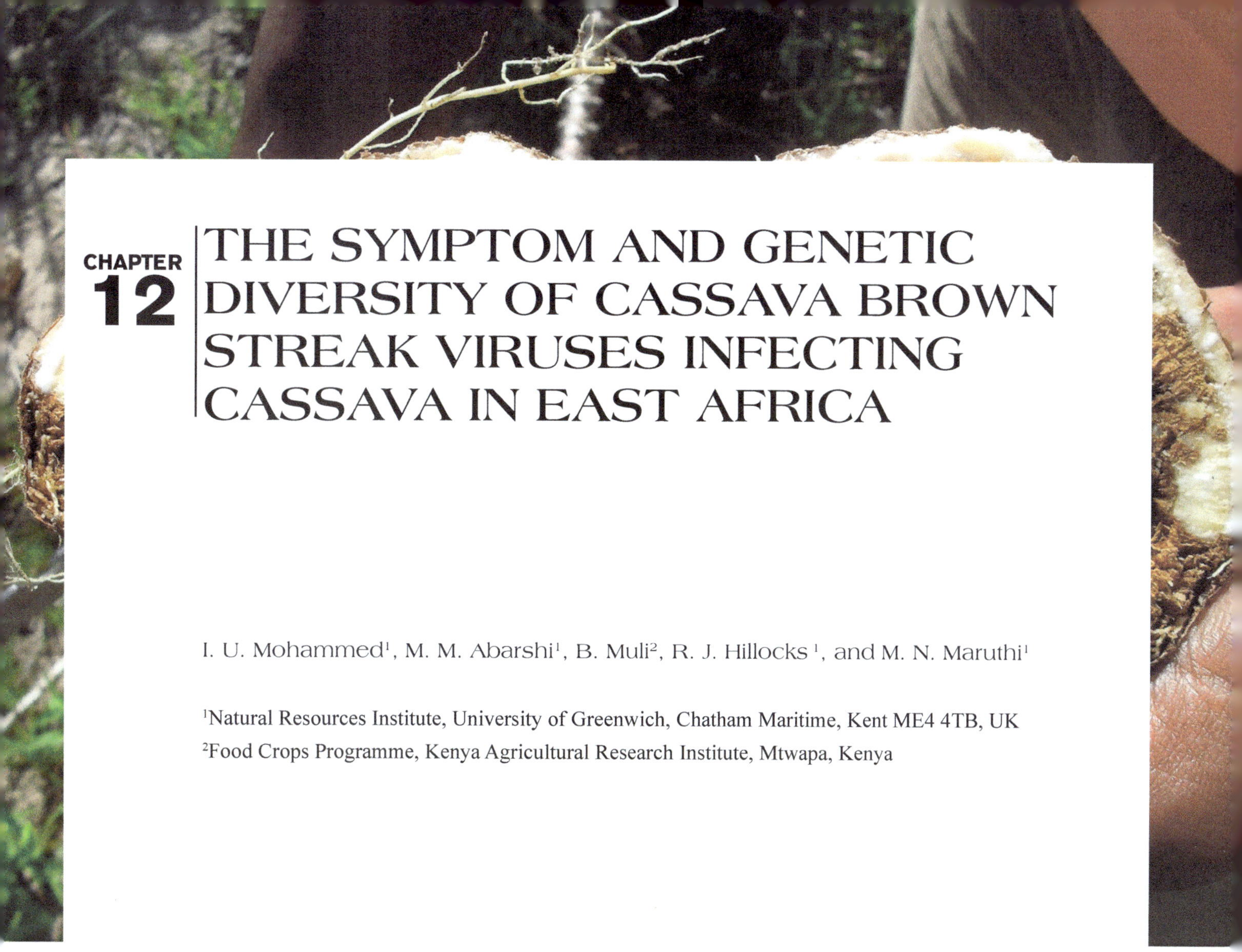

CHAPTER 12

THE SYMPTOM AND GENETIC DIVERSITY OF CASSAVA BROWN STREAK VIRUSES INFECTING CASSAVA IN EAST AFRICA

I. U. Mohammed[1], M. M. Abarshi[1], B. Muli[2], R. J. Hillocks [1], and M. N. Maruthi[1]

[1]Natural Resources Institute, University of Greenwich, Chatham Maritime, Kent ME4 4TB, UK
[2]Food Crops Programme, Kenya Agricultural Research Institute, Mtwapa, Kenya

ABSTRACT

The genetic and symptom diversity of six virus isolates causing cassava brown streak disease (CBSD) in the endemic (Kenya, Mozambique, and Tanzania) and the recently affected epidemic areas (Uganda) of eastern Africa was studied. Five cassava varieties; Albert, Colombian, Ebwanateraka, TMS60444 (all susceptible) and Kiroba (tolerant) were graft inoculated with each isolate. Based on a number of parameters including the severity of leaf and root symptoms, and the extent of virus transmission by grafting, the viruses were classified as either severe or relatively mild. These results were further confirmed by the mechanical inoculation of 13 herbaceous hosts in which the virulent isolates caused plant death in *Nicotiana clevelandii*and *N. benthamiana* whereas the milder isolates did not. Phylogenetic analysis of complete coat protein gene sequences of these isolates together with sequences obtained from 14 other field-collected samples from Kenya and Zanzibar, and reference sequences grouped them into two distinct clusters, representing the two species of cassava brown streak viruses. Put together, these results did not suggest the association of a hypervirulent form of the virus with the current CBSD epidemic in Uganda. Identification of the severe and milder isolates, however, has further implications for disease management and quarantine requirements.

INTRODUCTION

Cassava brown streak disease (CBSD) is endemic in areas along the Indian Ocean coast of eastern Africa, from the northeastern border of Kenya across the Tanzanian border down as far as the Zambezi River in Mozambique, and it was widespread around the shore of Lake Malawi. In the endemic areas, CBSD was confined to altitudes below 1,000 metres above sea level [1–3]. More recently, CBSD has been reported at midaltitude levels (1200–1500 meters above sea levels) in Democratic Republic Congo [4], Uganda [5], and the Lake zone areas of Tanzania [6, 7], which were not considered to be at risk by the disease previously. This is a serious concern because the disease incidences of up to 100% were recorded [8], and in sensitive varieties the disease causes rotting of tubers, reducing both the quality and quantity of tubers available for consumption [1, 2, 9]. A moderate infection by CBSD (10–30% damage to root surface area) decreases the market value of cassava tubers drastically by 90%, fetching under US $5 per tonne, as opposed

Citation: I. U. Mohammed, M. M. Abarshi, B. Muli, R. J. Hillocks, and M. N. Maruthi, The Symptom and Genetic Diversity of Cassava Brown Streak Viruses Infecting Cassava in East Africa, doi:10.1155/2012/795697.

to $55 for fresh healthy cassava root [10]. Severely diseased roots are completely destroyed and unfit for market or family use. Recent estimates indicate that CBSD causes economic losses of up to $100 million annually to the African farmer [11] and these are probably an underestimate, as the disease has since spread into new areas [5, 7]. The disease is now considered to be the most important cause of food insecurity in the coastal and lake zone areas of eastern Africa.

Based on complete genome sequences, CBSD is known to be caused by two distinct virus species. The coastal endemic virus is referred to as *Cassava brown streak virus* (CBSV), and the highland epidemic virus as *Cassava brown streak Uganda virus* (CBSUV). Both species belong to the genus *Ipomovirus*, family*Potyviridae* [12–16], and are transmitted by the whiteflies (*Bemisia tabaci* Gennadius) [17, 18].

The prominent symptoms of CBSD appear on leaves with varying patterns of chlorosis which enabled Nichols [2] to distinguish two types of CBSD isolates. Leaf chlorosis appear in a feathery pattern, first along the margins of the secondary veins, later affecting tertiary veins and may develop into chlorotic blotches. Alternatively, the chlorosis may not be clearly associated with the veins but appear in roughly circular patches between the main veins. There is considerable variation in the expression of foliar symptoms depending on variety, growing conditions (temperature, rainfall, and altitude), age of the plant, and the virus isolate involved in causing the symptoms. Some cultivars show marked foliar symptoms but without or delayed root symptoms (example var. Kiroba), and *vice versa*. Symptoms of the disease are more difficult to recognize in older plants as the lower leaves with prominent symptoms senesce and fall off. New leaves produced from these plants often do not show symptoms, especially at high temperatures. Symptoms can be also transient when a period of active growth produces symptom-free tissues [19]. However, it is difficult to interpret these observations precisely because they have been made in field situations with varying agroclimatic conditions, on cassava varieties with differing virus resistance levels and crop age, and possibly infected with different virus strains, which on their own or in combinations affected symptom development. The main objectives of this study were, therefore, to study symptom diversity of CBSV and CBSUV isolates under uniform environmental conditions, and to identify whether the CBSD epidemic was associated with a severe form of the virus. Accordingly, six CBSD isolates from endemic (Mozambique, Kenya, Tanzania, and Zanzibar) and epidemic areas (Uganda) were each inoculated to five cassava varieties with varying levels of virus tolerance. The effect of each virus on plant growth and symptom development were recorded under standard conditions in a glasshouse. Symptom diversity was also investigated by inoculating 13 species of experimental host-plants. The genetic diversity of the virus isolates was estimated by cloning and sequencing complete coat protein (CP) genes.

MATERIALS AND METHODS

Cassava Varieties and CBSD Isolates

Stem cuttings of five disease-free cassava varieties Ebwanateraka (collected from Uganda) and Albert and Kiroba (both collected from Tanzania) were collected from farmer's fields. Variety Colombian was obtained from the University of Bristol, UK, and TMS60444 from the International Laboratory for Tropical Agricultural Biotechnology (ILTAB), St. Louis, USA. Plants were grown at 28 ± 5°C, 50–60% relative humidity in the quarantine glasshouse at the Natural Resources Institute (NRI), UK, and observed for cassava mosaic disease (CMD) and CBSD symptoms. Plants were virus-indexed using reverse transcription polymerase chain reaction (RT-PCR) tests and the absence of two cassava brown streak viruses (CBSVs) using primers designed in this study (see below) as well as for cassava mosaic begomoviruses (CMB) [20,21]. Symptomless plants in which no virus was detected was further cultivated through the micropropagation of nodal buds using tissue culture techniques [22]. PCR was used on tissue-cultured plants to further confirm the absence of CBSVs and CMBs. The resulting virus-free plants were used in subsequent virus inoculation experiments.

The six CBSD virus isolates used in this study were collected as stem cuttings of unknown cassava varieties in farmer's fields (Table 1, Figure 1) from disease endemic areas in Nampula, Mozambique (CBSV-[MZ:Nam1-1:07]); Naliendele, Tanzania (CBSV-[TZ:Nal3-1:07]); Zanzibar, Tanzania (CBSV-[TZ:Zan6-2:08]); Mwalumba, Kenya (CBSUV-[KE:Mwa16-2:08]); Kibaha, Tanzania (CBSUV-[TZ:Kib10-2:03]), and from the epidemic area of Kabanyoro, Uganda (CBSUV-[UG:Kab4-3:07]) [23, 24]. The identity of the viruses was confirmed using RT-PCR, cloning, and sequencing of the complete CP genes (see below).

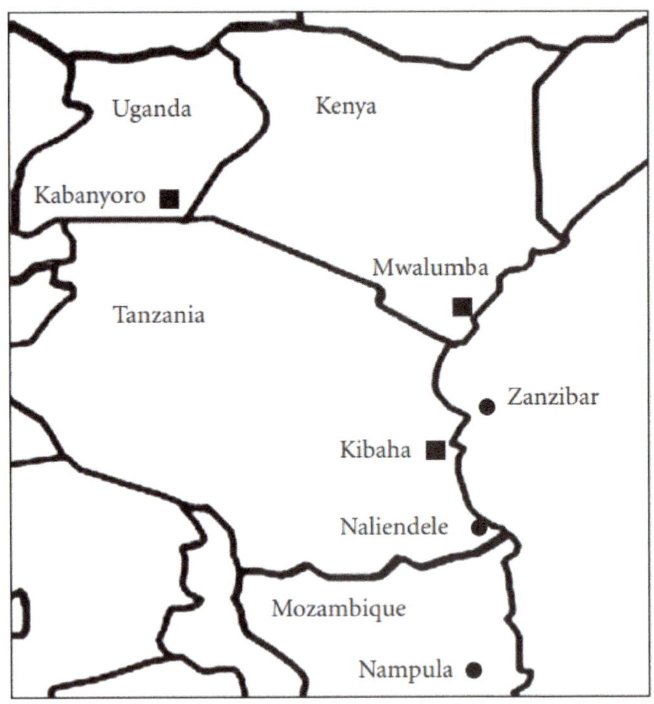

Figure 1: A sketch map of eastern Africa showing the collection sites of CBSV (circles) and CBSUV (squares) isolates used in symptom diversity studies.

Table 1: List of CBSD isolates sequenced in this study; those used in the symptom diversity study are in bold

Isolate name/abbreviation	Place and country of collection	Collection date	Accession numbe
CBSUV-[KE:Den1-2:08]	Denyenye, Kenya	October 2008	HM346937
CBSUV-[KE:Kil18-2:08]	Kilifi, Kenya	October 2008	HM346938
CBSUV-[KE:Kil20-1:08]	Kilifi, Kenya	October 2008	HM346939
CBSUV-[KE:Kil20-3:08]	Kilifi, Kenya	October 2008	HM346940
CBSUV-[KE:Dia3-1:08]	Diani, Kenya	October 2008	HM346941
CBSUV-[KE:Nyu5-4:08]	Nyumbasita, Kenya	October 2008	HM346942
CBSUV-[KE:Shi6-1:08]	Shirazi, Kenya	October 2008	HM346943
CBSUV-[KE:Shi7-1:08]	Shirazi, Kenya	October 2008	HM346944
CBSUV-[KE:Mri8-1:08]	Mrima, Kenya	October 2008	HM346945
CBSUV-[KE:Kik11-5:08]	Kikonde, Kenya	October 2008	HM346946
CBSUV-[KE:Kik10-1:08]	Kikonde, Kenya	October 2008	HM346947
CBSUV-[KE:Mba12-1:08]	Mwabandari, Kenya	October 2008	HM346948
CBSUV-[KE:Mwa16-2:08]	Mwalumba, Kenya	October 2008	HM346949
CBSUV-[KE:Chu21-1:08]	Chumani, Kenya	October 2008	HM346950
CBSUV-[KE:Nam2-1:08]	Namulonge, Uganda	December 2004	HM346951
CBSUV-[UG:Kab4-3:07]	Kabanyoro, Uganda	May 2007	HM346952
CBSV-[MZ:Nam1-1:07]	Nampula, Mozambique	November 2007	HM346953
CBSV-[TZ:Nal3-1:07]	Naliendele, Tanzania	November 2007	HM346954
CBSUV-[TZ:Kib10-2:03]	Kibaha, Tanzania	March 2003	HM346955
CBSV-[TZ:Zan6-2:08]	Zanzibar, Tanzania	October 2008	HM346956
CBSV-[TZ:Zan8-2:08]	Zanzibar, Tanzania	October 2008	HM346957
CBSV-[TZ:Zan7-1:08]	Zanzibar, Tanzania	October 2008	HM346958
CBSV-[TZ:Zan13-1:08]	Zanzibar, Tanzania	October 2008	HM346959
CBSV-[TZ:Zan11-1:08]	Zanzibar, Tanzania	October 2008	HM346960

Graft Inoculation of Virus Isolates and Recording Symptom Severity

The six CBSD isolates were grafted onto two-month-old healthy cassava plants of five cassava varieties: Ebwanateraka, Albert, Kiroba, Colombian, and TMS60444. Plants were kept in constant environment at 28 ± 5°C and 50–60% relative humidity for symptom development. Various parameters were recorded at weekly intervals for determining CBSD symptom severity on cassava leaves and roots, herbaceous hosts (see below), the rate of graft-transmission, the sprouting of the infected cassava cuttings and virus titres in infected plants. Ten cuttings were made for each virus-variety combination (6 viruses × 5 varieties = 30 treatments) which resulted in a total of 300 cuttings. Numbers of cuttings that sprouted was recorded to measure the effect of CBSD on sprouting young cuttings.

Leaf symptoms severity was scored on 3-month-old plants using a five point scale where 1 = no visible CBSD symptoms, 2 = mild foliar symptoms on some leaves, 3 = pronounced foliar symptoms but no die-back, 4 = pronounced foliar symptoms which might include slight die-back of terminal branches, and 5 = severe foliar symptoms and plant die-back [9, 25]. Root symptoms were recorded about 18 months after planting by horizontally cutting the tubers for every 1-2 cm.

Sap-Inoculation of Herbaceous Host-Plants

Thirteen herbaceous species/varieties were tested for their response to CBSVs through sap-inoculations (Table 2). The 0.06 M potassium phosphate buffer was prepared (80.2 mL of 0.6 M K_2HPO_4 + 19.8 mL of 0.6 M KH_2PO_4 + 900 mL of SDW) and pH was adjusted to 7.4 and autoclaved. For each isolate, a cassava leaf showing clear CBSD symptom was collected and ground separately in ~20 mL of the potassium phosphate buffer using pestle and mortar. The leaf debris was separated from the sap by squeezing through sterile muslin cloth. Fully-open young leaves of herbaceous plants were sprinkled with fine 600 mesh carborundum powder and the plant sap was applied gently using a cotton wool pad stroking from petiole to the leaf tip. Virus inoculated leaves were rinsed thoroughly using a jet of water 10 min after the application of sap and the plants were kept for symptom development. Plants inoculated with buffer alone served as controls.

Table 2: Herbaceous hosts inoculated with CBSD isolates

Species/variety	Number of plants infected/ inoculated for each isolate						Total number of infected/ inoculated plants[1] (%)
	CBSUV-			CBSV-			
	[UG:Kab4-3:07]	[KE:Mwa16-2:08]	[TZ:Kib10-2:03]	[TZ:Zan6-2:08]	[MZ:Nam1-1:07]	[TZ:Nal3-1:07]	
Chenopodium quinoa	0/10	0/10	0/10	0/10	0/10	0/10	0/60 (0.0)
Cucurbita maxima	0/10	0/10	0/10	0/10	0/10	0/10	0/60 (0.0)
Datura metel	0/10	0/10	0/10	0/10	0/10	0/10	0/60 (0.0)
Datura stramonium	4/10	2/10	2/10	3/10	9/10	4/10	24/60 (40.0)
Solanum lycopersicum	0/10	0/10	0/10	0/10	0/10	0/10	0/60 (0.0)
Ipomoea batatas	0/10	0/10	0/10	0/10	0/10	0/10	0/60 (0.0)
Nicotiana benthamiana	40/40	5/40	40/40	20/40	40/40	40/40	185/240 (77.0)
Nicotiana clevelandii	10/10	10/10	10/10	10/10	10/10	10/10	60/60 (100.0)
Nicotiana glutinosa	20/40	13/40	23/40	12/40	37/40	40/40	145/240 (60.0)
Nicotiana hesperis	0/10	0/10	0/10	0/10	0/10	0/10	0/60 (0.0)
Nicotiana tabacum nn	19/20	17/20	20/20	20/20	20/20	20/20	116/120 (96.6)
Nicotiana tabacum NN	10/10	10/10	10/10	7/10	9/10	10/10	56/60 (93.3)
Nicotiana rustica	18/20	17/20	15/20	20/20	20/20	20/20	110/120 (91.7)
Total number of infected/inoculated plants[2] (%)	121/210 (57.6)	74/210 (35.2)	120/210 (57.1)	92/210 (43.8)	145/210 (69.0)	144/210 (68.6)	696/1260 (55.2)

[1]Number of infected plants for each cassava variety.
[2]Number of infected plants for each virus isolate.

Detection of CBSVs and Estimation of Virus Titres

Total nucleic acids were extracted separately from cassava leaves infected with each virus isolate using the modified cetyl trimethyl ammonium bromide (CTAB) method [21, 26, 27]. For the purposes of designing virus-specific primers, the 12 complete sequences of CBSVs that were available in gene bank database European Molecular Biology Laboratory (EMBL) were aligned and the primers were designed to the most conserved regions in the 3′ terminal region of the genome. A single forward degenerate primer CBSVF2 (5′ GGR CCA TAC ATY AAR TGG TT 3′) common to CBSVs was designed in the middle of the conserved HAM1h protein about 250 bases upstream of the 5′ end of CP. The two reverse primers CBSVR7 (5′ CCC TTT GCA AAR CTR AAA TARC 3′) and CBSVR8 (5′ CCA TTR TCT YTC CAM ADC TTC 3′) specific to CBSUV and CBSV, respectively, were designed in the conserved regions of 3′ untranslated region (UTR). Viral cDNAs were prepared from two samples of the graft-inoculated cassava plants for each isolate using the OligodT primer and the RT-PCR was carried out as described by Abarshi et al. [27]. For estimating relative concentrations of virus particles, cDNAs were diluted serially: 10^{-1}, 10^{-2}, 10^{-3}, 10^{-4} and 10^{-5}, and the viral genomes amplified using RT-PCR.

Cloning and Sequencing of the Virus Coat Protein Gene

The CBSVF2 was used in combination with another degenerate primer CBSVR1 (5′ AAY ARA AAG GAT ATG GAG AAA G 3′) to amplify the complete CP of CBSVs following the protocols of Abarshi et al. [27]. CBSVR1 was designed to the conserved region of 3′ UTR and together with CBSVF2, these primers amplified approximately 1,600 bp fragment encompassing the partial HAM1 gene, partial 3′ UTR, and complete CP of CBSVs. The RT-PCR amplicons obtained were cloned into pGEMT Easy vector (Promega, UK) and sequenced. For each sample two clones were sequenced in both directions where possible. Sequences were edited and aligned using the software package MEGA4 [28]. BLAST search analysis was carried out to confirm the identity of the sequences. Maximum parsimony analysis and heuristic search were used to generate the most parsimonious phylogenetic tree. The reliability of the tree was estimated by performing 1,000 bootstrap repetitions. The CP sequences of all CBSD isolates were compared with those reference sequences obtained from the EMBL database (accession numbers: FN434109, FJ185044,FN433931, FN433933, FN433932, FJ039520, FN433930, FN434436 and GQ329864).

CBSV Genetic Diversity in Field Samples

In order to correlate symptom variation observed with the genetic diversity of the viruses in the field, a mini survey was carried out in October 2008 for CBSD in the coastal regions of Kenya north and south of Mombasa city [23]. The survey extended about 120 km to the south towards the Tanzanian border along the main road A14, and to the north of Mombasa along B8 up to the town of Kilifi. Cassava leaves showing CBSD symptoms were collected in farmers' fields neighbouring the highways at every 15–20 km intervals. Five CBSD affected leaf samples were also collected for analysis from farmer fields on the Island of Zanzibar during the same period. Genomes of CBSVs were amplified from field-collected samples using CBSVF2 and CBSVR1 primers and the PCR products were cloned and sequenced as above.

RESULTS

Parameters Measured to Estimate CBSD Symptom Severity on Cassava

Efficiency of Graft Transmission —

Of the 30 plants graft inoculated for each virus and variety combination, all plants inoculated with CBSV-[TZ:Nal3-1:07] and CBSV-[MZ:Nam1-1:07] resulted in infections (Table 3). The efficiency of infection varied for the remaining isolates as only 60% of the plants became infected with the epidemic isolate CBSUV-[UG:Kab4-3:07]. No significant differences were observed amongst the varieties in the rate of CBSD infection, which varied from 77 to 87%.

Table 3: The rate of graft transmission of six CBSD isolates to different cassava varieties

Cassava variety	Number of plants infected/grafted with each virus isolate						Total number of infected/grafted plants[1] (%)
	CBSUV-				CBSV-		
	[UG:Kab4-3:07]	[KE:Mwa16-2:08]	[TZ:Kib10-2:03]	[TZ:Zan6-2:08]	[MZ:Nam1-1:07]	[TZ:Nal3-1:07]	
Albert	4/5	3/5	4/5	4/5	5/5	5/5	25/30 (83.3)
Kiroba	3/5	4/5	4/5	5/5	5/5	5/5	26/30 (86.6)
Ebwanateraka	3/5	4/5	3/5	3/5	5/5	5/5	23/30 (76.6)
Colombian	3/5	4/5	3/5	4/5	5/5	5/5	24/30 (80.0)
TMS 60444	2/5	4/5	3/5	4/5	5/5	5/5	23/30 (76.6)
Total number of infected/grafted plants[2] (%)	15/25 (60.0)	19/25 (76.0)	17/25 (68.0)	20/25 (80.0)	25/25 (100)	25/25 (100)	121/150 (80.7)

[1]Number of infected plants for each cassava variety.
[2]Number of infected plants for each virus isolate.

Sprouting of the Infected Cuttings —

Amongst the isolates, maximum number of cuttings were sprouted from the epidemic isolate CBSUV-[UG:Kab4-3:07] (96%) and the least number of cuttings from CBSV-[TZ:Nal3-1:07] (74%) (Table 4). No significant differences were observed amongst the varieties except for TMS60444 from which only 67% of the cuttings were sprouted.

Table 4: The effects of CBSD infections on the sprouting of cassava stem cuttings

Cassava variety	Number of cuttings sprouted/planted when infected with each isolate[3]						Total number of sprouted/planted cuttings[1] (%)
	CBSUV-				CBSV-		
	[UG:Kab4-3:07]	[KE:Mwa16-2:08]	[TZ:Kib10-2:03]	[TZ:Zan6-2:08]	[MZ:Nam1-1:07]	[TZ:Nal3-1:07]	
Albert	9/10	9/10	8/10	9/10	9/10	9/10	53/60 (88.3)
Kiroba	10/10	8/10	10/10	9/10	8/10	6/10	51/60 (85.0)
Ebwanateraka	10/10	7/10	10/10	8/10	9/10	10/10	54/60 (90.0)
Colombian	10/10	10/10	8/10	10/10	9/10	10/10	57/60 (95.0)
TMS 60444	9/10	10/10	10/10	5/10	4/10	2/10	40/60 (66.6)
Total number of sprouted/planted cuttings[2] (%)	48/50 (96.0)	44/50 (88.0)	46/50 (92.0)	41/50 (82.0)	39/50 (78.0)	37/50 (74.0)	255/300 (85.0)

[1]Number of sprouted and fully grown plants for each cassava variety.
[2]Number of sprouted and fully grown plants for each virus isolate.
[3]All 10 cuttings were obtained from plants infected with viruses and showing typical CBSD symptoms.

Leaf Symptom Severity Scores —

Mean maximum leaf symptom severity score of 3.8 was recorded for CBSV-[MZ:Nam1-1:07] and the mean minimum score of 1.9 for CBSUV-[UG:Kab4-3:07] (Table 5, Figure 2). The symptom severity score for each variety varied. When a multiple comparison ANOVA was carried out, significantly high differences ($P < 0.05$) were observed for virus-variety interactions on symptom score (data not shown).

Figure 2: (a) Typical leaf and root symptoms expressed by the CBSUV-[UG:Kab4-3:07] and CBSV-[MZ:Nam1-1:07] in varieties Albert and Kiroba. (b) Typical symptoms observed on *N. clevelandii* plants 5-6 weeks after inoculated with sap extracted from infected cassava ...

Table 5: Mean symptom severity scores for each CBSD isolate on different cassava varieties (on a 0–5 scale using the procedure of [9])

Cassava variety	Mean symptom severity scores for each virus isolate						Mean symptom severity[1]
	CBSUV-			CBSV-			
	[UG:Kab4-3:07]	[KE:Mwa16-2:08]	[TZ:Kib10-2:03]	[TZ:Zan6-2:08]	[MZ:Nam1-1:07]	[TZ:Nal3-1:07]	
Albert	1.9	2.9	2.2	2.8	4.0	3.9	3.0
Kiroba	1.9	2.0	2.0	2.4	3.0	2.7	2.3
Ebwanateraka	1.9	2.6	2.1	2.8	4.0	4.0	3.0
Colombian	1.9	2.9	2.1	2.9	4.0	4.0	3.0
TMS 60444	2.1	2.9	2.7	3.1	4.0	4.0	3.1
Mean symptom severity[2]	1.9	2.7	2.2	3.0	3.8	3.7	2.8

[1]Mean symptom severity for each variety.
[2]Mean symptom severity for each virus isolate.

Relative Virus Concentrations —

In a serial dilution of viral cDNA from 10^{-1} to 10^{-5} folds, virus was detectable at 10^{-5} dilutions only from CBSV-[MZ:Nam1-1:07] and CBSV-[TZ:Nal3-1:07] isolates. For the remaining four isolates virus was not detectable at 10^{-3} or greater dilutions.

Root Symptoms —

Typical necrosis and dry rotting of infected tubers was recorded for all virus-variety combinations except for Kiroba infected with CBSUV-[UG:Kab4-03:07], which did not produce root symptoms (Figure 2).

CBSD Symptom Phenotype on Cassava

CBSD symptoms were variable but two recognisable patterns emerged, which are associated with the virus species involved.

CBSUV Symptom Pattern —

Initial symptoms of plants infected with CBSUV isolates appeared as faint yellow spots on the affected leaves which later developed into bright yellow patches of usually irregular to occasionally circular shape. The yellow patches were vividly defined especially in susceptible varieties (e.g., Albert, Figure 2). The symptoms were not always associated with veins and not uniformly distributed throughout the leaflet leaving some parts of the leaf unaffected. As the symptoms developed further, most of the symptomatic leaf turned bright yellow while some areas remained green.

CBSV Symptom Pattern —

Initial symptoms of plants infected with CBSV isolates appeared as faint yellow streaks usually along the tertiary veins which later developed into severe chlorosis and feathery yellowing extending to secondary and primary veins. The yellowing of veins was mostly even, spreading uniformly throughout the affected leaf which unlike CBSUV, symptoms did not develop into concentric bright yellow patches (Figure 2). Necrotic spots were seen on sensing leaves which also appeared completely yellow before leaf fall.

CBSD Symptom Severity on Herbaceous Hosts

All six CBSD isolates infected *Datura stramonium, Nicotiana benthamiana, N. clevelandii, N. glutinosa, N. tabacum nn, N. tabacum NN,* and *N. rustica* with varying rates of infection (Table 2). All plants of *N. clevelandii* were infected with each isolate. Most but not all plants of *N. tabacum* nn, *N. tabacum* NN, and*N. rustica* were also infected with each isolate. Time taken for first symptom expression on these hosts varied for each isolate and it depended on the plant species infected. Amongst the isolates, CBSV-[MZ:Nam1-1:07] produced symptoms on all hosts within a week of inoculation, which is closely followed by CBSV-[TZ:Nal3-1:07]. Symptom expression ranged from week 1–4 for the remaining five isolates (data not shown). Of the plant species, *N. clevelandii* was most susceptible, showing symptoms on all plants between weeks 1 to 3.

Symptom severity on herbaceous plants varied especially on *N. clevelandii* and *N. benthamiana.* Plants infected with CBSV-[TZ:Nal3-1:07] and CBSV-[MZ:Nam1-1:07] were severely stunted and subsequently wilted by developing leaf necrosis (Figure 2(b)). Most of these plants died usually within four weeks after virus inoculation. Plants infected with the remaining isolates developed various patterns of chlorosis, vein clearing, leaf malformation, and stunting but not necrosis or death. Symptoms on other hosts also varied but in general included leaf chlorosis, mosaic, and mottling. Local lesions were seen on *N. tabacum* nn, chlorosis/mosaic patterns in *N. tabacum* NN, and vein clearing in *N. benthamiana* by all the isolates.

Detection of CBSVs and Virus Diversity

All six CBSD isolates were detected by RT-PCR using novel primers. CBSV and CBSUV were distinguished using virus-specific primers; CBSVF2 & CBSVR7 and CBSVF2 and CBSUVR8, which specifically amplified CBSV (345 bp) and CBSUV (440 bp), respectively (Figure 3). No amplifications were obtained from RNA extracted from virus-free plants (healthy). PCR products were sequenced to confirm the specificity of the primers to respective viruses (data shown). CBSVs were detected in all five Zanzibar samples from Tanzania and all but two (Mwatundo and Mwajambo) Kenyan samples. The amplified products contained ~1600 nucleotide sequences upon sequencing. Clones from most samples yielded unique consensus sequences, and those with more than one unique sequence are shown in Table 1 (e.g., three distinct sequences were obtained from Kilifi, and two each from Shirazi and Kikonde).

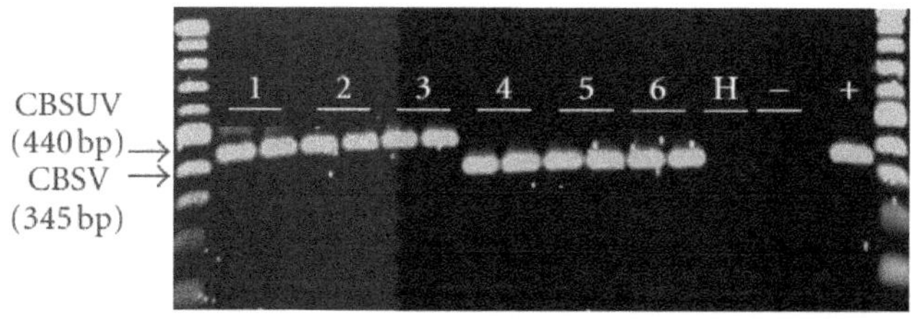

Figure 3: RT-PCR amplification of CBSUV and CBSV genomes using newly designed primers (CBSVF2 in combination with CBSVR7 and CBSVR8). 1 = CBSUV-[UG:Kab4-3:07], 2 = CBSUV-[TZ:Kib10-2:03], 3 = CBSUV-[KE:Mwa16-2:08], 4 = CBSV-[TZ:Zan6-2:08], 5 = CBSV-[MZ:Nam1-1:07], ...

The deduced amino acid (aa) sequences for the complete CP of CBSV and CBSUV isolates consisted of 378 and 367 bases, respectively. These were used to estimate their genetic relationships together with the reference sequences of CBSV, CBSUV, *cucumber vein yellowing virus* (CVYV), *squash vein yellowing virus* (SqVYV), and *sweet potato mild mottle virus* (SPMMV). The most parsimony analysis grouped the CP aa sequences into two major clusters: CBSV and CBSUV (Figure 4). Amongst the field-collected samples, all Kenyan sequences belonged to the CBSUV group and all Zanzibar sequences grouped with CBSV. Members within each group were conserved with an average aa similarity of 95.8% for CBSV group, and a 96.6% similarity for CBSUV group. The two groups were similar to each other by only 80.3%. The epidemic isolate CBSUV-[UG:Kab4-03:07] was highly similar (CP aa similarities between 96.9% to 99.2%) to the previously described isolates from Uganda, Kenya, and Malawi.

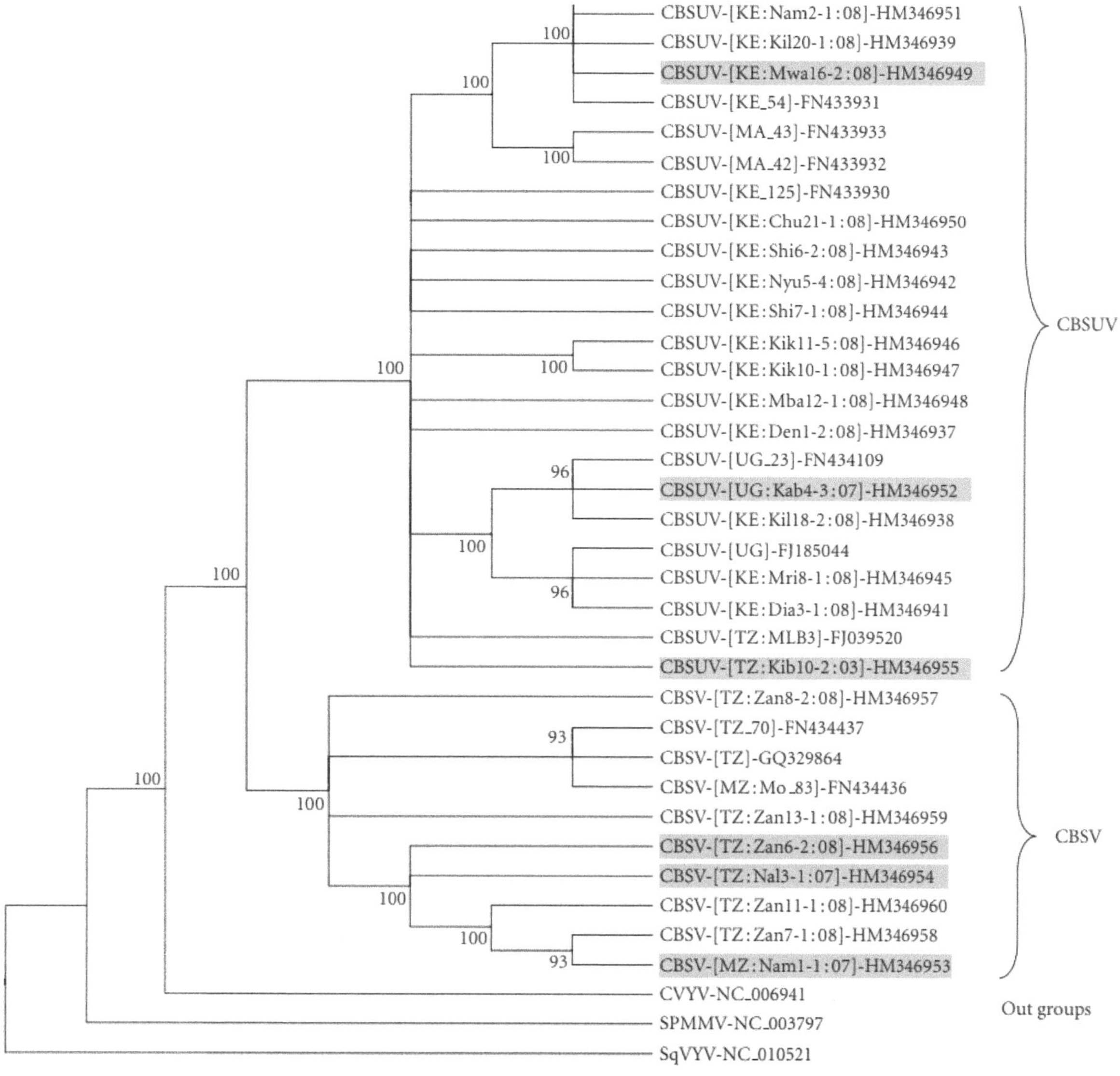

Figure 4: The most parsimonious tree showing the relationship between the two cassava brown streak viruses based on coat protein gene amino acid sequences. The isolates for which symptom diversity was studied are shaded in grey.

DISCUSSION

Until recently, research on CBSD diversity/severity has largely been restricted to observations in the field on cassava plants of different age, genetic makeup, and grown in different agroecological zones with varying environmental conditions and possibly infected with different virus strains, all of which can independently or in combination influence symptom development. This made the comparison of the field observations between the various studies particularly difficult and the question of whether a severe form of

CBSD is associated with the latest epidemic in Uganda has remained unanswered. Inoculation of herbaceous host plants by various researchers provided somewhat uniform conditions for symptom diversity studies [29] but until recently no such comparison has been made with isolates from the coastal endemic and inland epidemic areas involving the two different species of CBSVs [15, 16]. It was particularly difficult to conclude whether the severe CBSD symptoms observed in the fields of coastal Mozambique and Tanzania [9], for example, or the relatively milder leaf symptoms seen in Uganda (severity score of 2.0, [5]) were due to the effect of virus isolate or the tolerance/susceptibility of the cassava varieties being grown in those regions. In order to answer these questions, experiments were carried out in controlled environmental conditions in a glasshouse using a standard range of CBSD isolates from both the endemic and epidemic regions to determine the virulence of the isolates. This was particularly relevant to understand if the new outbreaks of CBSD at high altitudes in Uganda and the lake zone areas of Tanzania were due to the prevalence of a severe form of the virus, similar to those observed during the course of the CMD pandemic in Uganda in the early 1990s.

In order to investigate this, a number of parameters were used to assess the severity levels between one epidemic and five endemic CBSD isolates including the symptoms on leaf and roots of five infected cassava varieties, the effect of virus on sprouting of cassava stem cuttings, the rate of graft transmission, and virus titres in infected leaves as well as symptom severity on herbaceous host plants. Amongst the isolates examined, the endemic isolates CBSV-[MZ:Nam1-1:07] and CBSV-[TZ:Nal3-1:07] produced the most severe symptoms with mean symptom severity scores of 3.7-3.8 on a five-point scale [9]. In comparison, the epidemic CBSUV-[UG:Kab4-3:07] isolate produced relatively mild symptoms with a mean leaf severity score of 1.9. These differences were further confirmed upon the inspection of root symptoms in which CBSV-[MZ:Nam1-1:07] and CBSV-[TZ:Nal3-1:07] infections resulted in root necrosis in all five cassava varieties tested including the tolerant variety Kiroba, however, this was not by CBSUV-[UG:Kab4-3:07] (Figure 2). The severity of CBSVs can also be estimated by their ability to affect the young growing buds of infected cassava plants [2, 30]. Using these earlier observations as cues, the differences in the severity levels of the epidemic and endemic isolates were further demonstrated when a significantly higher number of cuttings failed to sprout from the severe endemic isolates compared to the milder epidemic isolate. Between 22 and 26% of the cuttings failed to sprout when infected with CBSV-[MZ:Nam1-1:07] or CBSV-[TZ:Nal3-1:07] while only 4% of the cuttings were similarly affected by the infection of CBSUV-[UG:Kab4-3:07] (Table 4). These observations were further supported by the higher rates of virus transmission by grafting of the endemic severe isolates which is probably due to high virus titre (about 1000-times higher virus titre in the two severe endemic isolates CBSV-[MZ:Nam1-1:07] or CBSV-[TZ:Nal3-1:07] compared to the epidemic isolate CBSUV-[UG:Kab4-3:07]). A notable difference observed between this and earlier studies, however, is the infection of the cassava variety Albert by all isolates of this study. In graft-inoculation experiments, Winter et al. [16] failed to infect Albert by the CBSD isolates from Kenya, Uganda, and Malawi. While the differences between these two similar studies could not be explained at this stage, these results nonetheless have great implications for developing disease management strategies since Albert once considered resistant to CBSD in Kenya, Uganda and Malawi has now been proven susceptible. In southern Tanzania, growing of Albert has been largely abandoned due to its susceptibility to the disease (R. J. Hillocks, unpublished).

The differences in symptoms were also observed on infected herbaceous hosts. Compared to the previously reported *N. benthamiana* [15, 16], *N. clevelandii* in particular were highly susceptible to both CBSV and CBSUV in our conditions, and this could be an excellent differential host for separating severe and mild isolates. On *N. clevelandii*, the severe isolates CBSV-[MZ:Nam1-1:07] and CBSV-[TZ:Nal3-1:07] produced symptoms early, caused severe stunting of infected plants, leaf necrosis, and often plant death. The remaining isolates including CBSUV-[UG:Kab4-3:07] caused various forms of leaf chlorosis, the symptoms were less severe and nonlethal (Figure 2). To correlate the symptom diversity observed to the genetic diversity of CBSD isolates, the complete CP aa sequences of 24 isolates were compared with those reference sequences available in gene bank databases. Similar to the results obtained in previous studies [15, 16, 23], our virus isolates grouped into two clusters based on the two described species: CBSV and CBSUV (Figure 4). Based on aa sequence identities, the Ugandan epidemic isolate was highly similar to those from Kenya and Malawi, suggesting that CBSUV-[UG:Kab4-3:07] may have originated form one or both these countries.

Put together, these collective observations on symptom diversity as well as genetic differences did not indicate the association of a severe form of CBSD in Uganda. These results are indeed consistent with studies on another epidemic isolate (Namulonge) from Uganda [16] and especially agree with field observations in which the maximum average severity recorded at the onset of CBSD in Uganda was only 2.0 [5]. In the absence of a particularly virulent virus in Uganda, our results, however, raise serious questions as to the factors responsible for the current outbreaks of CBSD in eastern African countries. The possible explanations for this are the presence of unusually high populations of whitefly vectors (*B. tabaci*) on cassava that may be responsible for the rapid spread of the virus in the field. The recent widespread introduction of CMD-resistant varieties that are particularly susceptible to CBSD or the combination of both could be a factor in disease outbreak. Recent surveys in Uganda indeed confirmed these possibilities, where more than 70% of the cassavas grown in 23 districts were CMD-resistant improved varieties, all of which are susceptible to CBSD. These varieties also support high whitefly numbers, in excess of 200 adults per top five leaves. Although such elite cassava has not been introduced in high quantities to the Lake Zone Tanzania, the high susceptibility of local land races grown in the region and the sudden development of unusually high whitefly populations on cassava there is ensuring the spread of CBSD [6, 7]. Identification of severe forms of CBSVs in CBSD endemic regions is particularly worrying because the spread of these isolates into areas of high whitefly population has greater potential to cause even more severe damage to cassava production than yet encountered. Our results emphasize the need for exercising strict quarantine measures for preventing further spread of CBSD between country borders and have also identified the need for developing cassava varieties with broad spectrum resistance to both viruses.

ACKNOWLEDGMENTS

I. U. Mohammed and M. M. Abarshi are funded by the Government of Kebbi State, Nigeria to undertake this research. The authors thank Professor Mike Thresh for his comments on the paper, and Mr. Haji Saleh for assistance in collecting cassava samples from Zanzibar.

REFERENCES

1. Storey HH. Virus diseases on East African plants - VII. A progress report on studies of diseases of cassava. *East African Journal*. 1936;2:34–39.
2. Nichols RFJ. The brown streak disease of cassava; distribution, climatic effects and diagnostic symptoms. *East African Agricultural Journal*. 1950;15:154–160.
3. Hillocks RJ, Raya MD, Thresh JM. Distribution and symptom expression of *Cassava brown streak virus*disease in southern Tanzania. *African Journal of Root and Tuber Crops*. 1999;3:57–62.
4. Mahungu NM, Bidiaka M, Tata H, et al. Cassava brown streak-like symptoms in Democratic Republic of Congo. *Roots*. 2003;8:8–9.
5. Alicai T, Omongo CA, Maruthi MN, et al. Re-emergence of cassava brown streak disease in Uganda.*Plant Disease*. 2007;91(1):24–29.
6. Jeremiah SC, Legg JP. *Cassava brown streak virus* disease: farmers' perspectives on a new outbreak of this disease from the Lake zone of Tanzania. 2008, http://www.youtube.com/watch?v=nCJdws9CnUw.
7. Legg JP, Jeremiah SC, Obiero HM, et al. Comparing the regional epidemiology of the cassava mosaic and*Cassava brown streak virus* pandemics in Africa. *Virus Research*. 2011;159(2):161–170. [PubMed]
8. Hillocks RJ, Raya MD, Mtunda K, Kiozia H. Effects of brown streak virus disease on yield and quality of cassava in Tanzania. *Journal of Phytopathology*. 2001;149(7-8):389–394.
9. Hillocks RJ, Raya M, Thresh JM. The association between root necrosis and above-ground symptoms of brown streak virus infection of cassava in southern Tanzania. *International Journal of Pest Management*.1996;42(4):285–289.
10. McSween S, Walker T, Salegua V, Pitoro R. *Research Report*. 1E. Mozambique: Mozambican Institute of Agricultural Research; Economic impact on food security of varietal tolerance to cassava brown streak disease in costal Mozambique. Research Paper Series, 2006.
11. IITA. Agronomy of Cassava. IITA Research Guides, Training Program, IITA, Ibadan, Nigeria, 2005.
12. Monger WA, Seal S, Isaac AM, Foster GD. Molecular characterization of the *Cassava brown streak virus* coat protein. *Plant Pathology*. 2001;50(4):527–534.
13. Monger WA, Alicai T, Ndunguru J, et al. The complete genome sequence of the Tanzanian strain of*Cassava brown streak virus* and comparison with the Ugandan strain sequence. *Archives of Virology*.2010;155(3):429–433.
14. Mbanzibwa DR, Tian Y, Mukasa SB, Valkonen JPT. *Cassava brown streak virus* (Potyviridae) encodes a putative Maf/HAM1 pyrophosphatase implicated in reduction of mutations and a P1 proteinase that suppresses RNA silencing but contains no HC-Pro. *Journal of Virology*. 2009;83(13):6934–6940.
15. Mbanzibwa DR, Tian YP, Tugume AK, et al. Genetically distinct strains of *Cassava brown streak virus*in the Lake Victoria basin and the Indian Ocean coastal area of East Africa. *Archives of Virology*.2009;154(2):353–359.
16. Winter S, Koerbler M, Stein B, Pietruszka A, Paape M, Butgereitt A. Analysis of *Cassava brown streak virus*es reveals the presence of distinct virus species causing cassava brown streak disease in East Africa.*Journal of General Virology*. 2010;91(5):365–376.
17. Maruthi MN, Hillocks RJ, Mtunda K, et al. Transmission of *Cassava brown streak virus* by *Bemisia tabaci*. *Journal of Phytopathology*. 2005;153(5):307–312.
18. Mware B, Narla R, Amata R, et al. Efficiency of *Cassava brown streak virus* transmission by two whitefly species in coastal Kenya. *Journal of General and Molecular Virology*. 2009;1:40–45.
19. Jennings DL. Observations on virus diseases of cassava in resistant and susceptible varieties. II. Brown streak disease. *Empire Journal of Experimental Agriculture*. 1960;28:261–269.
20. Deng D, Mcgrath PF, Robinson DJ, Harrison BD. Detection and differentiation of whitefly-transmitted geminiviruses in plants and vector insects by the polymerase chain reaction with degenerate primers. *Annals of Applied Biology*. 1994;125(2):327–336.
21. Maruthi MN, Colvin J, Seal S, Gibson G, Cooper J. Co-adaptation between cassava mosaic geminiviruses and their local vector populations. *Virus Research*. 2002;86(1-2):71–85.
22. Frison EA. Sanitation techniques for cassava. *Tropical Science*. 1994;34:146–153.
23. Mbanzibwa DR, Tian YP, Tugume AK, et al. Evolution of cassava brown streak disease-associated viruses. *Journal of General Virology*. 2011;92(4):974–987.
24. Patil BL, Ogwok E, Wagaba H, et al. RNAi-mediated resistance to diverse isolates belonging to two virus species involved in Cassava brown streak disease. *Molecular Plant Pathology*. 2010;12(1):31–41.
25. Hahn SK, Isoba JCG, Ikotun T. Resistance breeding in root and tuber crops at the International Institute of Tropical Agriculture (IITA), Ibadan, Nigeria. *Crop Protection*. 1989;8(3):147–168.
26. Lodhi MA, Ye GN, Weeden NF, Reisch BI. A simple and efficient method for DNA extraction from grapevine cultivars and *Vitis* species. *Plant Molecular Biology Reporter*. 1994;12(1):6–13.
27. Abarshi MM, Mohammed IU, Wasswa P, et al. Optimization of diagnostic RT-PCR protocols and sampling procedures for the reliable and cost-effective

detection of *Cassava brown streak virus*. *Journal of Virological Methods*. 2010;163(2):353–359.

28. Tamura K, Dudley J, Nei M, Kumar S. MEGA4: molecular evolutionary genetics analysis (MEGA) software version 4.0. *Molecular Biology and Evolution*. 2007;24(8):1596–1599.

29. Bock KR. Studies on *Cassava brown streak virus* disease in Kenya. *Tropical Science*. 1994;34:134–145.

30. Hillocks RJ, Jennings DL. Cassava brown streak disease: a review of present knowledge and research needs. *International Journal of Pest Management*. 2003;49(3):225–234.

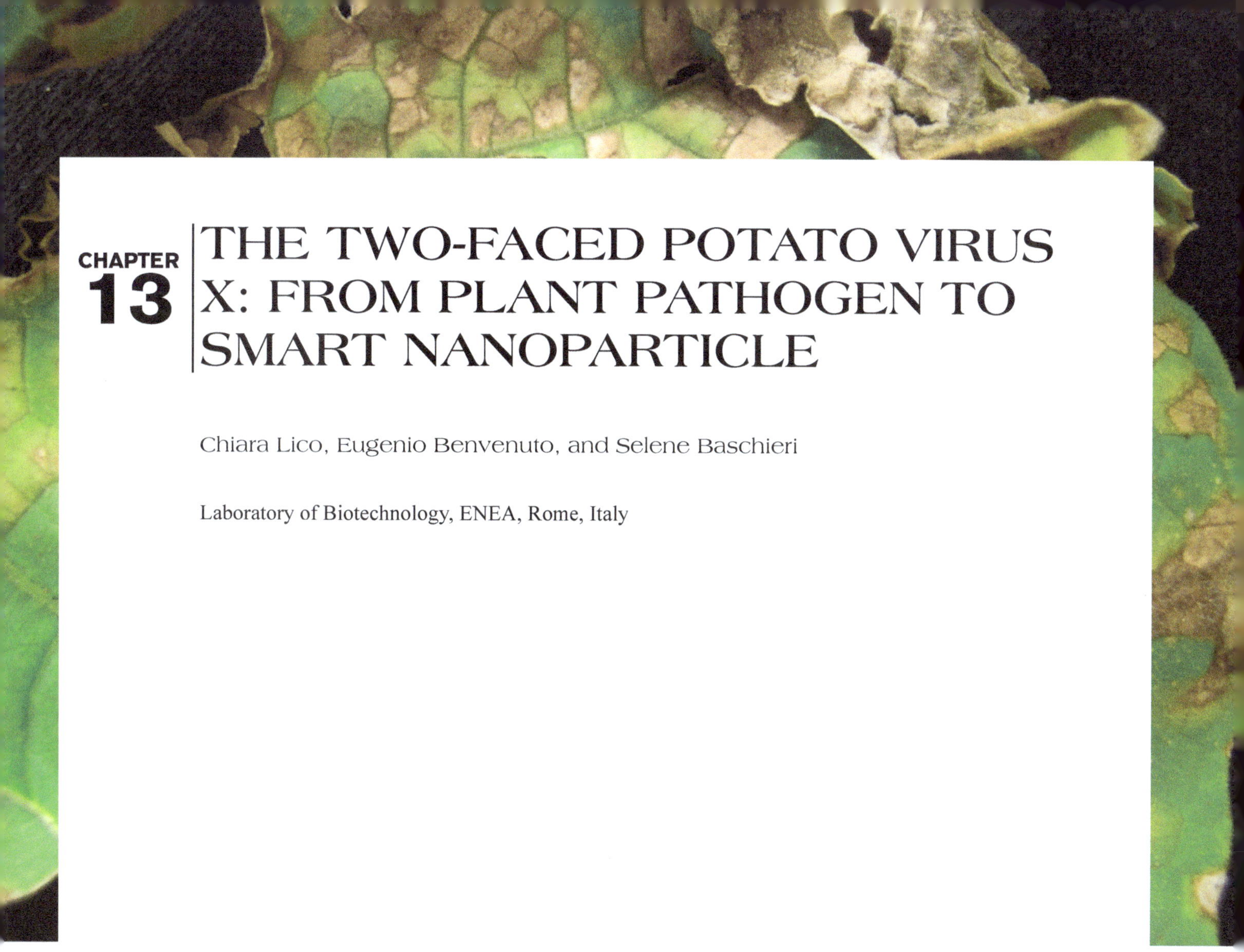

CHAPTER 13

THE TWO-FACED POTATO VIRUS X: FROM PLANT PATHOGEN TO SMART NANOPARTICLE

Chiara Lico, Eugenio Benvenuto, and Selene Baschieri

Laboratory of Biotechnology, ENEA, Rome, Italy

ABSTRACT

Potato virus X (PVX) is a single-stranded RNA plant virus, historically investigated in light of the detrimental effects on potato, the world's fourth most important food commodity. The study of the interactions with cells, and more generally with the plant, both locally and systemically, significantly contributed to unveil the mechanisms underlying gene silencing, fundamental not only in plant virology but also in the study of gene expression regulation. Unraveling the molecular events of PVX infection paved the way for the development of different viral expression vectors and consequential applications in functional genomics and in the biosynthesis of heterologous proteins in plants. Apart from that, the ease of manipulation and the knowledge of the virus structure (particle dimensions, shape and physicochemical features) are inspiring novel applications, mainly focused on nanobiotechnology. This review will lead the reader in this area, spanning from fundamental to applied research, embracing fields from plant pathology to vaccine and drug-targeted delivery, imaging and material sciences. Due to the versatile moods, PVX holds promise to become an interesting nanomaterial, in view to create the widest possible arsenal of new "bio-inspired" devices to face evolving issues in biomedicine and beyond.

INTRODUCTION TO POTATO VIRUS X MOLECULAR PATHOLOGY

Potato virus X (PVX) belongs to the genus *Potexvirus* of the *Flexiviridae* family that groups viruses phylogenetically related by replication mechanisms, structural proteins and genome type and organization (Martelli et al., 2007). PVX is mechanically transmitted and its main hosts are herbaceous plants, especially*Solanaceae*. It has been included among the pathogens with important economic impact and its effects can be worsened by co-infection with other viruses, in particular Potato virus Y (PVY; a phenomenon also known as "synergism"; Syller, 2012).

Citation: Chiara Lico, Eugenio Benvenuto and Selene Baschier, The Two-Faced Potato Virus X: From Plant Pathogen to Smart Nanoparticle, doi: 10.3389/fpls.2015.01009.

Potato virus X has a simple filamentous flexible structure of about 500 nm in length and 15 nm in diameter (Figure (Figure1A).1A). The virion has helical symmetry and a deeply grooved, highly hydrated surface (Parker et al., 2002). It is made of a single-stranded positive-sense RNA genome of approximately 6.4 kb (Huisman et al., 1988; Skryabin et al., 1988), wrapped in approximately 1300 units of a single coat protein (CP) type, with 8.9 CP units per helix turn (Kendall et al., 2008).

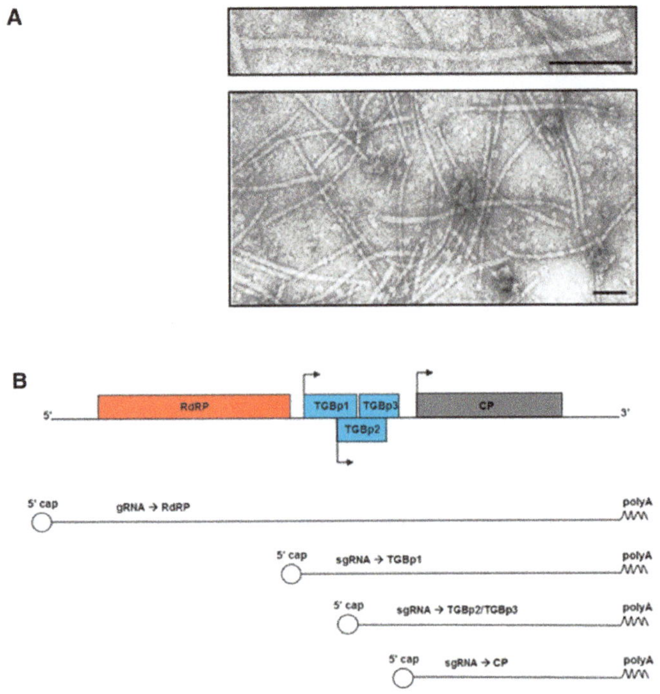

Figure 1: (A) Transmission electron micrographs of purified PVX nanoparticles (personal unpublished data). Bars: 100 nm. (B) Schematic diagram of PVX genome, genomic and subgenomic RNAs (gRNA and sgRNA, respectively), and the proteins they encode. RdRP: RNA-dependent ...

The genome is capped at the 5′-end and poly-adenylated at the 3′-terminus. It contains five open reading frames (ORFs) encoding five proteins: the RNA-dependent RNA Polymerase (RdRP), the movement proteins encoded by three overlapping ORFs that form the Triple Gene Block module (TGBp1, TGBp2, and TGBp3), and the CP (Figure (Figure1B;1B; Sonenberg et al., 1978; Morozov et al., 1987, 1991). The majority of these proteins are multifunctional and involved in several different steps of the infection process, together with numerous host factors. The RdRP, for example, contains a methyl transferase type 1 domain involved in viral RNAs capping (Rozanov et al., 1992), a helicase domain fundamental for viral replication (Davenport and Baulcombe, 1997), and a replicase domain (Longstaff et al., 1993); while TGBp1 has RNA-binding, ATPase (Leshchiner et al., 2006) and RNA-helicase (Kalinina et al., 2002) activities. Moreover, TGBp1 contributes to: (a) increase the size exclusion limit of the plasmodesmata (PD; Angell et al., 1996; Howard et al., 2004), (b) translationally activate the progeny virions (Karpova et al., 2006), and (c) counteract gene silencing (Voinnet et al., 2000), the plant innate mechanism of immune resistance.

When PVX enters plant cells, the particles unpack thanks to the phosphorylation events mediated by cellular enzymes on Ser/Thr residues located at the solvent-exposed N-terminus of the CP (Atabekov et al., 2001). The RdRP is directly synthetized from the viral RNA genome, exploiting the host translational machinery, to subsequently give rise to subgenomic RNAs and viral proteins (Batten et al., 2003). Later on, within infected cells, the replication complexes become part of structures called X-bodies that are formed next to the nucleus (Tilsner et al., 2012). X-bodies are isolated environments delimited by remodeled cellular membranes and organized in "compartments" that incorporate both viral and host components, coordinating the different phases of infection (replication, protein synthesis, assembly, and movement). The primary function of X-bodies is to protect and allow viral replication through the anchoring of RdRP in arrays to membranes, yet being sites where ribonucleoprotein (RNP) complexes (involved in cell-to-cell transfer of infection) and complete virus particles are assembled. TGBp1 has a central role in the formation of X-bodies as it participates to the remodeling of host actin filaments, endoplasmic reticulum (ER) and Golgi membranes, and to the recruitment of the viral TGBp2 and TGBp3. Within X-bodies, aggregates of TGBp1 and actin microfilaments wrap in spirals to form the typical "beaded sheets," electron microscopy-visible structures in PVX infected cells (Linnik et al., 2013). The spirals are surrounded by naked viral RNAs and fine membrane hoops/granular vesicles that contain TGBp2, TGBp3, ribosomes, and viral replicase, probably representing replication sites. The TGBp2 and TGBp3 membrane-spanning proteins contribute individually and/or synergistically to the formation of these vesicles by remodeling the ER membranes. At the periphery of X-bodies, assembled viral particles accumulate.

In the initial/middle phases of the infection, TGBp2 and TGBp3 exploit the dynamic ER/actin network to direct the transfer of RNP complexes, made of viral RNA and CP, from the perinuclear to the cortical ER (Tilsner et al., 2012). It has been recently suggested that later on the TGBp2 and TGBp3 remodeled ER-membranes form caps of the PD orifice and that these caps, harboring the RdRP and viral RNA, become the viral replication sites. According to this "co-replication model," as soon as they are synthetized, RNP complexes are inserted by the TGBp1 into the PD for cell-to-cell transfer (Tilsner et al., 2013).

As for viral spread throughout the plant, the CP is necessary both in cell-to-cell and in systemic movement, but recently published data indicate that virus particle assembly is dispensable for local while essential for long-distance transfer (Santa Cruz et al., 1998; Betti et al., 2012). The CP has a pivotal role in virus particle morphology and stability, and strongly influences the efficiency of infection spreading (Lico et al., 2006). Unfortunately, atomic-resolution data defining its folding are not available. A recent revision of the protein model indicates that it contains seven α-helices and six β-strands and confirms that the N-terminus is exposed on the viral surface (Baratova et al., 2004; Lukashina et al., 2009, 2012).

Potato virus X has been used as model to study plant-virus interactions and to elucidate the plant-defense mechanisms and viral countermeasures such as suppression of silencing (Ruiz et al., 1998; Ratcliff et al., 1999). The plant is not a passive host during viral infection and viral RNA sensing induces the activation of defense responses that involve a complex set of proteins. The post-transcriptional RNA gene-silencing (PTGS) is a homology-dependent gene inactivation strategy that is triggered by the unusual presence in the plant cell of double stranded (ds) RNAs (Tenllado and Díaz-Ruíz, 2001). In the case of a viral infection, the PTGS response is activated by highly structured regions in the viral genomes or in the viral replication intermediates. These abnormal RNA structures are digested by a plant cell dsRNA-specific RNase, named Dicer, into 21–24 nucleotides RNA species (small interfering RNAs, siRNAs; Hamilton and Baulcombe, 1999). The siRNAs are then exploited through a sequence complementarity mechanism to recruit the homologous RNA into an RNA-induced silencing complex (RISC), where it is digested (Voinnet, 2005).

Also viral proteins contribute to trigger plant innate defense. In particular, the CP has been shown to work as pathogen avirulence effector of the Rx protein-mediated resistance (Bendahmane et al., 1995). Folding of the Rx protein is modified by CP binding and consequently nucleotide-binding site leucine-rich domains are exposed thus determining the activation of the signaling cascade and cell death (Moffett et al., 2002; Rairdan and Moffett, 2006).

Like other plant viruses, PVX has evolved mechanisms to counteract plant innate defense through one of its multifunctional protein, the TGBp1. The protein inhibits the formation of the effector complexes involved in RNA silencing (Bayne et al., 2005) by inducing, through the proteasome pathway, the degradation of the Argonaute 1 nuclease, one of the RISC components (Chiu et al., 2010).

Also the PVX-*Potyviruses* synergistic interaction that results in the enhancement of pathogenicity of PVX in*Nicotiana benthamiana* plants is somehow related to virus-induced gene silencing through the involvement of the *Potyviruses* P1/Helper component proteinase (HC-Pro; González-Jara et al., 2005; Pacheco et al., 2012).

To bring over these fundamental studies aimed to clarify the complex mechanisms underlying the plant-virus interaction, a number of viral expression vectors have been developed (Lico et al., 2008; Hefferon, 2012;Gleba et al., 2014; Peyret and Lomonossoff, 2015). First generation viral expression vectors usually harbor the cDNA form of the complete viral genome, sometimes engineered to easily insert foreign sequences as additional ORFs, or in substitution of a viral one, in association to a strong promoter (Chapman et al., 1992). These vectors can be directly used to infect the plant hosts, when the viral genome is under the control of a plant specific promoter, or used as template to generate *in vitro* an infectious transcript. In second-generation vectors, viral components are separately inserted into binary vectors and used to transform independently different *Agrobacterium* cells. Bacterial cells are then mixed together and used to co-infiltrate plant leaves (Gleba et al., 2014). Most of these vectors have found application in different technological fields.

In functional genomics studies, they are used to vehicle an endogenous gene fragment, triggering the specific suppression of the corresponding sequence in the genome (Ruiz et al., 1998; Baulcombe, 1999). This reverse genetics technique known as virus-induced gene silencing, is a high-throughput approach to the analysis of plant gene functions.

In "molecular farming," PVX-based vectors are currently being used to vehicle and induce the expression in plants of foreign genes encoding high added-value biomolecules (Komarova et al., 2010).

Last, but only in the order of appearance list, these vectors appear extremely appealing also for nanotechnologies (Lico et al., 2013), as described in the following paragraphs.

PVX NANOPARTICLES FOR SUBUNIT VACCINE DELIVERY

Subunit vaccines are formulations based on isolated pathogens components (proteins or peptides) that allow the activation of highly specific and protective immune responses. Nowadays, these vaccines are typically produced by recombinant DNA technologies. Compared to traditional inactivated or attenuated vaccines, subunit vaccines guarantee selectivity, specificity, low toxicity, stability and reduced risk of undesired side effects (Rappuoli, 2007). Main limitation of these vaccines is that isolated proteins or peptides being small (<10 nm) are *per se* unable to stimulate complete immune responses (innate, antibody, and cell-mediated). Indeed, it has been established that the efficiency of antigen uptake by antigen presenting cells (APCs) is strictly related to antigen sizes, and the larger surfaces of particulate antigens improve the interaction with APCs (Bachman and Jennings, 2010). To increase subunit vaccine immunogenicity, it is thus important to arrange isolated antigens in larger particles (20–200 nm). This is possible by entrapping them with adjuvants or by favoring self-assembly in supramolecular structures, such as in the case of virus-like particles generated by the self-assembly of viral capsid proteins (Rosenthal et al., 2014). A further possibility consists in delivering the antigen in association to nanoparticles, such as liposomes.

Many attempts have also been made to increase subunit vaccines immunogenicity and stability using genetically engineered plant virus nanoparticles as carriers for their delivery (Lico et al., 2007, 2013; Lebel et al., 2015; Streatfield et al., 2015). As for PVX, this provides for the fusion of the sequence encoding the immunogenic peptides or proteins in frame to the 5′-end of the gene encoding the CP in viral expression vectors. The N-terminus of the protein has indeed been demonstrated to be exposed on the external surface of the virion (Baratova et al., 1992). The modified viral expression vectors are then used to induce infection onset in plants and produce on large scale chimeric virus particles (CVPs) displaying on each CP subunit (approx. 1300 per virion) the (poly)peptide of interest (Figure2). CVPs are then extracted from plant tissues and purified to be used in immunization trials.

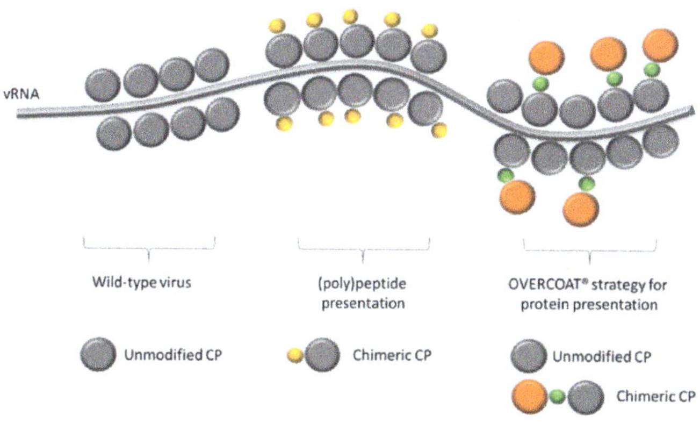

Figure 2: Schematic representation of a PVX nanoparticle, showing the viral RNA (vRNA) and the shell composition of: a wild-type virion; a chimeric virus particle exposing on every single CP unit (in gray) a copy of the heterologous (poly)peptide (in yellow); a ...

Because the CP is involved in mediating the interaction with the plant environment, substantial modifications are not always tolerated in that they may interfere with both viral encapsidation and/or spread. These issues have been narrowly explored in the past years and some basic rules have been defined, whose compliance increases the chance of getting genetically stable infectious CVPs. Clearly, the shorter the exogenous (poly)peptide, the higher the probability that CP folding is preserved and compatible with efficient assembly. But length is not the unique parameter to be considered. Studies performed by fusing peptides (variable in length and amino acid composition) to a N-terminally truncated CP (a natural PVX mutant lacking the first 21 amino acids) have indicated that also isoelectric point, Ser/Thr and Trp content in the heterologous sequence may affect viral fitness (Lico et al., 2006). Recently, the insertion of a heterologous sequence between amino acids 23 and 24 at the N-terminus of the CP has also been reported (Vaculik et al., 2015). Nonetheless, up to now no CP domains other than the N-terminus have been identified that can harbor fusion peptides/proteins. Notably, the fusion to the C-terminus does not give rise to CVPs able to move throughout the plant (Cerovska et al., 2012). In fact, the C-terminus of the protein is crucial for the interaction with the viral RNA, hence for virion assembly (Nemykh et al., 2008).

On these bases, PVX particles displaying peptides of immunological relevance have been successfully produced in plants and their efficacy verified in animal models. PVX CVPs displaying the linear 2F5 epitope (2F5e) of the gp41 protein of Human Immunodeficiency Virus type 1 (HIV-1) envelope are able to induce the production of epitope-specific IgA and IgG, when administered intranasally or intraperitoneally in normal mice without the need to co-administer adjuvants. Notably, human dendritic cells pulsed with these CVPs are able to activate a 2F5e-specific human primary antibody response in severe combined immunodeficient mice engrafted with human lymphocytes. Finally, the antibodies induced by immunization with the CVPs are able to neutralize HIV-1 particles in viral neutralization assays (Marusic et al., 2001).

Potato virus X CVPs displaying a Major Histocompatibility Complex class I-restricted peptide of the nucleoprotein of influenza virus have also been engineered, taking into account mechanisms and limitations imposed by both plant virus biology and immune response. The synthetic DNA fragment encoding the influenza peptide was designed to be translated in an amino acid sequence optimized to: (i) obtain assembly of CVPs able to move systemically; (ii) minimize the risk of occurrence of unexpected mutations during viral replication; (iii) avoid detrimental effects of N-terminal modifications (such as acetylation) and improper C-terminal context on peptide processing by APCs. These rationally designed CVPs resulted to be genetically stable and triggered epitope-specific $CD8^+$ T cells without the need of adjuvant co-delivery (Lico et al., 2009).

Overall, these data indicate that appropriately designed PVX CVPs are excellent epitope carriers endowed of adjuvant properties, probably not only because of the complex particulate structure but also because of the presence of the viral RNA that may trigger Toll-Like Receptor 7 on APCs (Jobsri et al., 2015). Altogether, these features account for the inherent ability of PVX to trigger a whole immune response, ideal condition to achieve complete protection from pathogens through vaccination.

Because one of the factors that may limit the construction of PVX CVPs is the length of the fused (poly)peptide, a "smart" procedure based on the OVERCOAT® PVX vector has also been devised to avoid possible steric hindrance of chimeric CPs during particle assembly (Santa Cruz et al., 1996). To this aim, a 16 amino acid linker peptide (2A) derived from the Foot and Mouth Disease Virus (FMDV) has been inserted between the foreign sequence and the CP. During translation, this sequence may induce a ribosomal skip with a frequency that is dependent on the nature of the 18–30 amino acids in the region immediately upstream the 2A cleavage site (Minskaia and Ryan, 2013; Minskaia et al., 2013). The incomplete FMDV 2A-mediated "cleavage" results in the production of a heterogeneous population of CPs including original and engineered version, both contributing to the assembly of fully functional CVPs able to spread throughout the plant (Figure2). By this strategy, PVX particles were constructed displaying on the surface different (poly)peptides of immunological interest. In the case of the rotavirus major inner capsid protein (VP6;O'Brien et al., 2000) and the tuberculosis ESAT-6 antigen (Zelada et al., 2006), data demonstrated only the structural stability of the CVPs and the correct display of the foreign sequences. Subsequent papers demonstrated also the ability of these PVX CVPs to induce antibody responses in animal models (rabbits and mice). This is the case of the CVPs tailored to display the D2 peptide from *Staphylococcus aureus*fibronectin-binding protein (Brennan et al., 1999), two peptides from the E2 glycoprotein of the classical swine fever virus (Marconi et al., 2006) or the R9 peptide from the hypervariable region I of hepatitis C virus (Uhde-Holzem et al., 2010).

Alternative approaches to the construction of PVX CVPs, even if at higher costs, imply standard chemical protein conjugation procedures. In this case, the success depends on the presence of free amine groups (Lys residues) and carboxylate groups (Asp and Glu residues). These amino acids are all present in PVX CP. Different bioconjugation protocols have been tested to functionalize PVX, revealing that the carboxylates groups are either not solvent-exposed or not accessible. On the other hand, N-Hydroxysuccinimide (NHS)-ester based and Cu(I)-catalyzed azide-alkyne 1,3-dipolar cycloaddition ("CLICK") reactions on amine groups are both feasible (Steinmetz et al., 2010).

Potato virus X particles displaying a peptide derived from the extracellular domain of the Human epidermal growth factor receptor 2 (HER2) have been constructed by conjugation to Lys on the CP surface through a two-step reaction via a heterofunctional NHS-(poly)ethyleneglycole $(PEG)_4$-maleimide linker (Shukla et al., 2014a). HER2 is an antigen usually expressed on breast cancer cells and known to induce the activation of B lymphocytes and antibody production. The chemically derivatized PVX particles were tested in mice and demonstrated to be effective in inducing the overcoming of HER2 immunological tolerance by Enzyme Linked ImmunoSorbent Assay and confocal microscopy using HER2-positive human cancer cells.

PVX NANOPARTICLES FOR IMAGING, DRUG DELIVERY, AND DIAGNOSIS

Pioneering works in vaccine technology exploiting plant viruses such as PVX have worked as trailblazer in the field of nanotechnology. Indeed, the encouraging results prompted scientists in nanomedicine to broaden the exploration and exploitation of PVX particles as tools for targeted drug delivery or for diagnosis. In this perspective, the evaluation of the "biocompatibility" of viral particles, their natural tropism and biodistribution are fundamental to enter clinical trials. Indeed, the intrinsic toxicity cannot be taken for granted even if, for the animal cell, plant viruses are expected to behave as unreplicative and biologically safe nano-objects.

In vitro and *in vivo* experiments have been recently carried out to evaluate the toxicity and teratogenicity potential of unmodified PVX and Tomato Bushy Stunt Virus (TBSV), the type member of another viral genus and family, different in shape and dimensions (Blandino et al., 2015). Haemolysis assay was performed *in vitro* to determine detrimental effect of virus nanoparticles on human erythrocytes while toxicity and teratogenicity were evaluated *in vivo* through the early embryo assay (EEA). Chicken embryo is considered an excellent model to perform this kind of studies even if the reliability of the results depends on critical factors such as the choice of adequate controls, number of eggs per dose, and volume and route of delivery. Overall, the results of this study demonstrated that both plant viruses have neither cytotoxic effects on human erythrocytes nor toxic/teratogenic effects on chicken embryos. These data represent a good prerequisite for safety evaluation in view of forthcoming use of these nanoparticles as functionalized nanocarriers.

Biodistribution studies *in vivo* have been performed with PVX particles conjugated to various fluorescent dyes and to PEG with different chain lengths. PEG is frequently used to create a shield around biopharmaceuticals to increase solubility and stability and reduce immunogenicity even if, as a side effect, this masking reduces the interactions with cells, such as in the case of some plant virus nanoparticles (Raja et al., 2003; Schlick et al., 2005; Steinmetz and Manchester, 2009).

Preliminary studies *in vitro*, with human cervical cancer (HeLa) cells and murine fibroblasts (BalbCl7), highlight somewhat contradictory effects of PEGylation on PVX. In fact, this procedure effectively prevents interaction with cells of particles labeled with OregonGreen 488 (O488), while it promotes interaction if particles are labeled with Alexafluor 647 (A647; Steinmetz et al., 2010), and this is probably due to some intrinsic feature of the dye.

The analysis of the biodistribution and clearance of A647-labeled and PEGylated PVX in mice evidences that the viral nanoparticles remain mainly trapped by the mononuclear phagocyte system (MPS) that speeds up their removal from the circulation and from tissues. Nonetheless, the flexible filamentous PVX has enhanced tumor homing and tissue penetration, as compared to the icosahedral Cowpea Mosaic virus (Shukla et al., 2013, 2014b). Similarly, PVX CVPs constructed by the OVERCOAT® procedure and carrying the mCherry fluorescent protein but not conjugated to PEG, were shown to accumulate in the liver and to co-localize with macrophages for final clearance from tissues via MPS, 7 days post-administration in mice (Shukla et al., 2014c). It has been recently shown that it is possible to significantly improve PVX half-life in the serum reducing tissue accumulation, renal clearance as well as uptake by immune cells through derivatization of the viral surface with a PEG chain tailored in length and conformation (brush vs. mushroom;Lee et al., 2015).

Beside these early studies aimed to gain insights on PVX nanoparticles-based delivery, advanced studies have already started to exploit the virus as scaffold for the conjugation of chemical moieties, constructing nano-bullets for molecular imaging and targeted drug delivery.

A647-labeled PVX nanoparticles, bioconjugated with a 12 amino acid peptide with affinity to the epidermal growth factor receptor (EGFR), are effective in detection and imaging of carcinoma cell lines that upregulate EGFR, and prefer partitioning to cancer cells rather than to macrophages (Chariou et al., 2015). Furthermore, PVX particles conjugated to the Herceptin (Trastuzumab) monoclonal antibody specific for the peptide HER2 have been demonstrated to be able to induce the apoptosis of HER2 positive cell lines (Esfandiari et al., 2015).

OTHER PVX NANOPARTICLES APPLICATIONS

Genetically or chemically modified PVX nanoparticles can be usefully applied also in fields other than vaccine and drug delivery or *in vivo* imaging.

A diagnostic kit for the early detection of the Sjogren's Syndrome (SjS) has been recently developed (patent application No. 102015000020005 Italian Patent Office), based on PVX CVPs displaying a SjS diagnostic peptide. It has indeed been observed that serum antibodies recognize much less efficiently this soluble peptide if compared to the same peptide displayed on the viral scaffold so that the sensitivity of the assay is noticeably increased, preserving the specificity.

Potato virus X CVPs, displaying an antimicrobial decapeptide derived from an anti-idiotypic antibody that acts as a functional internal image of a microbicidal broad-spectrum yeast killer toxin, were active against*Staphylococcus aureus* and *Candida albicans*, as well as against the plant pathogens *Erwinia carotovora*,*Botrytis cinerea*, and *Fusarium oxysporum*. Moreover, they effectively protected plants from *Pseudomonas syringae* attacks (Donini et al., 2005).

Potato virus X CVPs carrying CPs fused to a single chain antibody specific to *Diuron* herbicide were purified from plant extracts by *Diuron*-based immune-capture, demonstrating their potential as tools for *in situ*applications including remediation of contaminated soils and waterways (Smolenska et al., 1998).

The fusion of GFP or mCherry protein to the PVX CP through the 2A peptide strategy allowed to follow infection progression throughout the plant, providing a non-invasive tool to the study of viral multiplication and spread (Santa Cruz et al., 1996, 1998; Roberts et al., 1997; Tilsner et al., 2013; Shukla et al., 2014c).

Potato virus X provides also opportunities for nanochemistry applications such as the construction of catalytic systems operating like metabolons in nature. A proof of concept has been published, in which the CVPs were constructed through the OVERCOAT® system to display Lipase B, an enzyme derived from *Candida antarctica* known to be an efficient enantiospecific catalyst for chemical hydrolysis. The virus-anchored lipase molecules retained their catalytic activity toward the substrate p-nitrophenyl caproate in solution (Carette et al., 2007). Unlike manmade supported catalysts, these PVX-based biocatalysts were easily produced using plant machinery and, because the virus particle is made of many proteins, they intrinsically possessed multiple catalytic sites.

Recently, PVX nanoparticles have also been investigated as scaffold for the attachment of metal atoms in an attempt to develop new bioinorganic materials for nanoelectronics. Platinum atom clusters (Pt nanoparticles, Pt-NPs) were built selectively at one end of the virus helical structure by chemical reduction, yielding virions that resemble a push-pin with a platinum head and a virus needle (Drygin et al., 2013). These results highlight the specificity of the coordinate metal interaction with a complex polyvalent biological structure, thus providing insights on the mechanisms of Pt-NPs formation, important to understand the interaction of viral particles with ionic metals. Chimeric PVX particles have also been constructed displaying multiple copies of a peptide that is known to induce the formation of SiO_2. By the conditions adopted for synthesis, isolated SiO_2 spots of about 10 nm were formed along the surface of the chimeric virus. The not uniform SiO_2 coating allowed additional immunogold labeling of the particles. The possibility to combine silicification with immunolabeling provides a method to develop triple-hybrid complex structures, with controllable composition and functional properties (Van Rijn et al., 2015).

CONCLUSION

Overall, the knowledge of PVX, type-member of the *Potexvirus* genus, has evolved from basic molecular pathology to applications in biomedicine and material science.

The knowledge of plant-pathogen interactions and related silencing mechanisms has opened the way to the development of novel genetically-controlled strategies of plant infection, and through these to functional genomics and molecular farming.

By virtue of the innate versatility, PVX is now in the spotlight as a promising nanoobject for applications in vaccinology, drug targeting and delivery, diagnostic kits development and also as imaging reporter tool.

Last but not least, it is investigated as scaffold for enzymatic reactions or nanoelectronics.

For a long time, other plant viruses, mainly spherical or rigid rod-shaped, have been preferred to PVX for this kind of applications, but the data reviewed here clearly demonstrate that the flexuous mood of this nanoparticle is not a constraint rather a positive trait to be adequately exploited to respond to specific needs. Although many of the illustrated applications are still in their infancy, there is no reason to suppose they will not progress rapidly.

The shift of PVX from foe to friend has been made possible because this virus has been largely investigated in its innermost mechanisms. This cultural trail clearly witnesses how basic research can trigger a virtuous cycle of knowledge, removing existing interdisciplinary boundaries and opening up unimagined prospects, thus fuelling the development of innovative technologies.

REFERENCES

1. Angell S. M., Davies C., Baulcombe D. C. (1996). Cell-to-cell movement of Potato virus X is associated with a change in the size-exclusion limit of plasmodesmata in tricomere cells of *Nicotiana clevelandii*. Virology 216, 197–201. 10.1006/viro.1996.0046
2. Atabekov J. G., Rodionova N. P., Karpova O. V., Kozlovsky S. V., Novikov V. K., Arkhipenko M. V. (2001). Translational activation of encapsidated potato virus X RNA by coat protein phosphorylation. Virology 286, 466–474. 10.1006/viro.2001.1013
3. Bachman M. F., Jennings G. T. (2010). Vaccine delivery: a matter of size, geometry, kinetics and molecular patterns. Nat. Rev. Immunol. 10, 787–796. 10.1038/nri2868
4. Baratova L. A., Grebenshchikov N. I., Dobrov E. N., Gedrovich A. V., Kashirin I. A., Shishkov A. V., et al. (1992). The organization of potato virus X coat proteins in virus particles studied by tritium planigraphy and model building. Virology 188, 175–180. 10.1016/0042-6822(92)90747-D
5. Baratova L. A., Fedorova N. V., Dobrov E. N., Lukashina E. V., Kharlanov A. N., Nasonov V. V., et al. (2004). N-Terminal segment of potato virus X coat protein subunits in glycosylated and mediates formation of a bound water shell on the virion surface. Eur. J. Biochem. 271, 3136–3145. 10.1111/j.1432-1033.2004.04243.x
6. Batten J. S., Yoshinari S., Hemenway C. (2003). Potato virus X: a model system for virus replication, movement and gene expression. Mol. Plant Pathol. 4, 125–131. 10.1046/j.1364-3703.2003.00156.x
7. Baulcombe D. C. (1999). Fast forward genetics based on virus-induced gene silencing. Curr. Opin. Plant Biol. 2, 109–113. 10.1016/S1369-5266(99)80022-3
8. Bayne E. H., Rakitina D. V., Morozov S. Y., Baulcombe D. C. (2005). Cell-to-cell movement of Potato *Potexvirus* X is dependent on suppression of RNA silencing. Plant J. 44, 471–482. 10.1111/j.1365-313X.2005.02539.x
9. Bendahmane A., Köhn B. A., Dedi C., Baulcombe D. C. (1995). The coat protein of potato virus X is a strain-specific elicitor of Rx1-mediated virus resistance in potato. Plant J. 8, 933–941. 10.1046/j.1365-313X.1995.8060933.x
10. Betti C., Lico C., Maffi D., D'Angeli S., Altamura M. M., Benvenuto E., et al. (2012). Potato virus X movement in *Nicotiana benthamiana*: new details revealed by chimeric coat protein variants. Mol. Plant Pathol. 13, 198–203. 10.1111/j.1364-3703.2011.00739.x
11. Blandino A., Lico C., Baschieri S., Barberini L., Cirotto C., Blasi P., et al. (2015). *In vitro* and *in vivo*toxicity evaluation of plant virus nanocarriers. Colloids Surf. B Biointerfaces 129, 130–136. 10.1016/j.colsurfb.2015.03.039
12. Brennan F. R., Jones T. D., Longstaff M., Chapman S., Bellaby T., Smith H., et al. (1999).Immunogenicity of peptides derived from a fibronectin-binding protein of *S. aureus* expressed on two different plant viruses. Vaccine 17, 1846–1857. 10.1016/S0264-410X(98)00485-X
13. Carette N., Engelkamp H., Akpa E., Pierre S. J., Cameron N. R., Christianen P. C., et al. (2007). A virus-based biocatalyst. Nat. Nanotechnol. 2, 226–229. 10.1038/nnano.2007.76

14. Cerovska N., Hoffmeisterova H., Moravec T., Plchova H., Folwarczna J., Synkova H., et al. (2012).Transient expression of Human papillomavirus type 16 L2 epitope fused to N- and C-terminus of coat protein of Potato virus X in plants. J. Biosci. 37, 125–133. 10.1007/s12038-011-9177-z

15. Chapman S., Kavanagh T., Baulcombe D. (1992). Potato virus X as a vector for gene expression in plants. Plant J. 2, 549–557.

16. Chariou P. L., Lee K. L., Wen A. M., Gulati N. M., Stewart P. L., Steinmetz N. F. (2015). Detection and imaging of aggressive cancer cells using an epidermal growth factor receptor (EGFR)-targeted filamentous plant virus-based nanoparticle. Bioconjug. Chem. 26, 262–269. 10.1021/bc500545z[PMC free article]

17. Chiu M. H., Chen I. H., Baulcombe D. C., Tsai C. H. (2010). The silencing suppressor P25 of Potato virus X interacts with Argonaute1 and mediates its degradation through the proteasome pathway. Mol. Plant Pathol. 11, 641–649. 10.1111/j.1364-3703.2010.00634.x

18. Davenport G. F., Baulcombe D. C. (1997). Mutation of the GKS motif of the RNA-dependent RNA polymerase from potato virus X disables or eliminates virus replication. J. Gen. Virol. 78, 1247–1251. 10.1099/0022-1317-78-6-1247

19. Donini M., Lico C., Baschieri S., Conti S., Magliani W., Polonelli L., et al. (2005). Production of an engineered killer peptide in *Nicotiana benthamiana* by using a potato virus X expression system.Appl. Environ. Microbiol. 71, 6360–6367. 10.1128/AEM.71.10.6360-6367.2005

20. Drygin Y., Kondakova O., Atabekov J. (2013). Production of platinum atom nanoclusters at one end of helical plant viruses. Adv. Virol. 2013:746796. 10.1155/2013/746796

21. Esfandiari N., Arzanani M. K., Soleimani M., Kohi-Habibi M., Svendsen W. E. (2015). A new application of plant virus nanoparticles as drug delivery in breast cancer. Tumour. Biol. [Epub ahead of print].10.1007/s13277-015-3867-3

22. Gleba Y. Y., Tusé D., Giritch A. (2014). Plant viral vectors for delivery by *Agrobacterium*. Curr. Top. Microbiol. Immunol. 375, 155–192. 10.1007/82_2013_352

23. González-Jara P., Atencio H. A., Martínez-García B., Barajas D., Tenllado F., Díaz-Ruiz J. R. (2005).A single amino acid mutation in the Plum pox virus helper component-proteinase gene abolishes both synergistic and RNA silencing suppression activities. Phytopathology 95, 894–901. 10.1094/PHYTO-95-0894

24. Hamilton A. J., Baulcombe D. C. (1999). A species of small antisense RNA in posttranscriptional gene silencing in plants. Science 286, 950–952. 10.1126/science.286.5441.950

25. Hefferon K. L. (2012). Plant virus expression vectors set the stage as production platforms for biopharmaceutical proteins. Virology 433, 1–6. 10.1016/j.virol.2012.06.012

26. Howard A. R., Hepffer M. L., Ju H. J., Krishnamurthy K., Payton M. E., Verchot-Lubicz J. (2004).Potato virus X TGBp1 induces plasmodesmata gating and moves between cells in several host species whereas CP moves only in *N. benthamiana* leaves. Virology 328, 185–197. 10.1016/j.virol.2004.06.039

27. Huisman M. J., Linthorst H. J., Bol J. F., Cornelissen J. C. (1988). The complete nucleotide sequence of potato virus X and its homologies at the amino acid level with various plus-stranded RNA viruses.J. Gen. Virol. 69, 1789–1798. 10.1099/0022-1317-69-8-1789

28. Jobsri J., Allen A., Rajagopal D., Shipton M., Kanyuka K., Lomonossoff G. P., et al. (2015). Plant virus particles carrying tumour antigen activate TLR7 and Induce high levels of protective antibody.PLoS ONE 10:e0118096. 10.1371/journal.pone.0118096

29. Kalinina N. O., Rakitina D. A., Soloyev A. G., Schiemann J., Morozov S. Y. (2002). RNA helicase activity of the plant virus movement proteins encoded by the first gene of the triple gene block.Virology 296, 321–329. 10.1006/viro.2001.1328

30. Karpova O. V., Zayakina O. V., Arkhipenko M. V., Sheval E. V., Kiselyova O. I., Poljakov V. Y., et al. (2006). Potato virus X RNA-mediated assembly of single-tailed ternary 'coat protein-RNA-movement protein' complexes. J. Gen. Virol. 87, 2731–2740. 10.1099/vir.0.81993-0

31. Kendall A., McDonald M., Bian W., Bowles T., Baumgarten S. C., Shi J., et al. (2008). Structure of flexible filamentous plant viruses. J. Virol. 82, 9546–9554. 10.1128/JVI.00895-08

32. Komarova T. V., Baschieri S., Donini M., Marusic C., Benvenuto E., Dorokhov Y. L. (2010). Transient expression systems for plant-derived biopharmaceuticals. Expert. Rev. Vaccines 9, 859–876. 10.1586/erv.10.85

33. Lebel M. E., Chartrand K., Leclerc D., Lamarre A. (2015). Plant viruses as nanoparticle-based vaccines and adjuvants. Vaccines 3, 620–637. 10.3390/vaccines3030620

34. Lee K. L., Shukla S., Wu M., Ayat N. R., El Sanadi C. E., Wen A. M., et al. (2015). Stealth filaments: polymer chain length and conformation affect the *in vivo* fate of PEGylated potato virus X. Acta Biomater. 19, 166–179. 10.1016/j.actbio.2015.03.001

35. Leshchiner A. D., Solovyev A. G., Morozov S. Y., Kalinina N. O. (2006). A minimal region in the NTPase/helicase domain of the TGBp1 plant virus movement protein is responsible for ATPase activity and cooperative RNA binding. J. Gen. Virol. 87, 3087–3095. 10.1099/vir.0.81971-0

36. Lico C., Baschieri S., Marusic C., Benvenuto E. (2007). "Molecular farming for antigen (vaccine) production in plants," in Improvement of Crop Plants for Industrial End Uses, ed. Ranalli P., editor. (Dordrecht, Netherland: Springer;), 417–433.

37. Lico C., Capuano F., Renzone G., Donini M., Marusic C., Scaloni A., et al. (2006). Peptide display on Potato virus X: molecular features of the coat protein-fused peptide affecting cell-to-cell and phloem movement of chimeric virus particles. J. Gen. Virol. 87, 3103–3112. 10.1099/vir.0.82097-0

38. Lico C., Chen Q., Santi L. (2008). Viral vectors for production of recombinant proteins in plants. J. Cell. Physiol. 216, 366–377. 10.1002/jcp.21423

39. Lico C., Mancini C., Italiani P., Betti C., Boraschi D., Benvenuto E., et al. (2009). Plant-produced potato virus X chimeric particles displaying an influenza virus-derived peptide activate specific $CD8^{+}T$ cells in mice. Vaccine 27, 5069–5076. 10.1016/j.vaccine.2009.06.045

40. Lico C., Schoubben A., Baschieri S., Blasi P., Santi L. (2013). Nanoparticles in biomedicine: new insights from plant viruses. Curr. Med. Chem. 20, 3471–3487. 10.2174/09298673113209990035

41. Linnik O., Liesche J., Tilsner J., Oparka K. J. (2013). Unraveling the structure of viral replication complexes at super-resolution. Front. Plant Sci. 4:6. 10.3389/fpls.2013.00006

42. Longstaff M., Brigneti G., Boccard F., Chapman S., Baulcombe D. (1993). Extreme resistance to potato virus X infection in plants expressing a modified component of the putative viral replicase.EMBO J. 12, 379–386.

43. Lukashina E., Badun G., Fedorova N., Ksenofontov A., Nemykh M., Serebryakova M., et al. (2009).Tritium planigraphy study of structural alterations in the coat protein of Potato virus X induced by binding of its triple gene block 1 protein to virions. FEBS J. 276, 7006–7015. 10.1111/j.1742-4658.2009.07408.x

44. Lukashina E., Ksenofontov A., Fedorova N., Badun G., Mukhamedzhanova A., Karpova O., et al. (2012). Analysis of the role of the coat protein N-terminal segment in Potato virus X virion stability and functional activity. Mol. Plant Pathol. 13, 38–45. 10.1111/j.1364-3703.2011.00725.x

45. Marconi G., Albertini E., Barone P., De Marchis F., Lico C., Marusic C., et al. (2006). *In planta*production of two peptides of the Classical Swine Fever Virus (CSFV) E2 glycoprotein fused to the coat protein of potato virus X. BMC Biotechnol. 6: 29. 10.1186/1472-6750-6-29

46. Martelli G. P., Adams M. J., Kreuze J. F., Dolja V. V. (2007). Family *Flexiviridae*: a case study in virion and genome plasticity. Ann. Rev. Phytopathol. 45, 73–100. 10.1146/annurev.phyto.45.062806.094401

47. Marusic C., Rizza P., Lattanzi L., Mancini C., Spada M., Belardelli F., et al. (2001). Chimeric plant virus particles as immunogens for inducing murine and human immune responses against human immunodeficiency virus type 1. J. Virol. 75, 8434–8439. 10.1128/JVI.75.18.8434-8439.2001

48. Minskaia E., Nicholson J., Ryan M. D. (2013). Optimisation of the foot-and-mouth disease virus 2A co-expression system for biomedical applications. BMC Biotechnol. 13:67. 10.1186/1472-6750-13-67

49. Minskaia E., Ryan M. D. (2013). Protein coexpression using FMDV 2A: effect of “linker” residues.Biomed. Res. Int. 2013:291730. 10.1155/2013/291730

50. Moffett P., Farnham G., Peart J., Baulcombe D. C. (2002). Interaction between domains of a plant NBS-LRR protein in disease resistance-related cell death. EMBO J. 21, 4511–4519. 10.1093/emboj/cdf453

51. Morozov S. Y., Lukasheva L. I., Chernov B. K., Skryabin K. G., Atabekov J. G. (1987). Nucleotide sequence of the open reading frames adjacent to the coat protein cistron in the potato virus X genome.FEBS Lett. 213, 438–442. 10.1016/0014-5793(87)81538-7 [Cross Ref]

52. Morozov S. Y., Miroshnichenko N. A., Solovyev A. G., Fedorkin O. N., Zelenina D. A., Lukasheva L. I., et al. (1991). Expression strategy of the potato virus X triple gene block. J. Gen. Virol. 72, 2039–2042. 10.1099/0022-1317-72-8-2039

53. Nemykh M. A., Efimov A. V., Novikov V. K., Orlov V. N., Arutyunyan A. M., Drachev V. A., et al. (2008). One more probable structural transition in potato virus X virions and a revised model of the virus coat protein structure. Virology 373, 61–71. 10.1016/j.virol.2007.11.024

54. O’Brien G. J., Bryant C. J., Voogd C., Greenberg H. B., Gardner R. C., Bellamy A. R. (2000).Rotavirus VP6 expressed by PVX vectors in *Nicotiana benthamiana* coats PVX rods and also assembles into virus-like particles. Virology 270, 444–453. 10.1006/viro.2000.0314

55. Pacheco R., García-Marcos A., Barajas D., Martiáñez J., Tenllado F. (2012). PVX-potyvirus synergistic infections differentially alter microRNA accumulation in *Nicotiana benthamiana*. Virus Res. 165, 231–235. 10.1016/j.virusres.2012.02.012

56. Parker L., Kendall A., Stubbs G. (2002). Surface features of potato virus X from fiber diffraction.Virology 300, 291–295. 10.1006/viro.2002.1483

57. Peyret H., Lomonossoff G. P. (2015). When plant virology met *Agrobacterium*: the rise of the deconstructed clones. Plant Biotechnol. J. 13, 1121–1135. 10.1111/pbi.12412

58. Rairdan G. J., Moffett P. (2006). Distinct domains in the ARC region of the potato resistance protein Rx mediate LRR binding and inhibition of activation. Plant Cell 18, 2082–2093. 10.1105/tpc.106.042747

59. Raja K. S., Wang Q., Gonzalez M. J., Manchester M., Johnson J. E., Finn M. G. (2003). Hybrid virus-polymer materials. 1. Synthesis and properties of PEG-decorated cowpea mosaic virus.Biomacromolecules 4, 472–476. 10.1021/bm025740+

60. Rappuoli R. (2007). Bridging the knowledge gaps in vaccine design. Nat. Biotechnol. 25, 1361–1366. 10.1038/nbt1207-1361

61. Ratcliff F. G., MacFarlane S. A., Baulcombe D. C. (1999). Gene silencing without DNA. rna-mediated cross-protection between viruses. Plant Cell 11, 1207–1216. 10.1105/tpc.11.7.1207

62. Roberts A. G., Cruz S. S., Roberts I. M., Prior D., Turgeon R., Oparka K. J. (1997). Phloem unloading in sink leaves of *Nicotiana benthamiana*: comparison of a fluorescent solute with a fluorescent virus. Plant Cell 9, 1381–1396. 10.1105/tpc.9.8.1381

63. Rosenthal J. A., Chen L., Baker J. L., Putnam D., DeLisa M. P. (2014). Pathogen-like particles: biomimetic vaccine carriers engineered at the nanoscale. Curr. Opin. Biotechnol. 28, 51–58. 10.1016/j.copbio.2013.11.005

64. Rozanov M. N., Koonin E. V., Gorbalenya A. E. (1992). Conservation of the putative methyltransferase domain: a hallmark of the ‘Sindbis-like’ supergroup of

positive-strand RNA viruses.J. Gen. Virol. 73, 2129–2134. 10.1099/0022-1317-73-8-2129

65. Ruiz M. T., Voinnet O., Baulcombe D. C. (1998). Initiation and maintenance of virus-induced gene silencing. Plant Cell 10, 937–946. 10.1105/tpc.10.6.937

66. Santa Cruz S., Chapman S., Roberts A. G., Roberts I. M., Prior D. A., Oparka K. J. (1996). Assembly and movement of a plant virus carrying a green fluorescent protein overcoat. Proc. Natl. Acad. Sci. U.S.A. 93, 6286–6290. 10.1073/pnas.93.13.6286

67. Santa Cruz S., Roberts A. G., Prior D. A., Chapman S., Oparka K. J. (1998). Cell-to-cell and phloem-mediated transport of potato virus X. The role of virions. Plant Cell 10, 495–510. 10.1105/tpc.10.4.495

68. Schlick T. L., Ding Z., Kovacs E. W., Francis M. B. (2005). Dual-surface modification of the tobacco mosaic virus. J. Am. Chem. Soc. 127, 3718–3723. 10.1021/ja046239n

69. Shukla S., Ablack A. L., Wen A. M., Lee K. L., Lewis J. D., Steinmetz N. F. (2013). Increased tumor homing and tissue penetration of the filamentous plant viral nanoparticle Potato virus X. Mol. Pharm.10, 33–42. 10.1021/mp300240m

70. Shukla S., Wen A. M., Commandeur U., Steinmetz N. F. (2014a). Presentation of HER2 epitopes using a filamentous plant virus-based vaccination platform. J. Mater. Chem. B 2, 6249–6258. 10.1039/C4TB00749B [Cross Ref]

71. Shukla S., Wen A. M., Ayat N. R., Commandeur U., Gopalkrishnan R., Broome A. M., et al. (2014b). Biodistribution and clearance of a filamentous plant virus in healthy and tumor-bearing mice.Nanomedicine 9, 221–235. 10.2217/nnm.13.75

72. Shukla S., Dickmeis C., Nagarajan A. S., Fischer R., Commandeur U., Steinmetz N. F. (2014c).Molecular farming of fluorescent virus-based nanoparticles for optical imaging in plants, human cells and mouse models. Biomater. Sci. 2, 784–797. 10.1039/c3bm60277j [Cross Ref]

73. Skryabin K. G., Kraev A. S., Morozov S. Y., Rozanov M. N., Chernov B. K., Lukasheva L. I., et al. (1988). The nucleotide sequence of potato virus X RNA. Nucleic Acids Res. 16, 10929–10930. 10.1093/nar/16.22.10929

74. Smolenska L., Roberts I. M., Learmonth D., Porter A. J., Harris W. J., Wilson T. M., et al. (1998).Production of a functional single chain antibody attached to the surface of a plant virus. FEBS Lett.441, 379–382. 10.1016/S0014-5793(98)01586-5

75. Sonenberg N., Shatkin A. J., Ricciardi R. P., Rubin M., Goodman R. M. (1978). Analysis of terminal structures of RNA from potato virus X. Nucleic Acids Res. 5, 2501–2512. 10.1093/nar/5.7.2501

76. Steinmetz N. F., Manchester M. (2009). PEGylated viral nanoparticles for biomedicine: the impact of PEG chain length on VNP cell interactions *in vitro* and *ex vivo*. Biomacromolecules 10, 784–792. 10.1021/bm8012742

77. Steinmetz N. F., Mertens M. E., Taurog R. E., Johnson J. E., Commandeur U., Fischer R., et al. (2010). Potato virus X as a novel platform for potential biomedical applications. Nano Lett. 10, 305–312. 10.1021/nl9035753

78. Streatfield S. J., Kushnir N., Yusibov V. (2015). Plant-produced candidate countermeasures against emerging and reemerging infections and bioterror agents. Plant Biotechnol. J. 13, 1136–1159. 10.1111/pbi.12475

79. Syller J. (2012). Facilitative and antagonistic interactions between plant viruses in mixed infections.Mol. Plant Pathol. 13, 204–216. 10.1111/j.1364-3703.2011.00734.x

80. Tenllado F., Díaz-Ruíz J. R. (2001). Double-stranded RNA-mediated interference with plant virus infection. J. Virol. 75, 12288–12297. 10.1128/JVI.75.24.12288-12297.2001

81. Tilsner J., Linnik O., Wright K. M., Bell K., Roberts A. G., Lacomme C., et al. (2012). The TGB1 movement protein of Potato virus X reorganizes actin and endomembrane into the X-body, a viral replication factory. Plant Physiol. 158, 1359–1370. 10.1104/pp.111.189605

82. Tilsner J., Linnik O., Louveaux M., Roberts I. M., Chapman S. N., Oparla K. J. (2013). Replication and trafficking of a plant virus are coupled at the entrances of plasmodesmata. J. Cell Biol. 201, 981–995. 10.1083/jcb.201304003

83. Uhde-Holzem K., Schlösser V., Viazov S., Fischer R., Commandeur U. (2010). Immunogenic properties of chimeric potato virus X particles displaying the hepatitis C virus hypervariable region I peptide R9. J. Virol. Methods 166, 12–20. 10.1016/j.jviromet.2010.01.017

84. Vaculik P., Plchova H., Moravec T., Hoffmeisterova H., Cerovska N., Smahel M. (2015). Potato virus X displaying the E7 peptide derived from human papillomavirus type 16: a novel position for epitope presentation. Plant Cell Tiss. Organ Cult. 120, 671–680. 10.1007/s11240-014-0634-x [Cross Ref]

85. Van Rijn P., Van Bezouwen L. S., Fischer R., Boekema E. J., Böker A., Commandeur U. (2015).Virus-SiO2 and Virus-SiO2-Au hybrid particles with tunable morphology. Part. Part. Syst. Charact.32, 43–47. 10.1002/ppsc.201400068 [Cross Ref]

86. Voinnet O. (2005). Induction and suppression of RNA silencing: insights from viral infections. Nat. Rev. Genet. 6, 206–220. 10.1038/nrg1555

87. Voinnet O., Lederer C., Baulcombe D. C. (2000). A viral movement protein prevents spread of gene silencing signal in *Nicotiana benthamiana*. Cell 103, 157–167. 10.1016/S0092-8674(00)00095-7

88. Zelada A. M., Calamante G., de la Paz Santangelo M., Bigi F., Verna F., Mentaberry A., et al. (2006).Expression of tuberculosis antigen ESAT-6 in *Nicotiana tabacum* using a potato virus X-based vector.Tuberculosis 86, 263–267. 10.1016/j.tube.2006.01.003

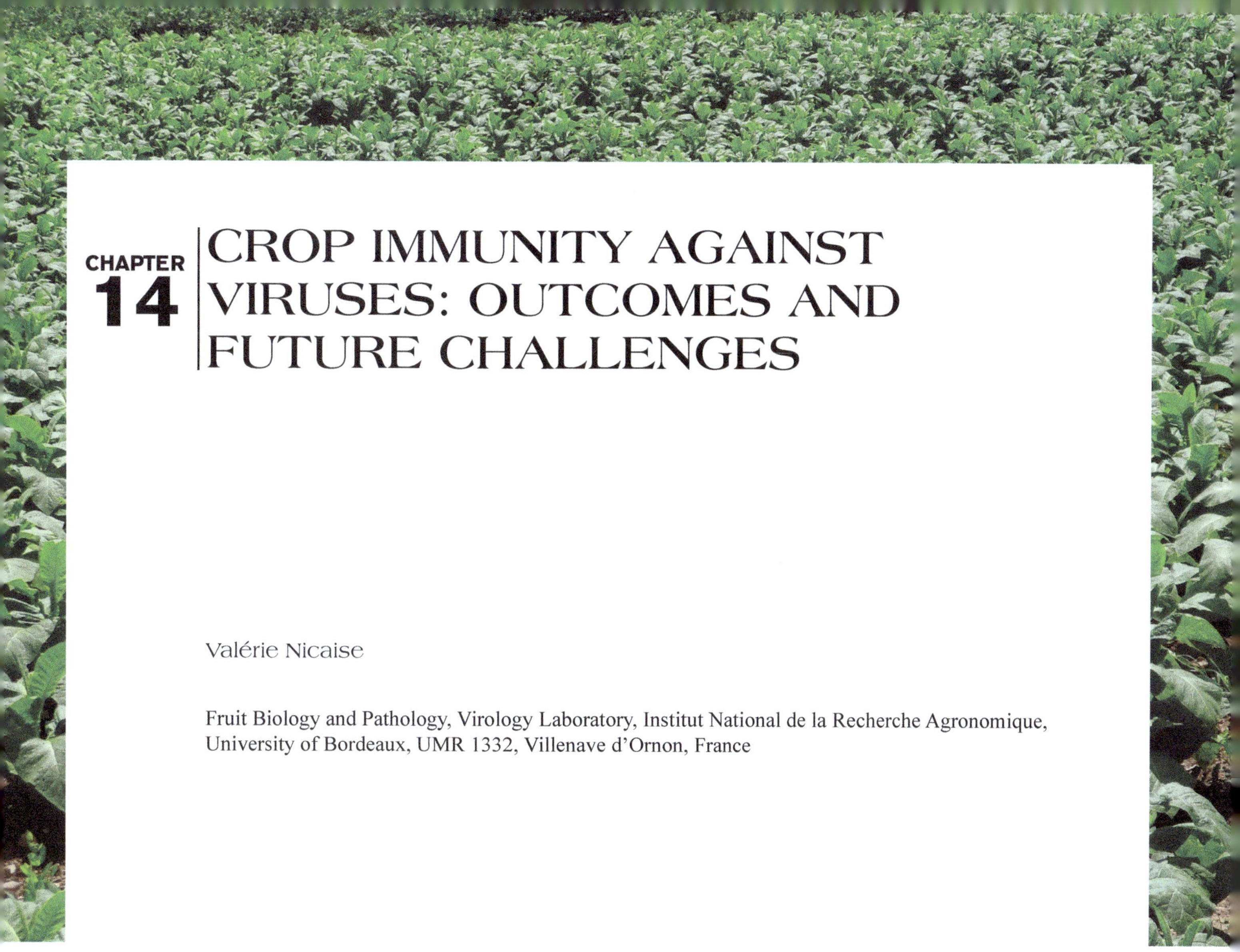

CHAPTER 14

CROP IMMUNITY AGAINST VIRUSES: OUTCOMES AND FUTURE CHALLENGES

Valérie Nicaise

Fruit Biology and Pathology, Virology Laboratory, Institut National de la Recherche Agronomique, University of Bordeaux, UMR 1332, Villenave d'Ornon, France

ABSTRACT

Viruses cause epidemics on all major cultures of agronomic importance, representing a serious threat to global food security. As strict intracellular pathogens, they cannot be controlled chemically and prophylactic measures consist mainly in the destruction of infected plants and excessive pesticide applications to limit the population of vector organisms. A powerful alternative frequently employed in agriculture relies on the use of crop genetic resistances, approach that depends on mechanisms governing plant–virus interactions. Hence, knowledge related to the molecular bases of viral infections and crop resistances is key to face viral attacks in fields. Over the past 80 years, great advances have been made on our understanding of plant immunity against viruses. Although most of the known natural resistance genes have long been dominant R genes (encoding NBS-LRR proteins), a vast number of crop recessive resistance genes were cloned in the last decade, emphasizing another evolutive strategy to block viruses. In addition, the discovery of RNA interference pathways highlighted a very efficient antiviral system targeting the infectious agent at the nucleic acid level. Insidiously, plant viruses evolve and often acquire the ability to overcome the resistances employed by breeders. The development of efficient and durable resistances able to withstand the extreme genetic plasticity of viruses therefore represents a major challenge for the coming years. This review aims at describing some of the most devastating diseases caused by viruses on crops and summarizes current knowledge about plant–virus interactions, focusing on resistance mechanisms that prevent or limit viral infection in plants. In addition, I will discuss the current outcomes of the actions employed to control viral diseases in fields and the future investigations that need to be undertaken to develop sustainable broad-spectrum crop resistances against viruses.

INTRODUCTION

As obligatory intracellular parasites, plant viruses depend on the host machinery to multiply and invade their hosts. In their simplest form, viruses consist of a DNA or RNA genomic segment encoding only few genes and encapsided into a protein shell, called the capsid. Because of natural physical barriers (cuticle, cell wall), viruses are delivered into plant cells through wounds or through the action of vectors (insects, nematodes, fungi) that feed on or infect the plants. Following entry into a host cell and genome decapsidation, the infectious cycle includes translation and replication of the viral genome, assembly of progeny virus particles, generalized

Citation: Valérie Nicaise, Crop immunity against viruses: outcomes and future challenges, doi: 10.3389/fpls.2014.00660.

invasion of the host through cell-to-cell and long-distance movements of viral particles or ribonucleoprotein complexes and finally, transmission to new hosts by vectors. In some cases, transmission to the following generation of host plant is also observed as a result of seed infection. In 2012, the International Committee on Taxonomy of Viruses reported 92 genera of plant viruses of which 82 were assigned in 21 different families (King et al., 2012).

Plant infection by viruses causes physiological disorders responsible for plant diseases of economic and agronomic significance in many crops. Widely employed in fields, the use of plant varieties carrying genetic resistances constitutes the most effective, economical and ecological measure to control viral infections. The last decades have seen substantial advances on the molecular dialog between viral pathogens and their plant hosts, bringing new strategies directly exploitable in crop improvement programs. Nevertheless, the spread of crop viral pests has increased dramatically in recent years. Globalization, trade and climate change, as well as reduced resilience in production systems due to decades of agricultural intensification have all played a part.

This review intends: (i) to provide a brief overview of severe virus-associated plant diseases and their impact on crop production, (ii) to summarize prophylactic strategies employed to control viral epidemics in fields, (iii) to bring an update on current knowledge about plant resistances against viruses, (iv) to present the technical approaches currently employed in crop improvement programs, and (v) to discuss how antiviral mechanisms based on PAMP-triggered immunity (PTI) could be the source of novel plant resistances in fields.

WHAT IS THE REAL IMPACT OF VIRAL DISEASES ON CROPS?

Management of plant virus diseases is a matter of vital importance and concern to farmers, horticulturists, foresters, manufacturers, as well as consumers. It is well-established that virus diseases in different crops cause enormous losses all over the world in terms of quantity and/or quality of products. Although it's very difficult to put a clear figure on the financial impact of plant viruses in agriculture, the yield losses that can be ascribed to plant viruses are estimated to cost worldwide more than $30 billion annually (Sastry and Zitter, 2014).

Amongst the most damaging virus-associated threats, strains of *Cassava mosaic begomoviruses* cause more than 25 million tons of losses per year in Africa, India, and Sri Lanka (Legg and Thresh, 2000; Calvert and Tresh, 2002; Thresh and Cooter, 2005). Because the Cassava crop represents the daily staple for more than 500 million people all over the world, epidemics are often associated to famine events (Legg, 1999; Legg and Thresh, 2000; Calvert and Tresh, 2002). *Potato leafroll polerovirus* is responsible for an annual potato loss of $100 millions in the US and #x000A3;30–50 millions in UK (Wale et al., 2008; Sastry and Zitter, 2014). The losses in citrus tree cultures attributed to *Citrus Tristeza closterovirus* (CTV) were estimated to over 100 million trees worldwide (Moreno et al., 2008; Harper, 2013). Barley yellow dwarf is the most widely distributed viral disease of cereals, affecting oats, rice, barley, maize, and wheat. It is caused by *Barley yellow dwarf luteovirus*, that costs producers in UK alone about £10 million a year in lost production (Ordon et al., 2009; Sastry and Zitter, 2014). It is estimated that the international costs of managing the Sharka disease [caused by *Plum pox potyvirus* (PPV)] since the 1970s have exceeded 10 billion euros (Cambra et al., 2006). Viruses affecting rice cultures result in yield losses estimated at more than $1.5 billion in South-East Asia alone (Abo and Sy, 1998; Hull, 2013; Sasaya et al., 2013). Discovered in Ghana in the middle of the 20th century, *Cacao swollen shoot badnavirus* is currently endemic in Togo, Ghana, and Nigeria. Over 200 million trees have already been eradicated, representing the most costly effort of any country in the world against a viral plant disease (Dzahini-Obiatey et al., 2010).

It's important to mention that diseases of perennial and fruit crops not only lead to the loss of the crop but the loss of time and cost in bringing the trees to bearing, the losses of other crops that could have been grown on the land during that time, and the differences in the value of the land with and without a productive orchard. Moreover, virus-associated losses in fields are highly under-estimated as some viral infections are asymptomatic alone but contribute in a synergic manner to damages due to attacks from other pathogens (Hull, 2013). More importantly, viruses have been described causing half of the reported emerging infectious diseases from plants (Anderson et al., 2004). Hence, viruses represent serious crop threats, responsible for considerable agronomical losses, global food problem, and human life overall all over the world.

PROPHYLACTIC MEASURES IN FIELDS

Because viral agents are obligate intracellular parasites, curative treatments of virus infections are impossible, making viral diseases very difficult to control in fields. Prophylactic control measures are therefore crucial in combating epidemics on crops. They consist mainly on combining cultural practices, biosecurity measures and organism-vector management (Figure1).

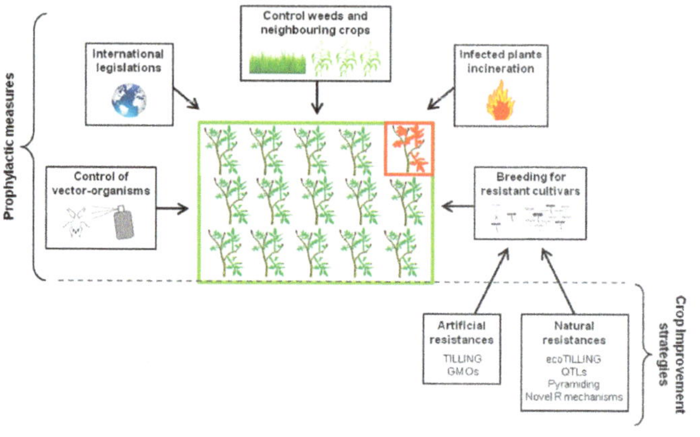

Figure 1: Prophylactic measures and main crop improvement strategies employed to control plant viral diseases.

Perform Regular Inspection for the Presence of Viral Pathogens

In this domain, the rising up of molecular biology techniques combined to continuous characterization of new etiological agents improved significantly the sensitivity, the specificity and the rapidity required to an accurate diagnosis of plant pathogenic viruses (Boonham et al., 2014). The reliability of the available diagnosis tests is a key point in viral disease management in fields, as infected plants need to be eradicated as fast as possible to minimize the virus spread.

Monitor Organism-Vector Populations

Plant viruses need to be transmitted by an organism-vector (insects, nematodes, zoosporic endoparasites) for their plant-to-plant spread. Hence, viral diseases can be efficiently controlled by limiting the populations of their vectors with the applications of appropriate pesticides. The use of non-host "trap plants" may be also considered to attract vectors to reduce the number of individuals feeding on the crop of interest and thus, the transmission of the disease (Bragard et al., 2013).

Set Up a Rigorous Control Program on Weeds and Other Host Plants in the Vicinity of the Field

Epidemics often arise from new viruses or new variants of classic viruses that spilled over from reservoir species to crops. Although this phenomenon results from a complex evolutionary process in which the main players are ecological factors, virus genetic plasticity and host factors, viral diseases can be controlled by managing the spatial structure and composition of field parcels, which impacts resistance durability (Elena et al., 2011; Fabre et al., 2012).

Respect the Phytosanitary Measures Decreed By Various International Commissions

Minimizing viral epidemics involves the respect of international legislations concerning worldwide trade of virus-free plant material, which applies to any development stage of a plant that can be carrier of viruses (seeds or fruit stones, grafts, rootstocks, seedlings, flowers,...), as well as manipulation of decontaminated horticultural tools.

Use Crop Cultivars That are Resistant to Viruses

The use of genetically resistant plants is one of the most efficient, sustainable and frequently employed strategies to control virus infections in fields. For centuries, it has involved plants selected by breeders for their agronomic proprieties combined to the absence of disease symptoms. However, from the middle of the 20th century, plant improvement programs capitalize strongly on the knowledge associated to plant–virus interactions to develop resistant varieties exploitable in agriculture.

WHAT DO WE KNOW ABOUT PLANT IMMUNITY AGAINST VIRUSES?

Faced with viral attacks, plants defend themselves through several resistance layers, that are complementary in terms of defense timing (at early or late infection steps), location (in the first infected leaf or in systemic tissues) and targeting the virus-derived molecules (the viral genome or the viral proteins; Figure2).

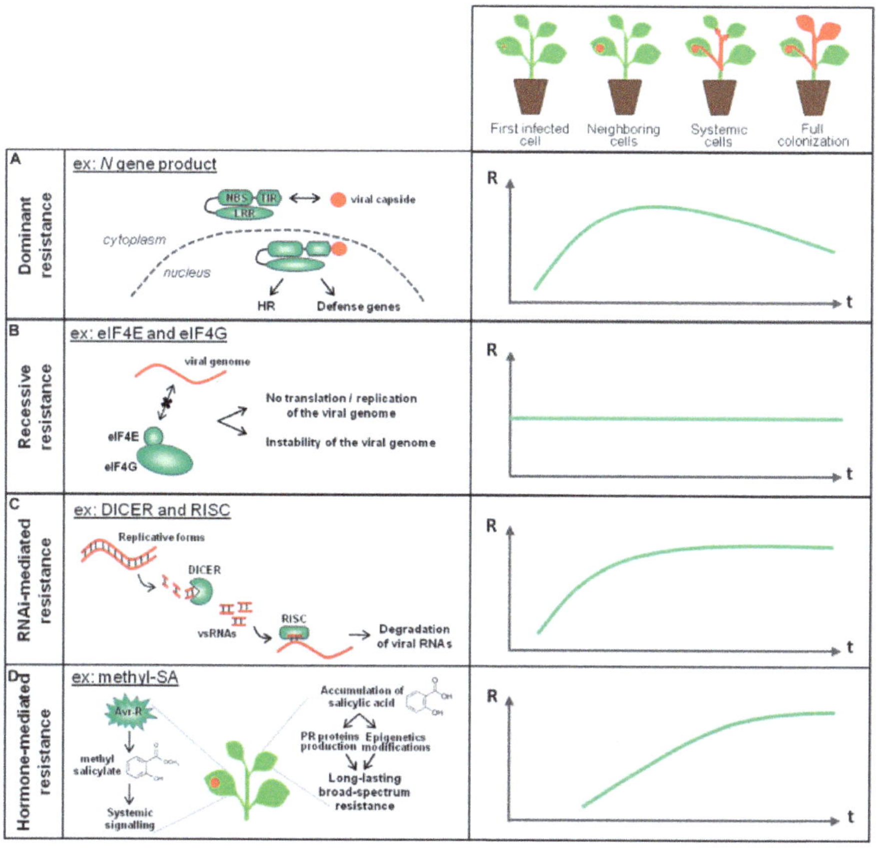

Figure 2: Known antiviral immune mechanisms in plants. Plant resistance mechanisms against viruses are complementary in terms of plant defense timing, location (from the first infected cell to the generalized colonization) and targeting the virus-derived molecules ...

Dominant Resistances

The majority of dominant resistance genes (*R* genes) identified in plant–virus interactions belong to the nucleotide binding site-leucine-rich repeat (NBS-LRR) class, that specifically recognize the viral avirulence (*avr*) gene products, through the establishment of the so-called "gene-for-gene" interaction. Although a direct physical interaction between the *avr* and *R* gene products had originally been suggested, the current understanding favors the more sophisticated "guard hypothesis" model (Soosaar et al., 2005). Many NBS-LRR proteins conferring resistance against viruses have been identified so far, and are classified on the basis of their N-terminal structure, that carries either a Toll–interleukin-1 receptor (TIR) domain or a coiled-coil (CC) domain (Moffett, 2009; De Ronde et al., 2014).

One of the best characterized example is the potato *Rx1* gene that encodes a typical CC-NBS-LRR protein and mediates resistance to *Potato potexvirus X* (PVX) through the recognition of the PVX capsid (CP). Rx1 CC domain has been shown to form a heterodimer with the cellular ranGTPase-activating protein 2 (ranGAP2), interaction required for Rx1 function (Rairdan et al., 2008; Tameling et al., 2010). Although direct interaction between ranGAP2 and PVX CP has not been detected, it has been proposed that ranGAP2 might nonetheless interact with the CP, causing a conformational change perceived by the rest of the protein and leading Rx to switch to an active state (Hao et al., 2013). In another well-characterized *R* gene example, the tobacco *N* gene product, that displays a TIR-NBS-LRR structure, interacts directly with the helicase domain of the replicase of *Tobacco mosaic tobamovirus* (TMV), in an ATP dependent manner (Figure2A;Ueda et al., 2006). Full resistance to TMV requires the N receptor-interacting protein 1 (NRIP1), which is recruited from the chloroplast to the cytoplasm and nucleus and interacts with both viral replicase and N factor (Caplan et al., 2008). In both *Rx1* and *N* cases, the *R* gene product is activated in the cytoplasmic compartment, while its nucleocytoplasmic distribution is required for full functionality (Slootweg et al., 2010;Tameling et al., 2010). The study of *N*- and *Rx*-mediated resistances have led to characterize the R signaling cascade in plant–virus interactions, that include rapid activation of MAP kinases and the action of a molecular chaperone complex composed of SGT1 (*Suppressor of G2 allele of SKP1*), HSP90 (*Heat shock protein*), and RAR1 (*Required for Mla resistance 1*; Azevedo et al., 2006; Botër et al., 2007; Kobayashi et al., 2010; Hoser et al., 2013), whose role seems to both stabilize R factors and mediate their degradation, thereby maintaining a tight cellular balance between defense signaling and attenuation (Kadota and Shirasu, 2012).

Immune events downstream of R protein activation are frequently associated with calcium ion influx, MAPK-mediated signaling, reactive oxygen species (ROS) production, salicylic acid (SA) accumulation, and extensive transcriptional reprogramming. In addition, the activation of *R* genes is most of the time associated with a hypersensitive response (HR), a phenomenon involving the programmed death of the infected and neighboring cells, with the consequences to confine the pathogen in the hypersensitive lesion and to prevent any further pathogen spread in the plant. A functional module that mediates HR against viruses (as well as non-viral pathogens) requires the interaction of two lipases, EDS1 (*Enhanced disease susceptibility 1*) and PAD4 (*Phytoalexin deficient 4*) with the protein SAG101 (*Senescence-associated gene 101*; Liu et al., 2002;Marathe et al., 2002; Wagner et al., 2013). In *Arabidopsis*, the EDS1/PAD4/SAG101 complex regulates*HRT*-mediated resistance against *Turnip crinkle carmovirus* (Zhu et al., 2011). Recent insights into R protein downstream signaling indicate that the resistance process and the HR-programmed cell death are distinct physiological pathways, although both can work in concert (Bendahmane et al., 1999; Bai et al., 2012). Meanwhile *R* gene mediated defense is taking place locally, it also induces defense signaling in distally located tissues, called systemic acquired resistance (SAR), a mechanism that has been demonstrated for both*N* and *Rx1* genes (Delaney et al., 1994; Liu et al., 2010) and that is described in the Section "Plant Hormone-Mediated Resistance" of this review.

In contrast with the structure of the classical *R* genes, the *RTM* genes (for *Restricted TEV Movement*) were the first cloned non-NBS-LRR dominant resistance genes against viruses. Genetic characterization of natural*Arabidopsis* accessions and mutants showed that at least five dominant *RTM* genes are involved in resistance to *Tobacco etch potyvirus* (TEV), *Lettuce mosaic potyvirus* (LMV), and PPV (Cosson et al., 2012). It has been recently proposed that RTM members may form a phloem-resident multiprotein complex involved in the resistance mechanisms to block the long-distance movement of potyviruses (Cosson et al., 2010a,b). No induction of HR or production of SA has been detected, in contrast with NBS-LRR mediated resistance responses (Mahajan et al., 1998). Another example of a non-NLS-LRR dominant resistance conferring resistance to TMV comes from the tomato *Tm-1* gene, which encodes a protein with a TIM-barrel like structure, that interacts directly with the viral replicase, strongly impairing the viral genome replication (Ishibashi and Ishikawa, 2013).

Resistance Recessives

Around one-half of the approximately 200 known resistance genes that target plant viruses are recessively inherited (Diaz-Pendon et al., 2004), suggesting that this form of resistance is more common for viruses than for other plant pathogens. The use of such genes is therefore a very important tool in breeding programs to control plant diseases caused by pathogenic viruses.

Recessive resistances are often achieved through the absence of appropriate host factors that are required for the virus to complete its biological cycle. Over the last decade, a large number of recessive resistance genes have been cloned from crop species and shown to encode eukaryotic initiation factors (eIFs) belonging to the eIF4E and eIF4G families (Figure2B; Truniger and Aranda, 2009; Wang and Krishnaswamy, 2012; Julio et al., 2014; Revers and Nicaise, 2014). These factors belong to the cellular 43S complex, that recruit both mRNAs and ribosomes before the process of translation (Pestova et al., 2001) and are hypothesized to act as susceptibility factors recruited during the replication/translation steps of the viral genome. Firstly highlighted for successful potyvirus infection, the role of eIF4E and eIF4G have been quickly extended to other plant virus families, involving bymoviruses, cucumoviruses, ipomoviruses, sobemoviruses, carmoviruses, and waikiviruses (Revers and Nicaise, 2014), suggesting that they contribute to a broad mechanism of plant susceptibility to viruses. An intriguing aspect of eIF4E- and eIF4G-mediated resistances is that they cover a diverse range of resistance phenotypes. Although in many investigated cases they govern complete qualitative resistance, they have also been shown in some instances to provide partial resistance or to be components of partial or polygenic resistances (Caranta et al., 1997; Nicaise et al., 2003; Acosta-Leal and Xiong, 2008; Charron et al., 2008). Interestingly, ectopic overexpression of an *eIF4E* resistance gene generates dominant potyvirus resistance in tomato and potato crops (Kang et al., 2007; Cavatorta et al., 2011), probably through the fact that the abundance of the "resistant eIF4E" makes the endogenous "susceptible eIF4E" inaccessible for its recruitment by viruses. Other components belonging to the translational machinery were demonstrated to be required for virus multiplication in *Arabidopsis thaliana* and*Nicotiana benthamiana*, including the translation initiation factor 4B (eIF4B), the translation elongation factors 1A and 1B (eEF1A and eEF1B) and the poly(A)-binding proteins (PABPs) 2, 4, and 8 (Sasvari et al., 2011; Patarroyo et al., 2012; Hwang et al., 2013; Li et al., 2014b). These results suggest that crop genes encoding these translation factors may lead to new resistance sources that need to be explored for viral disease control.

Other susceptible recessive resistance genes which are not encoding translation initiation factors have been identified through the analysis of mutants collections and/or natural cultivated or wild species. A positional cloning strategy exploring barley natural variability revealed recently the key role of the PDI5-1 (*Protein disulfide isomerase like 5-1*) protein in the recessive resistance to bymoviruses (Yang et al., 2014). Another recessive resistance gene named *ra* blocking vascular transport of *Potato potyvirus A* (PVA) was genetically characterized in potato (Hämäläinen et al., 2000). In addition, from the perspective of identifying new resistance sources, exploring *Arabidopsis* genetic diversity has provided original insights into the genes involved in plant–virus interactions, which could be used as potentially resistance sources against viruses. For example, the recessive resistance genes *rlm1* and *rpv1*, conferring resistance to LMV and PPV respectively, map in a genomic region containing no translation factor genes (Revers et al., 2003; Decroocq et al., 2006). The lack of co-segregation with *eIF4E* or *eIF4G* genes is also true for *dstm1* mediating TMV resistance (Serrano et al., 2008) and *sha3*, a major quantitative trait locus (QTL) contributing to systemic resistance against PPV (Pagny et al., 2012). Very recently, data revealed that the gene *rwm1* conferring resistance to*Watermelon mosaic potyvirus* in *Arabidopsis* acts at an early stage of infection by impairing viral accumulation in initially infected leaf tissues and encodes a conserved nucleus-encoded chloroplast phosphoglycerate kinase (Ouibrahim et al., 2014). In the case of *tom1* and *tom2A Arabidopsis* mutants, TMV accumulation is suppressed in single cells. After further characterization, it appears that both genes encode transmembrane proteins localized in the tonoplast and are required for tobamovirus replication (Ishibashi et al., 2012). Consistent with the fact that viral replication complexes associate with host intracellular membranes (Hull, 2013), prospective studies identified many membrane-associated components as key factors required for plant infection success, providing new candidates for novel genetic sources of crop resistances (Diaz et al., 2010; Hyodo et al., 2013; Barajas et al., 2014a,b).

RNA Interference Mediated-Resistance

Over the past decades, RNA interference (RNAi; also called gene silencing) has been recognized as an evolutionarily conserved process in most eukaryotes, that is triggered by double-stranded RNAs (dsRNAs). These dsRNAs are processed by ribonuclease III-type DICER-like (DCL) enzymes into small RNAs (sRNAs), 21–24 nucleotides in length, that are incorporated into an RNA-induced cytoplasmic silencing complex (RISC), whose key catalytic component corresponds to one member of the ARGONAUTE (AGO) protein family. Once integrated into the RISC, sRNAs base-pair to their target-mRNA and induce their cleavage (Bologna and Voinnet, 2014). In plants, silencing pathways are particularly diverse and partially overlapping. At least, three basic processes can be distinguished: cytoplasmic RNA silencing (or post-transcriptional gene silencing; PTGS) mediated by small interfering RNAs (siRNAs), silencing-mediated by plant-encoded microRNAs (miRNAs) and transcriptional gene silencing (TGS) mediated by siRNA-directed methylation of DNA and histone proteins. Key components of these RNA silencing pathways have been shown to have an important protective role against invading viral pathogens (Bologna and Voinnet, 2014).

Most plant viruses have RNA genome that commonly contain double-stranded secondary structure elements and/or produce dsRNA intermediates via the action of viral RNA-dependent RNA polymerases (RDRs) during the replication step. These molecules are targeted by the RNA-silencing machinery to produce virus-derived small RNAs (vsRNAs; Pumplin and Voinnet, 2013; Figure2C). What initially appeared counter-intuitive, infection from DNA viruses produce also dsRNAs, most likely via bidirectional convergent transcription (Aregger et al., 2012). The integration of vsRNAs in the RISC leads to the sequence-specific degradation of viral nucleic acids, as well as the generation of a mobile-silencing signal, which spreads between cells through plasmodesmata and over long-distances via the phloem, through a relay-amplification process involving host RDRs (Pumplin and Voinnet, 2013). This process activates RNA silencing in non-infected cells and is notably responsible for the plant recovery phenomenon. Given gene silencing induces immune mechanisms highly specific to the pathogen, it is commonly accepted that RNAi is classified into plant adaptive immunity (Voinnet, 2001; Waterhouse et al., 2001).

Encoded by multigene families, plant DCLs, RDRs and AGOs are often specialized in the production and function of the distinct sRNA classes. Despite its universality, current knowledge remains incompletely understood, as it relies mostly on reverse genetic studies conducted in *Arabidopsis*. This notwithstanding, it appears clear that specific RNAi-associated components are involved in silencing antiviral functions. Thus, DCL4 is the major enzyme for generating RNA virus-derived vsRNAs, even if DCL2 can substitute for DCL4 to some extent (Blevins et al., 2006; Deleris et al., 2006; Qu et al., 2008; Jakubiec et al., 2012). In contrast, all four DCLs (DCL1-4) produce DNA virus-derived vsRNAs (Akbergenov et al., 2006; Moissiard and Voinnet, 2006; Blevins et al., 2011). So far, only AGO1, AGO2, and AGO7 seem to contribute to antiviral RNAi (Morel et al., 2002; Qu et al., 2008; Shivaprasad et al., 2008; Azevedo et al., 2010; Harvey et al., 2011; Jaubert et al., 2011; Wang et al., 2011; Zhang et al., 2012). In particular, RNA viruses-derived vsRNAs seem to be strongly recruited by AGO1 and AGO2 proteins, an observation consistent with the hyper-susceptibility of *ago1 ago2* double mutants (Wang et al., 2011). Antiviral RNAi has also been shown to be dependent on one or more of RDR1, RDR2, and RDR6 for signal initiation and/or amplification (Diaz-Pendon et al., 2007; Qu et al., 2008; Qi et al., 2009; Vaistij and Jones, 2009; Garcia-Ruiz et al., 2010; Wang et al., 2011; Jiang et al., 2012). An increase of susceptibility against RNA viruses is also observed in*Arabidopsis* plants defected for the gene *HEN1* (for HUA ENHANCER 1) encoding a methyltransferase that protects siRNA and miRNA duplexes from degradation (Vogler et al., 2007; Zhang et al., 2012). Interestingly, the natural resistance genes *Ty-1* and *Ty-3* conferring resistance to *Tomato yellow leaf curl begomovirus* (TYLCV) has been recently shown to encode a tomato RDR. Although resistant plants do not show symptoms upon a challenge with TYLCV, low levels of virus are still detectable, a phenomenon characteristic of a virus tolerance more than a real resistance (Verlaan et al., 2013).

Viruses have evolved diverse mechanisms to avoid silencing-mediated resistance, most notably through silencing suppressor activities. Identified for almost all types of plant viruses, silencing suppressors target RNAi pathways at different points and through diverse mechanisms, including the impair of siRNA biogenesis, the defect of siRNA incorporation into the RISC, the degradation of AGOs, the trapping of sRNAs and the suppression of RNAi amplification (reviewed in Burgyán and Havelda, 2011; Pumplin and Voinnet, 2013; Bologna and Voinnet, 2014; Li et al., 2014a). In turn, increasing evidence suggest that plants have evolved by establishing specific defenses against RNA-silencing suppression by pathogens, providing yet another illustration of the never-ending molecular arms race between plant pathogens and their hosts (Pumplin and Voinnet, 2013; Sansregret et al., 2013).

Plant Hormone-Mediated Resistance

Plant hormones play important roles in regulating signaling networks involved in plant defenses. Upon pathogen attack, the quantity, composition and timing of the plant hormonal blend produced by the plant depends greatly on the lifestyle and infection strategy of the invading attacker. In the last decades, significant progress has been made in identifying the key components and understanding the role of phytohormones in plant responses to biotic stresses (Robert-Seilaniantz et al., 2011).

During the R-mediated resistance activation (see the Section "Dominant Resistances" in this review), cellular responses elicited at the infection site are emitted to distant non-infected tissues, resulting in a resistance or in reduced susceptibility state that can remain efficient during several weeks (Fu and Dong, 2013). This phenomenon is referred as the SAR (Figure2D). In the case of TMV-triggered HR, the response persists up to 3 weeks during the time plants are protected against not only TMV but also other pathogens (Ross, 1961). How SAR can be sustained for so long is not clear but epigenetic modifications, such as DNA methylation and chromatin remodeling, seem critical to maintain a SAR signal (Spoel and Dong, 2012). Moreover, during a viral infection (in a manner similar to non-viral infections), this long lasting and broad-spectrum disease resistance requires endogenous accumulation of SA, resulting in transcriptional reprogramming of a battery of genes encoding pathogenesis-related (PR) proteins (Tsuda et al., 2008; Yi et al., 2014). The signal emitted by the infection spot to protect the uninfected tissues against pathogen invasion may circulate as an heterocomplex, where methyl-SA binds to lipid derivatives and lipid-transport proteins and moves through the phloem to the rest of the plant. Thus, studies on TMV-infected tobacco plants revealed that MeSA participates to the perpetuation of SAR defense (Park et al., 2007; Dempsey and Klessig, 2012). Recent data suggest that the composition of the mobile immune signal in SAR relies on a complex network of cross-interacting signals (e.g., MeSA, glycerol-3-phosphate, the lipid-transfer protein DIR1 and the amino acid-derivative pipecolic acid), that differs depending on the plant species and the type of plant–pathogen interaction (Vlot et al., 2008; Dempsey and Klessig, 2012; Spoel and Dong, 2012). Jasmonic acid (JA) is also strongly involved in plant defense against viruses and the HR response initiated by Avr-R protein interactions results on a modulation of SA and JA. Although SA clearly acts as a positive regulator of plant resistance to viruses, the role of JA is controversial and remains to be fully elucidated. For example, JA seems to regulate negatively the local resistance to TMV in tobacco (Oka et al., 2013) but is essential for systemic resistance to TMV in *N. benthamiana* (Zhu et al., 2014). It is likely that a balance between endogenous JA and SA play a key role for determining the degree of resistance, in a similar way to pathosystems involving non-viral plant pathogens (Thaler et al., 2012). Interestingly, plant viruses have evolved targeting hormone pathways, often exploiting the antagonistic interactions between SA and JA pathways (Kazan and Lyons, 2014). Although their function in plant–virus interactions remains poorly understood, recent studies indicate that other plant hormones modulate antiviral resistance mechanisms, especially abscisic acid (Chen et al., 2013; Alazem et al., 2014; Seo et al., 2014), ethylene (Fischer and Dröge-Laser, 2004; Love et al., 2007; Zhang et al., 2009; Chen et al., 2013), and brassinosteroids (Ali et al., 2014).

CROP IMPROVEMENT AND VIRAL PEST MANAGEMENT: MODERN TECHNOLOGY TO THE RESCUE

Ultimately, the main objective of research on plant–virus interactions consists on the implementation of efficient antiviral resistances in crop plants. For long, genetics-mediated resistance strategies have involved exploiting plant natural variability by introgression of resistance genes through a classical breeding process. However, the onset of evolved pathogens able to overcome these resistance, sometimes very rapidly after resistance deployment (García-Arenal and McDonald, 2003), raised the problem of crop resistance durability and thus the urge to develop new breeding strategies. The last decades saw the emergence of new approaches combining modern technology and state-of-the-art knowledge on plant–virus interactions toward crop improvement programs, triggering a new green revolution in agriculture.

Technical Assets from Marker-Assisted Selection

The conventional breeding represents a laborious and time consuming process. In this context, the advent of DNA marker techniques such as random amplified polymorphic DNA (RAPD), restriction fragment length polymorphism (RFLP), amplified fragment length polymorphism (AFLP), and simple sequence repeats (SSRs) contributed, through what is called "Marker-assisted selection (MAS)," to make substantially easier, quicker, and more accurate the selection of resistant genotypes during the introgression steps (Collard and Mackill, 2008). Importantly, this progress has recently been driven by next generation sequencing- (NGS-) based technologies successfully employed for *de novo* whole genome shotgun (WGS) sequencing of reference genotypes and whole genome resequencing (WGRS) of several cultivars, land races, and wild relatives (Mascher and Stein, 2014). Despite the fact that the creation of improved varieties via the MAS strategy is still at its infancy (Collard and Mackill, 2008), MAS has an enormous potential and represents a great challenge of molecular breeding in the 21st century.

Exploiting Resistance-Associated QTLs

It is probable that only a small proportion of the natural biodiversity available for disease resistance has been exploited so far. Although more than 80% of reported plant resistances to viruses are monogenically controlled, most agronomic traits in crop plants do not segregate as single defined qualitative monogenic characters but as quantitative and polygenically controlled traits (Maule et al., 2007). Mapping QTL for quantitative resistance requires large sized progenies, nearly saturated genetic maps, as well as reliable and quantitative phenotyping procedures. To date, relatively few QTLs analyses have been performed in plant–virus interactions compared to other pathogens (Maule et al., 2007; Palloix and Ordon, 2011). It is a general assumption that the greater the number of mutations required for virus virulence, the more durable is the resistance. Consequently, the higher durability of polygenic resistances in plant–virus interactions is commonly hypothesized and has recently been validated in the case of pepper resistance to *Potato potyvirus Y* (Palloix et al., 2009; Quenouille et al., 2013). Notwithstanding their clear relevance, QTLs present particular technical challenges for their characterization as well as their incorporation into crops, leading them to be mostly dropped in favor of approaches using monogenic resistances or combining several known major genes.

Gene Pyramiding Strategy

The concept of transferring several characterized resistance genes into one plant is called "Gene Pyramiding." The dogma behind this strategy is that the probability of a pathogen mutating to "virulence against all resistance genes in the pyramid would be the product of the probabilities for each gene singly" (Mundt, 2014), thus making the probability of a virulent pathotype arising highly unlikely. Hence, pyramiding relies on resistance genes that have been previously characterized singly and whose functions are combined within the same plant. This strategy has been successfully applied to plant–virus interactions (Werner et al., 2005; Palloix et al., 2009; Shi et al., 2009) and could bring the opportunity to associate different types of antiviral plant resistances targeting different virus-derived molecules (proteins or nucleic acids) and various processes during virus cycle (replication/translation, plasmodesmata crossing, systemic colonization), with the possible consequence to reduce significantly the probability of resistance breaking by new virus variants (Quenouille et al., 2013). Despite arduous population sizes required for multiple resistance genes along with other agronomical traits, gene pyramiding is gaining considerable importance in plant–pathogen interactions and thus is representing an ambitious challenge for crop improvement programs.

Engineering Resistances

Genetic engineering directly manipulates the genome of a plant by inducing the expression of novel proteins or by modulating the expression of targeted genes. The genetically modified (GM) plants resistant to viral attacks are grouped into three major categories. First of all, resistance genes with a dominant determinism can be introduced to another plant species. The best examples in plant–virus interactions involve the successful transfer of *R* genes into solanaceous related plants (Whitham et al., 1996; Bendahmane et al., 1999; Baurès et al., 2008; Candresse et al., 2010) but it seems also possible across plant families (Seo et al., 2006). The opportunity to introduce *R* genes from sexually incompatible species is, however, weakened by the fact that most *R* genes display a limited-spectrum resistance. Alternative transgenic approaches have also been developed based on the integration of a viral protein or genomic region into the host plant, through a phenomenon called pathogen-derived resistance (PDR). Initially reported in transgenic tobacco plants expressing the TMV CP (Abel et al., 1986), the resistance observed was ultimately ascribed to the action of the transgene-encoded CP, which disturbs the disassembly of incoming TMV particles. In many other studies, resistance was obtained through the expression of partial or non-coding viral sequences. This kind of resistance, more efficient that the protein-mediated resistance, was later shown to be based on the RNAi mechanism (Tenllado et al., 2004). These observations led to a plethora of RNAi-mediated engineered resistances, using virus-derived double-stranded, hairpin RNAs, or artificial miRNAs (e.g., Qu et al., 2007;Hashmi et al., 2011; Shimizu et al., 2011; Yadav et al., 2011; Zhang et al., 2011a,b; Lin et al., 2012;Shekhawat et al., 2012; Lemgo et al., 2013; Odipio et al., 2013; Zhao and Song, 2014). In the cases where virus-resistant GM varieties have been deployed and/or commercialized (including transgenic squash, papaya, plum, potato, and bean), this strategy has so far proven to be remarkably efficient (Tricoli et al., 1995; Lius et al., 1997; Ferreira et al., 2002; Malinowski et al., 2006; Zagrai et al., 2008; Bravo-Almonacid et al., 2012; Aragão et al., 2013). In spite of engineering resistance success (especially RNAi-based resistant crops), public concerns over the potential ecological impact of GM crops and GM organisms overall have so far significantly limited their use, in particular in Europe.

The Tilling and Ecotilling Revolution

The early 2000s have seen the emergence of the TILLING (Targeting Induced Local Lesions IN Genomes) method, that consists of a classical mutagenesis step followed by the targeted search for plants carrying a mutation in a gene of interest (McCallum et al., 2000). By creating artificial polymorphism directly into crops, this technique allows potentially: (i) to replace a resistance allele overcome by viral strains, (ii) to generate novel resistances with a broader spectrum, or (iii) to create a new resistance gene based on knowledge acquired in heterologous systems. Such a strategy increase the natural allelic diversity by the identification of novel artificial alleles. Although this strategy requires the previous characterization of the gene conferring the resistance, the main advantage of TILLING is that it can be applied to any plant species, regardless of its genome size, ploidy level or method of propagation, and without introducing heterologous DNA as for GM plants. Similarly, the TILLING natural alternative (using natural germplasms collections instead of EMS mutants collections) is calling ecoTILLING and consists of exploiting the whole natural variability of a plant species (including wild-related and cultivated genotypes; Comai et al., 2004). Initially developed in *A. thaliana*, both TILLING and ecoTILLING have spread rapidly to other model plants (*Medicago truncatula*, *Lotus japonicus*, *Brachypodium distachyon*) and major crops (e.g., maize, soybean, sorghum, tomato, pepper, cucumber, pea, wheat, banana, bean, rice, barley, *Brassica napus*). It seems now clear that these strategies are emerging as major crop improvement tools, with especially successful examples of recent applications to antiviral protection (Nieto et al., 2007; Piron et al., 2010).

Chemical Treatments

Curative treatments for virus control in fields is impossible. However, chemical applications can prime plant defense, i.e., resistance mechanisms are switched on prior to future infections. Based on the findings that exogenous applications of plant hormones trigger an efficient and broad-spectrum resistance to viral and non-viral pathogens, many SAR-priming molecules have been characterized. Among a series of SA analogs (Gozzo and Faoro, 2013), benzothiadiazole (BTH) has been identified as the safest and most efficient SAR activator and was brought to the market with the common name of acibenzolar-*S*-methyl, that leads to plant protection against many pathogens including viruses (Friedrich et al., 1996; Lawton et al., 1996; Mandal et al., 2008; Takeshita et al., 2013; Trejo-Saavedra et al., 2013). This priming approach represents an environment-friendly and efficient way to control plant diseases by exploiting a natural phenomenon. Despite these positive aspects, so far it has not yet been met with enthusiastic favor by the farmers, as the treatment efficacy with BTH (as well as other SAR-priming molecules) depends on many variables: dose, plant species and cultivar, growth stage of plant, pathogen pressure, climatic conditions and timing of chemical applications (Gozzo and Faoro, 2013). Nevertheless, coupled to genetic strategies, chemical SAR priming is still a valuable method to increase plant resistance against viruses for pest management programs in fields.

Plant Vaccination

Plant vaccination gambles on cross-protection, a phenomenon whereby the inoculation of a virus into a host plant prevents the multiplication of a subsequent challenge virus. This strategy mainly relies on the manipulation of the primary virus, whose infection is weak (symptomless with low viral load), and that triggers virus-induced gene silencing (VIGS) targeting both primary and challenge viruses (Ziebell and Carr, 2010; Nishiguchi and Kobayashi, 2011). Primary viruses act as vaccines and are classified into two categories: the attenuated and the engineering viruses. An attenuated virus corresponds to a weak isolate that triggers cross-protection against virulent isolates of the same virus or closed related viruses (Ziebell and Carr, 2010; Nishiguchi and Kobayashi, 2011). Many examples of cross-protection have been identified since the discovery of this phenomenon by McKinney in 1920s (McKinney, 1926; Crowdy and Posnette, 1947;Fletcher, 1978; Wang, 1991; Wen et al., 1991; Hugues and Ollennu, 1994; Valkonen et al., 2002; Kurihara and Watanabe, 2003; Ichiki et al., 2005; You et al., 2005; Kosaka et al., 2006; Yoon et al., 2006; Ziebell et al., 2007; Nakazono-Nagaoka et al., 2009; Kurth et al., 2012). A great illustration of this method involves an attenuated isolate of *Zucchini yellow mosaic potyvirus*, that has since been registered as the pesticide CUBIO ZY-02 and successfully employed in virus disease control in Japan (Kosaka and Fukunishi, 1997; Kosaka et al., 2009). Moreover, tristeza disease caused by CTV is currently controlled by mild CTV isolates, which when inoculated into existing field trees, extend the productive life of orchards and enable a more graduate replanting of trees on CTV-tolerant rootstocks (Lee and Keremane, 2013). However, identifying effective attenuated viruses for each virus of economical importance might be very arduous. A possible solution came recently from the development of engineering vaccines based on viral vectors carrying a genomic fragment of the virulent virus of interest (Culver, 1996; Bazzini et al., 2006; Tamura et al., 2013; Taki et al., 2013). Given *Apple latent spherical cheravirus* (ALSV) is able to infect a broad range of herbaceous crops (e.g., tomato, lettuce, zucchini, spinach, soybean, tobacco) and fruit trees (e.g., apple, cherry, peach, plum, pear, and citrus) without causing any symptoms, this virus was identified as an excellent candidate for VIGS-mediated vaccination, that could be simply applied widely by replacing the insert with a sequence derived from a different virus. This strategy has been recently illustrated (Tamura et al., 2013; Taki et al., 2013) and represents a very promising technology in virus epidemics management in fields.

NEW RESEARCH LEAD: LET'S TALK ABOUT ANTIVIRAL PTI IN PLANTS

In an attempt to increase resistance durability in fields, crop improvement requires a continuous pipeline of new resistance genes. Given the importance of PAMP-triggered immunity (PTI) in the field of pathology over the last decades, PTI against plant viruses represents an unexplored question that needs to be addressed.

Plant Innate Immunity

Over the last decades, a concept revolutionizing the understanding of immunity emerged in plant pathology. This new concept stems from the ability for each organism to discriminate between self and non-self molecules through the action of pattern recognition receptors (PRRs) perceiving specific microbial molecular signatures, named pathogen-associated molecular patterns (PAMPs). Perception of PAMPs by these immune receptors induces a downstream signaling cascade including PRR association with positive regulators, phosphorylation events, successive activation of cytoplasmic kinases (including the MAP kinases) and defense-related transcription factors, as well as specific defense genes expression. This range of fast, efficient and multi-layered defense responses is referred as PTI (Nicaise et al., 2009; Macho and Zipfel, 2014). To counteract this defense strategy, successful pathogens deploy effectors proteins, the primary function of which is to evade/interfere with PTI. In turn, some plant cultivars have evolved R proteins to block these effectors, leading to effector-triggered immunity (ETI; Jones and Dangl, 2006). Because both PTI and ETI defense layers rely on cellular actors already present primary to the infection, they are commonly classified into plant innate immunity, in opposition to the concept of adaptive immunity (e.g., gene silencing), where the defense responses are acquired following an infection and are adapted to the pathogen.

It's All about Good Questions

To date, most current models about plant innate immunity exclude viruses, as they are not generally viewed as encoding typical PAMPs and effectors that would trigger PTI and ETI responses (Jones and Dangl, 2006;Boller and Felix, 2009; Schwessinger and Ronald, 2012; Dangl et al., 2013). The fact that virus biology differs greatly from other pathogens raises several questions of importance: could plants defend themselves against viral attacks through similar defense mechanisms employed against non-viral pathogens? And then, could the zig-zag model from Jones and Dangl (2006) be applied to plant–virus interactions?

Given that *R* gene-based immunity in plant–virus interactions involves specific recognition of viral avirulence factors by host-encoded NBS-LRR proteins and leads to HR and SAR (described previously in this review), recent hypotheses advance that it may correspond to antiviral ETI mechanisms (Moffett, 2009; Mandadi and Scholthof, 2013; De Ronde et al., 2014; Nakahara and Masuta, 2014). If so, it may suggest that viral avirulent factors perceived by R proteins correspond to pathogenic effectors, the primary function being to target PTI. Therefore, the concept of antiviral innate immunity as a whole lies mainly in whether it exists virus-associated PRR pathways and viral-encoded effectors interfering with PTI responses. Therefore, the question to assess in priority is whether plant viruses possess molecular features meeting the definition of PAMPs.

Do Viruses Encode Pamps?

Pathogen-associated molecular patterns are defined as highly conserved molecular signatures, characteristic of entire classes of microbes and with an essential function for these microbes (Medzhitov and Janeway, 1997), which implies that they are under strong evolutive pressure and thus, not easily modified by mutations. Many PAMPs from plant pathogens have been identified and so far, they are comprised of proteins, lipids, or carbohydrates (Boller and Felix, 2009). Virus genomes are highly variable even within the same viral family and display high mutation rates during the infection process (≥1 mutation per genome per round of replication), making the expression "viral populations" more suitable than "viruses" when speaking of infected hosts. This elevated mutation rate allows viruses to quickly adapt to changes in their host or environment. Therefore, the PAMP definition referring to small molecular motifs conserved within a class of pathogen is hardly applicable to viral protein motifs. By contrast, virus genomes and their replicative forms (such as dsRNAs) display structural characteristics distinguishing them from the nucleic acid sequences from prokaryotes and eukaryotes. Consequently, these very conserved nucleic acids-composed features could be perceived as non-self signals in plants, acting thus as viral PAMPs.

What do We Know about PTI in Animal–Virus Interactions?

In the animal field, the most prominent group of PRRs comprises the Toll-like receptors (TLRs), a family of a dozen type I-transmembrane proteins containing a domain with LRRs and a tail that contains a conserved region called Toll/IL-1 receptor (TIR) domain (Beutler, 2004), and sharing common recognition strategies and structural features with plant PRRs (Zipfel and Felix, 2005; Ronald and Beutler, 2010). The TLRs involved in antiviral PTI are localized in endosomes and perceive pathogen nucleic acids: TLR3 binds virus-derived dsRNA, TLR7 and TLR8 recognize viral single-stranded RNAs (ssRNA) and TLR9 mediates the perception of unmethylated CpG DNA from viruses (as well as bacteria and protozoa; Beutler, 2004). After PAMP perception, the activated receptor seems to form homo- and/or hetero-dimers with other TLRs. The TIR domains of these complexes recruit cytoplasmic adaptors, triggering a signaling cascade that mainly consist on the successive activation of cytoplasmic kinases including the MAP kinases, defense-related transcription factors and the expression of specific defense genes (Beutler, 2004; Kawai and Akira, 2008;Sandig and Bulfone-Paus, 2012). In addition, subsequent studies revealed that TLR-independent recognition of viruses can be also accomplished by cytosolic PRRs, including the two RNA helicases RIG-I (*Retinoic acid-inducible gene I*) and MDA5 (*Melanoma differentiation-associated gene 5*), both highly activated by dsRNAs, as well as 5′ triphosphate-RNAs for RIG-I (Kawai and Akira, 2008). In the arm race between host and microbes, successful animal viruses have evolved to suppress and/or hijack PTI responses, leading to host colonization. This overcoming of defense mechanisms is orchestrated by virulence effectors encoded by virus genomes. Among the strategies employed by viruses to suppress host immunity, inactivation of cytoplasmic kinases, inhibition of PRR complexes, stabilization of immunity-related negative regulators and inactivation of transcription factors are particularly noteworthy (Schröder and Bowie, 2007; Yokota et al., 2010).

On the Trail Of PTI against Plant Viruses

The existence of animal PRRs specific to viral PAMPs raises

the question of the existence of PRRs perceiving plant viruses. Although no PTI pathways against plant viruses has been formally characterized to date, typical PTI cellular responses are observed during plant–virus interactions, including ion fluxes (Shabala et al., 2010, 2011; Otulak and Garbaczewska, 2011), ROS production (Allan et al., 2001; Love et al., 2005; Díaz-Vivancos et al., 2008), ethylene signaling (de Laat and van Loon, 1982; Love et al., 2005), and callose deposition (Iglesias and Meins, 2000; Li et al., 2012; Zavaliev et al., 2013). Key components of plant PTI pathways are also involved in antiviral defense mechanisms. SA, known to regulate innate immunity including PTI responses (Tsuda et al., 2008; Yi et al., 2014), plays an important role in plant–virus interactions (Carr et al., 2010). MAPK4, that regulates negatively plant PTI mechanisms (Gao et al., 2008;Kong et al., 2012), represses soybean defense against *Bean pod mottle comovirus* (Liu et al., 2011). The receptor-like kinases (RLKs) BAK1 (BRI1-Associated receptor Kinase 1) and BKK1 (BAK1-like Kinase 1), key activators of plant PRRs after PAMP perception (Kim et al., 2013), are required for *Arabidopsis*resistance to plant viruses (Yang et al., 2010; Kørner et al., 2013). Moreover, the protein NIK1, another RLK highly similar to BAK1 and BKK1, has been strongly associated with *Arabidopsis* defense against the*Cabbage leaf curl geminivirus* (CaLCuV). Strikingly, NIK1 is cleaved by the viral protein NSP during the infection, suppressing its kinase activity and thus enhancing susceptibility to CaLCuV infection (Fontes et al., 2004; Carvalho et al., 2008). Overall, these data support the hypothesis that plants may defend themselves against viruses through similar PTI pathways employed against non-viral pathogens.

Accumulating data indicate that: (i) colonization strategies employed by plant and animal viruses inside host cells are highly similar (Thivierge et al., 2005; Taliansky et al., 2010; Nicholson and White, 2014), (ii) plant and animal PTI mechanisms display impressive structural and functional similarities (Zipfel and Felix, 2005;Dardick and Ronald, 2006; Ronald and Beutler, 2010), (iii) effectors from animal and plant pathogens target similar PTI components (Schröder and Bowie, 2007; Yokota et al., 2010; Deslandes and Rivas, 2012) and, (iv) plant defense against viruses involves many components belonging to PTI signaling (described above). Therefore, the apparent universality of PTI strategies identified so far in eukaryotic organisms enable an extrapolation of antiviral pathways from animal toward plant models.

Thus, strongly based on the understanding of plant immune pathways and the knowledge on antiviral mechanisms from the animal field, I propose here a model of plant immunity against viruses (Figure3), that does not invalidate previous models but instead tries (i) to bring the concept of PTI as a novel plant antiviral mechanism, and (ii) to integrate the concept of innate (PTI and ETI) and adaptive (gene silencing) antiviral immunities, both triggered by similar virus-derived elicitor molecules, in order to address the important question of plant defense against viruses in a more holistic way.

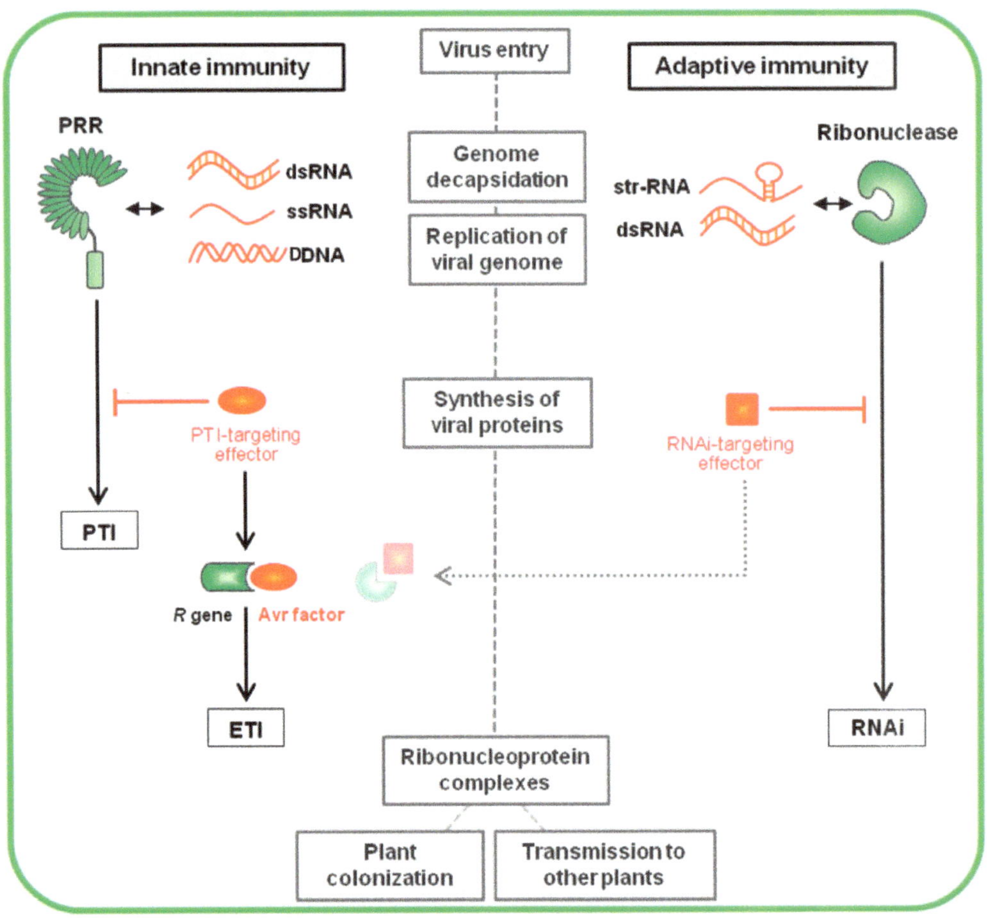

Figure 3: Plant innate immunity vs. Plant adaptive immunity. Following entry into a host cell and genome decapsidation, the

virus genome is liberated and replicated, leading to the accumulation of pathogenic nucleic acids that are perceived as viral PAMPs by specific ...

Thus, following the virus entry into a plant cell, viral nucleic acids are perceived as PAMPs by specific intracellular PRRs triggering PTI pathways (Figure 3). Successful viruses encode specialized effectors that are able to suppress this defense layer. In turn, specific *R* gene products interact with these effectors (playing then the role of avirulence factors), leading to ETI. In the case of RNAi-based adaptive immunity, the viral elicitor molecules (mostly dsRNAs) are recognized by DCLs. This leads to virus degradation at the nucleic acid level, a process that can be suppressed by a second class of virus-encoded effectors, commonly named silencing suppressors. Interestingly, Sansregret et al. (2013) recently highlighted a resistance mechanism antagonizing the silencing suppressor P19, through a phenomenon of extreme resistance characterized by a SA- and ethylene-dependent process without microscopic cell death. Although the reliance of this mechanism on one or several R proteins remains to be established, this data suggest that the action of RNAi-suppressing effectors might be countered by ETI-like mechanisms.

It is important to note here that recent reviews refer DCLs themselves as PRRs perceiving viral nucleic acids and triggering immune responses equivalent to the zig-zag model first layer (Pumplin and Voinnet, 2013;Sansregret et al., 2013; Nakahara and Masuta, 2014). Of course, PTI and RNAi mechanisms are both induced by highly similar virus-derived molecules that both fit the definition of PAMPs, suggesting that these virus-derived molecules trigger two different pathways in parallel (PTI and RNAi), similarly as in the animal field (Yan and Chen, 2012). In line with this, DCL-mediated immunity should maybe be called PTI too. However, this idea faces a conceptual and semantic conflict, as the classical view of PTI lies on innate immunity, while silencing triggers an immunity that is classified as adaptive. This issue could be partially solved by raising the concepts of innate and adaptive PTIs, that both would be targeted by plant ETI. This notwithstanding, this situation reveals that these processes represents two distinct but tight and intricate branches of the whole plant immune arsenal against viruses, raising important conceptual questions that will need to be answered. Indeed, future investigations will need first to bring the clear proof of the existence of PTI pathways triggered by the recognition of viral PAMPs by plant PRRs. To this perspective, the identification of pathogenic effectors from successful plant viruses that target specifically PTI responses will be definitely an asset, as pathogenic effectors represent fantastic molecular tools commonly used to characterize innate immunity signaling.

Which advantage PTI could bring into Crop Improvement Programs?

So far, the only powerful strategy to control viral epidemics in field relies on the use of genetic resistances. Unfortunately, viruses evolve very quickly the ability to overcome the resistances employed by breeders. Hence, handling the extreme genetic plasticity of viruses represents a major challenge for the coming years. The durability of plant resistances, even more with viruses than any other plant pathogen, depends mostly on the resistance timing/efficiency and on the elicitor nature/conservation within pathogen families, in order to enable an efficient, sustainable, and broad-spectrum resistance.

Given their importance for pathogen survival, PAMPs are under a strong evolutive pressure and are highly conserved within pathogen families, this definition fitting perfectly with the characteristics of viral nucleic acids, whose features (e.g., dsRNAs replicative forms, highly structured genomes) are highly conserved within virus families and differ radically from prokaryotic or eukaryotic nucleic acids. It is these typical viral features that are perceived by specialized PRRs in animals (e.g., TLRs), triggering very fast and very efficient immune responses. Because the innate pathway is solely programmed by structural and nucleotide sequence genomic features, plant antiviral PTI should be a remarkably versatile mechanism because it could respond virtually to any plant virus. Moreover, timeliness of defense responses is a key parameter in resistance success, especially in the case of viruses, given their tremendous adaptation abilities. And yet, the first measurable PTI responses are within minutes after PAMP perception (Nicaise et al., 2009), making PTI the quickest plant defense identified so far. Hence, translating fundamental knowledge about PTI mechanisms for new crop resistance deployment must bring considerable advances in the control of viral epidemics in fields. Moreover, what is increasingly clear is that plant antiviral arsenal ensure a highly robust plant immune system, through different plant defense layers (R genes, SAR, RNAi, PTI, and ETI) that emerge to be finely coordinated (Voinnet, 2005; Alamillo et al., 2006; Boccara et al., 2014; Padmanabhan and Dinesh-Kumar, 2014). Consequently, addressing plant–virus interactions holistically by combining PTI-mediated resistance with other defense pathways (such as silencing) could provide highly robust resistance phenotypes targeting virus cycle as a whole.

CONCLUSION

In the world agriculture context, intensification of cultural practices, climate alterations, and extensive exchanges affecting global markets are associated with an increased incidence of plant viral diseases. In consequence, management of viral epidemics is a matter of global food security. Nowadays, antiviral measures employed in fields benefits greatly from the latest knowledge highlighted by plant virologists, geneticists and molecular biologists, as well as engineers and breeders, whose coordinated effort enables a deployment in crop improvement programs. This notwithstanding, the situation in fields remains worrying, given the emergence of numerous crop diseases ascribed to viruses, a statement that could be aggravated in the future through climate changes that might affect strongly virus and organism-vectors populations in terms of magnitude, locations and dynamics (Anderson et al., 2004; Gautam et al., 2013). In a coordinated manner with technological advances, fundamental research needs to explore new scientific leads, deciphering more and more thoroughly the intricate molecular dialog between a plant virus and its host. Hence, future challenges associated to the management of crop viral diseases will rely mainly on integrated research actions with a view to translating fundamental understanding toward applied programs, and thus reducing the gap existing between the laboratory and the field.

ACKNOWLEDGMENTS

This work was supported by the European Framework Programme 7, through a Marie-Curie Intra-European Fellowship (Grant number 327341).

REFERENCES

1. Abel P. P., Nelson R. S., De B., Hoffmann N., Rogers S. G., Fraley R. T., et al. (1986). Delay of disease development in transgenic plants that express the tobacco mosaic virus coat protein gene.*Science* 232 738–743 10.1126/science.3457472
2. Abo M. E., Sy A. A. (1998). Rice virus diseases: epidemiology and management strategies. *J. Sustain. Agric.* 11 113–134 10.1300/J064v11n02_09
3. Acosta-Leal R., Xiong Z. (2008). Complementary functions of two recessive R-genes determine resistance durability of tobacco "Virgin A Mutant" (VAM) to potato virus Y. *Virology* 379 275–283 10.1016/j.virol.2008.06.026
4. Akbergenov R., Si-Ammour A., Blevins T., Amin I., Kutter C., Vanderschuren H., et al. (2006).Molecular characterization of geminivirus-derived small RNAs in different plant species. *Nucleic Acids Res.* 34 462–471 10.1093/nar/gkj447
5. Alamillo J. M., Saénz P., García J. A. (2006). Salicylic acid-mediated and RNA-silencing defense mechanisms cooperate in the restriction of systemic spread of plum pox virus in tobacco. *Plant J.* 48217–227 10.1111/j.1365-313X.2006.02861.x
6. Alazem M., Lin K.-Y., Lin N.-S. (2014). The abscisic acid pathway has multifaceted effects on the accumulation of bamboo mosaic virus. *Mol. Plant Microbe Interact.* 27 177–189 10.1094/MPMI-08-13-0216-R
7. Ali S. S., Gunupuru L. R., Kumar G. B., Khan M., Scofield S., Nicholson P., et al. (2014). Plant disease resistance is augmented in uzu barley lines modified in the brassinosteroid receptor BRI1.*BMC Plant Biol.* 14:227 10.1186/s12870-014-0227-1
8. Allan A. C., Lapidot M., Culver J. N., Fluhr R. (2001). An early tobacco mosaic virus-induced oxidative burst in tobacco indicates extracellular perception of the virus coat protein. *Plant Physiol.*126 97–108 10.1104/pp.126.1.97
9. Anderson P. K., Cunningham A. A., Patel N. G., Morales F. J., Epstein P. R., Daszak P. (2004).Emerging infectious diseases of plants: pathogen pollution, climate change and agrotechnology drivers.*Trends Ecol. Evol.* 19 535–544 10.1016/j.tree.2004.07.021
10. Aragão F. J. L., Nogueira E. O., Tinoco M. L., Faria J. C. (2013). Molecular characterization of the first commercial transgenic common bean immune to the bean golden mosaic virus. *J. Biotechnol.* 16642–50 10.1016/j.jbiotec.2013.04.009
11. Aregger M., Borah B. K., Seguin J., Rajeswaran R., Gubaeva E. G., Zvereva A. S., et al. (2012).Primary and secondary siRNAs in geminivirus-induced gene silencing. *PLoS Pathog.* 8:e100294110.1371/journal.ppat.1002941
12. Azevedo C., Betsuyaku S., Peart J., Takahashi A., Noël L., Sadanandom A., et al. (2006). Role of SGT1 in resistance protein accumulation in plant immunity. *EMBO J.* 25 2007–2016 10.1038/sj.emboj.7601084
13. Azevedo J., Garcia D., Pontier D., Ohnesorge S., Yu A., Garcia S., et al. (2010). Argonaute quenching and global changes in Dicer homeostasis caused by a pathogen-encoded GW repeat protein. *Genes Dev.* 24 904–915 10.1101/gad.1908710
14. Bai S., Liu J., Chang C., Zhang L., Maekawa T., Wang Q., et al. (2012). Structure-function analysis of barley NLR immune receptor MLA10 reveals its cell compartment specific activity in cell death and disease resistance. *PLoS Pathog.* 8:e1002752 10.1371/journal.ppat.1002752
15. Barajas D., Martín I. F., Pogany J., Risco C., Nagy P. D. (2014a). Noncanonical role for the host Vps4 AAA+ ATPase ESCRT protein in the formation of tomato bushy stunt virus replicase. *PLoS Pathog.* 10:e1004087 10.1371/journal.ppat.1004087
16. Barajas D., Xu K., de Castro Martín I. F., Sasvari Z., Brandizzi F., Risco C., et al. (2014b). Co-opted oxysterol-binding ORP and VAP proteins channel sterols to RNA virus replication sites via membrane contact sites. *PLoS Pathog.* 10:e1004388 10.1371/journal.ppat.1004388 [PubMed]
17. Baurès I., Candresse T., Leveau A., Bendahmane A., Sturbois B. (2008). The Rx gene confers resistance to a range of potexviruses in transgenic *Nicotiana* plants. *Mol. Plant Microbe Interact.* 211154–1164 10.1094/MPMI-21-9-1154
18. Bazzini A. A., Asurmendi S., Hopp H. E., Beachy R. N. (2006). Tobacco mosaic virus (TMV) and potato virus X (PVX) coat proteins confer heterologous interference to PVX and TMV infection, respectively. *J. Gen. Virol.* 87 1005–1012 10.1099/vir.0.81396-0
19. Bendahmane A., Kanyuka K., Baulcombe D. C. (1999). The Rx gene from potato controls separate virus resistance and cell death responses. *Plant Cell* 11 781–792 10.1105/tpc.11.5.781
20. Beutler B. (2004). Innate immunity: an overview. *Mol. Immunol.* 40 845–859 10.1016/j.molimm.2003.10.005
21. Blevins T., Rajeswaran R., Aregger M., Borah B. K., Schepetilnikov M., Baerlocher L., et al. (2011).Massive production of small RNAs from a non-coding region of cauliflower mosaic virus in plant defense and viral counter-defense. *Nucleic Acids Res.* 39 5003–5014 10.1093/nar/gkr119
22. Blevins T., Rajeswaran R., Shivaprasad P. V., Beknazariants D., Si-Ammour A., Park H. S., et al. (2006). Four plant Dicers mediate viral small RNA biogenesis and DNA virus induced silencing.*Nucleic Acids Res.* 34 6233–6246 10.1093/nar/gkl886
23. Boccara M., Sarazin A., Thiébeauld O., Jay F., Voinnet O., Navarro L., et al. (2014). The *Arabidopsis*miR472-RDR6 silencing pathway modulates PAMP- and effector-triggered immunity through the post-transcriptional control of disease resistance genes. *PLoS Pathog.* 10:e100388310.1371/journal.ppat.1003883

24. Boller T., Felix G. (2009). A renaissance of elicitors: perception of microbe-associated molecular patterns and danger signals by pattern-recognition receptors. *Annu. Rev. Plant Biol.* 60 379–406 10.1146/annurev.arplant.57.032905.105346

25. Bologna N. G., Voinnet O. (2014). The diversity, biogenesis, and activities of endogenous silencing small RNAs in *Arabidopsis*. *Annu. Rev. Plant Biol.* 65 473–503 10.1146/annurev-arplant-050213-035728

26. Boonham N., Kreuze J., Winter S., van der Vlugt R., Bergervoet J., Tomlinson J., et al. (2014).Methods in virus diagnostics: from ELISA to next generation sequencing. *Virus Res.* 186 20–31 10.1016/j.virusres.2013.12.007

27. Botër M., Amigues B., Peart J., Breuer C., Kadota Y., Casais C., et al. (2007). Structural and functional analysis of SGT1 reveals that its interaction with HSP90 is required for the accumulation of Rx, an R protein involved in plant immunity. *Plant Cell* 19 3791–3804 10.1105/tpc.107.050427

28. Bragard C., Caciagli P., Lemaire O., Lopez-Moya J. J., MacFarlane S., Peters D., et al. (2013). Status and prospects of plant virus control through interference with vector transmission. *Annu. Rev. Phytopathol.* 51 177–201 10.1146/annurev-phyto-082712-102346

29. Bravo-Almonacid F., Rudoy V., Welin B., Segretin M. E., Bedogni M. C., Stolowicz F., et al. (2012).Field testing, gene flow assessment and pre-commercial studies on transgenic *Solanum tuberosum* spp. tuberosum (cv. Spunta) selected for PVY resistance in Argentina. *Transgenic Res.* 21 967–982 10.1007/s11248-011-9584-9

30. Burgyán J., Havelda Z. (2011). Viral suppressors of RNA silencing. *Trends Plant Sci.* 16 265–272 10.1016/j.tplants.2011.02.010

31. Calvert L. A., Tresh J. M. (2002). "The viruses and virus diseases of cassava," in *Cassava: Biology, Production and Utilization* eds Hillocks R. J., Thresh J. M., Bellotti A. C., editors. (Wallingford: CABI Publishing;).

32. Cambra M., Capote N., Myrta A., Llácer G. (2006). Plum pox virus and the estimated costs associated with sharka disease. *EPPO Bull.* 36 202–204 10.1111/j.1365-2338.2006.01027.x

33. Candresse T., Marais A., Faure C., Dubrana M. P., Gombert J., Bendahmane A. (2010). Multiple coat protein mutations abolish recognition of Pepino mosaic potexvirus (PepMV) by the potato rx resistance gene in transgenic tomatoes. *Mol. Plant Microbe Interact.* 23 376–383 10.1094/MPMI-23-4-0376

34. Caplan J. L., Mamillapalli P., Burch-Smith T. M., Czymmek K., Dinesh-Kumar S. P. (2008).Chloroplastic protein NRIP1 mediates innate immune receptor recognition of a viral effector. *Cell* 132449–462 10.1016/j.cell.2007.12.031

35. Caranta C., Lefebvre V., Palloix A. (1997). Polygenic resistance of pepper to potyviruses consists of a combination of isolate-specific and broad-spectrum quantitative trait loci. *Mol. Plant Microbe Interact.*10 872–878 10.1094/MPMI.1997.10.7.872

36. Carr J. P., Lewsey M. G., Palukaitis P. (2010). Signaling in induced resistance. *Adv. Virus Res.* 76 57–121 10.1016/S0065-3527(10)76003-6

37. Carvalho C. M., Santos A. A., Pires S. R., Rocha C. S., Saraiva D. I., Machado J. P. B., et al. (2008). Regulated nuclear trafficking of rpL10A mediated by NIK1 represents a defense strategy of plant cells against virus. *PLoS Pathog.* 4:e1000247 10.1371/journal.ppat.1000247

38. Cavatorta J., Perez K. W., Gray S. M., Van Eck J., Yeam I., Jahn M. (2011). Engineering virus resistance using a modified potato gene. *Plant Biotechnol. J.* 9 1014–1021 10.1111/j.1467-7652.2011.00622.x

39. Charron C., Nicolaï M., Gallois J. L., Robaglia C., Moury B., Palloix A., et al. (2008). Natural variation and functional analyses provide evidence for co-evolution between plant eIF4E and potyviral VPg. *Plant J.* 54 56–68 10.1111/j.1365-313X.2008.03407.x

40. Chen L., Zhang L., Li D., Wang F., Yu D. (2013). WRKY8 transcription factor functions in the TMV-cg defense response by mediating both abscisic acid and ethylene signaling in *Arabidopsis*. *Proc. Natl. Acad. Sci. U.S.A.* 110 E1963–E1971 10.1073/pnas.1221347110

41. Collard B. C. Y., Mackill D. J. (2008). Marker-assisted selection: an approach for precision plant breeding in the twenty-first century. *Philos. Trans. R. Soc. Lond. B Biol. Sci.* 363 557–572 10.1098/rstb.2007.2170

42. Comai L., Young K., Till B. J., Reynolds S. H., Greene E. A., Codomo C. A., et al. (2004). Efficient discovery of DNA polymorphisms in natural populations by Ecotilling. *Plant J.* 37 778–786 10.1111/j.0960-7412.2003.01999.x

43. Cosson P., Schurdi-Levraud V., Le Q. H., Sicard O., Caballero M., Roux F., et al. (2012). The RTM resistance to potyviruses in *Arabidopsis thaliana*: natural variation of the RTM genes and evidence for the implication of additional genes. *PLoS ONE* 7:e39169 10.1371/journal.pone.0039169

44. Cosson P., Sofer L., Le Q. H., Léger V., Schurdi-Levraud V., Whitham S. A., et al. (2010a). RTM3, which controls long-distance movement of potyviruses, is a member of a new plant gene family encoding a meprin and TRAF homology domain-containing protein. *Plant Physiol.* 154 222–232 10.1104/pp.110.155754

45. Cosson P., Sofer L., Schurdi-Levraud V., Revers F. (2010b). A member of a new plant gene family encoding a meprin and TRAF homology (MATH) domain-containing protein is involved in restriction of long distance movement of plant viruses. *Plant Signal. Behav.* 5 1321–1323 10.4161/psb.5.10.13244

46. Crowdy S. H., Posnette A. F. (1947). Virus diseases of cacao in West Africa II. Cross immunity experiments with viruses 1A, 1B and 1C. *Ann. Appl. Biol.* 34 403–411 10.1111/j.1744-7348.1947.tb06373.x

47. Culver J. N. (1996). Tobamovirus cross protection using a potexvirus vector. *Virology* 226 228–235 10.1006/viro.1996.0650

48. Dangl J. L., Horvath D. M., Staskawicz B. J. (2013). Pivoting the plant immune system from

dissection to deployment. *Science* 341 746–751 10.1126/science.1236011

49. Dardick C., Ronald P. (2006). Plant and animal pathogen recognition receptors signal through non-RD kinases. *PLoS Pathog.* 2:e2 10.1371/journal.ppat.0020002
50. Decroocq V., Sicard O., Alamillo J. M., Lansac M., Eyquard J. P., García J. A., et al. (2006). Multiple resistance traits control plum pox virus infection in *Arabidopsis thaliana. Mol. Plant Microbe Interact.* 19 541–549 10.1094/MPMI-19-0541
51. de Laat A. M. M., van Loon L. C. (1982). Regulation of ethylene biosynthesis in virus-infected tobacco leaves 1. *Plant Physiol.* 69 240–245 10.1104/pp.69.1.240
52. Delaney T. P., Uknes S., Vernooij B., Friedrich L., Weymann K., Negrotto D., et al. (1994). A central role of salicylic acid in plant disease resistance. *Science* 266 1247–1250 10.1126/science.266.5188.1247
53. Deleris A., Gallego-Bartolome J., Bao J., Kasschau K. D., Carrington J. C., Voinnet O. (2006).Hierarchical action and inhibition of plant Dicer-like proteins in antiviral defense. *Science* 313 68–71 10.1126/science.1128214
54. Dempsey D. A., Klessig D. F. (2012). SOS – too many signals for systemic acquired resistance?*Trends Plant Sci.* 17 538–545 10.1016/j.tplants.2012.05.011
55. De Ronde D., Butterbach P., Kormelink R. (2014). Dominant resistance against plant viruses. *Front. Plant Sci.* 5:307 10.3389/fpls.2014.00307
56. Deslandes L., Rivas S. (2012). Catch me if you can: bacterial effectors and plant targets. *Trends Plant Sci.* 17 644–655 10.1016/j.tplants.2012.06.011
57. Diaz A., Wang X., Ahlquist P. (2010). Membrane-shaping host reticulon proteins play crucial roles in viral RNA replication compartment formation and function. *Proc. Natl. Acad. Sci. U.S.A.* 107 16291–16296 10.1073/pnas.1011105107
58. Diaz-Pendon J. A., Li F., Li W.-X., Ding S.-W. (2007). Suppression of antiviral silencing by cucumber mosaic virus 2b protein in *Arabidopsis* is associated with drastically reduced accumulation of three classes of viral small interfering RNAs. *Plant Cell* 19 2053–2063 10.1105/tpc.106.047449
59. Diaz-Pendon J. A., Truniger V., Nieto C., Garcia-Mas J., Bendahmane A., Aranda M. A. (2004). Advances in understanding recessive resistance to plant viruses. *Mol. Plant Pathol.* 5 223–233 10.1111/J.1364-3703.2004.00223.X
60. Díaz-Vivancos P., Clemente-Moreno M. J., Rubio M., Olmos E., García J. A., Martínez-Gómez P., et al. (2008). Alteration in the chloroplastic metabolism leads to ROS accumulation in pea plants in response to plum pox virus. *J. Exp. Bot.* 59 2147–2160 10.1093/jxb/ern082
61. Dzahini-Obiatey H., Domfeh O., Amoah F. M. (2010). Over seventy years of a viral disease of cocoa in Ghana: from researchers' perspective. *Afr. J. Agric. Res.* 5 476–485 10.5897/AJAR09.625
62. Elena S. F., Bedhomme S., Carrasco P., Cuevas J. M., de la Iglesia F., Lafforgue G., et al. (2011). The evolutionary genetics of emerging plant RNA viruses. *Mol. Plant Microbe Interact.* 24 287–293 10.1094/MPMI-09-10-0214
63. Fabre F., Rousseau E., Mailleret L., Moury B. (2012). Durable strategies to deploy plant resistance in agricultural landscapes. *New Phytol.* 193 1064–1075 10.1111/j.1469-8137.2011.04019.x
64. Ferreira S. A., Pitz K. Y., Manshardt R., Zee F., Fitch M., Gonsalves D. (2002). Virus coat protein transgenic papaya provides practical control of papaya ringspot virus in Hawaii. *Plant Dis.* 86 101–105 10.1094/PDIS.2002.86.2.101
65. Fischer U., Dröge-Laser W. (2004). Overexpression of NtERF5, a new member of the tobacco ethylene response transcription factor family enhances resistance to tobacco mosaic virus. *Mol. Plant Microbe Interact.* 17 1162–1171 10.1094/MPMI.2004.17.10.1162
66. Fletcher J. T. (1978). The use of avirulent virus strain to protect plants against the effects of virulent strains. *Ann. Appl. Biol.* 89 110–114 10.1111/j.1744-7348.1978.tb02581.x
67. Fontes E. P. B., Santos A. A., Luz D. F., Waclawovsky A. J., Chory J. (2004). The geminivirus nuclear shuttle protein is a virulence factor that suppresses transmembrane receptor kinase activity.*Genes Dev.* 18 2545–2556 10.1101/gad.1245904
68. Friedrich L., Lawton K., Ruess W., Masner P., Specker N., Rella M. G., et al. (1996). A benzothiadiazole derivative induces systemic acquired resistance in tobacco. *Plant J.* 10 61–70 10.1046/j.1365-313X.1996.10010061.x
69. Fu Z. Q., Dong X. (2013). Systemic acquired resistance: turning local infection into global defense. *Annu. Rev. Plant Biol.* 64 839–863 10.1146/annurev-arplant-042811-105606
70. Gao M., Liu J., Bi D., Zhang Z., Cheng F., Chen S., et al. (2008). MEKK1, MKK1/MKK2 and MPK4 function together in a mitogen-activated protein kinase cascade to regulate innate immunity in plants. *Cell Res.* 18 1190–1198 10.1038/cr.2008.300
71. García-Arenal F., McDonald B. A. (2003). An analysis of the durability of resistance to plant viruses.*Phytopathology* 93 941–952 10.1094/PHYTO.2003.93.8.941
72. Garcia-Ruiz H., Takeda A., Chapman E. J., Sullivan C. M., Fahlgren N., Brempelis K. J., et al. (2010). *Arabidopsis* RNA-dependent RNA polymerases and dicer-like proteins in antiviral defense and small interfering RNA biogenesis during turnip mosaic virus infection. *Plant Cell* 22 481–496 10.1105/tpc.109.073056
73. Gautam H. R., Bhardwaj M. L., Kumar R. (2013). Climate change and its impact on plant diseases. *Curr. Sci.* 105 1685–1691.
74. Gozzo F., Faoro F. (2013). Systemic acquired resistance (50 years after discovery): moving from the lab to the field. *J. Agric. Food Chem.* 61 12473–12491 10.1021/jf404156x
75. Hämäläinen J. H., Kekarainen T., Gebhardt C., Watanabe K. N., Valkonen J. P. (2000). Recessive and

dominant genes interfere with the vascular transport of potato virus a in diploid potatoes. *Mol. Plant Microbe Interact.* 13 402–412 10.1094/MPMI.2000.13.4.402

76. Hao W., Collier S. M., Moffett P., Chai J. (2013). Structural basis for the interaction between the potato virus X resistance protein (Rx) and its cofactor Ran GTPase-activating protein 2 (RanGAP2). *J. Biol. Chem.* 288 35868–35876 10.1074/jbc.M113.517417
77. Harper S. J. (2013). Citrus tristeza virus: evolution of complex and varied genotypic groups. *Front. Microbiol.* 4:93 10.3389/fmicb.2013.00093
78. Harvey J. J. W., Lewsey M. G., Patel K., Westwood J., Heimstädt S., Carr J. P., et al. (2011). An antiviral defense role of AGO2 in plants. *PLoS ONE* 6:e14639 10.1371/journal.pone.0014639
79. Hashmi J. A., Zafar Y., Arshad M., Mansoor S., Asad S. (2011). Engineering cotton (*Gossypium hirsutum* L.) for resistance to cotton leaf curl disease using viral truncated AC1 DNA sequences. *Virus Genes* 42 286–296 10.1007/s11262-011-0569-9
80. Hoser R., Zurczak M., Lichocka M., Zuzga S., Dadlez M., Samuel M. A., et al. (2013).Nucleocytoplasmic partitioning of tobacco N receptor is modulated by SGT1. *New Phytol.* 200 158–171 10.1111/nph.12347
81. Hugues J. A., Ollennu L. A. A. (1994). Mild strain protection of cocoa in Ghana against cocoa swollen shoot virus—a review. *Plant Pathol.* 43 442–457 10.1111/j.1365-3059.1994.tb01578.x
82. Hull R. (2013). *Plant Virology* 5th Edn. New York: Academic Press.
83. Hwang J., Oh C. S., Kang B. C. (2013). Translation elongation factor 1B (eEF1B) is an essential host factor for tobacco mosaic virus infection in plants. *Virology* 439 105–114 10.1016/j.virol.2013.02.004
84. Hyodo K., Mine A., Taniguchi T., Kaido M., Mise K., Taniguchi H., et al. (2013). ADP ribosylation factor 1 plays an essential role in the replication of a plant RNA virus. *J. Virol.* 87 163–176 10.1128/JVI.02383-12
85. Ichiki T. U., Nagaoka E. N., Hagiwara K., Uchikawa K., Tsuda S., Omura T. (2005). Integration of mutations responsible for the attenuated phenotype of pepper mild mottle virus strains results in a symptomless cross-protecting strain. *Arch. Virol.* 150 2009–2020 10.1007/s00705-005-0551-8
86. Iglesias V. A., Meins F. (2000). Movement of plant viruses is delayed in a beta-1,3-glucanase-deficient mutant showing a reduced plasmodesmatal size exclusion limit and enhanced callose deposition. *Plant J.* 21 157–166 10.1046/j.1365-313X.2000.00658.x
87. Ishibashi K., Ishikawa M. (2013). The resistance protein Tm-1 inhibits formation of a tomato mosaic virus replication protein-host membrane protein complex. *J. Virol.* 87 7933–7939 10.1128/JVI.00743-13
88. Ishibashi K., Miyashita S., Katoh E., Ishikawa M. (2012). Host membrane proteins involved in the replication of tobamovirus RNA. *Curr. Opin. Virol.* 2 693–698 10.1016/j.coviro.2012.09.011
89. Jakubiec A., Yang S. W., Chua N. H. (2012). *Arabidopsis* DRB4 protein in antiviral defense against turnip yellow mosaic virus infection. *Plant J.* 69 14–25 10.1111/j.1365-313X.2011.04765.x
90. Jaubert M., Bhattacharjee S., Mello A. F. S., Perry K. L., Moffett P. (2011). ARGONAUTE2 mediates RNA-silencing antiviral defenses against potato virus X in *Arabidopsis*. *Plant Physiol.* 1561556–1564 10.1104/pp.111.178012
91. Jiang L., Qian D., Zheng H., Meng L. Y., Chen J., Le W. J., et al. (2012). RNA-dependent RNA polymerase 6 of rice (*Oryza sativa*) plays role in host defense against negative-strand RNA virus, rice stripe virus. *Virus Res.* 163 512–519 10.1016/j.virusres.2011.11.016
92. Jones J. D. G., Dangl J. L. (2006). The plant immune system. *Nature* 444 323–329 10.1038/nature05286
93. Julio E., Cotucheau J., Decorps C., Volpatti R., Sentenac C., Candresse T., et al. (2014). A eukaryotic translation initiation factor 4E (eIF4E) is responsible for the "va" tobacco recessive resistance to potyviruses. *Plant Mol. Biol. Rep.* 10.1007/s11105-014-0775-4
94. Kadota Y., Shirasu K. (2012). The HSP90 complex of plants. *Biochim. Biophys. Acta* 1823 689–97 10.1016/j.bbamcr.2011.09.016
95. Kang B. C., Yeam I., Li H., Perez K. W., Jahn M. M. (2007). Ectopic expression of a recessive resistance gene generates dominant potyvirus resistance in plants. *Plant Biotechnol. J.* 5 526–536 10.1111/j.1467-7652.2007.00262.x
96. Kawai T., Akira S. (2008). Toll-like receptor and RIG-I-like receptor signaling. *Ann. N. Y. Acad. Sci.*1143 1–20 10.1196/annals.1443.020
97. Kazan K., Lyons R. (2014). Intervention of phytohormone pathways by pathogen effectors. *Plant Cell*26 2285–2309 10.1105/tpc.114.125419
98. Kim B. H., Kim S. Y., Nam K. H. (2013). Assessing the diverse functions of BAK1 and its homologs in *Arabidopsis*, beyond BR signaling and PTI responses. *Mol. Cells* 35 7–16 10.1007/s10059-013-2255-3
99. King A. M. Q., Adams M. J., Carstens E. B., Lefkowitz E. (2012). *Virus Taxonomy: Classification and Nomenclature of Viruses: Ninth Report of the International Committee on Taxonomy of Viruses.*San Diego: Elsevier Academic Press.
100. Kobayashi M., Seo S., Hirai K., Yamamoto-Katou A., Katou S., Seto H., et al. (2010). Silencing of WIPK and SIPK mitogen-activated protein kinases reduces tobacco mosaic virus accumulation but permits systemic viral movement in tobacco possessing the N resistance gene. *Mol. Plant Microbe Interact.* 23 1032–1041 10.1094/MPMI-23-8-1032
101. Kong Q., Qu N., Gao M., Zhang Z., Ding X., Yang F., et al. (2012). The MEKK1-MKK1/MKK2-MPK4 kinase cascade negatively regulates immunity mediated by a mitogen-activated protein kinase kinase kinase in *Arabidopsis*. *Plant Cell* 24 2225–2236 10.1105/tpc.112.097253
102. Kørner C. J., Klauser D., Niehl A., Domínguez-Ferreras

A., Chinchilla D., Boller T., et al. (2013).The immunity regulator BAK1 contributes to resistance against diverse RNA viruses. *Mol. Plant Microbe Interact.* 26 1271–1280 10.1094/MPMI-06-13-0179-R

103. Kosaka Y., Fukunishi T. (1997). Multiple inoculation with three attenuated viruses for the control of cucumber virus disease. *Plant Dis.* 81 733–738 10.1094/PDIS.1997.81.7.733

104. Kosaka Y., Ryang B.-S., Katagiri N., Yasuhara H. (2009). "Development and diffusion of CUBIO ZY-02 a water-soluble formulation of an attenuated zucchini yellow mosaic virus isolate, for the control of cucumber mosaic disease," in *Proceedings of PSJ Biocontrol Workshop* Vol. 11 Yamagata: 63–72.

105. Kosaka Y., Ryang B.-S., Kobori T., Shiomi H., Yasuhara H., Kataoka M. (2006). Effectiveness of an attenuated *Zucchini yellow mosaic virus* isolate for cross-protecting cucumber. *Plant Dis.* 90 67–72 10.1094/PD-90-0067

106. Kurihara Y., Watanabe Y. (2003). Cross-protection in *Arabidopsis* against crucifer tobamovirus Cg by an attenuated strain of the virus. *Mol. Plant Pathol.* 4 259–269 10.1046/j.1364-3703.2003.00174.x

107. Kurth E. G., Peremyslov V. V., Prokhnevsky A. I., Kasschau K. D., Miller M., Carrington J. C., et al. (2012). Virus-derived gene expression and RNA interference vector for grapevine. *J. Virol.* 86 6002–6009 10.1128/JVI.00436-12

108. Lawton K. A., Friedrich L., Hunt M., Weymann K., Delaney T., Kessmann H., et al. (1996).Benzothiadiazole induces disease resistance in *Arabidopsis* by activation of the systemic acquired resistance signal transduction pathway. *Plant J.* 10 71–82 10.1046/j.1365-313X.1996.10010071.x

109. Lee R. F., Keremane M. L. (2013). Mild strain cross protection of tristeza: a review of research to protect against decline on sour orange in Florida. *Front. Microbiol.* 4:259 10.3389/fmicb.2013.00259

110. Legg J. P. (1999). Emergence, spread and strategies for controlling the pandemic of cassava mosaic virus disease in east and central Africa. *Crop Prot.* 18 627–637 10.1016/S0261-2194(99)00062-9

111. Legg J. P., Thresh J. M. (2000). Cassava mosaic virus disease in East Africa: a dynamic disease in a changing environment. *Virus Res.* 71 135–149 10.1016/S0168-1702(00)00194-5

112. Lemgo G. N. Y., Sabbadini S., Pandolfini T., Mezzetti B. (2013). Biosafety considerations of RNAi-mediated virus resistance in fruit-tree cultivars and in rootstock. *Transgenic Res.* 22 1073–1088 10.1007/s11248-013-9728-1

113. Li F., Huang C., Li Z., Zhou X. (2014a). Suppression of RNA silencing by a plant DNA virus satellite requires a host calmodulin-like protein to repress RDR6 expression. *PLoS Pathog.* 10:e100392110.1371/journal.ppat.1003921

114. Li Z., Gonzalez P. A., Sasvari Z., Kinzy T. G., Nagy P. D. (2014b). Methylation of translation elongation factor 1A by the METTL10-like See1 methyltransferase facilitates tombusvirus replication in yeast and plants. *Virology* 448 43–54 10.1016/j.virol.2013.09.012

115. Li W., Zhao Y., Liu C., Yao G., Wu S., Hou C., et al. (2012). Callose deposition at plasmodesmata is a critical factor in restricting the cell-to-cell movement of soybean mosaic virus. *Plant Cell Rep.* 31 905–916 10.1007/s00299-011-1211-y

116. Lin C. Y., Ku H. M., Chiang Y. H., Ho H. Y., Yu T. A., Jan F. J. (2012). Development of transgenic watermelon resistant to cucumber mosaic virus and watermelon mosaic virus by using a single chimeric transgene construct. *Transgenic Res.* 21 983–993 10.1007/s11248-011-9585-8

117. Liu J.-Z., Horstman H. D., Braun E., Graham M. A., Zhang C., Navarre D., et al. (2011). Soybean homologs of MPK4 negatively regulate defense responses and positively regulate growth and development. *Plant Physiol.* 157 1363–1378 10.1104/pp.111.185686

118. Liu P.-P., Yang Y., Pichersky E., Klessig D. F. (2010). Altering expression of benzoic acid/salicylic acid carboxyl methyltransferase 1 compromises systemic acquired resistance and PAMP-triggered immunity in *Arabidopsis*. *Mol. Plant Microbe Interact.* 23 82–90 10.1094/MPMI-23-1-0082

119. Liu Y., Schiff M., Marathe R., Dinesh-Kumar S. P. (2002). Tobacco Rar1, EDS1 and NPR1/NIM1 like genes are required for N-mediated resistance to tobacco mosaic virus. *Plant J.* 30 415–429 10.1046/j.1365-313X.2002.01297.x

120. Lius S., Manshardt R. M., Fitch M. M. M., Slightom J. L., Sanford J. C., Gonsalves D. (1997).Pathogen-derived resistance provides papaya with effective protection against papaya ringspot virus.*Mol. Breed.* 3 161–168 10.1023/A:1009614508659

121. Love A. J., Laval V., Geri C., Laird J., Tomos A. D., Hooks M. A., et al. (2007). Components of*Arabidopsis* defense- and ethylene-signaling pathways regulate susceptibility to cauliflower mosaic virus by restricting long-distance movement. *Mol. Plant Microbe Interact.* 20 659–670 10.1094/MPMI-20-6-0659

122. Love A. J., Yun B. W., Laval V., Loake G. J., Milner J. J. (2005). Cauliflower mosaic virus, a compatible pathogen of *Arabidopsis*, engages three distinct defense-signaling pathways and activates rapid systemic generation of reactive oxygen species. *Plant Physiol.* 139 935–948 10.1104/pp.105.066803

123. Macho A. P., Zipfel C. (2014). Plant PRRs and the activation of innate immune signaling. *Mol. Cell* 54263–272 10.1016/j.molcel.2014.03.028

124. Mahajan S. K., Chisholm S. T., Whitham S. A., Carrington J. C. (1998). Identification and characterization of a locus (RTM1) that restricts long-distance movement of tobacco etch virus in*Arabidopsis thaliana*. *Plant J.* 14 177–186 10.1046/j.1365-313X.1998.00105.x

125. Malinowski T., Cambra M., Capote N., Zawadzka B., Gorris M. T., Scorza R., et al. (2006). Field Trials of plum clones transformed with the plum pox virus coat protein (PPV-CP) gene. *Plant Dis.* 901012–1018 10.1094/PD-90-1012

126. Mandadi K. K., Scholthof K.-B. G. (2013). Plant immune responses against viruses: how does a virus cause disease? *Plant Cell* 25 1489–1505 10.1105/tpc.113.111658

127. Mandal B., Mandal S., Csinos A. S., Martinez N., Culbreath A. K., Pappu H. R. (2008). Biological and molecular analyses of the acibenzolar-S-methyl- induced systemic acquired resistance in flue-cured tobacco against tomato spotted wilt virus. *Phytopathology* 98 196–204 10.1094/PHYTO-98-2-0196

128. Marathe R., Anandalakshmi R., Liu Y., Dinesh-Kumar S. P. (2002). The tobacco mosaic virus resistance gene, N. *Mol. Plant Pathol.* 3 167–172 10.1046/j.1364-3703.2002.00110.x

129. Mascher M., Stein N. (2014). Genetic anchoring of whole-genome shotgun assemblies. *Front. Genet.*5:208 10.3389/fgene.2014.00208

130. Maule A. J., Caranta C., Boulton M. I. (2007). Sources of natural resistance to plant viruses: status and prospects. *Mol. Plant Pathol.* 8 223–231 10.1111/j.1364-3703.2007.00386.x

131. McCallum C. M., Comai L., Greene E. A., Henikoff S. (2000). Targeting Induced Local Lesions IN Genomes (TILLING) for plant functional genomics. *Plant Physiol.* 123 437–442 10.1104/pp.123.2.439

132. McKinney H. H. (1926). Virus mixtures that may not be detected in young tobacco plants.*Phytopathology* 16 883.

133. Medzhitov R., Janeway C. A. (1997). Innate immunity: the virtues of a nonclonal system of recognition. *Cell* 91 295–298 10.1016/S0092-8674(00)80412-2

134. Moffett P. (2009). Mechanisms of recognition in dominant R gene mediated resistance. *Adv. Virus Res.*75 1–33 10.1016/S0065-3527(09)07501-0

135. Moissiard G., Voinnet O. (2006). RNA silencing of host transcripts by cauliflower mosaic virus requires coordinated action of the four *Arabidopsis* Dicer-like proteins. *Proc. Natl. Acad. Sci. U.S.A.*103 19593–19598 10.1073/pnas.0604627103

136. Morel J.-B., Godon C., Mourrain P., Béclin C., Boutet S., Feuerbach F., et al. (2002). Fertile hypomorphic ARGONAUTE (ago1) mutants impaired in post-transcriptional gene silencing and virus resistance. *Plant Cell* 14 629–639 10.1105/tpc.010358

137. Moreno P., Ambrós S., Albiach-Martí M. R., Guerri J., Peña L. (2008). Citrus tristeza virus: a pathogen that changed the course of the citrus industry. *Mol. Plant Pathol.* 9 251–268 10.1111/j.1364-3703.2007.00455.x

138. Mundt C. C. (2014). Durable resistance: a key to sustainable management of pathogens and pests. *Infect. Genet. Evol.* 27C, 446–455 10.1016/j.meegid.2014.01.011

139. Nakahara K. S., Masuta C. (2014). Interaction between viral RNA silencing suppressors and host factors in plant immunity. *Curr. Opin. Plant Biol.* 20 88–95 10.1016/j.pbi.2014.05.004

140. Nakazono-Nagaoka E., Takahashi T., Shimizu T., Kosaka Y., Natsuaki T., Omura T., et al. (2009).Cross-protection against bean yellow mosaic virus (BYMV) and clover yellow vein virus by attenuated BYMV isolate M11. *Phytopathology* 99 251–257 10.1094/PHYTO-99-3-0251

141. Nicaise V., German-Retana S., Sanjuán R., Dubrana M., Mazier M., Maisonneuve B., et al. (2003).The eukaryotic translation initiation factor 4E controls lettuce susceptibility to the potyvirus lettuce mosaic virus. *Plant Physiol.* 132 1272–1282 10.1104/pp.102.017855.1

142. Nicaise V., Roux M., Zipfel C. (2009). Recent advances in PAMP-triggered immunity against bacteria: pattern recognition receptors watch over and raise the alarm. *Plant Physiol.* 150 1638–1647 10.1104/pp.109.139709

143. Nicholson B. L., White K. A. (2014). Functional long-range RNA–RNA interactions in positive-strand RNA viruses. *Nat. Rev. Microbiol.* 12 493–504 10.1038/nrmicro3288

144. Nieto C., Piron F., Dalmais M., Marco C. F., Moriones E., Gómez-Guillamón M. L., et al. (2007).EcoTILLING for the identification of allelic variants of melon eIF4E, a factor that controls virus susceptibility. *BMC Plant Biol.* 7:34 10.1186/1471-2229-7-34

145. Nishiguchi M., Kobayashi K. (2011). Attenuated plant viruses: preventing virus diseases and understanding the molecular mechanism. *J. Gen. Plant Pathol.* 77 221–229 10.1007/s10327-011-0318-x

146. Odipio J., Ogwok E., Taylor N. J., Halsey M., Bua A., Fauquet C. M., et al. (2013). RNAi-derived field resistance to cassava brown streak disease persists across the vegetative cropping cycle. *GM Crops Food* 5 1–4.

147. Oka K., Kobayashi M., Mitsuhara I., Seo S. (2013). Jasmonic acid negatively regulates resistance to tobacco mosaic virus in tobacco. *Plant Cell Physiol.* 54 1999–2010 10.1093/pcp/pct137

148. Ordon F., Habekuss A., Kastirr U., Rabenstein F., Kühne T. (2009). Virus resistance in cereals: sources of resistance, genetics and breeding. *J. Phytopathol.* 157 535–545 10.1111/j.1439-0434.2009.01540.x

149. Otulak K., Garbaczewska G. (2011). Cellular localisation of calcium ions during potato hypersensitive response to potato virus Y. *Micron* 42 381–391 10.1016/j.micron.2010.11.001

150. Ouibrahim L., Mazier M., Estevan J., Pagny G., Decroocq V., Desbiez C., et al. (2014). Cloning of the *Arabidopsis* rwm1 gene for resistance to watermelon mosaic virus points to a new function for natural virus resistance genes. *Plant J.* 79 705–716 10.1111/tpj.12586

151. Padmanabhan M. S., Dinesh-Kumar S. P. (2014). The conformational and subcellular compartmental dance of plant NLRs during viral recognition and defense signaling. *Curr. Opin. Microbiol.* 20 55–61 10.1016/j.mib.2014.05.003

152. Pagny G., Paulstephenraj P. S., Poque S., Sicard O., Cosson P., Eyquard J.-P., et al. (2012). Family-based linkage and association mapping reveals novel genes affecting plum pox virus infection in*Arabidopsis thaliana*. *New Phytol.* 196 873–86 10.1111/j.1469-8137.2012.04289.x

153. Palloix A., Ayme V., Moury B. (2009). Durability of plant major resistance genes to pathogens depends on the genetic background, experimental evidence and consequences for breeding strategies.*New Phytol.* 183 190–199 10.1111/j.1469-8137.2009.02827.x

154. Palloix A., Ordon F. (2011). "Advanced breeding for virus resistance in plants," in *Recent Advances in Plant Virology* eds Caranta C., Aranda M. A., Tepfer M., Lopez-Moya J. J., editors. (Norwich: Caister Academic Press;) 195–218.

155. Park S. W., Kaimoyo E., Kumar D., Mosher S., Klessig D. F. (2007). Methyl salicylate is a critical mobile signal for plant systemic acquired resistance. *Science* 318 113–116 10.1126/science.1147113

156. Patarroyo C., Laliberté J.-F., Zheng H. (2012). Hijack it, change it: how do plant viruses utilize the host secretory pathway for efficient viral replication and spread? *Front. Plant Sci.* 3:30810.3389/fpls.2012.00308

157. Pestova T. V, Kolupaeva V. G., Lomakin I. B., Pilipenko E. V, Shatsky I. N., Agol V. I., et al. (2001).Molecular mechanisms of translation initiation in eukaryotes. *Proc. Natl. Acad. Sci. U.S.A.* 98 7029–7036 10.1073/pnas.111145798

158. Piron F., Nicolaï M., Minoïa S., Piednoir E., Moretti A., Salgues A., et al. (2010). An induced mutation in tomato eiF4E leads to immunity to two potyviruses. *PLoS ONE* 5:e1131310.1371/journal.pone.0011313

159. Pumplin N., Voinnet O. (2013). RNA silencing suppression by plant pathogens: defence, counter-defence and counter-counter-defence. *Nat. Rev. Microbiol.* 11 745–760 10.1038/nrmicro3120

160. Qi X., Bao F. S., Xie Z. (2009). Small RNA deep sequencing reveals role for *Arabidopsis thaliana*RNA-dependent RNA polymerases in viral siRNA biogenesis. *PLoS ONE* 4:e497110.1371/journal.pone.0004971

161. Qu F., Ye X., Morris T. J. (2008). *Arabidopsis* DRB4, AGO1, AGO7, and RDR6 participate in a DCL4-initiated antiviral RNA silencing pathway negatively regulated by DCL1. *Proc. Natl. Acad. Sci. U.S.A.* 105 14732–14737 10.1073/pnas.0805760105

162. Qu J., Ye J., Fang R. (2007). Artificial microRNA-mediated virus resistance in plants. *J. Virol.* 816690–6699 10.1128/JVI.02457-06

163. Quenouille J., Montarry J., Palloix A., Moury B. (2013). Farther, slower, stronger: how the plant genetic background protects a major resistance gene from breakdown. *Mol. Plant Pathol.* 14 109–118 10.1111/j.1364-3703.2012.00834.x

164. Rairdan G. J., Collier S. M., Sacco M. A., Baldwin T. T., Boettrich T., Moffett P. (2008). The coiled-coil and nucleotide binding domains of the potato Rx disease resistance protein function in pathogen recognition and signaling. *Plant Cell* 20 739–751 10.1105/tpc.107.056036

165. Revers F., Guiraud T., Houvenaghel M.-C., Mauduit T., Le Gall O., Candresse T. (2003). Multiple resistance phenotypes to lettuce mosaic virus among *Arabidopsis thaliana* accessions. *Mol. Plant Microbe Interact.* 16 608–616 10.1094/MPMI.2003.16.7.608

166. Revers F., Nicaise V. (2014). "Plant resistance to infection by viruses," in *Encyclopedia of Life Sciences* ed. Wiley-Blackwell (Chichester: John Wiley & Sons Ltd;). 10.1002/9780470015902.a0000757.pub3

167. Robert-Seilaniantz A., Grant M., Jones J. D. G. (2011). Hormone crosstalk in plant disease and defense: more than just JASMONATE-SALICYLATE antagonism. *Annu. Rev. Phytopathol.* 49 317–343 10.1146/annurev-phyto-073009-114447

168. Ronald P. C., Beutler B. (2010). Plant and animal sensors of conserved microbial signatures. *Science*330 1061–1064 10.1126/science.1189468

169. Ross F. A. (1961). Systemic acquired resistance induced by localized virus infections in plants.*Virology* 14 340–358 10.1016/0042-6822(61)90319-1

170. Sandig H., Bulfone-Paus S. (2012). TLR signaling in mast cells: common and unique features. *Front. Immunol.* 3:185 10.3389/fimmu.2012.00185

171. Sansregret R., Dufour V., Langlois M., Daayf F., Dunoyer P., Voinnet O., et al. (2013). Extreme resistance as a host counter-counter defense against viral suppression of RNA silencing. *PLoS Pathog.*9:e1003435 10.1371/journal.ppat.1003435

172. Sasaya T., Nakazono-Nagaoka E., Saika H., Aoki H., Hiraguri A., Netsu O., et al. (2013). Transgenic strategies to confer resistance against viruses in rice plants. *Front. Microbiol.* 4:40910.3389/fmicb.2013.00409

173. Sastry S. K., Zitter T. A. (2014). "Management of virus and viroid diseases of crops in the tropics," in*Plant Virus and Viroid Diseases in the Tropics* Vol. 2*Epidemiology and Management* (Dordrecht: Springer;) 149–480 10.1007/978-94-007-7820-7_2

174. Sasvari Z., Izotova L., Kinzy T. G., Nagy P. D. (2011). Synergistic roles of eukaryotic translation elongation factors 1Bγ and 1A in stimulation of tombusvirus minus-strand synthesis. *PLoS Pathog.*7:e1002438 10.1371/journal.ppat.1002438

175. Schröder M., Bowie A. G. (2007). An arms race: innate antiviral responses and counteracting viral strategies. *Biochem. Soc. Trans.* 35 1512–1514 10.1042/BST0351512

176. Schwessinger B., Ronald P. C. (2012). Plant innate immunity: perception of conserved microbial signatures. *Annu. Rev. Plant Biol.* 63 451–482 10.1146/annurev-arplant-042811-105518

177. Seo J.-K., Kwon S.-J., Cho W. K., Choi H.-S., Kim K.-H. (2014). Type 2C protein phosphatase is a key regulator of antiviral extreme resistance limiting virus spread. *Sci. Rep.* 4 5905 10.1038/srep05905

178. Seo Y.-S., Rojas M. R., Lee J.-Y., Lee S.-W., Jeon J.-S., Ronald P., et al. (2006). A viral resistance gene from common bean functions across plant families and is up-regulated in a non-virus-specific manner. *Proc. Natl. Acad. Sci. U.S.A.* 103 11856–11861 10.1073/pnas.0604815103

179. Serrano C., González-Cruz J., Jauregui F., Medina C., Mancilla P., Matus J. T., et al. (2008). Genetic and

histological studies on the delayed systemic movement of tobacco mosaic virus in *Arabidopsis thaliana*. *BMC Genet.* 9:59 10.1186/1471-2156-9-59

180. Shabala S., Babourina O., Rengel Z., Nemchinov L. G. (2010). Non-invasive microelectrode potassium flux measurements as a potential tool for early recognition of virus-host compatibility in plants. *Planta* 232 807–815 10.1007/s00425-010-1213-y

181. Shabala S., Baekgaard L., Shabala L., Fuglsang A., Babourina O., Palmgren M. G., et al. (2011).Plasma membrane Ca^{2+} transporters mediate virus-induced acquired resistance to oxidative stress.*Plant. Cell Environ.* 34 406–417 10.1111/j.1365-3040.2010.02251.x

182. Shekhawat U. K. S., Ganapathi T. R., Hadapad A. B. (2012). Transgenic banana plants expressing small interfering RNAs targeted against viral replication initiation gene display high-level resistance to banana bunchy top virus infection. *J. Gen. Virol.* 93 1804–1813 10.1099/vir.0.041871-0

183. Shi A., Chen P., Li D., Zheng C., Hou A., Zhang B. (2009). Gene pyramiding of genes of soybean mosaic virus resistance. *Mol. Breed.* 23 113–124 10.1007/s11032-008-9219-x

184. Shimizu T., Nakazono-Nagaoka E., Akita F., Uehara-Ichiki T., Omura T., Sasaya T. (2011). Immunity to rice black streaked dwarf virus, a plant reovirus, can be achieved in rice plants by RNA silencing against the gene for the viroplasm component protein. *Virus Res.* 160 400–403 10.1016/j.virusres.2011.05.011

185. Shivaprasad P. V., Rajeswaran R., Blevins T., Schoelz J., Meins F., Hohn T., et al. (2008). The CaMV transactivator/viroplasmin interferes with RDR6-dependent trans-acting and secondary siRNA pathways in *Arabidopsis*. *Nucleic Acids Res.* 36 5896–5909 10.1093/nar/gkn590

186. Slootweg E., Roosien J., Spiridon L. N., Petrescu A.-J., Tameling W., Joosten M., et al. (2010).Nucleocytoplasmic distribution is required for activation of resistance by the potato NB-LRR receptor Rx1 and is balanced by its functional domains. *Plant Cell* 22 4195–4215 10.1105/tpc.110.077537

187. Soosaar J. L. M., Burch-Smith T. M., Dinesh-Kumar S. P. (2005). Mechanisms of plant resistance to viruses. *Nat. Rev. Microbiol.* 3 789–798 10.1038/nrmicro1239

188. Spoel S. H., Dong X. (2012). How do plants achieve immunity? Defence without specialized immune cells. *Nat. Rev. Immunol.* 12 89–100 10.1038/nri3141

189. Takeshita M., Okuda M., Okuda S., Hyodo A., Hamano K., Furuya N., et al. (2013). Induction of antiviral responses by acibenzolar-s-methyl against cucurbit chlorotic yellows virus in Melon.*Phytopathology* 103 960–965 10.1094/PHYTO-08-12-0188-R

190. Taki A., Yamagishi N., Yoshikawa N. (2013). Development of apple latent spherical virus-based vaccines against three tospoviruses. *Virus Res.* 176 251–258 10.1016/j.virusres.2013.06.015

191. Taliansky M. E., Brown J. W. S., Rajamäki M. L., Valkonen J. P. T., Kalinina N. O. (2010).Involvement of the plant nucleolus in virus and viroid infections. Parallels with animal pathosystems.*Adv. Virus Res.* 77 119–158 10.1016/B978-0-12-385034-8.00005-3

192. Tameling W. I. L., Nooijen C., Ludwig N., Boter M., Slootweg E., Goverse A., et al. (2010).RanGAP2 mediates nucleocytoplasmic partitioning of the NB-LRR immune receptor Rx in the Solanaceae, thereby dictating Rx function. *Plant Cell* 22 4176–4194 10.1105/tpc.110.077461

193. Tamura A., Kato T., Taki A., Sone M., Satoh N., Yamagishi N., et al. (2013). Preventive and curative effects of Apple latent spherical virus vectors harboring part of the target virus genome against potyvirus and cucumovirus infections. *Virology* 446 314–324 10.1016/j.virol.2013.08.019

194. Tenllado F., Llave C., Díaz-Ruíz J. R. (2004). RNA interference as a new biotechnological tool for the control of virus diseases in plants. *Virus Res.* 102 85–96 10.1016/j.virusres.2004.01.019

195. Thaler J. S., Humphrey P. T., Whiteman N. K. (2012). Evolution of jasmonate and salicylate signal crosstalk. *Trends Plant Sci.* 17 260–270 10.1016/j.tplants.2012.02.010

196. Thivierge K., Nicaise V., Dufresne P. J., Cotton S., Laliberté J.-F., Le Gall O., et al. (2005). Plant virus RNAs. Coordinated recruitment of conserved host functions by (+) ssRNA viruses during early infection events. *Plant Physiol.* 138 1822–1827 10.1104/pp.105.064105.1822

197. Thresh J. M., Cooter R. J. (2005). Strategies for controlling cassava mosaic virus disease in Africa.*Plant Pathol.* 54 587–614 10.1111/j.1365-3059.2005.01282.x

198. Trejo-Saavedra D. L., García-Neria M. A., Rivera-Bustamante R. F. (2013). Benzothiadiazole (BTH) induces resistance to pepper golden mosaic virus (PepGMV) in pepper (*Capsicum annuum* L.). *Biol. Res.* 46 333–340 10.4067/S0716-97602013000400004

199. Tricoli D. M., Carney K. J., Russell P. F., McMaster J. R., Groff D. W. (1995). Field-evaluation of transgenic squash containing single or multiple virus coat protein gene constructs for resistance to cucumber mosaic-virus. *Artif. Cells Blood Substit. Immobil. Biotechnol.* 13 1458–1465 10.1038/nbt1295-1458

200. Truniger V., Aranda M. A. (2009). Recessive resistance to plant viruses. *Adv. Virus Res.* 119–159 10.1016/S0065-3527(09)07504-6

201. Tsuda K., Sato M., Glazebrook J., Cohen J. D., Katagiri F. (2008). Interplay between MAMP-triggered and SA-mediated defense responses. *Plant J.* 53 763–775 10.1111/j.1365-313X.2007.03369.x

202. Ueda H., Yamaguchi Y., Sano H. (2006). Direct interaction between the tobacco mosaic virus helicase domain and the ATP-bound resistance protein, N factor during the hypersensitive response in tobacco plants. *Plant Mol. Biol.* 61 31–45 10.1007/s11103-005-5817-8

203. Vaistij F. E., Jones L. (2009). Compromised virus-induced gene silencing in RDR6-deficient plants.*Plant Physiol.* 149 1399–1407 10.1104/pp.108.132688

204. Valkonen J. P. T., Rajamäki M.-L., Kekarainen T. (2002). Mapping of viral genomic regions important

in cross-protection between strains of a potyvirus. *Mol. Plant Microbe Interact.* 15 683–692 10.1094/MPMI.2002.15.7.683

205. Verlaan M. G., Hutton S. F., Ibrahem R. M., Kormelink R., Visser R. G. F., Scott J. W., et al. (2013).The tomato yellow leaf curl virus resistance genes Ty-1 and Ty-3 are allelic and code for DFDGD-Class RNA-dependent RNA polymerases. *PLoS Genet.* 9:e1003399 10.1371/journal.pgen.1003399

206. Vlot A. C., Klessig D. F., Park S. W. (2008). Systemic acquired resistance: the elusive signal(s). *Curr. Opin. Plant Biol.* 11 436–442 10.1016/j.pbi.2008.05.003

207. Vogler H., Akbergenov R., Shivaprasad P. V, Dang V., Fasler M., Kwon M.-O., et al. (2007).Modification of small RNAs associated with suppression of RNA silencing by tobamovirus replicase protein. *J. Virol.* 81 10379–10388 10.1128/JVI.00727-07

208. Voinnet O. (2001). RNA silencing as a plant immune system against viruses. *Trends Genet.* 17 449–459 10.1016/S0168-9525(01)02367-8

209. Voinnet O. (2005). Induction and suppression of RNA silencing: insights from viral infections. *Nat. Rev. Genet.* 6 206–220 10.1038/nrg1555

210. Wagner S., Stuttmann J., Rietz S., Guerois R., Brunstein E., Bautor J., et al. (2013). Structural basis for signaling by exclusive EDS1 heteromeric complexes with SAG101 or PAD4 in plant innate immunity.*Cell Host Microbe* 14 619–630 10.1016/j.chom.2013.11.006

211. Wale S., Platt B., Cattlin N. D. (2008). *Diseases, Pests and Disorders of Potatoes: A Colour Handbook*. London: Manson Publishing Ltd.

212. Wang A., Krishnaswamy S. (2012). Eukaryotic translation initiation factor 4E-mediated recessive resistance to plant viruses and its utility in crop improvement. *Mol. Plant Pathol.* 13 795–803 10.1111/j.1364-3703.2012.00791.x

213. Wang H. L. (1991). Effectiveness of cross protection by a mild strain of zucchini yellow mosaic virus in cucumber, melon, and squash. *Plant Dis.* 75 203 10.1094/PD-75-0203

214. Wang X.-B., Jovel J., Udomporn P., Wang Y., Wu Q., Li W.-X., et al. (2011). The 21-nucleotide, but not 22-nucleotide, viral secondary small interfering RNAs direct potent antiviral defense by two cooperative argonautes in *Arabidopsis thaliana*. *Plant Cell* 23 1625–1638 10.1105/tpc.110.082305

215. Waterhouse P. M., Wang M. B., Lough T. (2001). Gene silencing as an adaptive defence against viruses. *Nature* 411 834–842 10.1038/35081168

216. Wen F., Lister R. M., Fattouh F. A. (1991). Cross-protection among strains of barley yellow dwarf virus. *J. Gen. Virol.* 72 791–799 10.1099/0022-1317-72-4-791

217. Werner K., Friedt W., Ordon F. (2005). Strategies for pyramiding resistance genes against the barley yellow mosaic virus complex (BaMMV, BaYMV, BaYMV-2). *Mol. Breed.* 16 45–55 10.1007/s11032-005-3445-2

218. Whitham S., McCormick S., Baker B. (1996). The N gene of tobacco confers resistance to tobacco mosaic virus in transgenic tomato. *Proc. Natl. Acad. Sci. U.S.A.* 93 8776–8781 10.1073/pnas.93.16.8776

219. Yadav J. S., Ogwok E., Wagaba H., Patil B. L., Bagewadi B., Alicai T., et al. (2011). RNAi-mediated resistance to cassava brown streak uganda virus in transgenic cassava. *Mol. Plant Pathol.* 12 677–687 10.1111/j.1364-3703.2010.00700.x

220. Yan N., Chen Z. J. (2012). Intrinsic antiviral immunity. *Nat. Immunol.* 13 214–222 10.1038/ni.2229

221. Yang H., Gou X., He K., Xi D., Du J., Lin H., et al. (2010). BAK1 and BKK1 in *Arabidopsis thaliana* confer reduced susceptibility to turnip crinkle virus. *Eur. J. Plant Pathol.* 127 149–156 10.1007/s10658-010-9581-5

222. Yang P., Lüpken T., Habekuss A., Hensel G., Steuernagel B., Kilian B., et al. (2014). PROTEIN DISULFIDE ISOMERASE LIKE 5-1 is a susceptibility factor to plant viruses. *Proc. Natl. Acad. Sci. U.S.A.* 111 2104–2109 10.1073/pnas.1320362111

223. Yi S. Y., Shirasu K., Moon J. S., Lee S. G., Kwon S. Y. (2014). The activated SA and JA signaling pathways have an influence on flg22-triggered oxidative burst and callose deposition. *PLoS ONE*9:e88951 10.1371/journal.pone.0088951

224. Yokota S.-I., Okabayashi T., Fujii N. (2010). The battle between virus and host: modulation of Toll-like receptor signaling pathways by virus infection. *Mediators Inflamm.* 2010 18432810.1155/2010/184328

225. Yoon J. Y., Ahn H. II, Kim M., Tsuda S., Ryu K. H. (2006). Pepper mild mottle virus pathogenicity determinants and cross protection effect of attenuated mutants in pepper. *Virus Res.* 118 23–30 10.1016/j.virusres.2005.11.004

226. You B.-J., Chiang C.-H., Chen L.-F., Su W.-C., Yeh S.-D. (2005). Engineered mild strains of papaya ringspot virus for broader cross protection in cucurbits. *Phytopathology* 95 533–540 10.1094/PHYTO-95-0533

227. Zagrai I., Capote N., Ravelonandro M., Cambra M., Zagrai L., Scorza R. (2008). Plum pox virus silencing of C5 transgenic plums is stable under challenge inoculation with heterologous viruses. *J. Plant Pathol.* 90 S1–S63.

228. Zavaliev R., Levy A., Gera A., Epel B. L. (2013). Subcellular dynamics and role of *Arabidopsis*β-1,3-glucanases in cell-to-cell movement of tobamoviruses. *Mol. Plant Microbe Interact.* 26 1016–1030 10.1094/MPMI-03-13-0062-R

229. Zhang G., Chen M., Li L., Xu Z., Chen X., Guo J., et al. (2009). Overexpression of the soybean GmERF3 gene, an AP2/ERF type transcription factor for increased tolerances to salt, drought, and diseases in transgenic tobacco. *J. Exp. Bot.* 60 3781–3796 10.1093/jxb/erp214

230. Zhang X., Li H., Zhang J., Zhang C., Gong P., Ziaf K., et al. (2011a). Expression of artificial microRNAs in tomato confers efficient and stable virus resistance in a cell-autonomous manner.*Transgenic Res.* 20 569–581 10.1007/s11248-010-9440-3

231. Zhang X., Sato S., Ye X., Dorrance A. E., Morris T. J., Clemente T. E., et al. (2011b). Robust RNAi-based

resistance to mixed infection of three viruses in soybean plants expressing separate short hairpins from a single transgene. *Phytopathology* 101 1264–1269 10.1094/PHYTO-02-11-0056

232. Zhang X., Singh J., Li D., Qu F. (2012). Temperature-dependent survival of turnip crinkle virus-infected *Arabidopsis* plants relies on an RNA silencing-based defense that requires DCL2, AGO2, and HEN1. *J. Virol.* 86 6847–6854 10.1128/JVI.00497-12
233. Zhao D., Song G. Q. (2014). Rootstock-to-scion transfer of transgene-derived small interfering RNAs and their effect on virus resistance in nontransgenic sweet cherry. *Plant Biotechnol. J.* [Epub ahead of print].10.1111/pbi.12243
234. Zhu F., Xi D.-H., Yuan S., Xu F., Zhang D.-W., Lin H.-H. (2014). Salicylic acid and jasmonic acid are essential for systemic resistance against tobacco mosaic virus in *Nicotiana benthamiana*. *Mol. Plant Microbe Interact.* 27 567–577 10.1094/MPMI-11-13-0349-R
235. Zhu S., Jeong R. D., Venugopal S. C., Lapchyk L., Navarre D., Kachroo A., et al. (2011). Sag101 forms a ternary complex with EDS1 and PAD4 and is required for resistance signaling against turnip crinkle virus. *PLoS Pathog.* 7:e1002318 10.1371/journal.ppat.1002318
236. Ziebell H., Carr J. P. (2010). Cross-protection: a century of mystery. *Adv. Virus Res.* 76 211–264 10.1016/S0065-3527(10)76006-1
237. Ziebell H., Payne T., Berry J. O., Walsh J. A., Carr J. P. (2007). A cucumber mosaic virus mutant lacking the 2b counter-defence protein gene provides protection against wild-type strains. *J. Gen. Virol.* 88 2862–2871 10.1099/vir.0.83138-0
238. Zipfel C., Felix G. (2005). Plants and animals: a different taste for microbes? *Curr. Opin. Plant Biol.* 8353–360 10.1016/j.pbi.2005.05.004

INDEX